Lecture Notes in Physics

Volume 879

Founding Editors

W. Beiglböck
J. Ehlers
K. Hepp
H. Weidenmüller

Editorial Board

B.-G. Englert, Singapore, Singapore
U. Frisch, Nice, France
P. Hänggi, Augsburg, Germany
W. Hillebrandt, Garching, Germany
M. Hjorth-Jensen, Oslo, Norway
R.A.L. Jones, Sheffield, UK
M. Lewenstein, Barcelona, Spain
H. von Löhneysen, Karlsruhe, Germany
M. S. Longair, Cambridge, UK
J.-F. Pinton, Lyon, France
J.-M. Raimond, Paris, France
A. Rubio, Donostia, San Sebastian, Spain
M. Salmhofer, Heidelberg, Germany
D. Sornette, Zurich, Switzerland
S. Theisen, Potsdam, Germany
D. Vollhardt, Augsburg, Germany
W. Weise, Garching, Germany and Trento, Italy
J.D. Wells, Geneva, Switzerland

For further volumes:
www.springer.com/series/5304

The Lecture Notes in Physics

The series Lecture Notes in Physics (LNP), founded in 1969, reports new developments in physics research and teaching—quickly and informally, but with a high quality and the explicit aim to summarize and communicate current knowledge in an accessible way. Books published in this series are conceived as bridging material between advanced graduate textbooks and the forefront of research and to serve three purposes:

- to be a compact and modern up-to-date source of reference on a well-defined topic
- to serve as an accessible introduction to the field to postgraduate students and nonspecialist researchers from related areas
- to be a source of advanced teaching material for specialized seminars, courses and schools

Both monographs and multi-author volumes will be considered for publication. Edited volumes should, however, consist of a very limited number of contributions only. Proceedings will not be considered for LNP.

Volumes published in LNP are disseminated both in print and in electronic formats, the electronic archive being available at springerlink.com. The series content is indexed, abstracted and referenced by many abstracting and information services, bibliographic networks, subscription agencies, library networks, and consortia.

Proposals should be sent to a member of the Editorial Board, or directly to the managing editor at Springer:

Christian Caron
Springer Heidelberg
Physics Editorial Department I
Tiergartenstrasse 17
69121 Heidelberg/Germany
christian.caron@springer.com

Christoph Scheidenberger · Marek Pfützner
Editors

The Euroschool on Exotic Beams, Vol. IV

 Springer

Editors
Christoph Scheidenberger
NuSTAR department
GSI Helmholtzzentrum
  für Schwerionenforschung GmbH
Darmstadt, Germany

Marek Pfützner
Faculty of Physics
University of Warsaw
Warsaw, Poland

ISSN 0075-8450   ISSN 1616-6361 (electronic)
Lecture Notes in Physics
ISBN 978-3-642-45140-9   ISBN 978-3-642-45141-6 (eBook)
DOI 10.1007/978-3-642-45141-6
Springer Heidelberg New York Dordrecht London

Library of Congress Control Number: 2014930224

© Springer-Verlag Berlin Heidelberg 2014
This work is subject to copyright. All rights are reserved by the Publisher, whether the whole or part of the material is concerned, specifically the rights of translation, reprinting, reuse of illustrations, recitation, broadcasting, reproduction on microfilms or in any other physical way, and transmission or information storage and retrieval, electronic adaptation, computer software, or by similar or dissimilar methodology now known or hereafter developed. Exempted from this legal reservation are brief excerpts in connection with reviews or scholarly analysis or material supplied specifically for the purpose of being entered and executed on a computer system, for exclusive use by the purchaser of the work. Duplication of this publication or parts thereof is permitted only under the provisions of the Copyright Law of the Publisher's location, in its current version, and permission for use must always be obtained from Springer. Permissions for use may be obtained through RightsLink at the Copyright Clearance Center. Violations are liable to prosecution under the respective Copyright Law.
The use of general descriptive names, registered names, trademarks, service marks, etc. in this publication does not imply, even in the absence of a specific statement, that such names are exempt from the relevant protective laws and regulations and therefore free for general use.
While the advice and information in this book are believed to be true and accurate at the date of publication, neither the authors nor the editors nor the publisher can accept any legal responsibility for any errors or omissions that may be made. The publisher makes no warranty, express or implied, with respect to the material contained herein.

Printed on acid-free paper

Springer is part of Springer Science+Business Media (www.springer.com)

Preface

We are pleased to present the continuation of the series Lecture Notes in Physics emerging from the Euroschool on Exotic Beams. This school, initiated in Leuven (Belgium) in 1993, has been running every year (with one exception in 1999). Based on lectures given at the Euroschool, the Lecture Notes provide an introduction for graduate students and young researchers to novel and exciting fields of physics with radioactive ion beams and their applications. The fourth volume in this series covers selected material presented in Euroschool lectures between 2007 and 2011.

Since the late 80s, the field of radioactive ion beams has been rapidly developing and substantially expanding. While many of its roots were founded in Europe, and also leadership of the field was for many years concentrated in Europe, there are meanwhile intense efforts worldwide to build and exploit dedicated second-generation radioactive beam facilities. The exciting physics of radioactive ions is mainly linked to the study of nuclear structure under extreme conditions of isospin, mass, spin and temperature. Radioactive ion beam science addresses problems in nuclear astrophysics, solid-state physics and fundamental interactions. Furthermore, important applications and spin-offs also originate from this basic research. The development of new production, acceleration and ion manipulation techniques and the construction of new detectors is also an important part of this science. A major aim is the development of a unified picture of the atomic nucleus, to understand the structure and dynamics of nuclei and to provide reliable predictions of nuclear structure properties within the "Terra Incognita", the regions in the nuclear chart which cannot be explored with present experimental techniques.

As with previous volumes, the present Lecture Notes do not comprise a complete overview of the field, but represent sample topics of theory and experiment. Since the appearance of the latest volume in 2009, and already before, many new subjects were covered in the lectures, and some of them are presented here. The topics have been selected by the editors to exhibit recent advances in the field and to complement previous Lecture Notes. None of them has been covered in previous volumes, all represent an active field of current research, and all authors are well known experts in their domains and are highly respected scientists. Their contributions to this book are not meant to be review-type articles rather they provide a modern introduction to a specific subject in a didactic way, given by practitioners at the forefront of scientific research. This approach has proved to be successful and the Euroschool Lecture Notes are popular among both students and scientists. The content of volumes I, II and III of this series is available on the Euroschool website http://www.euroschoolonexoticbeams.be. We hope that the present volume will be no less successful than the previous editions.

The Euroschool concept began as a European initiative. From the start the intention was to gather original questions, methods and results from the field of radioactive beams and exotic nuclei, and to bring them to the attention of students and young researchers working in this domain both within Europe and overseas. The school has, at all times, been open for European and international participants. Since 2001 the Euroschool has traveled to various locations and countries throughout Europe. Events took place in Finland (2001, 2011), France (2002, 2007), Spain (2003, 2010), United Kingdom (2004), Germany (2005), Italy (2006), Poland (2008), Belgium (2009), Greece (2012), and most recently in Russia (2013). The evident success of the Euroschool on Exotic Beams originates to a large extent from the excellence of the lecturers invited to share their knowledge with the students and the pleasant, informal atmosphere which generates a valuable forum for discussions. Despite some organizational changes which occurred during this time, the scope, format, spirit, and popularity among young participants has been maintained. It is our pleasure—and debt—to thank the sponsors for their support, which makes the Euroschool events possible:

- Demokritos—National Center for Scientific Research, Athens (Greece)
- ECT*, European Centre for Theoretical Studies in Nuclear Physics and Related Areas, Trento (Italy)
- GANIL—Grand Accelerateur National d'Ions Lourds, Caen (France)
- Gobierno de España—Ministerio de Economia y Competitividad—FNUC Network and CPAN Ingenio 2010, Madrid (Spain)
- GSI—Helmholtz Centre for Heavy Ion Research, Darmstadt (Germany)
- HIC-4-FAIR—Helmholtz International Center for FAIR, Darmstadt (Germany)
- IFIC-CSIC—Instituto de Fisica Corpuscular, Consejo Superior de Investigaciones Cientificas, Madrid (Spain)
- INFN—Instituto Nazionale di Fisica Nucleare (Italy)
- ISOLDE-CERN, Geneva (Switzerland)
- JINR—Joint Institute for Nuclear Research, Dubna (Russia)
- JYFL—University of Jyväskylä (Finland)
- KU Leuven—Instituut voor Kern- en Stralingsfysica, Leuven (Belgium)
- RuG-KVI—Kernfysisch Versneller Instituut, Groningen (The Netherlands)
- UCL—Centre de Recherche du Cyclotron, Louvain-la-Neuve (Belgium)
- University of Warsaw (Poland)
- USC—University of Santiago de Compostela (Spain)

Finally, we would like to thank all who have contributed to this volume. First of all to the authors who have given excellent lectures for the Euroschool students, and who have invested time and effort in preparing the contributions to this book in a comprehensive and pedagogical way. Secondly, we thank our colleagues on the Board of Directors of the Euroschool, who supported the development of this volume with interest and for their valuable ideas. Last, but not least, it is our pleasure to thank Dr. Chris Caron and his colleagues at Springer Verlag for the encouragement and continuous support in a fruitful collaboration.

University of Warsaw, Poland	Marek Pfützner
GSI Darmstadt and University of Giessen, Germany	Christoph Scheidenberger

Contents

1 Clustering in Light Nuclei; from the Stable to the Exotic 1
Martin Freer
   1.1 Clusters and Correlations in Context 1
   1.2 Clusters in First Principles Models 2
   1.3 Appearance of the Nuclear Cluster from the Mean-Field 3
   1.4 More Sophisticated Models of Clustering 11
      1.4.1 Bloch-Brink Alpha Cluster Model (ACM) 11
      1.4.2 Condensates and the THSR Wave-Function 13
      1.4.3 Microscopic Cluster Models 14
      1.4.4 Antisymmetrised Molecular Dynamics (AMD) and
             Fermionic Molecular Dynamics (FMD) 16
      1.4.5 Ab Initio Type Models 18
   1.5 Experimental Examples of Clustering 19
      1.5.1 The Example ^{8}Be 19
      1.5.2 The Structure of ^{12}C 20
   1.6 Experimental Techniques—Break-up and Resonant Scattering
       Reactions 22
      1.6.1 Resonant Scattering 22
      1.6.2 Break-up Measurements 24
   1.7 Beyond α-Clusters—Valence Neutrons and Molecules 26
      1.7.1 The Neutron-Rich Nucleus ^{10}Be 30
      1.7.2 More Complex Molecular States and the Extended Ikeda
             Diagram 32
   1.8 Summary and Conclusions 34
   References 35

**2 A Pedestrian Approach to the Theory of Transfer Reactions:
Application to Weakly-Bound and Unbound Exotic Nuclei** 39
Joaquín Gómez Camacho and Antonio M. Moro
   2.1 Introduction 39
   2.2 Theoretical Formalism 41

		2.2.1	Distorted Wave Born Approximation DWBA	43
		2.2.2	Adiabatic Distorted Wave Approximation ADWA	46
		2.2.3	Continuum Discretized Coupled Channels Born Approximation CDCC-BA	49
		2.2.4	Coupled Reaction Channels CRC	51
		2.2.5	Connection with the Faddeev Formalism	55
	2.3	Transfer to Unbound States		58
		2.3.1	Recent Applications to Weakly Bound Halo Nuclei	60
	2.4	Summary and Conclusions		62
	References			64
3	**What Can We Learn from Transfer, and How Is Best to Do It?**			67
	Wilton N. Catford			
	3.1	Motivation to Study Single-Nucleon Transfer Using Radioactive Beams		67
		3.1.1	Migration of Shell Gaps and Magic Numbers, Far from Stability	68
		3.1.2	Coexistence of Single Particle Structure and Other Structures	70
		3.1.3	Description of Single Particle Structure Using Spectroscopic Factors	70
		3.1.4	Disclaimer: What This Article Is, and Is Not, About	71
	3.2	Choice of the Reaction and the Bombarding Energy		72
		3.2.1	Kinematics and Measurements Using Normal Kinematics	72
		3.2.2	Differential Cross Sections: Dependence on Beam Energy and ℓ Transfer	73
		3.2.3	Choice of a Theoretical Reaction Model: The ADWA Description	77
		3.2.4	Comparisons: Other Transfer Reactions and Knockout Reactions	78
	3.3	Experimental Features of Transfer Reactions in Inverse Kinematics		80
		3.3.1	Characteristic Kinematics for Stripping, Pickup and Elastic Scattering	81
		3.3.2	Laboratory to Centre of Mass Transformation	85
		3.3.3	Strategies to Combat Limitations in Excitation Energy Resolution	86
	3.4	Examples of Light Ion Transfer Experiments with Radioactive Beams		89
		3.4.1	Using a Spectrometer to Detect the Beam-like Fragment	89
		3.4.2	Using a Silicon Array to Detect the Light (Target-like) Ejectile	90
		3.4.3	Choosing the Right Experimental Approach to Match the Experimental Requirements	94
		3.4.4	Using (d, p) with Gamma-Rays, to Study Bound States	97

		3.4.5	The Use of a Zero-Degree Detector in (d, p) and Related Experiments	102
		3.4.6	Simultaneous Measurements of Elastic Scattering Distributions	105
		3.4.7	Extending (d, p) Studies to Unbound States	106
		3.4.8	Simultaneous Measurement of Other Reactions Such as (d, t)	107
		3.4.9	Taking into Account Gamma-Ray Angular Correlations in (d, p)	108
		3.4.10	Summary	114
	3.5	Heavy Ion Transfer Reactions		115
		3.5.1	Selectivity According to $j_>$ and $j_<$ in a Semi-classical Model	115
		3.5.2	Examples of Selectivity Observed in Experiments	117
	3.6	Perspectives		118
	References			120
4	**Effective Field Theories of Loosely Bound Nuclei**			123
	U. van Kolck			
	4.1	Introduction		123
	4.2	Nuclear Physics Scales and Effective Field Theories		124
		4.2.1	Basic Ideas	126
		4.2.2	An Example: NRQED	133
		4.2.3	Summary	141
	4.3	QCD at Low Energies		141
		4.3.1	Building Blocks	141
		4.3.2	Chiral EFT	147
		4.3.3	Renormalization of Singular Potentials and Power Counting	158
		4.3.4	Summary	162
	4.4	Loosely Bound Systems		162
		4.4.1	Fine-Tuning	163
		4.4.2	Contact EFT	164
		4.4.3	Halo/Cluster EFT	173
		4.4.4	Summary	178
	4.5	Conclusions and Outlook		178
	References			179
5	**Direct Reactions at Relativistic Energies: A New Insight into the Single-Particle Structure of Exotic Nuclei**			183
	Dolores Cortina-Gil			
	5.1	Introduction		183
		5.1.1	First Experiments	186
	5.2	Knockout Reactions		189
		5.2.1	Extraction of Information in Knockout Reactions	189
		5.2.2	Experimental Needs and Relevant Observables	193

		5.2.3 Results of Knockout Measurements	201
	5.3	Quasi-Free Scattering Reactions with Rare Isotope Beams	220
		5.3.1 Status of the QFS Program with Exotic Rare Isotopes at R^3B	222
	5.4	Summary and Conclusions	226
	References		227
6	**Nuclear Charge Radii of Light Elements and Recent Developments in Collinear Laser Spectroscopy**		**233**
	Wilfried Nörtershäuser and Christopher Geppert		
	6.1	Introduction	233
	6.2	Atomic Theory: Isotope Shift and Charge Radii	234
		6.2.1 Mass Shift	235
		6.2.2 Field Shift	236
		6.2.3 Evaluation of Mass Shift and Field Shift Constants	239
	6.3	Nuclear Theory: Charge Radii Variations Along Isotopic Chains	248
		6.3.1 Spherical Nuclei	248
		6.3.2 Nuclear Deformation	249
		6.3.3 Clustering and Halos in Light Nuclei	251
	6.4	Measuring Charge Radii of Halo Isotopes	254
		6.4.1 The Challenge of Halo Nuclei	254
		6.4.2 Helium: Spectroscopy on Cold and Trapped Atoms	255
		6.4.3 Lithium: Doppler-Free Two-Photon Spectroscopy on Thermal Atoms	259
		6.4.4 Beryllium and Neon: High-Accuracy Measurements with Fast Beams of Ions	264
	6.5	Further Developments in Collinear Laser Spectroscopy	272
		6.5.1 Isotope Shift Determinations Using β-Asymmetry Detection	272
		6.5.2 Photon-Ion Coincidence Detection	275
	6.6	Towards the Limits: Improving Sensitivity with Cooled and Bunched Ion Beams	276
		6.6.1 Principle of a Radio-Frequency Quadrupole	276
		6.6.2 Applications of Ion Bunchers in CLS	278
		6.6.3 Optical Pumping in the Cooler and Buncher	282
	6.7	Future Prospects	285
	6.8	Conclusion	287
	References		288
7	**The Nuclear Energy Density Functional Formalism**		**293**
	T. Duguet		
	7.1	Introduction	293
		7.1.1 Generalities	293
		7.1.2 Nuclear Structure Theory	295
		7.1.3 Goal of the Present Lecture Notes	299
	7.2	Prelude	300

	7.2.1	Reference States and Bogoliubov Transformation	300
	7.2.2	Elements of Group Theory	301
	7.2.3	Collective Variable and Symmetry Breaking	303
7.3	Energy and Norm Kernels		304
	7.3.1	Norm Kernel	304
	7.3.2	Energy Kernel	305
	7.3.3	Pseudo-potential-based Energy Kernel	306
	7.3.4	Skyrme Parametrization	307
7.4	Single-Reference Implementation		315
	7.4.1	Equation of Motion	316
	7.4.2	One-Nucleon Addition and Removal Processes	318
	7.4.3	Effective Single-Particle Energies	320
	7.4.4	Equation of State of Infinite Nuclear Matter	323
	7.4.5	Symmetry Breaking and "Deformation"	327
	7.4.6	Connection to Density Functional Theory?	329
7.5	Multi-reference Implementation		330
	7.5.1	Symmetry-Restored Kernels	331
	7.5.2	Full Fledged MR Mixing	334
	7.5.3	Pseudo-potential-based Energy Kernel	335
	7.5.4	Other Observables	336
	7.5.5	Dynamical Correlations	336
	7.5.6	State-of-the Art Calculations	339
	7.5.7	Approximations to Full Fledged MR-EDF	339
	7.5.8	Pathologies of MR-EDF Calculations	340
	7.5.9	Towards Pseudo-potential-based Energy Kernels	343
	7.5.10	Towards Non-empirical Energy Kernels	344
7.6	Conclusions		345
References			347

Chapter 1
Clustering in Light Nuclei; from the Stable to the Exotic

Martin Freer

1.1 Clusters and Correlations in Context

The structure of nuclear matter is rich and varied. In one light the nucleus may behave like a liquid drop, with its shape and size corresponding to a balance between the long(ish) range attractive and short range repulsive behaviour of the nucleon-nucleon interaction and the charges of the constituent protons. This liquid drop displays collective properties such as vibrations where vibrational modes distort the nuclear surface; it can be encouraged to deform and then can be rotated—as the droplet spins it stretches which provides a mechanism for the determination of the equation-of-state of the fluid. At a critical angular momentum the droplet will fission. Similarly as the mass of a nucleus increases, typically so does the number of protons and hence the charge. The repulsive Coulomb energy should cause the nucleus to spontaneously fission when the number of protons is close to 100. However, it is at this point that another crucial feature contributes which allows nuclei to exist beyond that point—shell effects. Shell structure, which features for light and heavy nuclei alike, is associated with the quantal properties of the nucleus and marks a deviation from the constituent particles to a picture in which the particles are represented by standing waves. The associated quantum states are those of the nuclear shell model and give rise to a sequence of magic numbers which are associated with enhanced stability. A superposition of the macroscopic liquid drop and microscopic shell model-like behaviour is required to describe the stability of nuclei beyond the point at which the charged liquid drop should explode.

For light nuclei there is a similar interplay between the collective and single-particle nature, but here details of the nature of the interaction between the nucleons becomes increasingly important. Correlations become a dominant feature. The pairing interaction is evident in the nature of the drip-lines, which define the limits of stability on both the proton and neutron-rich side of the chart of nuclides (see

M. Freer (✉)
School of Physics and Astronomy, University of Birmingham, Birmingham, B15 2TT, UK
e-mail: M.Freer@bham.ac.uk

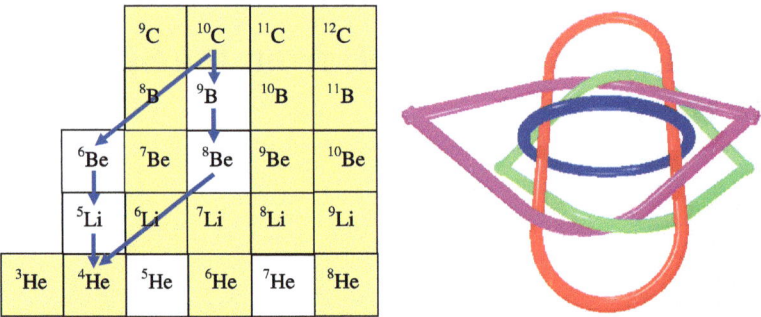

Fig. 1.1 Light nuclei. *Filled squares* are either stable or beta-decay, *unfilled particle* (neutron, proton or α) decay. The *arrows* show the paths corresponding to the removal of a proton or α-particle from ^{10}C. The diagram on the *right hand side* illustrates the 4th order Brunian knot

Fig. 1.1). For the helium nuclei, ^{4}He is stable, whereas ^{5}He is not. Similarly, ^{6}He and ^{8}He are stable and ^{7}He is not. The difference being that in addition to the ^{4}He core the stable isotopes have even numbers of neutrons, whereas the unstable ones do not. ^{6}He and ^{8}He are known as Borromean nuclei, as for example in the case of ^{6}He if a neutron is removed then the other two components dissociate; further if the α-particle is extracted then this leaves the unbound $2n$ system.

As an example of potentially exotic structures on the proton-rich side the ^{10}C nucleus sits at the head of a loop around unbound nuclei which include ^{9}B and ^{8}Be. ^{10}C may be thought of being composed of two protons and two α-particles and if any of the components are removed then the other three dissociate. This may be thought as a super-Borromean nucleus, or recognising that Borromean systems belong to a class of mathematical objects called Brunian knots then ^{10}C is a nucleus which is 4th order knot (as illustrated in Fig. 1.1).

These are rather extreme examples of correlations, but they are rather commonplace in light nuclei and have a determining role when it comes to the structure. These correlations can be spatial in addition to energy or momentum and then are referred to as clusters. The most prevalent cluster is the α-particle due to its remarkably high binding and inertness. This contribution examines some of the basic underlying principles behind the formation of clusters and examines some of the key areas experimentally where they strongly feature.

1.2 Clusters in First Principles Models

The formation of structures in nuclei that have large scale clustering is an intriguing phenomenon and is in part driven by correlations which stem from the details of the nucleon-nucleon interaction. For example, the *ab initio* Green's Function Monte Carlo (GFMC) calculations of ^{8}Be [1] predict the structure of nuclei based upon a starting point which is the nucleon-nucleon interaction expressed in terms of all two-body and three-body components. The two-body interactions are a parameterisation

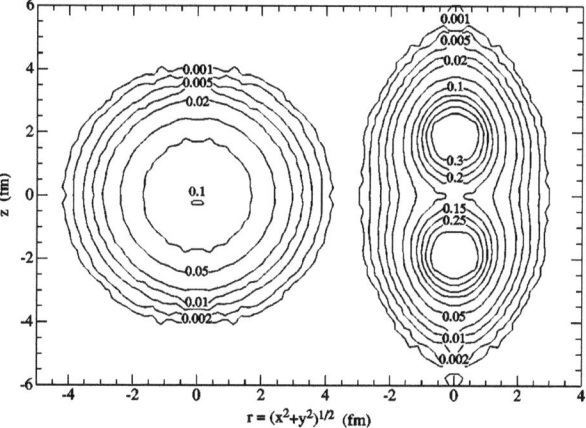

Fig. 1.2 The Green's Function Monte Carlo calculations of the density of ^{8}Be. The *left* and *right-hand images* are the densities calculated in the laboratory and intrinsic frames, respectively [1]. The 2α cluster structure is clearly evident

of the n–n force as determined from nucleon-nucleon scattering. It is not possible to determine the 3-body force in the same way, but is included through a parameterisation of terms such as the higher order pion exchange components devised by Fujita and Miyazawa [2]. In this manner the interaction is *ab initio* motivated rather than being grown from QCD. Given, that the model is one which contains the nucleonic degrees of freedom, it is somewhat remarkable that such an approach yields a ^{8}Be ground state, Fig. 1.2, that is clearly clustered [1]. At this point it is thus tempting to assert that the nucleus ^{8}Be corresponds to an α–α cluster structure in the ground state.

There have been many recent developments in the field of nuclear clusters including the ability to perform ab initio calculations of the light nuclei, such as the Green's Function Monte Carlo methods and Antisymmetrized Molecular Dynamics (Sect. 1.4.4) and Chiral Effective Field Theory (where nuclear properties are calculated on the lattice), the appearance of both experimental and theoretical evidence for molecular structures (Sect. 1.7) and the renewed focus on cluster states in nuclear synthesis, in particular the Hoyle-state in ^{12}C which may possess an α-condensate structure (Sect. 1.4.2). The following section attempts to provide a basic understanding of some of the underlying principles.

1.3 Appearance of the Nuclear Cluster from the Mean-Field

The possibility that the α-particles could be rearranged in some geometric fashion was realised even in the earliest days of the subject. An examination of the binding energy per nucleon of the light nuclei (Fig. 1.3—left-hand-side) shows that the nuclei which have even, and equal, number of protons and neutrons (so-called α-conjugate nuclei) are particularly stable, ^{8}Be, ^{12}C, ^{16}O, ^{20}Ne, …. Figure 1.3, right-hand-side, shows the binding energy per nucleon plotted against the energy of the first excited state for a variety of nuclei. The nucleus ^{4}He stands out as being both

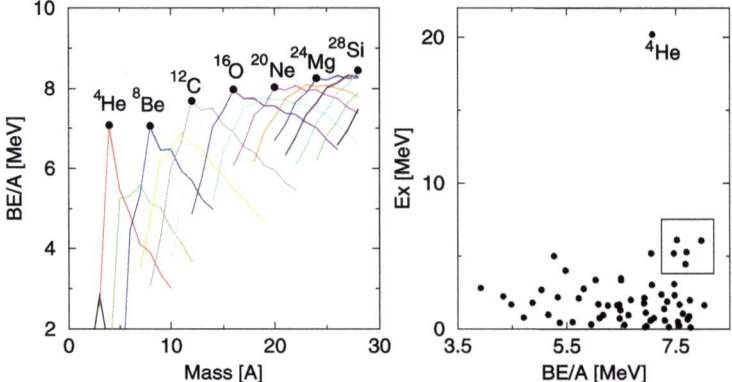

Fig. 1.3 (*Left panel*) Binding energy per nucleon of light nuclear systems (up to $A = 28$), the *lines* connect isotopes of each element. The "α-particle nuclei" are marked by the *circles*. (*Right panel*) Excitation energy of first excited states plotted versus binding energy per nucleon for nuclei up to $A = 20$. Good clusters should have both high binding energies and first excited states. The nucleus ^{4}He is clearly an outstanding cluster candidate. The *box* drawn includes nuclei which may also form clusters: ^{12}C, ^{14}O, ^{14}C, ^{15}N and ^{16}O

stable and inert. These systems were also considered by Hafstad and Teller [3], who characterised the binding energy with number of "bonds" or interactions between the α-particles (Fig. 1.4). The rather linear relationship pointed to an apparently constant α–α interaction and the inertness of the α-particle in the ground states of these nuclei (it should be noted that this view is not one which is currently held, where the cluster structure is believed to be eroded in most ground-states). In essence, what this reveals is that the binding energies of such $N\alpha$ nuclei (N being an integer representing the number of α-particles) can be described in terms of $N(BE_\alpha) + N \cdot B_{\alpha\alpha}$, where BE_α is the binding energy of the α-particle and $B_{\alpha\alpha}$ is the energy associated with the α–α interaction. In turn this may be indicative of the important of p–p, n–n and n–p correlation energies associated with occupation of common orbitals in nuclei with even and equal numbers of protons and neutrons (α-conjugate nuclei).

Earlier Morinaga had postulated, in a rather extreme prediction for the time, that it should be possible for the α-particles to arrange themselves in a linear fashion [4]. The idea that the cluster should not be manifest in the ground-state but emerge as the internal energy of the nucleus is increased was realised to be key in the 1960's [5]. For a nucleus to develop a cluster structure it must be energetically allowed. Asymptotically, when the nucleus is separated into its cluster components an energy equivalent to the mass difference between the host and the clusters must be provided. There is an additional contribution which is the interaction between clusters which is required to fully separate them. In other words, the cluster structure would expect to be manifest close to, and probably slightly below, the cluster decay threshold. This was the view reached by Ikeda and co-workers, and is summarised in the diagram in Fig. 1.5. The diagram illustrates that each new cluster degree of freedom arises as the cluster decay threshold is approached, or crossed. Thus, there is the gradual transition from the compact ground-state to the full liberation of the $N\alpha$

Fig. 1.4 Binding energy per nucleon of $A = 4n$ nuclei versus the number of α–α bonds. The analysis by Hafstad and Teller [3] suggested that the ground states of $A = 4n$ (n being an integer, i.e. 1, 2, 3 ...), α-conjugate, nuclei could be described by a constant interaction energy scaled by the number of bonds. For ^{8}Be there is one bond, ^{12}C—3, ^{16}O—6, ^{20}Ne—9, ^{24}Mg—12 and for structural reasons (the geometric packing of the α-particles) ^{28}Si—16

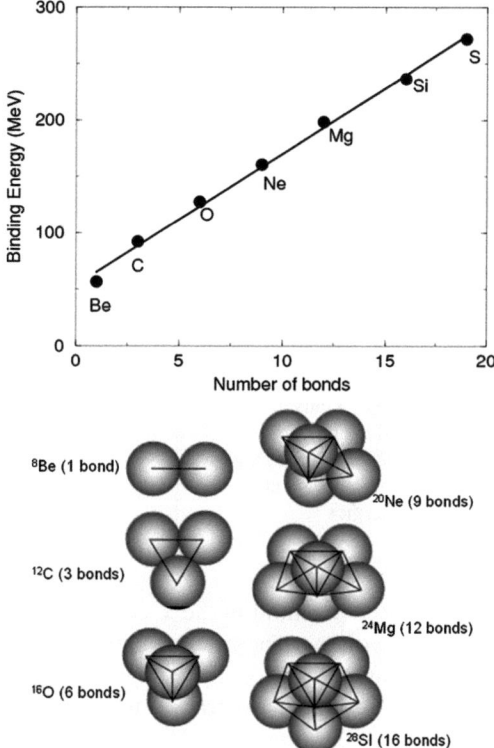

degree of freedom. Schematically, the diagram shows a linear arrangement of α-particles at the $N\alpha$-limit, though this need not be the most stable configuration. In fact, it may be argued that the linear structure has an inherent instability [7], though many have interpreted this limit as representing a linear structure.

There is a second key ingredient whose role greatly influences the possible geometric arrangements of the clusters—and that is symmetries. These symmetries can be thought of as arising from the packing of the α-particles, but have a deeper origin which relates to the quantal properties if the system. In order to illustrate this, we start with an analysis of a rather simple and schematic approach to the nuclear mean-field, but one which is nevertheless rather powerful. In the application of the harmonic oscillator (HO) to the nuclear problem, it is assumed that each nucleon moves within a parabolic potential (i.e. a linear restoring force) created by the mean-interaction of all of the other constituents. The solution of the Schrödinger equation then yields the well known energy levels

$$E = \hbar\omega(n + 3/2) \tag{1.1}$$

for the three dimensional nucleus, where oscillations can be along any of the three cartesian coordinate axes and n is the number of oscillator quanta. If the nucleus, or equivalently potential, is deformed, for example stretched along the z-axis, then

Fig. 1.5 The Ikeda picture [5], from [6]. The diagram shows how the cluster degree of freedom evolves as the excitation energy increases. The *numbers* indicate the excitation energies at which the cluster structures should appear—these are the binding energies of the cluster components in the parent nucleus. The important concept relayed by this diagram is that a cluster degree of freedom is only liberated close to a cluster decay threshold. Thus, for heavy systems the $N\alpha$ degree of freedom only appears at the highest energies

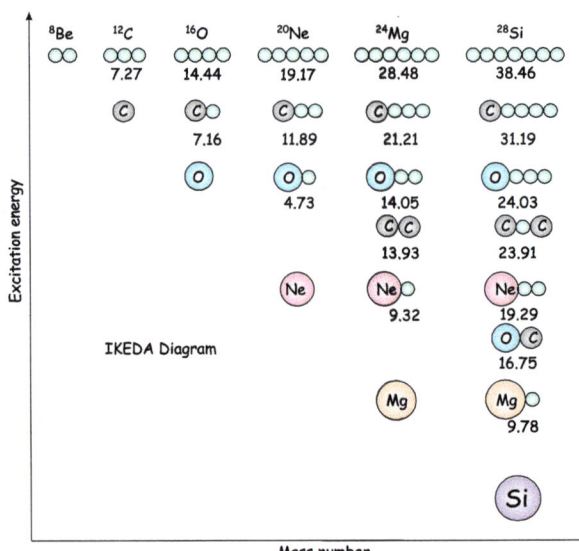

the size of the potential in the x and y-directions must shrink in order to conserve the nuclear volume. The extended potential in the z-direction lowers the oscillation frequency and, for an axially symmetric potential, is increased in the perpendicular direction. Thus, the degeneracy implicit in (1.1), is removed and

$$E = \hbar\omega_\perp n_\perp + \hbar\omega_z n_z + \frac{3}{2}\hbar\omega_0 \qquad (1.2)$$

where the characteristic oscillator frequencies for oscillations perpendicular ($\perp$) and parallel (z) to the deformation axis are now required. These are constrained such that $\omega_0 = (2\omega_\perp + \omega_z)$, and the quadrupole deformation is given by

$$\varepsilon = \varepsilon_2 = (\omega_\perp - \omega_z)/\omega_0. \qquad (1.3)$$

The total number of oscillator quanta is the sum of those on the parallel and perpendicular axes ($n_\perp + n_z$).

The characteristic energy levels of the deformed harmonic oscillator are shown in Fig. 1.6 [8]. The striking feature is the crossings of levels (regions of high degeneracy) which occur for axial deformations of ($\omega_\perp : \omega_z$) 2:1 and 3:1. In fact, such degeneracies occur whenever the ratios $\omega_x : \omega_y : \omega_z = a : b : c$ where a, b and c are simple integers. Here shell structure is generated and corresponding *deformed* magic numbers emerge. In fact, the magic numbers reveal some particularly interesting behaviour. If rather than examining the magic numbers the sequence of degeneracies is explored, then the sequence of spherical degeneracies (2, 6, 12, 20, ...) is repeated twice at a deformation of 2:1 and three times at 3:1. This pattern would indicate two interacting spherical harmonic oscillator potentials at 2:1 and three at 3:1, etc. Here the symmetry appears within the magic numbers. These ideas were articulated mathematically by Nazarewicz and Dobaczewski [10].

Fig. 1.6 The deformed harmonic oscillator. The shell structure which appears at $\varepsilon_2 = 0$ vanishes as the potential is deformed, but reappears at deformations of 2:1, 3:1, etc. It is at these shell closures that cluster structure appears. The *numbers in the circles* indicate the degeneracy of the level scheme at the crossing points, from Ref. [9]

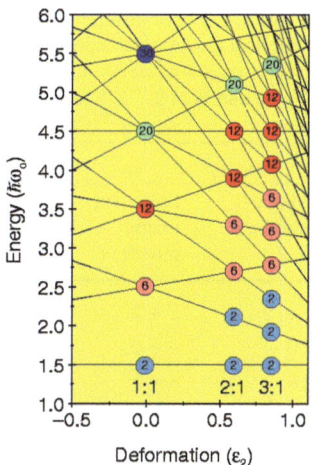

These symmetries have been explored elsewhere in detail in order to identify particular cluster partitions. Building on some of the earlier work of Bengtsson [11], Rae [12] focussed on the details of the deformed magic numbers in order to probe explicitly the cluster decompositions. These are shown in Table 1.1. Rae demonstrated that the deformed magic numbers could be expressed as the sums of spherical ones. This description locates at each deformation the associated cluster structure. At a deformation of 2:1 the superdeformed cluster states should be found in ^{8}Be ($\alpha + \alpha$), ^{20}Ne (^{16}O + α), ^{32}S (^{16}O + ^{16}O)... and at 3:1—hyperdeformation—^{12}C ($\alpha + \alpha + \alpha$), ^{24}Mg ($\alpha + {}^{16}$O + α), etc. Thus, the combination of the ideas of Rae and the Ikeda-picture permit the excitation-energy, deformation and single-particle configuration of cluster states to be determined.

The symmetries indicate a mapping between the shell structure and particular cluster states. However, the link runs deeper. We examine the rather trivial case of ^{8}Be. The levels which are labelled with degeneracy 2 are those with the oscillator quantum numbers $[n_\perp, n_z] = [0, 0]$ and $[0, 1]$. Each of these levels would be occupied by pairs of protons and pairs of neutrons with their spins coupled to zero. The density distributions of the particles is given by the square of the corresponding wave-functions, $\varphi_{0,0}$ and $\varphi_{0,1}$. The overall ^{8}Be density is given by $|\varphi_{0,0}|^2 + |\varphi_{0,1}|^2$. These three components are shown in Fig. 1.7. The feature which emerges is one in which the density is double humped corresponding to the localisation of the protons and neutrons within two "α-particles". Interestingly, the observed distribution is generated by particles moving in an axially deformed potential, this generates a clustered density distribution which then in turn creates the mean-field in which the particles move. This latter field is not identical to the first. Clearly, to provide stable solutions, self consistent approaches are required. Some of these are described later (e.g. Antisymmetrized Molecular Dynamics (AMD) and Fermionic Molecular Dynamics (FMD)).

The above operation can be also applied to the 3:1 deformed shell closure, where we consider the three lowest orbits which are labelled with degeneracy 2. These

Table 1.1 Relationship between the deformed magic numbers at deformations of 2:1 and 3:1 and spherical cluster decompositions from [12]. For example, at a deformation of 2:1 the neutron and proton magic numbers 4, 10 and 16 can be decomposed into the spherical neutron and proton magic numbers $2+2$, $8+2$ and $8+8$. Thus, one would expect at a deformation of 2:1 the cluster structures $\alpha+\alpha$, $\alpha+{}^{16}O$ and ${}^{16}O+{}^{16}O$ to appear

$\omega_\perp : \omega_z = 2:1$			$\omega_\perp : \omega_z = 3:1$		
Magic numbers at 2:1	Spherical magic numbers	Associated cluster configuration	Magic numbers at 3:1	Spherical magic numbers	Associated cluster configuration
4	$2+2$	$\alpha+\alpha$	6	$2+2+2$	$\alpha+\alpha+\alpha$
10	$8+2$	${}^{16}O+\alpha$	12	$2+8+2$	$\alpha+{}^{16}O+\alpha$
16	$8+8$	${}^{16}O+{}^{16}O$	18	$8+2+8$	${}^{16}O+\alpha+{}^{16}O$
28	$20+8$	${}^{40}Ca+{}^{16}O$	24	$8+8+8$	${}^{16}O+{}^{16}O+{}^{16}O$
40	$20+20$	${}^{40}Ca+{}^{40}Ca$	36	$8+20+8$	${}^{16}O+{}^{40}Ca+{}^{16}O$

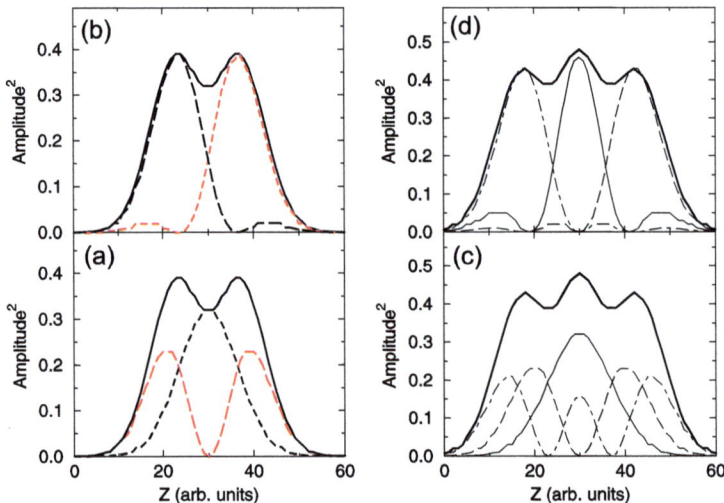

Fig. 1.7 The density corresponding to the HO configurations for (**a**) ^{8}Be and (**c**) ^{12}C. In (**a**) the square of the $(n_x, n_y, n_z) = (0, 0, 0)$ and $(0, 0, 1)$ orbits are plotted as is their sum (*solid line*). The square of the $(0, 0, 0)$, $(0, 0, 1)$ and $(0, 0, 2)$ orbits together with their sum (*solid line*) are shown in (**c**). Parts (**b**) and (**d**) show the separation into the two and three-centered components, respectively. These show the individual α-particle densities

are the $[n_\perp, n_z] = [0, 0]$, $[0, 1]$ and $[0, 2]$ HO levels. Figures 1.7 and 1.8 shows the densities which correspond to these three orbits. What can be clearly observed is that at the deformation of 3:1 there is a three humped structure. In other words, it is possible to see the evidence for the systems division into three centers. As with the ^{8}Be case, it is possible to project out the "α-particles" by appealing to the point symmetries of a three centered systems. If we employ the wave-functions containing

Fig. 1.8 The density of the three HO configurations associated with placing α-particles (pairs of protons and neutrons) in the orbits in Fig. 1.6 with degeneracy 2, at deformations of 2:1, 3:1 and 4:1. The densities correspond to the linear structures in the 2α, 3α and 4α systems ^{8}Be, ^{12}C and ^{16}O, respectively. In each case the presence of the α-particles is clear

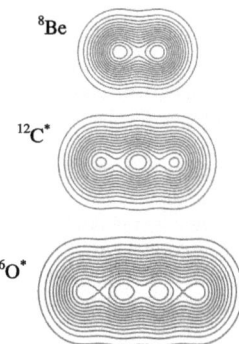

these symmetries we can equate the number of nodes in the multi-centered wavefunctions with those in the harmonic oscillator wave-functions under consideration;

$$\psi_{0,0} = \frac{1}{2}\phi_{\alpha(-)} + \frac{1}{\sqrt{2}}\phi_{\alpha(0)} + \frac{1}{2}\phi_{\alpha(+)} \tag{1.4}$$

$$\psi_{0,1} = \frac{1}{\sqrt{2}}\phi_{\alpha(-)} - \frac{1}{\sqrt{2}}\phi_{\alpha(+)} \tag{1.5}$$

$$\psi_{0,2} = -\frac{1}{2}\phi_{\alpha(-)} + \frac{1}{\sqrt{2}}\phi_{\alpha(0)} - \frac{1}{2}\phi_{\alpha(+)}. \tag{1.6}$$

These can be solved for the three α-particle like wave-functions $\phi_{\alpha(-,0,+)}$. The resulting α-particle densities are shown in Figs. 1.7 and 1.8. The greater overlap of the "α-particles" means that the central α-particle has additional higher order components (quantified in [9]).

Such an analysis may be performed universally across the deformed harmonic oscillator level scheme where ever shell structures arise and similar conclusions emerge; namely 2-fold clustering at a deformation at 2:1 and 3 at 3:1, etc. What is evident is that the cluster symmetries which are found in the HO are present *both* in degeneracies and densities. Figure 1.8 shows these symmetries for the first α-particle states appearing at deformations of 2:1, 3:1 and 4:1. Given the influence of the harmonic oscillator on more sophisticated nuclear models these cluster symmetries might be expected to be pervasive. The competition between the mean-field and clustering degrees of freedom is of great interest if the tendency of nuclei to fall either a shell-model or cluster-like description is to be probed. Itagaki and co-workers have recently explored this partition for a range of nuclei, e.g. Refs. [13–15].

Although more sophisticated models allow a more realistic description of the nucleus to be arrived at, the ideas developed here remain the leading order terms in our understanding of these nuclear states.

An example of this latter point may be found in a variety of calculations for ^{24}Mg. Figure 1.9 shows a compilation of calculations for ^{24}Mg. The central panel shows a Nilsson Strutinsky (NS) potential energy surface which is a macro-microscopic calculation which reveals a series of minima in the surface associated with meta-stable

Fig. 1.9 Comparison of a range of calculations of ^{24}Mg. The central panel shows a Nilsson Strutinsky calculation for the potential energy surface, the calculations around the outside show densities predicted by the Alpha Cluster Model (ACM), Hartree-Fock (HF) and Harmonic Oscillator (HO)

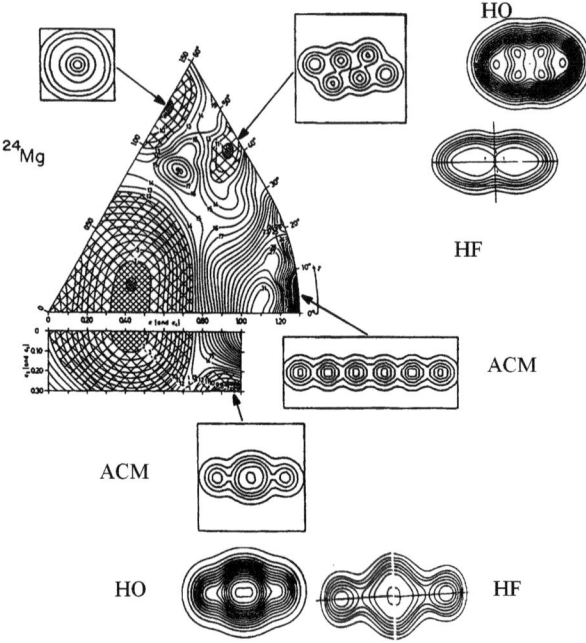

configurations. These may be linked directly with the appearance of shell structures in the deformed HO and also with Alpha Cluster Model (ACM) calculations in which the ^{24}Mg nucleus is described in terms of geometric arrangements of 6 α-particles—there is a one-to-one mapping between minima in the potential energy surface and the configurations found in the Alpha Cluster Model. In addition the densities for two structures found in Hartree-Fock (HF) calculations are shown— which bear a close resemblance to the structures found in the ACM. Finally, it is possible to extract from the NS calculations the underlying single-particle configuration and then this may be used to calculate the densities one would expect in the case of the harmonic oscillator (HO). Remarkably, these HO densities exhibit symmetries, or equivalently patterns, which are very strongly allied to those of the ACM. In other words, the symmetries that are associated with the arrangements of the α-particle clusters are pervasive in the mean-field type models. Thus, even if the α-particles themselves are not explicitly present within the nucleus their geometrical symmetries leave an imprint.

It should be noted that in the case of deformed states discussed here there exist two reference frames. The first is the intrinsic frame in which the coordinate system may be aligned with the deformation axis. In this case angular momentum of individual nucleons is not a good quantum number, only its projection onto the deformation axis. The second frame is the laboratory frame, which is the reference frame of the shell model—here angular momentum is a good quantum number. In calculations such as Hartree-Fock (HF) or Hartee-Fock-Bogoliubov (HFB), the latter including pairing, it is necessary to project out from the intrinsic states, states of good angular momentum (projection after variation). In the HF case this projection is performed using the Peierls-Yoccoz procedure [16] and for the more complex case a technique introduced by Blatt [17]. The majority of the cluster structures presented in the present review correspond to intrinsic states. It is of course within this framework in which collective rotational energies have a natural description. In the case of light nuclei in which SU(3) symmetry is respected it is often possible to deduce the relationship between the intrinsic and laboratory descriptions, i.e. the shell model limit corresponding to various cluster structures [10].

1.4 More Sophisticated Models of Clustering

The deformed harmonic oscillator provides a very good basis for distilling the underlying behavior of light nuclei, but is schematic. If one is to make progress towards a more detailed understanding and the ability to reproduce experimental observables such as transition rates, radii and energies then models of greater sophistication are required. Historically many models have taken as a starting point an implicit assumption of the existence of clustering and developing an interaction between α-particles. In more recent times it has been realized that the α-particles within the nucleus cannot be considered to be truly inert, but that interactions will distort, polarize and modify the internal structure and that the real degrees of freedom are those of the nucleons. This section explores some of the developments of models and their merits.

1.4.1 Bloch-Brink Alpha Cluster Model (ACM)

The Alpha Cluster Model was first conceived of by Margenau [18] and then developed by Brink [19] drawing on the work of Bloch. Within the nuclear shell model the ^{4}He nucleus is constructed from $2p + 2n$ all within the $0s_{1/2}$ orbital. The principle construction of the alpha particle model is to build on this idea and that quartets are produced from pairs of protons and neutrons which are coupled to a total angular momentum of zero, i.e. they may be represented by a relative $0s$-state. A collection of such quartet states may be modeled within the harmonic oscillator framework using

$$\phi_i(\mathbf{r}) = \sqrt{\frac{1}{b^3 \pi^{3/2}}} \exp\left[\frac{-(\mathbf{r} - \mathbf{R}_i)^2}{2b^2}\right]. \tag{1.7}$$

Here $\mathbf{R}_i$ is the vector describing the location of the ith quartet, and $b = (\hbar/m\omega)^{1/2}$ is a scale parameter which determines the size of the α-particle. The overall wavefunction of the system formed from the collection of α-particles must then be antisymmetrized in recognition that the true degrees of freedom are fermionic. Corre-

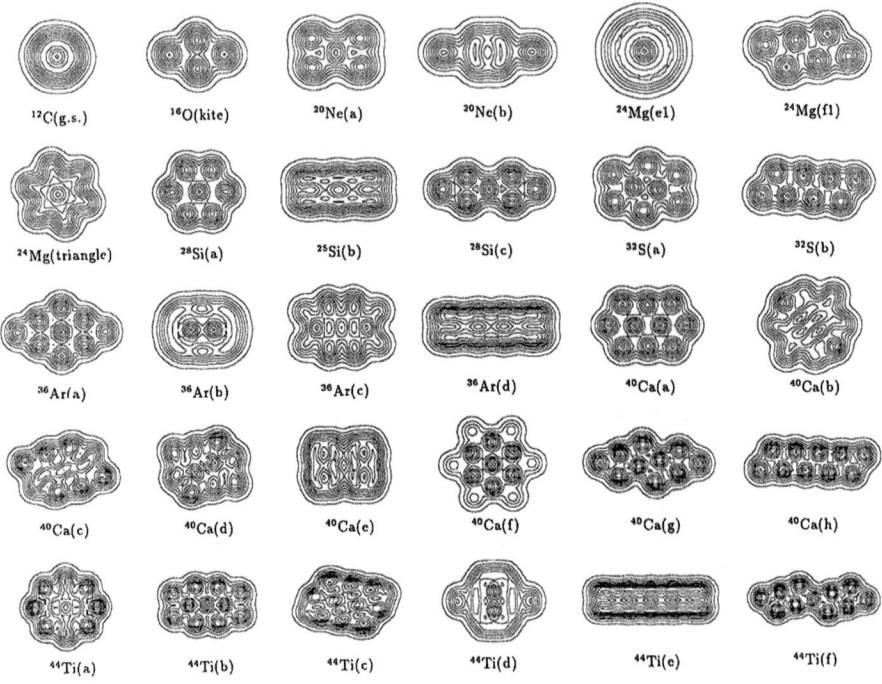

Fig. 1.10 Alpha Cluster Model (ACM) calculation for 2D structures in a range of light nuclei from Ref. [23]. See original work for further details

spondingly, the $N\alpha$ wave-function is then created using a Slater determinant

$$\Phi_\alpha(\mathbf{R}_1, \mathbf{R}_2, \ldots, \mathbf{R}_N) = K\mathscr{A} \prod_{i=1}^{N} \phi_i(\mathbf{R}_i) \qquad (1.8)$$

$\mathscr{A} \prod_{i=1}^{N} \phi_i(\mathbf{R}_i)$ being the Slater determinant wave-function ($\mathscr{A}$ is the antisymmetrization operator accounting for the Pauli Exclusion Principle) and K a normalisation constant. The antisymmetrizer recognizes that the wave-function is actually composed of the fermionic degrees of freedom, albeit the femions are embedded in the clusters. At short distances this will serve to break the α-particles. The Hamiltonian describing the total energy of the $N\alpha$-system is

$$H = \sum_{i=1}^{A} T_i + \frac{1}{2} \sum_{i \neq j} \left[v(\mathbf{r}_i - \mathbf{r}_j) + v_c(\mathbf{r}_i - \mathbf{r}_j) \right] - T_{c.m.} \qquad (1.9)$$

$T_{c.m.}$ is the center-of-mass energy and the α–α interactions are governed by the effective nucleon-nucleon potential $v(\mathbf{r}_i - \mathbf{r}_j)$ and Coulomb interaction $v_c(\mathbf{r}_i - \mathbf{r}_j)$.

The optimal arrangement of the α-particles is arrived at variationally, where the parameters which are optimised are the locations and size of the α-particles. This model has been applied extensively to light cluster systems by for example Brink [19], to the nucleus ^{16}O [20], a series of rather comprehensive set of calculations of the structure of ^{24}Mg by Marsh and Rae [21] (Fig. 1.9), linear arrangements of α-particles by Merchant [22] and finally a series of wide ranging calculations by Zhang et al. [23, 24], some of which are shown in Fig. 1.10. As was observed in Fig. 1.10, where the clusters were constrained to lie within a plane, many of the cluster structures are crystalline in nature.

As pointed out earlier there is a very strong mapping between the spatial symmetries found in these calculations and those which may be found in the densities associated with the deformed harmonic oscillator. In fact it is possible to deduce, in the limit that the separation of the α-particles tends to zero, the corresponding harmonic oscillator configurations. It is these oscillator configurations that then produce densities which emulate the patterns that are found in the Alpha Cluster Model.

1.4.2 Condensates and the THSR Wave-Function

An intriguing possibility that the Alpha Cluster Model raises is that there may be a class of states in nuclei in which the separation of the α-particles is such that the internal structure of the α-particle is no longer so important. The conditions necessary to achieve this require that the nuclear radius is sufficiently large. Such a condition may be achieved close to the α-decay threshold, where in a state which is only weakly bound an α-particle may significantly tunnel into the barrier increasing the nuclear volume. Perhaps the best candidate for such behaviour is the 7.65 MeV, 0^+, Hoyle-state in ^{12}C. From electron inelastic scattering measurements it is understood that the volume associated with the Hoyle state is some 3 to 4 times that of the ground-state. A further possibility then arises; if the state may be described by a collection of identical bosons is it possible for them to adopt bosonic symmetries and behave as an atomic Bose-Einstein condensate? In order to describe such a possibility, the Bloch-Brink wave-function (Sect. 1.4.1) has been adapted by Tohsaki, Horiuchi, Schuck and Röpke (THSR) to reflect the possible character of the state [25–27]. The condensed wave-function has the form

$$\langle \mathbf{r}_1, \ldots, \mathbf{r}_N | \Phi_{n\alpha} \rangle$$
$$= \mathscr{A}\left[\phi_\alpha(\mathbf{r}_1, \mathbf{r}_2, \mathbf{r}_3, \mathbf{r}_4)\phi_\alpha(\mathbf{r}_5, \mathbf{r}_6, \mathbf{r}_7, \mathbf{r}_8)\phi_\alpha(\mathbf{r}_{N-3}, \ldots, \mathbf{r}_N)\right] \quad (1.10)$$

here the construction is for N nucleons grouped into quartets described by ϕ_α. The wave-function of the α-particle is given by

$$\phi_\alpha(\mathbf{r}_1, \mathbf{r}_2, \mathbf{r}_3, \mathbf{r}_4) = e^{-\mathbf{R}^2/B^2} \phi(\mathbf{r}_1 - \mathbf{r}_2, \mathbf{r}_1 - \mathbf{r}_3, \ldots) \quad (1.11)$$

Fig. 1.11 The calculated inelastic form factor for electron inelastic scattering from the 0^+ ground state to the 0_2^+ excited state [29], compared with the experimental data from [30–32]

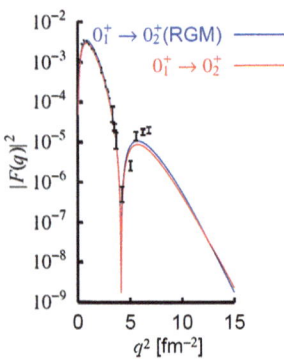

where $[\mathbf{R} = \mathbf{r}_1 + \mathbf{r}_2 + \mathbf{r}_3 + \mathbf{r}_4]/4$ is the c.o.m. coordinate of one α-particle and $\phi(\mathbf{r}_1 - \mathbf{r}_2, \mathbf{r}_1 - \mathbf{r}_3, \ldots)$ is a Gaussian wave-function

$$\phi(\mathbf{r}_1 - \mathbf{r}_2, \mathbf{r}_1 - \mathbf{r}_3, \ldots) = \exp\left(-[\mathbf{r}_1 - \mathbf{r}_2, \mathbf{r}_1 - \mathbf{r}_3, \ldots]^2/b^2\right) \quad (1.12)$$

as in the ACM b is the size parameter of the *free* α-particle and B ($\gg b$) is the parameter which describes the size of the common Gaussian distribution of the three α-particles. In the limit that $B \to \infty$ then the antisymmetrization operator $\mathscr{A}$ ceases to be important and the wave-function (1.10) becomes the product of Gaussians, i.e. a wave-function describing a free α-particle gas [28]. The important feature is that in the limit that the volume becomes small the antisymmetrization takes over and the wave-function respects the internal fermionic degrees of freedom. In this way the wave-function is very similar to that of the Alpha Cluster Model, but possesses an additional variational degree of freedom.

One of the main successes of this model is that it manages to reproduce the form factor for the electron elastic excitation to the Hoyle-state without any arbitrary normalisation [29] (see Fig. 1.11). There is remarkable agreement with the experimental data, which would confirm the nature of the Hoyle-state as being both spatially extended and strongly influenced by an internal α-particle structure.

1.4.3 Microscopic Cluster Models

The Alpha Cluster Model produces a rather good picture of the nature of states within $A = 4n$ nuclei which condense out into collections of α-particles. However, although it antisymmetrizes the α-particles, their individual constituents are ignored, i.e. the internal excitations of the cluster. For clusters such as α-particles this may be a good approximation, but for other clusters this is not the case. Such shortcomings are addressed within the generator coordinate method (GCM) (also within the resonating group method (RGM)) [33–42]. Moreover, this approach permits reactions between the asymptotic clusters to be studied, as has been performed extensively by Baye and Descouvement (e.g. Refs. [43–46]).

1 Clustering in Light Nuclei; from the Stable to the Exotic

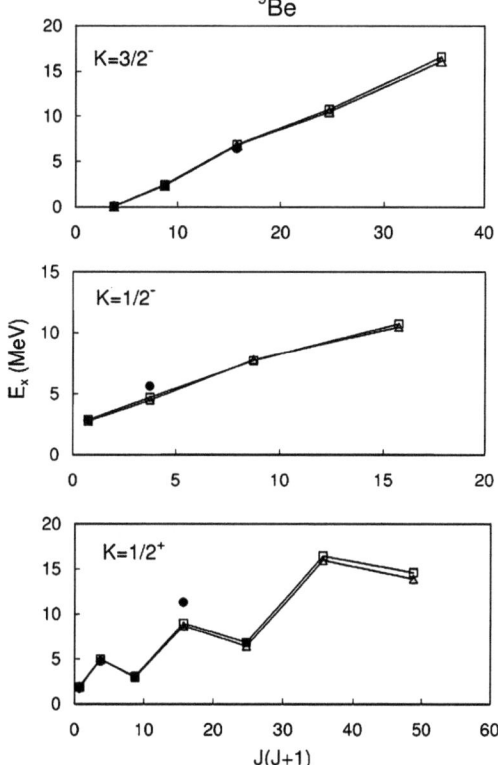

Fig. 1.12 The GCM calculations for ^{9}Be showing the three rotational bands associated with the $K^\pi = 3/2^-$ (π-configuration), $K^\pi = 1/2^+$ (σ-configuration) and $K^\pi = 1/2^-$ bands, from Ref. [47]. The experimental data are the *filled circles* and the *squares* and *circles* are the calculations for two different types of interaction

Within the RGM formalism the wave-function describing the A nucleons, separated into two clusters with A_1 and A_2 constituents, may be written as,

$$\Psi(\mathbf{r}_1, \mathbf{r}_2, \ldots, \mathbf{r}_A) = F(\mathbf{R}_{cm})\hat{A}\{\phi_1(\xi_1)\phi_2(\xi_2)g(\mathbf{R})\} \quad (1.13)$$

here $F(\mathbf{R}_{cm})$ describes the motion of the center of mass of the nucleus, ϕ_i represent antisymmetrized internal states of the two clusters (whose internal coordinates are described by ξ_i), $g(\mathbf{R})$ is a function of the relative motion of the two clusters (so that the relative coordinate $\mathbf{R}$ is given by $(1/A_1)\sum_{i=1}^{A_1}\mathbf{r}_i - (1/A_2)\sum_{j=1}^{A_2}\mathbf{r}_j$) and $\hat{A}$ is the antisymmetrization operator which exchanges nucleons between the two clusters. The great advantage of this approach is the fact that the constituents of the clusters are fully antisymmetrized and that the center-of-mass of the system is correctly treated so that the quantum numbers produced have a realistic meaning in terms of the asymptotic fragments. The above corresponds to the single-channel form of the RGM, if excitations of the cluster cores are required then so is a multi-channel approach.

An impressive demonstration of the GCM can be found in the calculations of the structure of the microscopic structure of ^{9,10,11}Be isotopes using $2\alpha + Xn$ configurations by Descouvemet [47]. The calculations for ^{9}Be reproduce almost perfectly the rotational bands in this system. In particular, the Coriolis decoupling of the $K^\pi = 1/2^+$ band is found (see Fig. 1.12). These GCM calculations reproduce the characteristics of the molecular states in the nuclei ^{9,10,11}Be. In this instance the neutrons reside in molecular orbits whereby they are exchanged between the two α-particle cores—π-orbit for the ground state band and σ for the excited states (see Sect. 1.7).

In recognition of this molecular behaviour, some approaches employ such orbitals explicitly in defining the basis states for the calculation of the structural properties. For example, this molecular-orbit (MO) approach has been used to calculate the properties of the neutron-rich beryllium [48–50] and carbon isotopes [51]. Here the molecular orbits are formed from linear combinations of p-orbitals based around α-particle centers. The MO framework also allows collisions between two nuclei to be considered, for example in the generalized two-center cluster model (GTCM), using a basis function of the form

$$\Phi_{m,n}^{J^\pi K} = \hat{P}_K^{J^\pi} \cdot \mathcal{A}\{\psi_L(\alpha)\psi_R(\alpha)\phi(m)\phi(n)\}, \tag{1.14}$$

the formation of resonances in ^{10}Be from ^{6}He+^{4}He has recently been considered [52]. Here $\psi_{L,R}(\alpha)$ is the wave-function of the left/right (L/R) α-particle and $\phi(m,n)$ are the molecular wave-functions of the neutrons. $\hat{P}_K^{J^\pi}$ and $\mathcal{A}$ are the parity projection and antisymmetrisation operators ensuring states have good angular momentum (J), angular momentum projection (K) and parity (π). Rather interestingly, these calculations indicate that in the inelastic scattering reaction ^{4}He+^{6}He $\Rightarrow$ ^{4}He+^{6}He(2^+) an avoided crossing which takes place between different molecular configurations that a Landau-Zener type transition [53, 54] is responsible for the inelastic scattering in the $L=1$ channel. In other words the formation of molecular configurations in the scattering process can have a marked impact on the elastic and inelastic scattering processes.

1.4.4 Antisymmetrised Molecular Dynamics (AMD) and Fermionic Molecular Dynamics (FMD)

The AMD approach, which has been comprehensively reviewed recently by Kanada-En'yo and Horriuchi [55], has many important advantages over microscopic cluster models, but the most significant is that there are no assumptions made about the cluster or the relative coordinates between clusters. The model is one in which the nucleonic degrees of freedom are explicitly included and the A-nucleon wave-function is then antisymmetrised again via a Slater determinant:

$$\Phi_{AMD}(\mathbf{Z}) = \frac{1}{\sqrt{A!}}\mathcal{A}\{\varphi_1, \varphi_2, \ldots, \varphi_A\}. \tag{1.15}$$

In this way the model resembles the Bloch-Brink cluster model, but contains as degrees of freedom the nucleons and releases the constraint that α-particles be preformed. Consequently, clusters emerge without being imposed. The φ_i are Gaussian

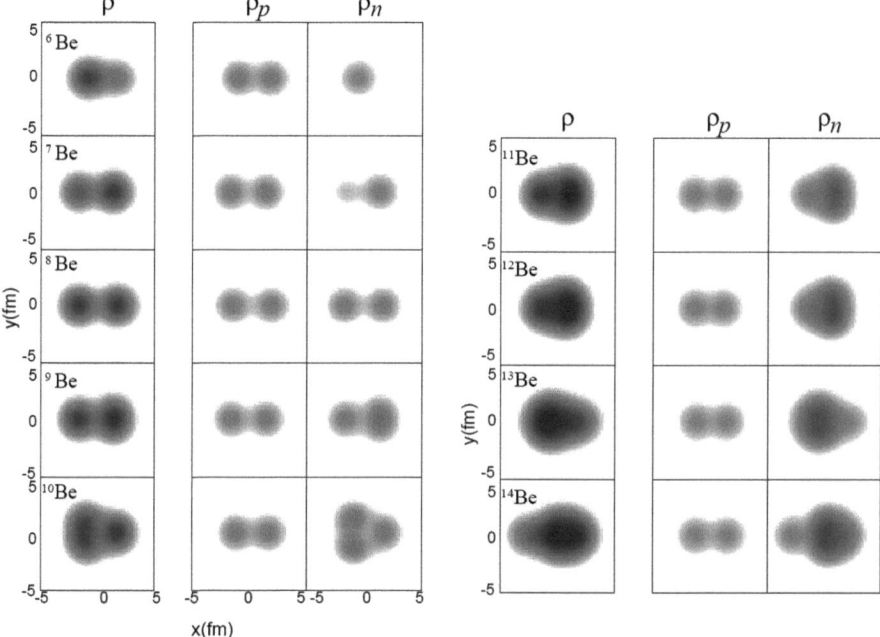

Fig. 1.13 The density distributions of the ground-states of the beryllium isotopes calculated within the framework of the AMD. The *first column* shows the total nucleon density (ρ) and the *middle* and *right-hand columns* the proton (ρ_p) and neutron densities (ρ_n). From [55]

wave-packets in space, $\phi_{\mathbf{X}_i}(\mathbf{r}_j) \propto \exp(-\nu(\mathbf{r}_j - \mathbf{X}_i/\sqrt{\nu})^2)$, but also possess spin (χ_i) and isospin character (τ_i): $\varphi_i = \phi_{\mathbf{X}_i}\chi_i\tau_i$. The wave-function is parameterized in terms of a complex set of variables $\mathbf{Z}$ describing the spin and geometry of the wave-function. The energy of the system is computed, variationally, utilizing an *effective* nucleon-nucleon interaction (see Ref. [55] for more details). The flexibility of this approach allows a suitable description of cluster and shell-model type systems, alike, and the structure emerges naturally from the details of the nucleon-nucleon interaction under the guidance of the Pauli Exclusion Principle.

An example of the appearance of the precipitation of clusters from the nucleon-nucleon interaction within the framework of the AMD is shown in Fig. 1.13 for the beryllium isotopes $^{6-14}$Be. All isotopes possess a proton distribution which is prolate and clustered. The role of the neutrons is clear. When the neutron number is the same as that of the protons (^{8}Be) the separation of the proton-cores is maximal (maximum clustering), whereas neutrons in more spherical distributions cause the separation of the proton centers to be reduced. This model has been widely applied, but with a particular focus on the Li, Be, B and C isotopes, see Ref. [55] and references therein. In general the model reproduces well both experimental binding energies, transition rates, radii and moments. Figure 1.14 shows some examples of the rather close agreement between the AMD calculations and the experimental electric quadrupole moments and electromagnetic transition rates.

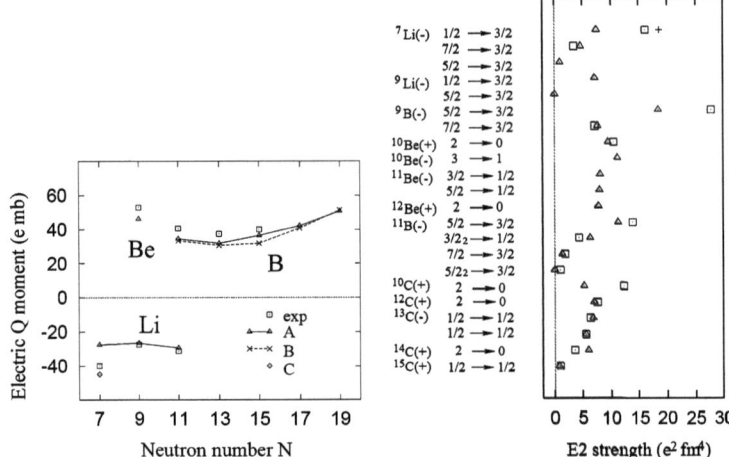

Fig. 1.14 (*Left*) Electric quadrupole moments for Li, Be and B isotopes. The *squares* are experimental data and other symbols are the AMD calculations with slightly different interactions or constructions. (*Right*) E2 transition strengths for Li, Be, B and C isotopes. The *squares* are the experimental data points, the other symbols are the AMD calculations. See Ref. [55] for further details

An alternate approach to AMD which contains an additional degree of freedom, namely each nucleon is represented by two Gaussian wave-packets, is fermionic molecular dynamics (FMD) [56]. Moreover, the interaction employed (Unitary Correlation Operator Method—UCOM) includes a tensor component. The features of these calculations essentially coincide with those of the AMD, but the variable Gaussian width should allow, in principle, a better description of shell-model like states and should potentially provide a better description of weakly bound states. The recent calculations for the structure of the 7.65 MeV state in ^{12}C are of particular note [57].

1.4.5 Ab Initio Type Models

Ultimately, it is important to be able to push beyond models which either employ assumptions of preformed clusters or effective interactions. The Green's Function Monte Carlo (GFMC) method, described earlier, uses realistic two-body interactions with a parameterization of the 3-body force. Not only does this method reproduce the properties of light nuclei up to $A = 12$ rather precisely, but, as shown in Fig. 1.2, also indicates the emergence of cluster like structures in nuclei such as ^{8}Be [1].

Another approach which attempts to extend beyond the shell model is the no-core shell model (NCSM) in which realistic interactions are used but with a set of basis states which are harmonic oscillator states [58]. This approach provides an analytic basis for the construction of the many-body Slater determinants. The downside is that HO wave-functions do not have the appropriate asymptotic behavior (as a function of r), which means that they tend not to be a good description of weakly bound systems, and also all states of the system end up being effectively bound due to the nature of the potential. As its name suggests the interaction between all nucleons is taken into account (rather than the valence nucleons beyond the closed shell) and it may use a variety of interactions including those used in the GFMC approach (the Argonne potentials) and those from Effective Field Theory (EFT).

In this latter case the interaction is grown from QCD by including various types of exchange processes which in leading order include one pion exchange terms. Higher order corrections include more complex processes for example next to leading order (NLO) includes 2 pion exchange and terms which correspond to pions being radiated and absorbed by a single nucleon which interacts with a second via pion exchange (called renormalisation of 1 pion exchange). Current models extend to N3LO (next to, next to, next to leading order) which amongst other components would include 3 pion exchange components [59, 60] and even N4LO.

Calculations of the states of ^{12}C using the no-core shell model [61, 62] struggle to reproduce the excitation of the 7.65 MeV, 0^+, Hoyle state without an extension of the basis to include excitations to HO levels at very high energies (large $\hbar\omega$). The Hoyle-state has long been known to possess a cluster-like structure and the failure of the NCSM to capture the detail of this state without a significant expansion of the basis is thus not surprising. In fact, this may be taken as a signature of clusterization.

Finally, a rather promising development is the use of chiral EFT interactions in lattice based calculations. A series of calculations of the structure of the states in ^{12}C, including the Hoyle-state, have been performed. These point to both clusterization and a rather different structure of the ^{12}C ground and excited states [63–65]. The lattice spacing used in these calculations remains rather coarse, but further optimization has the potential for providing great insight into the structure of light clustered systems and their reactions.

1.5 Experimental Examples of Clustering

1.5.1 The Example ^{8}Be

The ground-state of ^{8}Be is unbound to 2α decay by 92 keV, and has a lifetime of $\sim 10^{-16}$ s. It has a first excited 2^+ state at 3.03 MeV with a width of 1.51 MeV and a 4^+ state at 11.35 MeV with a width of 3.5 MeV. These three states have an energy separation which is consistent with a rotational behaviour given by $\hbar^2 J(J+1)/2\mathscr{I}$, where $\mathscr{I}$ is the moment of inertia. The value for the moment of inertia that one extracts is consistent with the picture of two touching α-particles, an essentially super-deformed nucleus. Indeed the Green's Function Monte Carlo calculations [1] reproduce the spectrum of excited states which reinforces this interpretation. There appears to be little doubt that clusterisation is a dominant factor in the structure of the ^{8}Be nucleus.

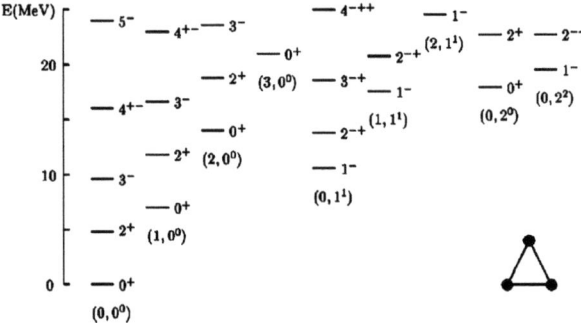

Fig. 1.15 Spectrum of the energy levels of an equilateral triangle configuration. The *bands* are labeled by $(v_1; v_2^l)$ [66]

1.5.2 The Structure of ^{12}C

If the structure of the ^{12}C ground state is influenced by clustering or the symmetries thereof, then the system can be constructed from a variety of geometric arrangements of three α-particles. It might be expected that the compact equilateral-triangle arrangement is the lowest energy configuration. Such an arrangement possesses a D_{3h} point group symmetry. The corresponding rotational and vibrational spectrum is described by a form [66]

$$E = E_0 + Av_1 + Bv_2 + CL(L+1) + D(K \pm 2l)^2 \quad (1.16)$$

where $v_{1,2}$ are vibrational quantum numbers, and v_2 is doubly degenerate; $l = v_2, v_2 - 2, \ldots, 1$ or 0, L is angular momentum, M its projection on a laboratory fixed axis and K a body-fixed axis [66]. A, B, C and D are adjustable parameters. The spectrum of states predicted by the choice $A = 7.0$, $B = 9.0$, $C = 0.8$ and $D = 0.0$ MeV is shown in Fig. 1.15.

The ground state band, $(v_1; v_2^l) = (0, 0^0)$, contains no vibrational modes and coincides well with the observed experimental spectrum. Here the states correspond to different values of K ($K = 3n$, $n = 0, 1, 2 \ldots$) and L. For $K = 0$, $L = 0, 2, 4$ etc., which is a rotation of the plane of the triangle about a line of symmetry, whereas for $K > 0$ $L = K, K+1, K+2, \ldots$. In the present case, $K = 0$ or 3 is plotted with the parity being given by $(-1)^K$. The $K = 0$ states coincide well with the well-known 0^+ (ground-state), 2^+ (4.4 MeV) and 4^+ (14.1 MeV) states. The $K = 3$ states correspond to a rotation about an axis which passes through the center of the triangle, with each of the α-particles carrying one unit of angular momentum. The first state has spin and parity 3^- and coincides with the 9.6 MeV, 3^-, excited state. The next such state would be $K = 6$, $J^\pi = 6^+$. A prediction of this model is that there should be a 4^- state almost degenerate with the 4^+ state. A recent measurement involving studies of the α-decay correlations indicated that the 13.35 MeV unnatural-parity state possessed $J^\pi = 4^-$ [67]. The close degeneracy with the 14.1 MeV 4^+ state would appear to confirm the D_{3h} symmetry. The rotational properties of these states are given by

$$E_{J,K} = \frac{\hbar^2 J(J+1)}{2\mathscr{I}_{Be}} - \frac{\hbar^2 K^2}{4\mathscr{I}_{Be}} \quad (1.17)$$

where $\mathscr{I}_{Be}$ is the moment of inertia corresponding to two touching α-particles which can be determined from the ^{8}Be ground-state rotational band [3].

Historically, one of the pre-eminent tests of our understanding of the structure of light nuclei lies in the nature of the second excited state in ^{12}C. This system resides at the limits of many of the *ab-initio* approaches. This state has character $J^\pi = 0^+$ and lies at $E_x = 7.65$ MeV. It is known as the Hoyle-state as it was predicted by Fred Hoyle [68, 69] as a solution to the discrepancy between the observed and predicted abundance of ^{12}C. ^{12}C is synthesized in the triple-α process, whereby the two α-particles briefly fuse to make ^{8}Be and at sufficient densities there is a finite probability of capturing a third α-particle to form ^{12}C. The 7.65 MeV state serves as a *doorway* resonance, substantially enhancing the reaction-rate. Without this resonance, or even if its energy were slightly different, the abundance of carbon would be dramatically reduced as would that of carbon based life-forms.

In the description illustrated in Fig. 1.15 the 0^+ state at 7.65 MeV corresponds to a vibrational mode ($v_1 = 1$). The coupling of rotational modes would then produce a corresponding 2^+ state at 4.4 MeV above 7.65 MeV, i.e. 12.05 MeV. There is no known 2^+ state at this energy, pointing to the more complex structure of this state. If the 7.65 MeV state in ^{12}C has a structure similar to that of the ground-state then a 2^+ state close to 12 MeV is expected. The closest state which has been reported with these characteristics is at 11.16 MeV [70]. This state was observed in the ^{11}B(^{3}He, d)^{12}C reaction, but has not been observed in measurements subsequently. A re-measurement of this reaction using the K600 spectrometer at iThemba in South Africa demonstrates that the earlier observation of a state at 11.16 MeV was an experimental artifact and no such state exists [71]. This introduces an interesting set of possibilities which lie at the heart of uncovering the structure of the Hoyle-state. If the Hoyle-state is more deformed than the ground-state, and the system behaves in a rotational fashion, then the 2^+ state would be lower in energy and an alternative possibility is that the Hoyle-state possesses no collective excitations. It has been suggested that due to the close proximity of the Hoyle-state close to the 3α-decay threshold, bound only by the presence of the Coulomb barrier, that the system obtains a bosonic rather than fermionic identity and that the α-particle bosons behave like a weakly interacting bosonic gas or even a bosonic condensate [25]. The resolution of the structure may follow from the identification of the 2^+ excitation—or otherwise.

Recent studies of the ^{12}C(α, α') [72–74] and ^{12}C(p, p') [75, 76] reactions indicate the presence of a 2^+ state close to 9.6–9.7 MeV with a width of 0.5 to 1 MeV. The state is only weakly populated in these reactions, presumably due to its underlying cluster structure, and is broad. Consequently, its distinction from other broad-states and dominant collective excitations (e.g. the 9.6 MeV, 3-) makes its unambiguous identification challenging. Further, and perhaps definitive, evidence for such an excitation comes from measurements of the ^{12}C($\gamma, 3\alpha$) reaction performed at the HIGS facility, TUNL [77] in the US. Here a measurable cross section for this process was observed in the same region of 9–10 MeV which cannot be attributed to known states in this region. Furthermore, the angular distributions of the α-particles are consistent with an $L = 2$ pattern, indicating a dominant 2^+ component. Based

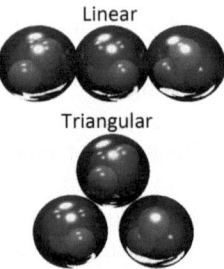

Fig. 1.16 Different arrangements of α-particles. The closest possibility fitting the experimental data is the triangular arrangement

on a rather simple description of this state in terms of three α-particles with radii given by the experimental charge radius (see Fig. 1.16 for possible arrangements), it is possible to use the 2 MeV separation between the Hoyle-state and the proposed 2^+ excitation to draw some conclusions as to the arrangements of the clusters. This would indicate that rather than a linear arrangement of the three clusters, a more appropriate description would be a loose arrangement of the α-particles in something approaching a triangular structure.

A natural extension of such a conclusion is that there should also be a collective 4^+ state. Using the simple $J(J+1)$ scaling, a 4^+ excitation close to $E_x(^{12}C) = 14$ MeV would be expected. Recent measurements of the two reactions $^9\text{Be}(\alpha, 3\alpha)n$ and $^{12}\text{C}(\alpha, 3\alpha)^4\text{He}$ have been performed [78]. These measurements indicate a candidate state close to 13.3 MeV with a width estimated to be 1.7 MeV. It is believed that this is not a contaminant and is observed with similar properties in all spectra. Angular correlation measurements made using the ^{12}C target are not definitive, but indicate a 4^+ assignment.

1.6 Experimental Techniques—Break-up and Resonant Scattering Reactions

A determination of the structure of light nuclei above the particle decay threshold, where gamma-decay ceases to be dominant, is challenging. In order to characterize the nature of excited states, the energies, total and partial widths and spins and parities should be determined. There are few experimental techniques which permit all of these quantities to be determined simultaneously.

1.6.1 Resonant Scattering

One approach which recently has found greater favor is thick target resonant scattering [79]. Here a beam passes through a thick target loosing energy as it traverses the medium. By far the majority of the interactions are with the atomic electrons slowing the beam, however occasionally a nuclear interaction takes place. The cross section for resonant capture reaches hundreds of millibarns. The resonance in the

Fig. 1.17 Resonant scattering using a thick helium target

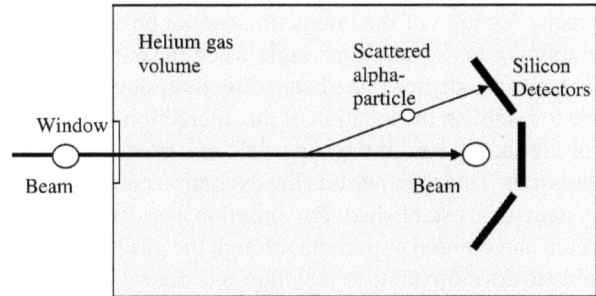

target-beam composite system then can decay either back into the entrance channel or into other final states. The fact that the beam energy is continuously varying in the medium means that the center-of-mass energy is scanned—this results in a technique which is considerably more efficient than the traditional excitation function measurements, where the beam energy must be re-tuned for each data point. For elastic resonant scattering involving a projectile and an α-particle target the cross section is given by

$$\sigma(E) = \pi \lambda^2 \frac{2J+1}{2J_1+1} \frac{\Gamma_\alpha^2}{(E-E_r)^2 + (\Gamma/2)^2} \quad (1.18)$$

where J is the spin of the resonance, J_1 is the spin of the projectile, E_r the energy of the resonance and Γ and Γ_α the total and α-partial widths, respectively. The cross section thus scales linearly with J and quadratically with Γ_α—the greater the degree of clusterization the larger the partial width and the larger the cross section. Resonant elastic scattering from an α-particle target is thus ideally matched to the study of cluster states. For inverse kinematics, where the beam is heavier than the target, the resolution with which it is possible to reconstruct excited states can exceed the energy resolution of the detection system.

The experimental approach is illustrated in Fig. 1.17. The beam, of energy typically a few MeV/u, passes through a window, which is typically Havar or Mylar of thickness 5 µm, to contain the helium gas with pressures up to about 1 atmosphere. The beam loses energy and undergoes energy-loss straggling as it passes through the window and the target gas. This leads to a loss in resolution. As the beam traverses the gas volume it again decelerates until finally it is stopped. The range of the beam is adjusted via the variation of the gas pressure, such that the beam stops immediately in front of the detectors. Of course if the range exceeds the distance to the detectors and the beam is sufficiently intense the detectors will be destroyed. Any interaction with an α-particle along the path of the beam has the potential to result in elastic scattering—either resonant or non-resonant. These two processes will interfere with each other. For center-of-mass angles close to 180 degrees the α-particles will be emitted in the same direction as the beam and since typically the beam has a mass and charge in excess of α-particles, the scattered α-particles have a lower energy loss in the gas and thus can reach the detectors. The main drawback for this approach arises when a helium gas target is extended and in this instance the precise location of the interaction cannot be determined. This means that there is an ambiguity in the emission angle when the α-particle is detected in the silicon array. Only at zero degrees (the beam direction) does this problem vanish. Here it is possible to establish the location of the interaction within the gas volume and thus correct for the energy loss of the α-particle as it traversed the gas and hence the energy upon emission. This then permits the excitation energy of the composite target+projectile system to be established. For emission away from zero degrees the path length of the beam and emitted α-particle through the gas is harder to establish—though it is possible to develop iterative techniques to address this. The excitation energy resolution away from zero degrees tends to be correspondingly degraded.

The study of resonances in the ^{18}O + α system by Rogachev et al. [80] is shown in Fig. 1.18. This shows the energy spin systematics of the resonances observed in ^{22}Ne obtained using this technique [80]. The systematics of the energies in the bands are compared with those for ^{20}Ne and show a similar rotational trend, but for each rotational level the states are split into two components. It is possible that the states observed have a molecular structure in which two neutrons are exchanged between α-particle and ^{16}O cores.

1.6.2 Break-up Measurements

The utility of break-up reactions in the study of nuclear clustering has been reviewed in Ref. [81]. In this approach, states above particle decay channels with a particular type of cluster structure are observed to decay into the cluster components. The argument being, that if the states have large cluster widths then they are more likely to decay in a manner respecting this structure and hence the break-up spectrum is most strongly populated by cluster states. The reaction populating such states may range from inelastic scattering to transfer. Figure 1.19 shows the sequence of states populated in the ^{12}C(^{24}Mg, ^{12}C + ^{12}C)^{12}C inelastic scattering reaction. The experimental technique employed is akin to invariant mass spectroscopy and has been termed resonant particle spectroscopy. It involves the simultaneous detection of the two decay products (in this case two ^{12}C nuclei) using detectors which are capable of measuring both the energy and emission angles of the particles. If the detection system is capable of also determining the mass of the fragments then the momentum of the two fragments may be established. Using the principles of conservation of momentum it is possible to calculate the momentum of the ^{24}Mg nucleus before decay and hence its kinetic energy, $E(^{24}\text{Mg})$. The excitation energy then follows

$$E_x = E\left(^{12}\text{C}_1\right) + E\left(^{12}\text{C}_1\right) - E\left(^{24}\text{Mg}\right) - Q_{bu} \quad (1.19)$$

where Q_{bu} is the breakup threshold, which in this instance is -13.93 MeV. Momentum conservation also permits the energy of the recoil to be calculated and hence the three-body reaction Q-value to be calculated. In this way it is also possible to select events in which the decay proceeds only to the ground states of the three final-state

Fig. 1.18 Excited states of ^{22}Ne, populated in ^{18}O + α scattering. The energy above threshold is given as $E_{c.m.}$. *Upper figure top panel*: excitation functions of resonant elastic scattering at different *cm*-angles as indicated. *Upper figure lower panel*: Details of the upper part with enhanced regions, showing the fits used to determine the spin values as indicated. *Lower figure*: plots of the observed excitation energies as function of spin ($J(J + 1)$) for ^{22}Ne compared to corresponding values of ^{20}Ne, from Ref. [80]

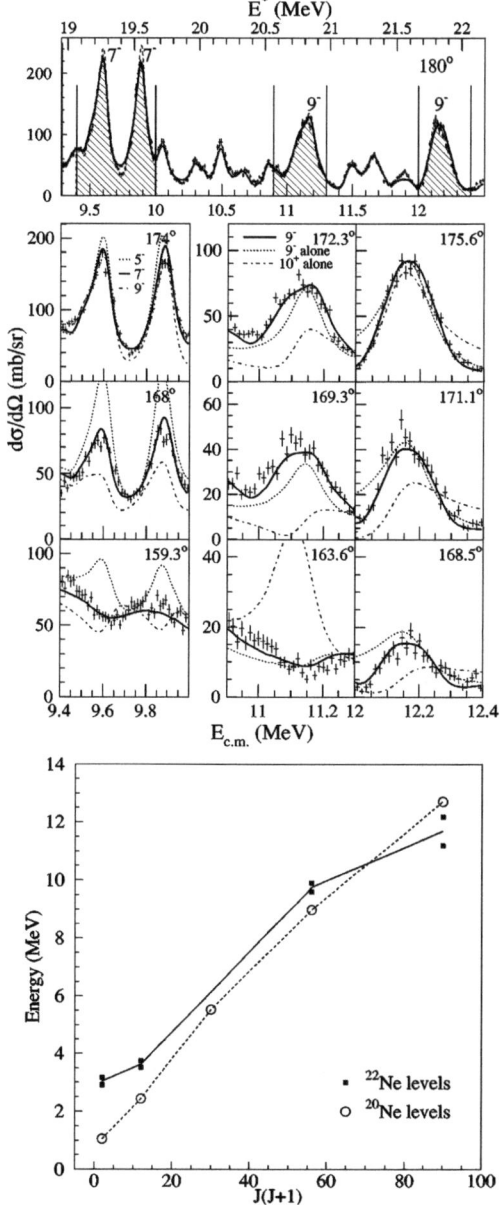

^{12}C nuclei. This is important otherwise the excitation energy spectrum contains an ambiguity corresponding to decays proceeding to the ^{12}C first excited state (2^+, 4.4 MeV).

A second advantage of being able to determine the fact that all three final state ^{12}C nuclei were produced in their 0^+ ground-states is that the technique of angular

Fig. 1.19 States in ^{24}Mg decaying into two ^{12}C nuclei populated in the ^{12}C(^{24}Mg, ^{12}C + ^{12}C)^{12}C inelastic scattering reaction. The *horizontal axis* is the excitation energy of the ^{24}Mg nucleus and the *vertical axis* represents the counts per bin. The *inset* shows the energy-spin systematics of the states which appear to follow a rotational behaviour consistent with a ^{12}C + ^{12}C cluster structure

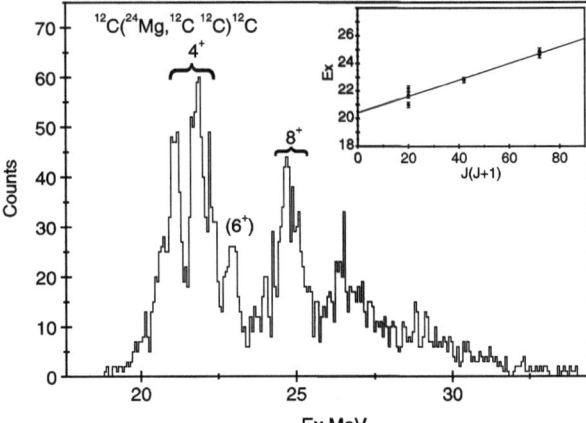

correlations may then be utilized. If all initial and final state nuclei are spin zero, then the mathematical form describing the angular distribution of the decay products is essentially that described by Legendre polynomials [82]. Given that there are two center-of-mass frames, the first associated with the inelastic scattering of the ^{24}Mg* nucleus and the second describing the decay of the ^{24}Mg* nucleus into two ^{12}C nuclei, there are two sets of angles and it is the correlation between these two processes which reveals the spin of the decaying ^{24}Mg* excited state. The angular correlation technique in principle permits quasi model independent spin determinations.

The breakup technique thus allows excitation energies and spins to be determined. However, it is often difficult to achieve excitation energy resolutions less than 100 keV and hence measuring the natural widths of states is challenging and in order to know the partial widths the excitation probability must also be determined which is also challenging. In some instances this has been overcome for example using a spectrometer to measure the recoil particle, e.g. [83], in order to determine the nature of the excitation energy spectrum prior to decay.

1.7 Beyond α-Clusters—Valence Neutrons and Molecules

Alpha-conjugate nuclei are clearly a very small subset of all those which exist in nature and in this instance that the clusterisation arises from the rather special properties that stem from the common orbitals in the mean-field limit. As has been observed this gives rise to particular symmetries which pervade both the mean-field and cluster model limits and may be interpreted as spatially localized clusters. When one moves away from such even N, Z, $N = Z$ nuclei then some of the energetic advantage associated with the α-particle are lost and the symmetries disturbed. The important question is, does clustering vanish at this point or does it remain influential even at the drip-lines? As described earlier on, there is evidence of the importance of correlations, or clustering, even at the drip-lines. The properties of ^{6}He may be traced both to the reliance of the α-particle and the effect of correlations between the neutrons [84]—the removal of one of the neutrons leaves the unbound ^{5}He. The presence of the α-particle also affects the binding of the two neutrons; the di-neutron is unbound. Understanding the behaviour of such finely balanced nuclei right at the drip-lines can give a deep insight into the intricacies of the strong nuclear force. In this instance it is the α-particle which forms part of the building block. Similarly the nuclei ^{6,7}Li possess $\alpha + d$ and $\alpha + t$ structures, respectively. The ^{6}Li ground-state spin of 1^+ would correspond to the $J^\pi = 0^+$ α-particle with a deuteron ($J^\pi = 1^+$) with a relative motion described by $L = 0$ (ignoring the small D-state component).

The first significant attempt to deal with the additional degrees of freedom that valence nucleons bring to systems was that of Hafstad and Teller [3]. This seminal piece of work set the ground rules for this field. These authors considered the sequence of nuclei, ^{5}He, ^{9}Be, ^{13}C and ^{17}O. The binding energies of these $4n + 1$ nuclei ($n = 1, 2, 3, \ldots$) depend on the α–α interaction energy, but also the character of the valence neutrons. The binding energy of the ^{5}He nucleus reflects the α–n interaction, whereas the $\alpha + n + \alpha$ nucleus ^{9}Be whilst containing similar terms in the Hamiltonian was recognised as having a contribution from an *exchange* interaction. Here, the systems were described in terms of the covalent exchange of neutrons between the α-cores. Again the building blocks are the α-particles and the neutrons are shared between the cores. This is highly reminiscent of the exchange of electrons in covalently bound atomic molecules. For example, the H_2^+ molecule is formed from two protons with a covalently exchanged electron. The electrons reside in single center s-orbitals and the covalent bond is formed from their linear combination:

$$\psi_\pm = \frac{1}{\sqrt{2}}(\varphi_1 \pm \varphi_2). \quad (1.20)$$

This generates two molecular wave-functions, one with no intermediate node (bonding) and a higher energy state with an internal node (anti-bonding). The development of atomic orbitals from symmetry adapted linear combinations (SALCs), is also widely used in molecular physics.

The exchange of neutrons between α-particle cores is a rather important concept which allows a detailed understanding of the structure of the beryllium isotopes to be developed [85–89]. The appearance of nuclear molecules is reviewed in [6]. The nucleus ^{9}Be demonstrates this beautiful piece of physics rather well. The $N = Z$ isotope ^{8}Be is unstable against α-decay, held together only by the Coulomb barrier for a period of $\sim 10^{-16}$ seconds. The only stable beryllium isotope is ^{9}Be. The additional neutron is exchanged between the cores just as electrons are exchanged between atoms in covalent atomic molecules. Thus, such states have been coined *nuclear molecules*. It is the delocalisation of the neutron which lowers its kinetic energy giving an enhanced binding energy for the ^{9}Be system compared to ^{8}Be. It is inferred from the neutron separation energy in ^{9}Be that the magnitude of the binding is approximately 1.6 MeV [86].

In the formation of such molecular states in the beryllium isotopes, the single-center wave-functions are those that the neutrons occupy in ^{5,6}He, i.e. $p_{3/2}$. Thus, one might expect the neutron to reside in covalent orbits, which are the analogues

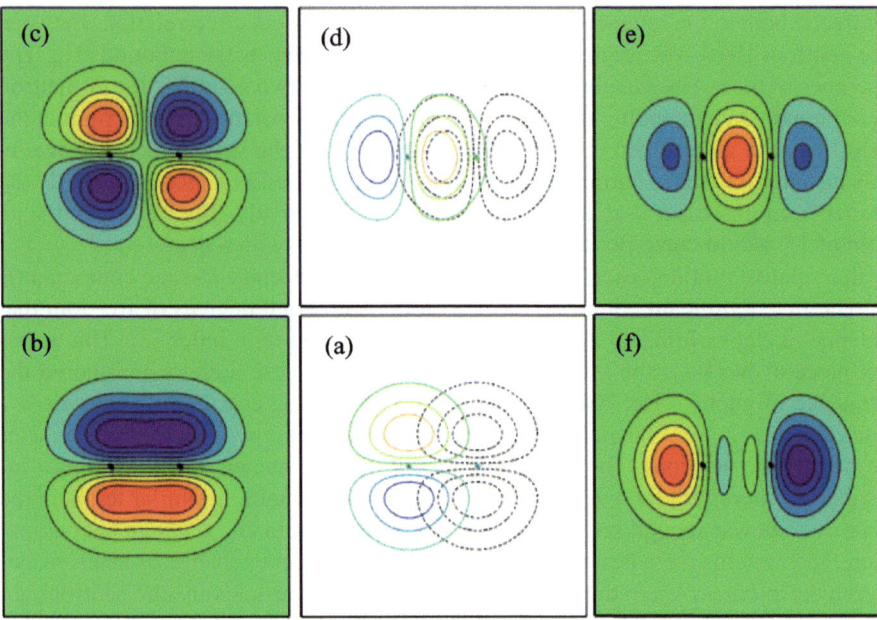

Fig. 1.20 Molecular orbitals associated with linear combinations of HO orbitals $[n_\perp, n_z] = [1, 0]$ and $[0, 1]$ orbits, equivalent to p-states. Here the z-direction is aligned with the separation axis of the two centers indicated by the *black dots*. (**a**) Shows the overlap of the two individual wave–functions. Diagrams (**b**) and (**c**) are the result of forming linear combinations: (**b**) corresponds to the binding π-state, and (**c**) to the anti-binding state. Diagram (**d**) shows the overlap of the two $(0, 0, 1)$ orbits, forming the σ-configurations, and (**e**) and (**f**) the two linear combinations

of those observed in carbon and oxygen molecules, namely σ and π-orbits, which are formed in the exchange of p-electrons.

To illustrate this, the possible linear combinations of the equivalent HO orbitals $[n_\perp, n_z] = [1, 0]$ and $[0, 1]$ are shown in Fig. 1.20. Note that there are two possible orientations of the dumbbell-like orbitals—either parallel or perpendicular to the axis separating the α-particles (though phases may vary). The linear combination shown in part (b) corresponds to the π-type structure for the valence neutron, and (e) to the σ-orbital. The notation σ and π corresponds to the projection of the angular momentum of the molecular orbit onto the symmetry axis of the molecule. If the linear combination of the p-orbits is considered, then for the orientation shown in Fig. 1.20a, this would correspond to $l = 1$ components along the separation axis and hence π-type orbits. For the alternate case, Fig. 1.20d, the projection of the orbital angular momentum of the two p-orbitals is perpendicular to the separation axis and thus the σ association (no angular momentum).

Figure 1.21 shows the energy evolution of the energy levels of the two-center shell model, where the Schrödinger equation is solved for two shell model potentials as a function of their separation—from infinite separation to zero. This model is one which is appropriate for the description of the merger of two nuclei (with

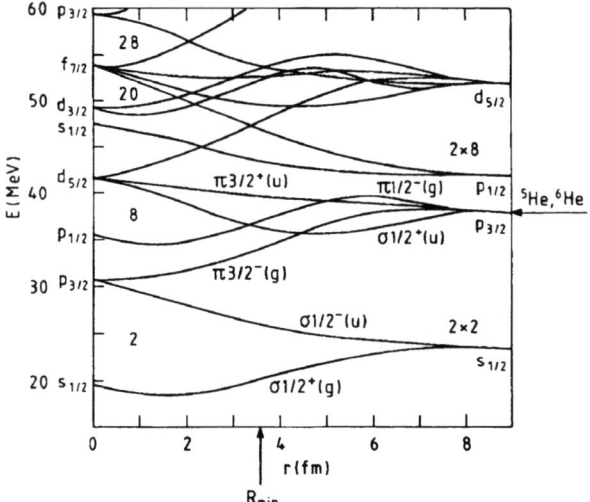

Fig. 1.21 The energy levels of the two center shell model from [87, 88]. The separation of the two potentials is defined in terms of the distance r. The present calculation corresponds to the energy-levels associated with the fusion of two ^{4}He nuclei. The separation at which the interaction potential reaches a minimum is marked, R_{min}—this would correspond to the ^{8}Be ground state

zero impact parameter forming a composite system) and traces the evolution of the initially degenerate energy levels in the two separate potentials to those of the merged system. As the separation varies, then the energy levels are essentially those of the prolate deformed nucleus and will strongly overlap with those found in the deformed shell, Nilsson, model (as illustrated in Fig. 1.22). The separation of the two potentials appropriate for two α-particles in the ground state of ^{8}Be is marked $R_{min} \sim 3.5$ fm in Fig. 1.21—the point at which the α–α potential attains its minimum. At this separation the two lowest energy orbits available for the neutron to follow are marked $\pi 3/2^-$ and $\sigma 1/2^+$. In fact the two levels are almost degenerate. These two orbits are analogues of the Nilsson orbitals from the $1p_{3/2}$ and $1d_{5/2}$ levels, with projections of the total angular momentum $K^\pi = 3/2^-$ and $1/2^+$, respectively.

A natural conclusion is that if such a description of ^{9}Be is correct then the ground state of ^{9}Be should be the head of a rotational band associated with $K^\pi = 3/2^-$. There should also be a second band linked with a $K^\pi = 1/2^+$ configuration and both bands should have a similar rotational gradient as that of the ^{8}Be ground state. In fact one would expect the $K^\pi = 1/2^+$ band to be slightly more deformed than the ground state band as the valence neutron in the σ-configuration intercedes between the two α-particles enhancing the deformation. Figure 1.23 shows the experimental situation for the nuclei ^{8}Be, ^{9}Be and ^{10}Be. The data indeed confirms the prediction; aside from the fact that the $K = 1/2$ band possesses Coriolis decoupling. For such bands, an additional term is introduced with an associated Coriolis decoupling parameter a,

$$E_J = \frac{\hbar^2}{2\mathscr{I}}[J(J+1) + (-)^{J+1/2}a(J+1/2)] \qquad (1.21)$$

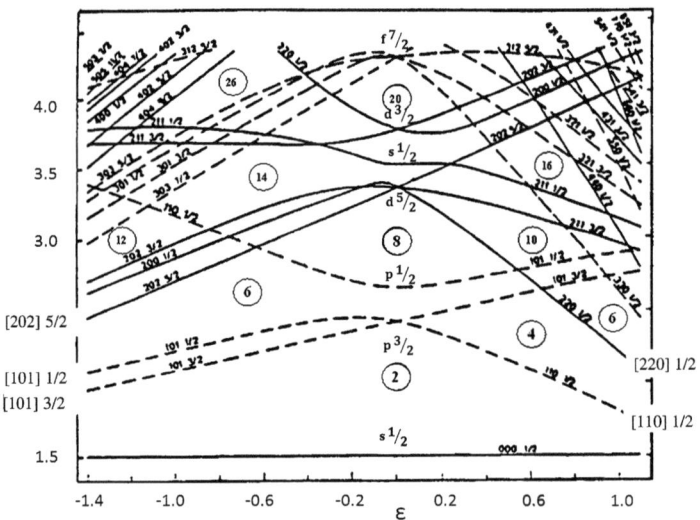

Fig. 1.22 The Nilsson single-particle energy levels. The parameter ε corresponds to the deformation of the potential. The magic numbers are labelled as are some of the key Nilsson orbits [90]

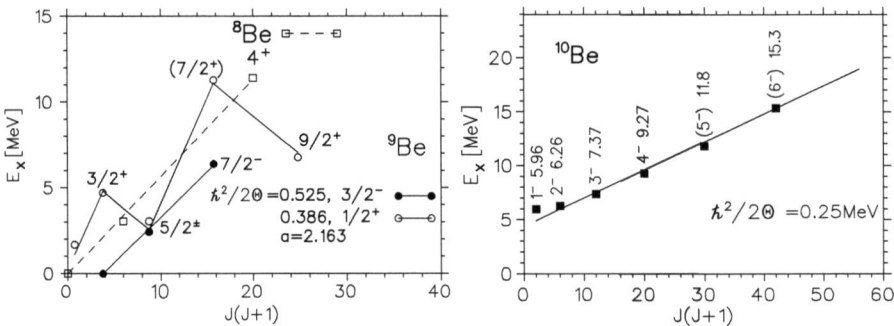

Fig. 1.23 Rotational bands of ^{8}Be, ^{9}Be (*left side*) and ^{10}Be (*right side*). The excitation energies are plotted as a function of angular momentum $J(J+1)$. The Coriolis decoupling parameter, a, for the $K = 1/2$ band is indicated. From Ref. [91]

$\mathscr{I}$ being the moment of inertia. It should be noted that the experimental moment of inertia for the $K = 1/2$ band is indeed larger than for the $K = 3/2$ ground-state band (as indicated in the left hand part of the figure).

1.7.1 The Neutron-Rich Nucleus ^{10}Be

The addition of two neutrons to the two α-particles results in the formation of ^{10}Be. The AMD calculations for the nucleus are shown in Fig. 1.24 [92]. The contour

1 Clustering in Light Nuclei; from the Stable to the Exotic   31

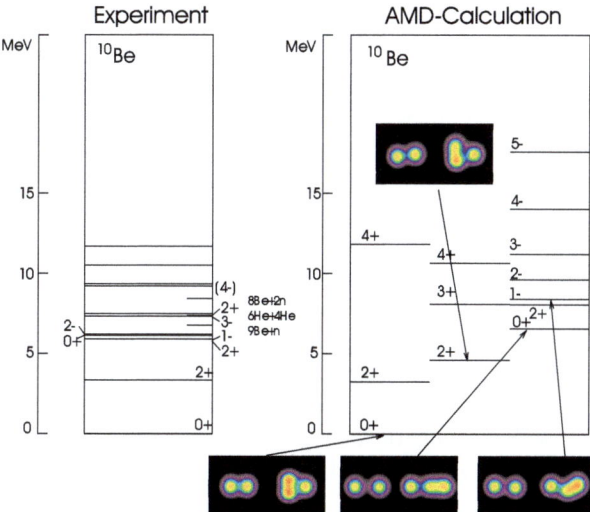

Fig. 1.24 *Left side*: Experimental level scheme of ^{10}Be, and, *right side*: that calculated with the spin parity projected AMD model [92]. The density plots of the intrinsic states are shown in special panels: for protons on the *left side* of each plot and for neutrons on the *right side*, respectively. The proton densities represent the positions of the α-particles. The neutron densities show for the ground state a density distribution characteristic of π-binding. The higher lying 0_2^+ state is well reproduced with a larger α-α distance (seen by the density of protons) as compared to the ground state, it shows the σ^2 configuration for the neutrons. The density of the 1^- state shows a mixture of σ-π orbitals with a distorted neutron density

plots show the density of the protons (left side) and neutrons (right side). In the case of the protons the α-particle structure can be clearly be seen. In the first 0^+ state (ground-state) the separation of the "α-particles" is smaller than that corresponding to the next 0^+ state (0_2^+). In the molecular picture this can be understood in terms of the orbitals of the valence neutrons. In the ground-state the neutrons occupy the π-orbital, forming a bridge between the two centers, whilst for the second 0^+ state the neutrons intercede between the two α-particles in a σ-orbital. The effect of the Pauli Exclusion Principle is to make it energetically unfavorable for the valence neutrons and those in the α-particle to overlap and hence the two α-particles are forced apart in order to minimize the energy of the configuration.

The 0_2^+ state should thus be the more deformed of the two—in fact could be the most deformed nuclear state yet seen in nature—experimentally it is found at 6.1793 MeV. The gamma-decay of this state is suppressed (it possesses a lifetime of the order of 1 ps)—an isomeric behavior that may be understood in terms of the small overlap of its structure and that of the more compact ground state. The excited state at 7.542 MeV (2^+) is believed the first rotational member of the associated band. This state lies very close to the α-decay threshold (7.409 MeV) and thus its decay to this channel is strongly suppressed by the Coulomb and ($L = 2$) centrifugal barriers. Nevertheless, the α-decay has been found to correspond to a very large reduced width [93], representative of the large degree of clusterisation associated with the state. This is however a single, and very challenging, measurement and needs confirming.

The 4^+ member of the same band would lie in the region of 10–11 MeV. There are a number of possible states which could correspond to the molecular band; 10.15 MeV and 10.57 MeV. The spin of the latter state is unknown, whereas the former has been associated with spins 3^- [94] and 4^+ [95]. The latter assignment was also found in a measurement of the resonant scattering of ^{6}He+^{4}He [96]. Recent re-measurements of resonant scattering verify the 4^+ assignment [97]. The energy and width of the state are consistent with the interpretation of an extremely deformed rotational band with a well-developed cluster structure.

Determination of the structure the ground-state of ^{10}Be cannot be readily made using particle spectroscopy techniques. A recent set of measurements of the electromagnetic transition strengths, $B(E2)$, between the 2^+ and 0^+ ground-state for ^{10}Be and ^{10}C, together with the isobaric analogue state in ^{10}B [98, 99] (made with an unprecedented precision) provide a significant benchmark against which the character of the state may be fixed.

The observations made for the nuclei ^{9}Be and ^{10}B may be extended to more complex $2\alpha + Xn$ systems such as ^{11,12}Be where the valence neutrons can be thought of as occupying combinations of σ and π-orbitals. The interactions between these valence particles will perturb the zeroth order molecular picture, but it is understood that some of the molecular characteristics are retained [6].

1.7.2 More Complex Molecular States and the Extended Ikeda Diagram

1.7.2.1 Asymmetric Cores

Based upon the concept that neutrons may be exchanged between α-particles it has been proposed that it may also be possible to form covalent structures from non-α cores in other systems. The next best two centered case corresponds to cores formed from an α-particle and an ^{16}O nucleus. ^{16}O possesses a closed shell, but not quite the degree of inertness of the α-particle (it has a first excited state of 6.05 MeV, 0^+, compared with 20.2 MeV). Nuclei formed from these two components produce neon isotopes. The nucleus ^{20}Ne is known to have a well-developed $\alpha + ^{16}$O cluster structure [55, 100, 101], the asymmetric structure giving rise to two rotational bands of $K^\pi = 0^\pm$ character [102]. The question as to what happens to valence neutrons introduced into this system was addressed by von Oertzen [103]. When the neutron orbits the α-particle it lies in a p-orbital (negative parity), when orbiting the closed shell ^{16}O it resides in the sd-shell (positive parity, associated with the $5/2^+$ ground-state).

The two orbitals which are aligned with the intrinsic deformation of the α-^{16}O system are linked to the harmonic oscillator levels $[n_x, n_y, n_z] = [0, 0, 1]$ and $[0, 0, 2]$. These are associated with the Nilsson orbitals with projections $K^\pi = 1/2^-$

Fig. 1.25 The covalent exchange of a neutron between the ^{16}O and α cores that occurs in the neon isotopes, from Ref. [104]

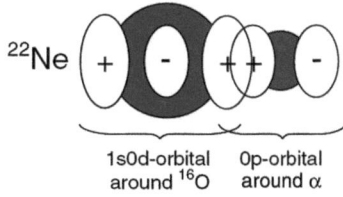

and $1/2^+$; both have σ-character. The strong overlap of these two orbitals in the region between the cores gives rise to the molecular binding effect, illustrated in Fig. 1.25. The resulting hybridized orbital gives rise to parity doublet bands [103]. For a complete description of the molecular bands that appear in the neutron-rich isotopes see Ref. [6].

1.7.2.2 More than Two Centers

The obvious extension from $2\alpha + Xn$ systems is to nuclei composed of 3α-particles—carbon isotopes. In this instance the α-particle cores may adopt a number of different arrangements. Two possible limits are a triangular and linear arrangements. This creates a greater spectrum of molecular states and hence complexity. There are a number of theoretical predictions for the appearance and characteristics of such states [6, 105, 106]. From the experimental perspective there is no definitive evidence for their existence [6], though measurements of ^{13}C and ^{14}C indicate possible rotational bands with the right characteristics.

1.7.2.3 The Extended Ikeda Diagram

The possibility that beyond α-conjugate nuclei there exists a series of states whose properties are strongly influenced by the underlying α-particle cluster structure, where the valence particles have the imprint of molecular exchange of the valence particles opens up some exciting possibilities. The present state-of-play is that only a few of these possibilities have been characterized. Understanding the conditions under which these states might appear is important and one element is the threshold energy—in most cases the states will not be close to the ground-state. Motivated by the Ikeda Diagram for α-conjugate cases (Fig. 1.5) von Oertzen has devised an *extended Ikeda Diagram* which is shown in Fig. 1.26. The diagram charts the expected location of these exotic states in terms of the constituent particle decay thresholds.

There are a number of intriguing possibilities in terms of new structures, for example the evolution of the behavior of the Beryllium isotopes ^{8}Be to ^{12}Be with increasing numbers of neutrons is indicated, as is the evolution of the 3α-systems.

Perhaps the one that captures the imagination most is that characterized by the structure $\alpha + 2n + {}^{16}\text{O} + 2n + \alpha$, in ^{28}Mg. This state has been called *nuclear water* due to the similarity with the atomic H_2O, however due to the nature of the valence orbitals it more closely resembles CO_2. Identification of such a structure would be an experimental tour-de-force.

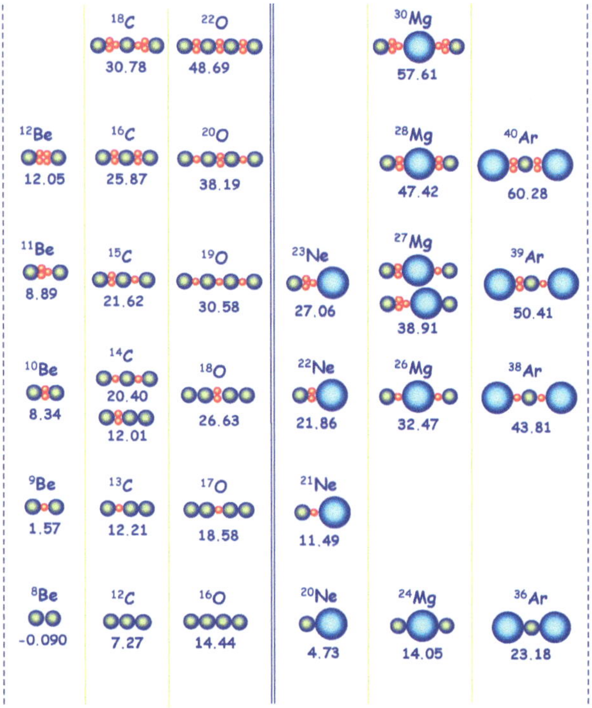

Fig. 1.26 The modified Ikeda Diagram proposed by von Oertzen. The *left hand side* shows the case for structures composed of α-particles (*blue*) and neutrons (*red*). The *right hand side* shows the case for larger cores of ^{16}O nuclei (*larger blue spheres*). The numbers shown are the excitation energies at which the cluster structures are expected to appear and correspond to the cluster binding energies

1.8 Summary and Conclusions

The history of clustering reaches to the earliest days of nuclear science when some of the first models captured nuclear properties in terms of constituent α-particles. Though the initial pictures have been found to be overly simplistic there are a number of cases where nuclei appear to have a behavior which reflects a well-developed α-particle structure. Key examples of these states are the ^{8}Be ground-state and the 7.65 MeV, 0$^+$, Hoyle-state in ^{12}C. It is nuclei such as this which have become the touchstones for the development of state-of-the-art nuclear models. Much of nuclear science has moved from this territory to the drip-lines—the limits of isospin stability. It is here that there is a significant increase in the number of neutrons, for example. It is in such systems that there is a co-existence of the boson and fermionic degrees of freedom and the valence neutrons can be thought of as being covalently exchanged between the α-particle cores. Though systems such as ^{9}Be and ^{10}Be are well characterized in these terms, the precise influence at the drip-lines has yet to be established. There is no doubt that correlations at the drip-lines play a defining role, but the question for the future is if they precipitate clusterization.

Acknowledgements The author would like to acknowledge his many colleagues who have worked on the development of both experimental and theoretical ideas contained in this review.

References

1. R.B. Wiringa, S.C. Pieper, J. Carlson, V.R. Pandharipande, Phys. Rev. C **62**, 014001 (2000)
2. J. Fujita, H. Miyazawa, Prog. Theor. Phys. **17**, 360 (1957)
3. L.R. Hafstad, E. Teller, Phys. Rev. **54**, 681 (1938)
4. H. Morinaga, Phys. Rev. **101**, 254 (1956)
5. K. Ikeda, N. Tagikawa, H. Horiuchi, Prog. Theor. Phys. Suppl., extra number, 464 (1968)
6. W. von Oertzen, M. Freer, Y. Kanada En'yo, Phys. Rep. **432**, 43 (2006)
7. N. Itagaki, W. von Oertzen, S. Okabe, Phys. Rev. C **74**, 067304 (2006)
8. A. Bohr, B.R. Mottelson, *Nuclear Structure*, vol. II (Benjamin, Reading, 1975)
9. M. Freer, R.R. Betts, A.H. Wuosmaa, Nucl. Phys. A **587**, 36 (1995)
10. W. Nazarewicz, J. Dobaczewski, Phys. Rev. Lett. **68**, 154 (1992)
11. T. Bengtsson, M.E. Faber, G. Leander, P. Moller, M. Ploszajczak, I. Ragnarsson, S. Aberg, Phys. Scr. **24**, 200 (1981)
12. W.D.M. Rae, Int. J. Mod. Phys. **3**, 1343 (1988)
13. N. Itagaki, S. Aoyama, S. Okabe, K. Ikeda, Phys. Rev. C **70**, 054307 (2004)
14. N. Itagaki, H. Masui, M. Ito, S. Aoyama, Phys. Rev. C **71**, 064307 (2005)
15. H. Masui, N. Itagaki, Phys. Rev. C **75**, 054309 (2007)
16. R.E. Peierls, J. Yoccoz, Proc. Phys. Soc. Lond. A **70**, 381 (1957)
17. J.M. Blatt, Austr. Math. Soc. **1**, 465 (1960)
18. H. Margenau, Phys. Rev. C **59**, 37 (1941)
19. D.M. Brink, in *Proceedings of the International School of Physics "Enrico Fermi", Course 36*, Varenna, 1965, ed. by C. Bloch (Academic Press, New York, 1966), p. 247
20. W. Bauhoff, H. Schultheis, R. Schultheis, Phys. Rev. C **29**, 1046 (1984)
21. S. Marsh, W.D.M. Rae, Phys. Lett. B **180**, 185 (1986)
22. A.C. Merchant, W.D.M. Rae, Nucl. Phys. A **549**, 431 (1992)
23. J. Zhang, W.D.M. Rae, Nucl. Phys. A **564**, 252 (1993)
24. J. Zhang, W.M.D. Rae, A.C. Merchant, Nucl. Phys. A **575**, 61 (1994)
25. A. Tohsaki et al., Phys. Rev. Lett. **87**, 192501 (2001)
26. Y. Funaki, A. Tohsaki, H. Horiuchi, P. Schuck, G. Röpke, Phys. Rev. C **67**, 051306 (2003)
27. T. Yamada, P. Schuck, Phys. Rev. C **69**, 024309 (2004)
28. G. Röpke, P. Schuck, Mod. Phys. Lett. A **21**, 2513 (2006)
29. Y. Funaki, A. Tohsaki, H. Horiuchi, P. Schuck, G. Röpke, Eur. Phys. J. A **28**, 259 (2006)
30. I. Sick, J.S. McCarthy, Nucl. Phys. A **150**, 631 (1970)
31. A. Nakada, Y. Torizuka, Y. Horikawa, Phys. Rev. Lett. **27**, 745 (1971) and 1102 (Erratum)
32. P. Strehl, Th.H. Schucan, Phys. Lett. B **27**, 641 (1968)
33. J.A. Wheeler, Phys. Rev. **52**, 1083 and 1107 (1937)
34. D.L. Hill, J.A. Wheeler, Phys. Rev. **89**, 1102 (1953)
35. J.J. Griffin, J.A. Wheeler, Phys. Rev. **108**, 311 (1957)
36. K. Wildermuth, W. McClure, in *Cluster Representations of Nuclei*. Springer Tracts in Modern Physics, vol. 41 (Springer, Berlin, 1966)
37. K. Wildermuth, Y.C. Tang, in *A Unified Theory of the Nucleus* (Academic Press, New York, 1977)
38. A. Arima, H. Horiuchi, K. Kubodera, N. Takigawa, Adv. Nucl. Phys. **5**, 345 (1972) (Plenum, New York, 1972). Edited by Baranger M. and Vogt E.
39. H. Furutani, H. Kanada, T. Kaneko, S. Nagata, H. Nishioka, S. Okabe, S. Saito, T. Sakuda, M. Seya, Prog. Theor. Phys. Suppl. **68**, 193 (1980)
40. Y.C. Tang, M. LeMere, D.R. Thompson, Phys. Rep. **47**, 167 (1978)
41. K. Langanke, H. Friedrich, Adv. Nucl. Phys. **17**, 223 (1987) (Plenum, New York, 1987). Edited by Negele J.W. and Vogt E.
42. K. Langanke, Adv. Nucl. Phys. **21**, 85 (1994) (Plenum, New York, 1994). Edited by Negele J.W. and Vogt E.
43. D. Baye, P.-H. Heenen, Nucl. Phys. A **233**, 304 (1974)
44. D. Baye, P.-H. Heenen, M. Liebert-Heinemann, Nucl. Phys. A **291**, 230 (1977)

45. D. Baye, P. Descouvement, in *Proceedings of the fifth International Conference on "Clustering aspects in nuclear and subnuclear systems"*, Kyoto, Japan, 25th–29th July ed. by K. Ikeda, K. Katori, Y. Suzuki (1988), p. 103 [Supplement to the Journal of the Physical Society of Japan, vol. 58 (1989)]
46. D. Baye, in *Proceedings of the Sixth International Conference on "Clusters in Nuclear Structure and Dynamics"*, Strasbourg, France, 6th–9th September 1994, ed. by F. Haas (1994), p. 259
47. P. Descouvement, Nucl. Phys. A **699**, 463 (2002)
48. N. Itagaki, S. Hirose, T. Otsuka, S. Okabe, K. Ikeda, Phys. Rev. C **65**, 044302 (2002)
49. N. Itagaki, S. Okabe, Phys. Rev. C **61**, 044306 (2000)
50. N. Itagaki, S. Okabe, K. Ikeda, Phys. Rev. C **62**, 034301 (2000)
51. N. Itagaki et al., Phys. Rev. C **64**, 014301 (2001)
52. M. Ito, Phys. Lett. B **636**, 293 (2006)
53. L.D. Landau, E.M. Lifshitz, *Quantum Mechanics*, 3rd edn. (Elsevier, Butterworth–Heinemann, Amsterdam/Stoneham, 1958)
54. H. Nakamura, *Nonadiabatic Transition* (World Scientific, Singapore, 2000) and references therein
55. Y. Kanada-En'yo, H. Horiuchi, Prog. Theor. Phys. **142**, 205 (2001)
56. R. Roth, T. Neff, H. Hergert, H. Feldmeier, Nucl. Phys. A **745**, 3 (2004)
57. M. Chernykh, H. Feldmeier, T. Neff, von P. Neumann-Cosel, A. Richter, Phys. Rev. Lett. **98**, 032501 (2007)
58. P. Navratil et al., J. Phys. G **36**, 083101 (2009)
59. E. Epelbaum, Part. Nucl. Phys. **57**, 654 (2006)
60. E. Epelbaum, H.-W. Hammer, U.-G. Meißner, Rev. Mod. Phys. **81**, 1773 (2009)
61. P. Navrátil, J.P. Vary, B.R. Barrett, Phys. Rev. Lett. **84**, 5728 (2000)
62. P. Navrátil, V. Gueorguiev, J.P. Vary, W.E. Ormand, A. Nogga, Phys. Rev. Lett. **99**, 042501 (2007)
63. E. Epelbaum, H. Krebs, T.A. Lähde, D. Lee, U.-G. Meißner, Phys. Rev. Lett. **110**, 112502 (2013)
64. E. Epelbaum, H. Krebs, T.A. Lähde, D. Lee, U.-G. Meißner, Phys. Rev. Lett. **109**, 252501 (2012)
65. E. Epelbaum, H. Krebs, D. Lee, U.-G. Meißner, Phys. Rev. Lett. **106**, 192501 (2011)
66. R. Bijker, F. Iachello, Phys. Rev. C **61**, 067305 (2000)
67. M. Freer et al., Phys. Rev. C **76**, 034320 (2007)
68. F. Hoyle, Astrophys. J. Suppl. Ser. **1**, 12 (1954)
69. C.W. Cook et al., Phys. Rev. **107**, 508 (1957)
70. G.M. Reynolds, D.E. Rundquist, M. Poichar, Phys. Rev. C **3**, 442 (1971)
71. F.D. Smit et al., Phys. Rev. C **86**, 03701 (2012)
72. M. Itoh et al., Nucl. Phys. A **738**, 268 (2004)
73. M. Itoh et al., Phys. Rev. C **84**, 054308 (2011)
74. M. Freer et al., Phys. Rev. C **86**, 034320 (2012)
75. M. Freer et al., Phys. Rev. C **80**, 041303(R) (2009)
76. W.R. Zimmerman, N.E. Destefano, M. Freer, M. Gai, F.D. Smit, Phys. Rev. C **84**, 027304 (2011)
77. W.R. Zimmerman et al., Phys. Rev. Lett. **110**, 152502 (2013)
78. M. Freer et al., Phys. Rev. C **83**, 034314 (2011)
79. K.P. Artemov et al., Sov. J. Nucl. Phys. **52**, 406 (1990)
80. G.V. Rogachev et al., Phys. Rev. C **64**, 051302 (2001)
81. M. Freer, A.C. Merchant, J. Phys. G **23**, 261 (1997)
82. M. Freer, Nucl. Instrum. Methods A **383**, 463 (1996)
83. P.J. Haigh, Phys. Rev. C **79**, 014302 (2009)
84. S. Aoyama, S. Mukai Kato, K. Ikeda, Prog. Theor. Phys. **93**, 99 (1995)
85. M. Seya, M. Kohno, S. Nagata, Prog. Theor. Phys. **65**, 204 (1981)
86. W. von Oertzen, Z. Phys. A **354**, 37 (1996)

87. W. von Oertzen, Z. Phys. A **357**, 355 (1997)
88. W. von Oertzen, Nuovo Cimento A **110**, 895 (1997)
89. N. Itagaki, S. Okabe, Phys. Rev. C **61**, 044306 (2000)
90. S.G. Nilsson, Mat. Fys. Medd. Dan. Vid. Selsk. **29**, 16 (1955)
91. H.G. Bohlen, W. von Oertzen, A. Blazevic, B. Gebauer, M. Milin, T. Kokalova, Ch. Schulz, S. Thummerer, A. Tumino, in *Proceedings of the International Symposium on Exotic Nuclei*, Lake Baikal, Russia, 2001, ed. by Yu.E. Penionzhkevich, E.A. Cherepanov (World Scientific, Singapore, 2002), p. 453
92. Y. Kanada-En'yo, H. Horiuchi, A. Dóte, J. Phys. G **24**, 1499 (1998)
93. J.A. Liendo et al., Phys. Rev. C **65**, 034317 (2002)
94. N. Curtis et al., Phys. Rev. C **64**, 044604 (2001)
95. M. Milin et al., Nucl. Phys. A **753**, 263 (2005)
96. M. Freer et al., Phys. Rev. Lett. **96**, 042501 (2006)
97. G. Rogachev, (2012), Private communication
98. E.A. McCutchan et al., Phys. Rev. C **86**, 057306 (2012)
99. E.A. McCutchan et al., Phys. Rev. C **86**, 014312 (2012)
100. H. Horiuchi, K. Ikeda, Prog. Theor. Phys. A **40**, 277 (1968)
101. Y. Kanada En'yo, H. Horiuchi, Prog. Theor. Phys. **93**, 115 (1995)
102. P.A. Butler, W. Nazarewicz, Rev. Mod. Phys. **68**, 350 (1996)
103. W. von Oertzen, Eur. Phys. J. A **11**, 403 (2001)
104. M. Kimura, Phys. Rev. C **75**, 034312 (2007)
105. P. McEwan, M. Freer, J. Phys. G **30**, 1 (2004)
106. N. Itagaki, T. Otsuka, K. Ikeda, S. Okabe, Phys. Rev. Lett. **92**, 014301 (2004)

Chapter 2
A Pedestrian Approach to the Theory of Transfer Reactions: Application to Weakly-Bound and Unbound Exotic Nuclei

Joaquín Gómez Camacho and Antonio M. Moro

2.1 Introduction

Let us consider a simplified semiclassical time-dependent picture of transfer. Imagine a nucleus, such as ^{11}Be, that is traveling towards a target, say ^{208}Pb. When ^{11}Be is far away from the target, we can see its structure. It is made of 4 protons and 7 neutrons, that interact with each other, and are subject to Pauli principle. However, one of the neutrons is very weakly bound, so it is described by an extended halo wavefunction. The weakly bound neutron (assuming that we can identify it, because all neutrons are identical), spends a large fraction of the time well separated of the rest, although some fraction of the time it interacts strongly with the other protons and neutrons, and hence it is strongly correlated with them. In fact, due to this correlation, we suspect that the ground state of ^{11}Be, in which our nucleus presently is, is a combination of two components, in each one of them the halo neutron is coupled to a different state of the ^{10}Be core, that is how we call the other 4 protons and 6 neutrons.

Now the ^{11}Be nucleus begins to approach the ^{208}Pb target, in what is called the incident channel. The strong Coulomb force of the target begins to act, and repels the 4 protons, but does not affect in principle the 7 neutrons. However, as 6 of the neutrons are strongly correlated with the protons, the ^{10}Be core moves as a whole, and separates from the halo neutron. As the nuclei get closer, the nuclear force acts, and as a result the halo neutron excites, and also the ^{10}Be core can excite. So,

---

J. Gómez Camacho (✉)
Centro Nacional de Aceleradores, Universidad de Sevilla/Junta de Andalucía/CSIC, 41092 Seville, Spain
e-mail: gomez@us.es

J. Gómez Camacho · A.M. Moro
Departamento de FAMN, Universidad de Sevilla, Apartado 1065, 41080 Seville, Spain

A.M. Moro
e-mail: moro@us.es

the ^{11}Be nucleus, due to the Coulomb and nuclear interaction with the target, ends up in a combination of the initial ground state, other bound excited states and the continuum of break-up states.

Eventually, the two nuclei are sufficiently close so that the halo neutron has the possibility of populating an unoccupied neutron state in the ^{208}Pb target. This transition is the result of many-body dynamics, involving (at least) the interaction of the neutron with the ^{10}Be core, the neutron with the ^{208}Pb target, and the ^{10}Be core with the ^{208}Pb target. Indeed, the transition will be more probable if the initial state of the halo neutron, bound to the ^{10}Be core, has a significant overlap with the final state of the neutron, in a certain bound state with the ^{208}Pb target. This overlap is not only a spatial overlap, occurring when the projectile and target are sufficiently close, but also overlap in the momentum space. The neutron will not jump easily to a state where its momentum (or its velocity) is very different from the initial one. Looking from a coordinate frame fixed in the target, the initial neutron has a momentum distribution that is centered at the projectile momentum divided by the projectile mass, with a certain spread given by the bound wavefunction in momentum space. The final wavefunction, bound to the target, is centered at zero, and has a spread given by this momentum distribution of this wavefunction. The tails of these two distributions overlap, and this favours transfer.

Once the transfer occurs, the remaining nuclei, ^{10}Be and ^{209}Pb in this case, have to fly apart. As they separate, in the so called outgoing channel, coupling effects can occur. Both ^{10}Be and ^{209}Pb can suffer excitations, produced by the Coulomb and nuclear forces. They can break-up, starting from bound states, or can bind, starting from break-up states. In addition, both in the incident and in the outgoing channels, complicated processes such as fusion can occur. These complicated processes do not need to be taken into account explicitly, but at least the loss of flux to the transfer channel has to be considered.

This semiclassical time-dependent description of a transfer reaction is what we think that is happening, but we cannot observe it directly. The only things that we can observe are the outgoing particles and, in some cases, gamma rays. So, if in our detectors we observe ^{11}Be, with the proper energy, we will conclude that we observe elastic scattering. This does not mean that ^{11}Be was in its ground state all the time. It could have excited, broken up, even transferred one neutron, but in the end it finished in the ground state.

If in our detectors we observe ^{10}Be, then we can infer that neutron transfer, or neutron break-up, has occurred. If it is neutron break-up, we expect to see a continuum of kinetic energies of ^{10}Be, centered at an energy that is 10/11 times the incident energy of ^{11}Be. However, if it is transfer, the ^{10}Be will appear with a definite energy, that is determined, for each scattering angle, by the Q-value of the reaction and kinematic factors. A well defined experiment, with sufficient energy resolution, should be able to determine individually the transfer probability, or cross section, for each scattering angle and for each excitation energy of the final nuclei ^{209}Pb and ^{10}Be. Even if there is not sufficient energy resolution to separate the ground and excited states of ^{209}Pb and ^{10}Be, the measurement in coincidence of gamma rays can allow to separate individual states. A thorough and up-to-date review of the

Fig. 2.1 Co-ordinate representation of a transfer reaction, before and after transfer occurs

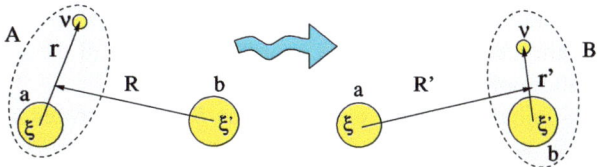

experimental procedures to measure transfer reactions is presented in the paper by W. Catford, in this volume.

A measurement of transfer reactions can give us unique information about the structure of the initial (^{11}Be) or final (^{209}Pb) composite nuclei, which are represented, respectively, by the capital letters A, B in Fig. 2.1. In particular, it can tell us which part of its wavefunction can be described it terms of the "cores" (^{10}Be and ^{208}Pb respectively), represented by the lowercase a, b in Fig. 2.1 and a neutron. Nevertheless, to obtain this structure information, one needs to deal with the complex dynamics of a quantum mechanical three-body problem, with Coulomb and nuclear forces in their full glory. To rescue the "beauty" of nuclear structure, we have to tame the "beast" of nuclear reactions. This paper is a beginners' guide to rescue the fragile beauty of weakly bound and unbound nuclear structure, taming the formidable beast of nuclear reactions in its full three-body power.

2.2 Theoretical Formalism

Let us denote generically a transfer reaction as:

$$A + b \rightarrow a + B.$$

In general, this is a very complex many-body reaction. However, we can visualize it considering that, initially, we have a nucleus A, which can be described as composed of two clusters $a + v$. During the reaction, the nucleus A is fragmented into the clusters $a + v$. The cluster a survives in the final state, while the cluster v gets attached to the nucleus b to form the composite system B. For example, in the reaction ^{10}Be$(d, p)^{11}$Be, b corresponds to ^{10}Be, A is the deuteron, a is the proton, B is ^{11}Be and v is the neutron.

We can describe the quantum mechanical state of the nucleus A as

$$\Phi_A = C_{av}^A \phi_a \phi_v \varphi_{av}(\mathbf{r}) + \Phi_A^C. \tag{2.1}$$

In this simplified notation, ϕ_a and ϕ_v represent the internal wavefunctions of clusters a and v, $C_{av}^A \varphi_{av}(\mathbf{r})$ represents the overlap function, which can be written in terms of a normalized relative wavefunction $\varphi_{av}(\mathbf{r})$ and a spectroscopic amplitude C_{av}^A. The product of these three terms is implicitly coupled to the angular momentum of nucleus A.

Notice that not all the state Φ_A can be described as two clusters a, v with a certain relative motion; Φ_A^C represents the part of the state that has a more complex configuration.

Similarly, the state of the nucleus B can be written as

$$\Phi_B = C_{bv}^B \phi_b \phi_v \varphi_{bv}(\mathbf{r}') + \Phi_B^C. \tag{2.2}$$

The transition matrix element that describes the transfer process will be a complex many-body matrix element, that can be written as

$$T(aB, Ab) = \langle \Phi_B \phi_a \chi_{aB} | \mathcal{T} | \Phi_A \phi_b \chi_{Ab} \rangle \tag{2.3}$$

where χ_{Ab} describes the relative motion of A and b, and similarly for χ_{aB}, and $\mathcal{T}$ is the adequate many-body T-matrix operator. The transition matrix element, squared, is proportional to the differential transfer cross section, which is the magnitude directly measured in the experiment. The scattering formalism relating transition matrix elements and scattering observables can be seen, for example in Ref. [1].

The following approximations allow us to reduce the many-body problem to a three-body problem:

- The terms Φ_A^C and Φ_B^C, corresponding to complex configurations of A and B, do not contribute significantly to transfer.
- The normalized overlap functions $\varphi_{bv}(\mathbf{r}')$ and $\varphi_{av}(\mathbf{r})$ can be approximated by the eigenstates of two-body Hamiltonians with interactions V_{bv} and V_{av}, respectively. They will be represented by some real mean-field interactions.
- During the collision process the interactions between the clusters a, b, and v are completely described by two-body interactions V_{bv}, V_{av} and U_{ab}, that cannot alter the internal states of the clusters. In our description of transfer, we do not consider explicitly processes that lead to the excitations of the clusters b and a, so the interaction between them is represented by an effective optical potential, complex in general, that we denote by U_{ab}.

With these approximations, the transfer matrix elements can be described as:

$$T(aB, Ab) = C_{bv}^{B*} C_{av}^A T^{(3)}(aB, Ab), \tag{2.4}$$

where the three-body matrix element can be expressed in the post form

$$T^{(3)}(aB, Ab) = \langle \chi_{aB}^{(-)}(\mathbf{R}') \varphi_{bv}(\mathbf{r}') | V_{av} + U_{ab} - U_{aB} | \Psi^{(+)}(\mathbf{R}, \mathbf{r}) \rangle. \tag{2.5}$$

Here, $\Psi^{(+)}(\mathbf{R}, \mathbf{r})$ is the exact solution of the 3-body problem of a, b, v with the corresponding interactions, with boundary conditions given by a plane wave with the incident momentum in the beam direction, on the A–b co-ordinate $\mathbf{R}$, times the bound wavefunction $\varphi_{av}(\mathbf{r})$, plus outgoing waves in all open channels. U_{aB} is a suitable potential, arbitrary at this stage, that is used to construct the two-body relative wavefunction $\chi_{aB}^{(-)}(\mathbf{R}')$. This wavefunction has boundary conditions given by a plane wave, with the final momentum in the direction of the detector, on the a–B co-ordinate $\mathbf{R}'$, plus incoming waves.

Equivalently, one can use the prior form,

$$T^{(3)}(aB, Ab) = \langle \Psi^{(-)}(\mathbf{R}', \mathbf{r}') | U_{ab} + V_{bv} - U_{Ab} | \chi_{Ab}^{(+)}(\mathbf{R}) \phi_{av}(\mathbf{r}) \rangle, \quad (2.6)$$

where the three-body and two-body wavefunctions have similar meanings as before, *mutatis mutandis*. Namely, $\Psi^{(-)}(\mathbf{R}', \mathbf{r}')$ is the exact solution of the 3-body problem of a, b, v with the corresponding interactions, with boundary conditions given by a plane wave with the final momentum in the detector direction, on the B–a relative co-ordinate $\mathbf{R}'$, times the bound wavefunction $\varphi_{bv}(\mathbf{r}')$, plus incoming waves in all open channels. U_{Ab} is a suitable potential, arbitrary at this stage, that is used to construct the two-body relative wavefunction $\chi_{Ab}^{(+)}(\mathbf{R})$. This wavefunction has boundary conditions given by a plane wave, with the initial momentum in the direction of the beam, on the b–A co-ordinate $\mathbf{R}$, plus incoming waves.

It should be noticed that the previous expressions are exact, assuming a 3-body model for the transfer process. Consequently, post and prior expressions give identical results, provided that the exact three-body wavefunction ($\Psi^{(-)}(\mathbf{R}', \mathbf{r}')$ or $\Psi^{(+)}(\mathbf{R}, \mathbf{r})$) is used to evaluate the transition amplitude. In general, this equivalence will break down when these exact wavefunctions are replaced by approximated ones.

2.2.1 Distorted Wave Born Approximation DWBA

The Distorted Wave Born Approximation (DWBA) [1–5] can be obtained assuming that the three-body wavefunction can be approximated by

$$\Psi^{(+)}(\mathbf{R}, \mathbf{r}) \simeq \chi_{Ab}^{(+)}(\mathbf{R}) \varphi_{av}(\mathbf{r}). \quad (2.7)$$

Thus, the transition matrix element becomes, in *post* representation,

$$T^{(3)}(aB, Ab) \simeq T_{\text{post}}^{\text{DWBA}}(aB, Ab)$$
$$= \langle \chi_{aB}^{(-)}(\mathbf{R}') \varphi_{bv}(\mathbf{r}') | V_{\text{post}} | \chi_{Ab}^{(+)}(\mathbf{R}) \varphi_{av}(\mathbf{r}) \rangle, \quad (2.8)$$

where

$$V_{\text{post}} \equiv V_{av} + U_{ab} - U_{aB}. \quad (2.9)$$

This approximation can be considered as the leading term of an expansion of the transition amplitude in terms of V_{post}. Thus, the accuracy of the DWBA approximation depends strongly on how the auxiliary potential U_{aB} is chosen. Not only this. The choice of this potential, which was arbitrary in the exact expression of $T^{(3)}(aB, Ab)$ in the post form, becomes very important in the DWBA approximation.

An equivalent derivation can be done starting from the *prior* expression of transfer, approximating

$$\Psi^{(-)}(\mathbf{R}', \mathbf{r}') \simeq \chi_{aB}^{(-)}(\mathbf{R}')\varphi_{bv}(\mathbf{r}'). \tag{2.10}$$

Thus, the transition matrix element becomes, in *prior* representation,

$$T^{(3)}(aB, Ab) \simeq T_{\text{prior}}^{\text{DWBA}}(aB, Ab)$$
$$= \langle \chi_{aB}^{(-)}(\mathbf{R}')\varphi_{bv}(\mathbf{r}') | V_{\text{prior}} | \chi_{Ab}^{(+)}(\mathbf{R})\varphi_{av}(\mathbf{r}) \rangle, \tag{2.11}$$

where

$$V_{\text{prior}} \equiv V_{bv} + U_{ab} - U_{Ab}. \tag{2.12}$$

It can be formally demonstrated that the *prior* and *post* expressions of DWBA give exactly the same result. Hence, the choice of one of another representation in DWBA is done by computational convenience, determined by the range of the interactions. In many situations, an appropriate choice of the auxiliary potential produces a certain cancellation of the *remnant term* ($U_{ab} - U_{aB}$ or $U_{ab} - U_{Ab}$ in the post and prior representations, respectively). In those situations, the transition amplitude is mostly determined by the interaction V_{av} (post) or V_{bv} (prior) and it results numerically advantageous to choose the representation for which this interaction is of shorter range.

The accuracy of DWBA depends on the choice of the auxiliary potentials in the incident channel (U_{Ab}) and in the outgoing channel (U_{aB}). These could be, in principle, any function of the co-ordinate $\mathbf{R}$ and $\mathbf{R}'$, respectively. Two approaches are usually taken:

- *The microscopic approach.* The auxiliary potential in the outgoing channel U_{aB} is taken as the expectation value, in the final bound state $\varphi_{bv}(\mathbf{r}')$, of the sum of the interactions $U_{ab} + V_{av}$. Explicitly,

$$U_{aB}(\mathbf{R}') = \int d^3\mathbf{r}' |\varphi_{bv}(\mathbf{r}')|^2 (U_{ab} + V_{av}). \tag{2.13}$$

Similarly, U_{Ab} is taken as the expectation value, in the initial bound state, of the sum of the interactions $U_{ab} + V_{bv}$,

$$U_{Ab}(\mathbf{R}) = \int d^3\mathbf{r} |\varphi_{av}(\mathbf{r})|^2 (U_{ab} + V_{bv}). \tag{2.14}$$

In practical applications of DWBA, it is very convenient that the auxiliary potentials are central, so that they depend on the value of the radial co-ordinate ($U_{Ab}(R)$, $U_{aB}(R')$) and not on its direction. This is achieved considering only the monopole part of the folding interaction, or, equivalently, averaging the folding potential over all the magnetic substates.

The microscopic approach has the advantage of being completely determined by the two-body interactions between the fragments. From the formal point of view, this approach would be the natural one to follow, in order to choose U_{Ab} so that the term $U_{ab} + V_{bv} - U_{Ab}$ is minimal, for the bound state φ_{av}.

On the negative side, it is not trivial that the interaction U_{Ab}, calculated according to Eq. (2.14), reproduces accurately the elastic scattering on the Ab channel. The interactions U_{ab}, V_{av}, and V_{bv} should be taken as complex interactions, in order to reproduce elastic scattering or transfer, but in this case V_{av} and V_{bv} can not be used to obtain bound states, unless the interactions are explicitly energy dependent. Finally, this approach excludes completely any effect of break-up channels on the three-body wavefunction. Hence, this approach would be valid when the three-body scattering wavefunctions are dominated by their elastic component, either in the incident or in the exit channels.

- *The phenomenological approach.* The auxiliary potential in the incident channel U_{Ab} is obtained by fitting the elastic scattering data on the Ab channel. The auxiliary potential in the exit channel, U_{aB}, is obtained by fitting the elastic scattering on the aB channel. This approach has the advantage of allowing for a consistent description of transfer reactions, as well as of elastic scattering in the incident and outgoing channels. It takes into account, through the use of optical potentials, the effect of complex reaction processes, such as fusion, that can remove flux from the elastic and from the transfer channels. Furthermore, the effect of some three-body reactions, such as break-up, which also remove flux from elastic and transfer channels, are approximately taken into account because the optical potentials fit the experimental elastic cross sections, which are affected by all these dynamic processes. On the negative side, it is not always possible to find the elastic data for the outgoing channel. If the final state of nucleus B is not in its ground state, but on an excited state, it will not be possible to measure the corresponding elastic scattering. This is particularly true if the final state is in the continuum. Moreover, the optical potentials reproduce typically the asymptotic wavefunctions, which determines the scattering amplitudes and differential cross sections. It does not necessarily reproduce the wavefunctions in the internal radial range that is relevant for the transfer matrix elements.

The coupling scheme assumed in the DWBA method is schematically depicted in Fig. 2.2(a) for the $^{10}\text{Be}(d, p)^{11}\text{Be}$ reaction. The solid arrow indicates that only transfer from the ground state of the deuteron to the $p + {}^{11}\text{Be}$ channel is explicitly included. The effect of breakup channels of the deuteron (represented by the shaded area) is completely neglected in the afore-mentioned microscopic approach, and only partially taken into account in the phenomenological approach, through its effect on the elastic wavefunction.

In general, DWBA has been, and still is, a key approach to describe transfer reactions, and it has been used extensively to extract spectroscopic information on nuclear structure, in particular spectroscopic amplitudes (see e.g. [4, 6–9]). However, this method is based on a rather crude approach to the three-body problem, and is expected to be accurate only when the elastic scattering, in the incident and outgoing channels, is dominant. For the case of exotic nuclei, which are frequently weakly bound, break-up channels can play a very important role in the three-body

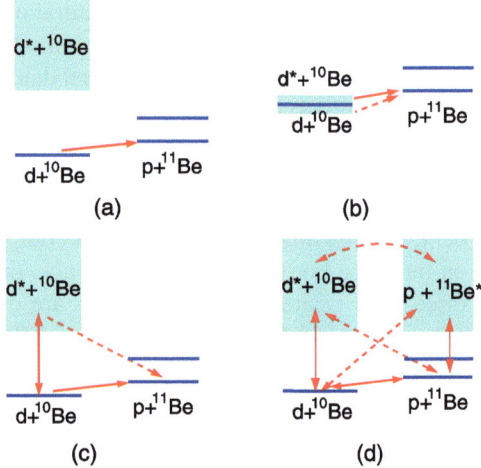

Fig. 2.2 Comparison of different coupling schemes discussed in this work for the reaction ^{10}Be$(d,p)^{11}$Be: (**a**) DWBA, (**b**) ADWA, (**c**) CDCC-BA and (**d**) CRC

dynamics. Hence, it is important, in order to extract reliable spectroscopic information from transfer reactions with exotic nuclei, to check the validity of the DWBA method by comparing it with other approaches that take into account the role of break-up channels.

2.2.2 Adiabatic Distorted Wave Approximation ADWA

The DWBA approach, as mentioned previously, relies heavily on the assumption that the elastic channel dominates the reaction. This does not only imply that the dominant cross sections is elastic, but also that, during the collision process, the three-body wavefunction can be approximated by the elastic component. Note that these two facts are not equivalent. There can be dynamic situations in which elastic cross section dominates, meaning that the asymptotic three-body wavefunction, at large distances, is dominated by the elastic component. However, this does not mean that at short projectile-target distances, which give the main contribution to the transfer matrix element, the elastic component should be dominant. Dynamic polarization effects make that the composite projectile can be strongly distorted at short distances, even when asymptotically the energy matching conditions make the elastic channel dominant.

Moreover, the phenomenological DWBA approach relies on the use of optical potentials, usually taken as local, angular momentum-independent potentials, chosen to reproduce elastic scattering. This means that the optical potentials will reasonably reproduce the phase shifts, for all partial waves, in the elastic channel. In other words, the phenomenological DWBA approach reproduces the elastic wavefunction asymptotically, at large projectile-target distances. It is not obvious that the elastic wavefunction used in the phenomenological DWBA approach reproduces correctly the elastic component of the wavefunction, in the radial range relevant for the transfer T-matrix elements.

Due to these limitations, it would be desirable to have an alternative formulation, which maintains the relative simplicity of DWBA, and whose ingredients can be completely determined from experiment. This is achieved by the Adiabatic Distorted Wave Approximation (ADWA), which was initially formulated by Johnson and Soper [10]. This approach is formulated in principle for (d, p), or (d, n) reactions, although it could be applied to other weakly bound composite systems. It relies on the fact that the composite projectile has a relatively low binding energy (2.22 MeV in the case of the deuteron), and so, if the collision energy is relatively high, we can expect that, during the collision process, the relative proton-neutron co-ordinate does not change significantly; it is "frozen". Under this situation, the relevant interaction that determines accurately the projectile-target wavefunction is not the phenomenological deuteron-target interaction that would reproduce elastic scattering, but the sum of the interactions of each one of the fragments of the projectile (proton and neutron in the deuteron case) with the target.

In the *adiabatic approximation* [10] (also called *sudden approximation* by some authors) the three-body wavefunction can be written as

$$\Psi^{(+)}(\mathbf{R},\mathbf{r}) \simeq \chi_{Ab}^{(+)}(\mathbf{R},\mathbf{r})\varphi_{av}(\mathbf{r}), \tag{2.15}$$

where $\chi_{Ab}^{(+)}(\mathbf{R},\mathbf{r})$ is the solution of a two-body scattering problem, on the co-ordinate $\mathbf{R}$, in which the interaction is given by

$$U_{Ab}(\mathbf{R},\mathbf{r}) = U_{ab}(R_{ab}) + V_{bv}(r'). \tag{2.16}$$

Indeed, the potential that describes the scattering wavefunction, although two-body, is not central and so the calculation of the adiabatic wavefunction, for each value of the a–v separation $\mathbf{r}$ is very complicated, but it has been done [11, 12]. Besides, the adiabatic approximation to the three-body wavefunction is not accurate for large values of $\mathbf{r}$, where one would expect to see outgoing waves, instead of the exponential decay given by the bound two-body wavefunction $\varphi_{av}(\mathbf{r})$.

Fortunately, these shortcomings of the adiabatic wavefunctions are not important, if one is only interested in evaluating the matrix element involved in transfer. These are dominated by the $V_{av}(r)$ interaction (the proton-neutron interaction, in the deuteron case) which has a short range. Note that, even if the a–v wavefunction $\varphi_{av}(\mathbf{r})$ has a relatively long range, which is the case for weakly bound halo systems, $V_{av}(r)$ has a much shorter range. Hence, for the purpose of evaluating the transfer matrix element, one can calculate the adiabatic wavefunction using the potential evaluated at $\mathbf{r} = 0$. This leads to the Johnson and Soper approximation [10], in which

$$\Psi^{(+)}(\mathbf{R},\mathbf{r}) \simeq \chi_{Ab}^{(+)}(\mathbf{R})\varphi_{av}(\mathbf{r}), \tag{2.17}$$

where $\chi_{Ab}^{(+)}(\mathbf{R})$ is the solution of a two-body scattering problem, on the co-ordinate $\mathbf{R}$, in which the interaction is given by

$$U_{Ab}^{JS}(R) = U_{ab}(R) + U_{bv}(R). \tag{2.18}$$

Note that in this expression the $b-v$ interaction V_{bv}, which would in general be energy dependent, and would be responsible for the bound state $\varphi_{bv}(\mathbf{r}')$, is replaced by the optical potential U_{bv} that describes the $b-v$ interaction at the same incident energy per nucleon. This is justified by the adiabatic approximation; the transfer process dynamics is consistent with freezing the $a-v$ co-ordinate, that then scatters from the b target with an interaction that is the sum of U_{bv} and U_{ab} interactions at the same energy per nucleon.

Several refinements and corrections have been performed within the ADWA formalism. For example, a finite-range version of the adiabatic potential was proposed by Johnson and Tandy [13]:

$$U_{Ab}^{JT}(R) = \frac{\langle \varphi_{av}(\mathbf{r}) | V_{av}(U_{ab}+U_{bv}) | \varphi_{av}(\mathbf{r}) \rangle}{\langle \varphi_{av}(\mathbf{r}) | V_{av} | \varphi_{av}(\mathbf{r}) \rangle}. \quad (2.19)$$

However, for the purpose of the analysis of (d, p) and (p, d) reactions, the simplest Johnson-Soper expression given by Eq. (2.18) is by far the most widely used. Here, we will outline its advantages and disadvantages. On the positive side, the ADWA approach ingredients are completely determined by experiments. These ingredients are the proton-target and neutron-target optical potentials, evaluated at half of the deuteron incident energy, as well as the well known proton-neutron interaction.

The adiabatic approximation is equivalent to neglect the excitation energy of the states of the projectile [10]. The adiabatic wavefunction takes into account the excitation to breakup channels, assuming that these states are degenerate in energy with the projectile ground state, as illustrated in Fig. 2.2(b). Therefore, the ADWA approach takes into account, approximately, the effect of deuteron break-up on the transfer cross section, within the adiabatic approximation. So, it should be well suited to describe deuteron scattering at high energies, around 100 MeV per nucleon. Systematic studies [14–16] have shown that ADWA is superior to standard DWBA for (d, p) scattering at high energies.

On the negative side, the ADWA approach does not consistently describe elastic scattering and nucleon transfer. Although physically one considers that elastic scattering, transfer and break-up should be closely related, so that the increase of flux in one channel should reduce the flux in the others, this connection is not present in ADWA. On the other hand, the arguments leading to ADWA are strongly associated with the assumption that the transfer is governed by a short range operator. So, it is not obvious that the approximations remain valid for other weakly bound systems, like ^{11}Be. Even in the case of (d, p) scattering, the transfer matrix element is determined not only by the $n-p$ interaction, but also by the proton-target and neutron-target interactions, that define the *remnant* term. It is not clear a-priori the role of these terms, that would have contributions of three-body configurations in which proton and neutron are not so close together. A promising alternative, that avoids the presence of these remnant terms, has been proposed by Timofeyuk and Johnson [17].

2.2.3 Continuum Discretized Coupled Channels Born Approximation CDCC-BA

In scattering of weakly bound nuclei, coupling to break-up channels can play an important role. DWBA may not be sufficiently accurate, as the three-body wavefunction is not dominated by the elastic channels. ADWA requires to assume the adiabatic approximation for the composite projectile, which may not be accurate if the collision energy is not sufficiently high. Besides, the simple Johnson-Soper expression requires to assume that the transfer operator is of short range, which may not be accurate beyond (d, p) reactions.

A more accurate approach for transfer is obtained if the three-body wavefunction is approximated in terms of a basis of the states of the relative motion of the $a + v$ sub-system, i.e.

$$\Psi^{(+)}(\mathbf{R}, \mathbf{r}) \approx \Psi^{(+)\text{CDCC}}(\mathbf{R}, \mathbf{r}) = \sum_{i=0}^{N} \chi_{Ab,i}^{(+)}(\mathbf{R}) \varphi_{av,i}(\mathbf{r}). \quad (2.20)$$

Here, the index i corresponds to a set of $a + v$ states explicitly included in a coupled channels calculation ($\varphi_{av,i}(\mathbf{r})$), which would correspond in general to a given spin and spin projection ($i = 0$ denotes the ground state of the $a + v$ system). This basis of states should include other possible bound states of the $a + v$ system, if present, as well as a suitable discrete representation of the two-body continuum states. In actual calculations, this continuum must be truncated in excitation energy and limited to a finite number of partial waves ℓ associated to the relative co-ordinate $\mathbf{r}$. Normalizable states representing the continuum should be obtained for each ℓ value. This can be achieved making use of a pseudo-state basis and diagonalizing the $a + v$ Hamiltonian [18]. Alternatively, continuum states of the $a + v$ Hamiltonian can be obtained, and normalizable states (bins) can be obtained by averaging these continuum states over a certain energy interval [19].

The model wavefunction given by Eq. (2.20) must verify the Schrödinger equation: $[H - E]\Psi^{(+)\text{CDCC}}(\mathbf{R}, \mathbf{r}) = 0$. To determine the radial coefficients $\chi_{Ab,i}^{(+)}(\mathbf{R})$, one multiplies this equation on the left by each of the internal wavefunctions $\varphi_{av,i}(\mathbf{r})^*$ and integrates along the coordinate $\mathbf{r}$. This gives rise to a set of coupled differential equations:

$$[E - \varepsilon_{av}^i - \hat{T}_{Ab} - U_{Ab}^{ii}(\mathbf{R})]\chi_{Ab,i}^{(+)}(\mathbf{R}) = \sum_{j \neq i}^{N} U_{Ab}^{ij}(\mathbf{R})\chi_{Ab,j}^{(+)}(\mathbf{R}), \quad (2.21)$$

where U_{Ab}^{ij} are the transition potentials defined as

$$U_{Ab}^{ij}(\mathbf{R}) = \int d\mathbf{r}\, \varphi_{av,i}^*(\mathbf{r})(U_{ab} + U_{bv})\varphi_{av,j}(\mathbf{r}). \quad (2.22)$$

The coupled channels solution $\chi_{Ab,i}^{(+)}(\mathbf{R})$ corresponds to the outgoing waves in all different channels i, for boundary conditions given by a plane wave in the initial bound state $i = 0$. The potentials U_{ab} and U_{bv} are to be understood as effective interactions (complex in general) describing the elastic scattering of the corresponding $a+b$ and $b+v$ sub-systems, at the same energy per nucleon as in the incident projectile. In particular, U_{bv} will be described in general by a complex optical potential, and will differ from the interaction V_{bv} used to generate the bound state wavefunction of the $b+v$ system.

Note that, without any loss of generality, we can introduce an arbitrary auxiliary potential $U_{Ab}(R)$, so that Eq. (2.21) can be written as

$$\left[E - \varepsilon_{av}^i - \hat{T}_{Ab} - U_{Ab}(R)\right]\chi_{Ab,i}^{(+)}(\mathbf{R}) = \sum_{j}^{N} V_{\text{prior}}^{ij}(\mathbf{R})\chi_{Ab,j}^{(+)}(\mathbf{R}), \qquad (2.23)$$

where $V_{\text{prior}}^{ij}(\mathbf{R})$ are the matrix elements of $V_{\text{prior}} = U_{ab} + U_{bv} - U_{Ab}$ between the states of the A system.

Once the CDCC wavefunction (2.20) is obtained, it can be inserted into Eq. (2.5) to give:

$$T^{(\text{CDCC})}(aB, Ab) = \langle \chi_{aB}^{(-)}(\mathbf{R}')\varphi_{bv}(\mathbf{r}')|V_{\text{post}}|\Psi^{(+)\text{CDCC}}(\mathbf{R}, \mathbf{r})\rangle. \qquad (2.24)$$

with V_{post} given by Eq. (2.9). To clarify the link between the CDCC-BA and DWBA methods it is convenient to rewrite this expression as:

$$T^{(\text{CDCC})}(aB, Ab) = \langle \chi_{aB}^{(-)}(\mathbf{R}')\varphi_{bv}(\mathbf{r}')|V_{\text{post}}|\chi_{Ab,0}^{(+)}(\mathbf{R})\varphi_{av,0}(\mathbf{r})\rangle$$
$$+ \sum_{i=1}^{N} \langle \chi_{aB}^{(-)}(\mathbf{R}')\varphi_{bv}(\mathbf{r}')|V_{\text{post}}|\chi_{Ab,i}^{(+)}(\mathbf{R})\varphi_{av,i}(\mathbf{r})\rangle. \qquad (2.25)$$

The first term in this expression corresponds to the *direct transfer*, proceeding directly from the ground state of the projectile (e.g. the deuteron, in a (d, p) reaction), whereas the second term accounts for the *multi-step* transfer occurring via the excited states of the projectile (p–n continuum states in the case of the deuteron). These two types of processes correspond, respectively, to the solid and dashed lines in Fig. 2.2(c) for the ^{10}Be$(d, p)^{11}$Be case. Clearly, the multi-step process going through the breakup channels are omitted in the DWBA calculation. At most, the DWBA considers the effect of these channels on the elastic scattering if a suitable choice of the entrance optical potential is made. The adiabatic approximation includes in principle both mechanisms, but under the assumption that the excited (breakup) channels of the projectile are degenerate with the ground state [Fig. 2.2(b)]. The advantage of the CDCC-BA approach is that all relevant bound and continuum states in the $a+v$ system are explicitly included in the calculation.

Some early comparisons between these three methods can be found in Refs. [18, 20–22] and the main results are also summarized in Ref. [19]. Due to numerical limitations, these first studies where done using a zero-range approximation of the V_{av} potential. Overall, they find that the ADWA model describes well the direct transfer contribution. However, the multi-step contribution, which are completely absent in DWBA, are described very inaccurately by the adiabatic approximation. At low energies ($E_d < 20$ MeV) the discrepancy between the ADWA and CDCC-BA calculation can be understood because at these energies the adiabatic approximation itself is questionable. However, even at medium energies ($E_d \approx 80$ MeV) there are situations in which transfer through breakup channels is found to be very significant, and therefore the ADWA method did not work well either. In these situations, the CDCC-BA should be better used instead. The disadvantage of the CDCC-BA calculations is that, in principle, a large basis of internal states has to be included, making this approach much more demanding numerically.

Finite-range effects have been studied within the adiabatic approximation in Refs. [23, 24] and were found to be small (<10 %) at energies below 20–30 MeV/u but become more and more important as the incident energy increases. This limitation should be also taken into account in the analysis of experimental data.

2.2.4 Coupled Reaction Channels CRC

It was stated that Eqs. (2.5) and (2.6) provide the exact solution to the 3-body scattering problem, provided that $\Psi^{(+)}(\mathbf{R}, \mathbf{r})$ (in the post form) or $\Psi^{(-)}(\mathbf{R}', \mathbf{r}')$ (in the prior form) correspond to the exact three-body wavefunctions with the appropriate boundary conditions. However, in practical calculations, these exact solutions are not available and thus they need to be replaced by approximated ones, such as the factorized form used in the DWBA method, the adiabatic wavefunction or the CDCC expansion. In all these approximations, the three-body wavefunction is restricted to configurations corresponding to either the initial or the final channel. For example, in the post representation, the initial state is a solution of the three-body Schrödinger equation

$$[\hat{T} + V_{av} + U_{bv} + U_{ab} - E]\Psi^{(+)}(\mathbf{r}, \mathbf{R}) = 0, \qquad (2.26)$$

where $\hat{T}$ stands for the full kinetic energy operator. Asymptotically, the solution of this equation is of the form

$$\Psi^{(+)}(\mathbf{r}, \mathbf{R}) \to \varphi_{av}(\mathbf{r})e^{i\mathbf{K}\cdot\mathbf{R}} + \text{outgoing waves}, \qquad (2.27)$$

where the "outgoing waves" contain contributions from all open channels. This includes elastic and breakup channels, but also rearrangement channels of the $a + b$ and $v + b$ pairs, if they are present. In principle, the eigenstates of the $a + v$ Hamiltonian form a complete set and hence the expansion Eq. (2.20) should contain all the relevant channels. In particular, the asymptotic part of (2.20) should contain information from all open channels, including rearrangement channels. However, rearrangement channels corresponding to the $b + v$ system would behave asymptotically as a product of the bound wavefunction $\varphi_{bv}(\mathbf{r}')$ times a plane wave in the aB co-ordinate. Although these states could be in principle expressed in the $\varphi_{av}(\mathbf{r})$ basis, this would require a very large number of energies and angular momenta [19]. In other words, any finite CDCC approximation will describe poorly the contribution from rearrangement channels.

A heuristic way of incorporating rearrangement channels is provided by the Coupled-Reaction-Channels (CRC) framework [1, 5, 25–27]. We present a brief derivation here that highlights its connection with the other methods discussed in this work and, in particular, with the CDCC-BA method of Sect. 2.2.3. The idea of the CRC method is to use a model wavefunction which incorporates explicitly contributions from several mass partitions. For simplicity, let us assume that we wish to consider explicitly excited states (bound or unbound) of the incoming partition plus some excited states of the aB partition. Then, we may use the following *ansatz*:

$$\Psi^{(+)}(\mathbf{R},\mathbf{r}) \approx \Psi^{(+)\text{CRC}}(\mathbf{R},\mathbf{r}) = \sum_i \chi^{(+)}_{Ab,i}(\mathbf{R}) \varphi_{av,i}(\mathbf{r})$$

$$+ \sum_j \chi^{(+)}_{aB,j}(\mathbf{R}') \varphi_{bv,j}(\mathbf{r}'). \quad (2.28)$$

This wavefunction can be interpreted as a generalization of the CDCC expansion of Eq. (2.20). The radial functions $\chi^{(+)}_{Ab,i}(\mathbf{R})$ and $\chi^{(+)}_{aB,j}(\mathbf{R}')$ are obtained by substituting the model wavefunction (2.28) into the Schrödinger equation:

$$[H - E]\Psi^{(+)\text{CRC}} = 0. \quad (2.29)$$

To get the equations satisfied by $\chi^{(+)}_{Ab,i}(\mathbf{R})$ we replace in this equation $\Psi^{(+)\text{CRC}}$ by the *ansatz* (2.28), multiply on the left by each of the functions $\varphi^*_{av,i}(\mathbf{r})$ and integrate along $\mathbf{r}$, giving rise to the system of equations:

$$\sum_{i'} \langle \varphi_{av,i} | H - E | \chi^{(+)}_{Ab,i'} \varphi_{av,i'} \rangle + \sum_j \langle \varphi_{av,i} | H - E | \chi^{(+)}_{aB,j} \varphi_{bv,j} \rangle = 0. \quad (2.30)$$

Now, recall that H can be written in two different forms, depending on whether one chooses the representation of the initial or final channel, namely,

$$H = \hat{T}_{Ab} + H_{av} + U_{Ab}(R) + V_{\text{prior}} \quad \text{(prior representation)} \quad (2.31a)$$

$$= \hat{T}_{aB} + H_{bv} + U_{aB}(R') + V_{\text{post}} \quad \text{(post representation)}, \quad (2.31b)$$

where $H_{av} = \hat{T}_{av} + V_{av}$ and $H_{bv} = \hat{T}_{bv} + V_{bv}$ are the internal Hamiltonians of the $a+v$ and $b+v$ systems, and $U_{Ab}(R)$ and $U_{aB}(R')$ are auxiliary potentials, to be specified later. The prior and post interactions are $V_{\text{prior}} = V_{Ab} - U_{Ab}(R)$ and $V_{\text{post}} = V_{aB} - U_{aB}(R')$, with $V_{Ab} \equiv V_{bv} + U_{ab}$, $V_{aB} \equiv V_{av} + U_{ab}$.

The first term in (2.30) is a matrix element between internal states of the initial partition; hence, the natural choice for H to evaluate this part is the prior representation:

$$[E - \varepsilon_{av}^i - \hat{T}_{Ab} - U_{Ab}^i(R)]\chi_{Ab,i}^{(+)}(\mathbf{R}) = \sum_{i'} \langle \varphi_{av,i}|V_{\text{prior}}|\varphi_{av,i'}\rangle \chi_{Ab,i'}^{(+)}(\mathbf{R})$$
$$+ \sum_j \langle \varphi_{av,i}|H - E|\chi_{aB,j}^{(+)}\varphi_{bv,j}\rangle, \quad (2.32)$$

where we have used the fact that $\langle \varphi_{av,i}|H_{av}|\varphi_{av,i'}\rangle = \varepsilon_{av}^i \delta_{i,i'}$ and that the kinetic energy operator $\hat{T}_{Ab}$ does not depend on the **r** coordinate. Note that a superscript i has been added to the auxiliary potential $U_{Ab}^i(R)$ to indicate that this potential can be taken differently for each of the equations above. For example, this potential could be taken as the monopole part of the cluster-folded potential $\langle \varphi_{av,i}|V_{Ab}|\varphi_{av,i}\rangle$.

Likewise, to get the equations for $\chi_{aB,j}^{(+)}(\mathbf{R}')$, we use the post form of the Hamiltonian in the second term of Eq. (2.30), and project the Schrödinger equation onto the functions $\varphi_{bv,j}^*(\mathbf{r}')$, giving rise to:

$$[E - \varepsilon_{bv}^j - \hat{T}_{aB} - U_{aB}^j(R')]\chi_{aB,j}^{(+)}(\mathbf{R}') = \sum_{j'} \langle \varphi_{bv,j}|V_{\text{post}}|\varphi_{av,j'}\rangle \chi_{aB,j'}^{(+)}(\mathbf{R}')$$
$$+ \sum_i \langle \varphi_{bv,j}|H - E|\chi_{Ab,i}^{(+)}\varphi_{bv,i}\rangle. \quad (2.33)$$

The set of equations (2.32) and (2.33) constitute the CRC equations for the three-body problem at hand. The first set of equations (2.32) correspond to the functions $\chi_{Ab,i}^{(+)}(\mathbf{R})$, which describe the relative motion between the projectile and the target for each state of the projectile i. The source term (RHS in this equation) shows that these functions are affected by two kinds of couplings. The first term, corresponds to couplings between the state i and other states of the same mass partition (i'), i.e., inelastic scattering. These coupling potentials are more explicitly given by:

$$\langle \varphi_{av,i}|V_{\text{prior}}|\varphi_{av,i'}\rangle = \int \varphi_{av,i}^*(\mathbf{r})(V_{Ab} - U_{Ab}^i)\varphi_{av,i'}(\mathbf{r})d\mathbf{r}. \quad (2.34)$$

The second term in the RHS of (2.32) describes the couplings between the states of the initial (Ab) partition and the second partition (aB). Explicitly,

$$\langle \varphi_{av,i}|H - E|\chi_{Ab,j}^{(+)}\varphi_{bv,j}\rangle = \int \varphi_{av,i}^*(\mathbf{r})(H - E)\chi_{aB,j}^{(+)}(\mathbf{R}')\varphi_{bv,j}(\mathbf{r}')d\mathbf{r}. \quad (2.35)$$

We see that, in this case, we cannot extract the $\chi_{aB,i}^{(+)}(\mathbf{R}')$ function from the integral, as we did for the first term. The reason is that this function depends on the variable $\mathbf{R}'$, which is a function of both **r** and **R**. This kind of couplings are said to be non-local, because they depend on the values of $\chi_{aB,j}^{(+)}(\mathbf{R}')$ in all the configuration space, and not just in a single point **R**. In the evaluation of this matrix element, we need to replace the Hamiltonian by either its prior or post form. Since these matrix elements are between states of two different pair Hamiltonians (H_{av} and H_{bv}) the choice is not as clear as in the case of the local matrix elements. In either case, we will get matrix elements of the interaction potentials, but also terms involving the overlaps $\langle \varphi_{av,i} | \varphi_{bv,j} \rangle$, which give a non-zero contribution because these states are not orthogonal (they are eigenstates of different Hamiltonians). These are the so-called *non-orthogonality* terms referred in the literature in the context of the CRC formalism. A more detailed discussion of these terms can be found elsewhere [1, 27].

The same kind of couplings are present in the CRC equations for $\chi^{(+)}_{aB,j}$, Eq. (2.33). That is, transfer channels are induced by couplings with the states of the initial partition but they are also indirectly affected by the couplings with other excited states of the final partition. These two kinds of couplings are depicted in Fig. 2.2(d). In particular, we see that, in this scheme, the elastic scattering will be modified by the coupling to the inelastic channels, as in the CDCC method, but also by couplings with the rearrangement channels. In many situations, however, it is assumed that the latter are small and thus a good approximation to the first set of equations can be obtained by just neglecting these couplings altogether, i.e.

$$\left[E - \varepsilon^i_{av} - \hat{T}_{Ab} - U^i_{Ab}(R) \right] \chi^{(+)}_{Ab,i}(\mathbf{R})$$
$$\approx \sum_{i'} \langle \varphi_{av,i} | V_{Ab} - U^i_{Ab}(R) | \varphi_{av,i'} \rangle \chi^{(+)}_{Ab,i'}(\mathbf{R}). \qquad (2.36)$$

Within our three-body model, the interaction V_{Ab} corresponds to the sum of the interactions between the projectile constituents (a and v) and the target b. In general, these are complicated operators, depending on the energy and angular momentum but, for the purpose of solving (2.36) they are typically approximated by some optical potentials describing the elastic scattering of each constituent by the target at the same incident energy per nucleon. That is, ones makes the approximation: $V_{Ab} \approx U_{bv} + U_{ab}$. Moreover, the auxiliary potential U^i_{Ab} is taken to minimize the difference $V_{Ab} - U^i_{Ab} = U_{bv} + U_{ab} - U^i_{Ab}$, which is just the term V_{prior} defined in Eq. (2.12). For example, a possible choice would be the monopole term of the expected value,

$$U^i_{Ab}(R) = \langle \varphi_{av,i} | U_{bv} + U_{ab} | \varphi_{av,i} \rangle. \qquad (2.37)$$

With this choice, the set of equations (2.36) are nothing else but the CDCC equations of Eq. (2.21). If we insert the approximated solutions $\chi^{(+)}_{Ab,i}(\mathbf{R})$ of (2.36) into (2.33) we get a first order approximation for the functions $\chi^{(+)}_{aB,j}(\mathbf{R}')$, from which the scattering amplitude for transfer can be obtained. This corresponds to the CDCC-BA approximation discussed in previous sections. It can be demonstrated (see e.g. [5]) that the scattering amplitude obtained from the asymptotics of $\chi^{(+)}_{aB,j}(\mathbf{R}')$ is entirely equivalent to the solution of the integral form given by Eq. (2.5) in which the exact wavefunction is approximated by its CDCC counterpart. Therefore, this first order solution of the CRC equations is just the CDCC-BA approximation discussed in Sect. 2.2.3.

The process could be continued iteratively, by inserting the $\chi^{(+)}_{aB,i}(\mathbf{R}')$ into the first set of equations (2.32), thus providing an improved approximation to the $\chi^{(+)}_{Ab,i}(\mathbf{R})$

functions and so on. An early comparison between the CDCC-BA and the full fledged CRC calculation can be found in [25] for the ^{16}O$(d, p)^{17}$O$(2s, 3.27$ MeV$)$ at several deuteron energies. At low energies ($E_d < 40$ MeV) the effect of the proton channel on the elastic cross section is significant but decreases rapidly as the energy increases, being negligible for $E_d > 40$ MeV. At these energies, the CDCC-BA is found to be accurate.

2.2.5 Connection with the Faddeev Formalism

The CRC method is based on a heuristic *ansatz* for the three-body wavefunction, rather than on a rigorous treatment of the three-body scattering problem. Such a rigorous solution exists and was provided many years ago by Faddeev [28]. The idea is to express the three-body wavefunction $\Psi^{(+)}$ as a sum of three components, each of them expressed in a definite Jacobi set of coordinates $\{\mathbf{r}_i, \mathbf{R}_i\}$, with $i = 1, 2, 3$ (using our previous notation, $\mathbf{r}_1 \equiv \mathbf{r}$ and $\mathbf{R}_1 \equiv \mathbf{R}$; $\mathbf{r}_2 \equiv \mathbf{r}'$ and $\mathbf{R}_2 \equiv \mathbf{R}'$):

$$\Psi^{(+)} = \Psi^{(1)}(\mathbf{r}_1, \mathbf{R}_1) + \Psi^{(2)}(\mathbf{r}_2, \mathbf{R}_2) + \Psi^{(3)}(\mathbf{r}_3, \mathbf{R}_3), \tag{2.38}$$

verifying the triad of equations

$$[E - \hat{T} - V_{av}]\Psi^{(1)} = V_{av}\left(\Psi^{(2)} + \Psi^{(3)}\right) \tag{2.39a}$$

$$[E - \hat{T} - V_{bv}]\Psi^{(2)} = V_{bv}\left(\Psi^{(1)} + \Psi^{(3)}\right) \tag{2.39b}$$

$$[E - \hat{T} - V_{ab}]\Psi^{(3)} = V_{ab}\left(\Psi^{(1)} + \Psi^{(2)}\right). \tag{2.39c}$$

Note that adding the three equations one recovers the original Schrödinger equation (2.26). The advantage of this decomposition is that each equation contains only one pair interaction and, therefore, the asymptotic form of the associate Faddeev component can only contain bound states supported by that interaction. Thus, $\Psi^{(1)}$ contains the v–a short range correlations, in particular, the v–a bound states, $\Psi^{(2)}$ contains the v–b short range correlations, including v–b bound states, and $\Psi^{(3)}$ contains the a–b short range correlations, and would include the a–b bound states, assuming that they were relevant for the reaction.

At first sight, the link between this set of equations and the methods discussed in the preceding sections, is not obvious. This connection became more clear after the work of Austern, Kawai and Yahiro [29, 30]. They use the alternative but equivalent set of Faddeev equations

$$\left[E - \hat{T} - V_{av} - \mathscr{P}(V_{bv} + V_{ab})\mathscr{P}\right]\widetilde{\Psi}^{(1)} = V_{av}\left(\widetilde{\Psi}^{(2)} + \widetilde{\Psi}^{(3)}\right) \tag{2.40a}$$

$$[E - \hat{T} - V_{bv}]\widetilde{\Psi}^{(2)} = (V_{bv} - \mathscr{P}V_{bv}\mathscr{P})\widetilde{\Psi}^{(1)} + V_{bv}\widetilde{\Psi}^{(3)} \tag{2.40b}$$

$$[E - \hat{T} - V_{ab}]\widetilde{\Psi}^{(3)} = (V_{ab} - \mathscr{P}V_{ab}\mathscr{P})\widetilde{\Psi}^{(1)} + V_{ab}\widetilde{\Psi}^{(2)} \tag{2.40c}$$

where $\mathscr{P}$ is a projector defined as

$$\mathscr{P} = \sum_{i=1}^{N} |\varphi_{av,i}\rangle\langle\varphi_{av,i}|, \qquad (2.41)$$

with N denoting a finite number of states of the $a+v$ system. In the present context, this projector corresponds to the model-space spanned by the CDCC wavefunction. The properties of this projector are discussed in [29]; it selects low angular momenta ℓ associated with the a–v relative coordinate. Note that the individual components $\{\widetilde{\Psi}^{(i)}\}$ will differ from the original ones $\{\Psi^{(i)}\}$ but the total wavefunction, obtained as the sum of the three components, will be the same. We can identify $\mathscr{P}V_{bv}\mathscr{P} \simeq U_{bv}$ and $\mathscr{P}V_{ab}\mathscr{P} \simeq U_{ab}$ with the effective complex interactions that would describe on-shell elastic matrix elements when the v–b and a–b energies per nucleon are close to the projectile incident energy per nucleon. This identification is possible because we are projecting on states in which the v–a relative energy and angular momentum is not large.

Addition of (2.40b) and (2.40c) gives:

$$[E - \hat{T} - V_{av} - U_{bv} - U_{ab}]\widetilde{\Psi}^{(1)} = V_{av}\widetilde{\Psi}^{(2+3)} \qquad (2.42a)$$

$$[E - \hat{T} - V_{bv} - V_{ab}]\widetilde{\Psi}^{(2+3)} = \left[V_{bv} + V_{ab} - \mathscr{P}(V_{bv} + V_{ab})\mathscr{P}\right]\widetilde{\Psi}^{(1)} \qquad (2.42b)$$

with $\Psi^{(2+3)} \equiv \Psi^{(2)} + \Psi^{(3)}$. As the model-space is augmented, the term $[V_{bv} + U_{ab} - \mathscr{P}(V_{bv} + V_{ab})\mathscr{P}]$ appearing in the RHS of (2.42b) becomes smaller and smaller, thus suppressing the component $\Psi^{(2+3)}$. Under these circumstances, the RHS in the first equation can be neglected giving the zero-th order approximation

$$[E - \hat{T} - V_{av} - U_{bv} - U_{ab}]\widetilde{\Psi}^{(1)} \approx 0. \qquad (2.43)$$

This corresponds to the CDCC approximation. It becomes clear that the accuracy of the CDCC-BA approximation depends on the ability of the CDCC expansion to represent the full solution, at least within the region of $\{\mathbf{r}, \mathbf{R}\}$ for which the transition operator is important. This is expected to happen when the short range correlations between b–v and a–b are not important. In particular, the pair interactions U_{bv} and U_{ab} should not support any bound states. In fact, it was argued by Austern and collaborators [29] that the apparent success of the CDCC method to describe elastic and breakup reactions is largely due to the use of complex optical potentials to represent the U_{bv} and U_{ab} interactions.

Recently, it has become possible to solve the Faddeev equations for a number of nuclear reactions [31–34], thus providing a very useful assessment for more approximate methods. In Ref. [35], a systematic comparison was performed for (d,p) reactions on ^{10}Be, ^{12}C and ^{48}Ca targets, at several incident energies, and it was found that CDCC-BA and Faddeev agree very well at small incident energies (a few MeV per nucleon) but the agreement progressively deteriorates with increasing incident energies. Moreover, as the energy increases, the absolute cross section drops down due to the less favorable energy/momentum matching conditions [19]. So, at these energies, small discrepancies in the absolute cross section imply a sizable relative uncertainty. This suggests that transfer reactions at high energies ($E > 100$ MeV) are not suitable for the extraction of structure information, such as spectroscopic factors. In Ref. [31] the two methods were compared for the $^{11}\text{Be}(p,d)^{10}\text{Be}$ reaction at 38.4 MeV/u and it was found that the former underestimates the Faddeev result by about 30 %. This relatively large discrepancy could be partially due to the absence of an imaginary part in one of the fragment-target interactions (p–n), which favours the coupling with rearrangement channels and makes less justified the approximation (2.43), as pointed out in Ref. [29]. Further calculations and benchmarks comparisons, for other systems and energies, are clearly needed to establish the limits of validity of the CDCC-BA and more approximated methods (DWBA, ADWA).

The Faddeev method can be also linked to the CRC formalism. For this purpose, we introduce a projector extended to the bound states, and possibly the narrow resonances, of the $a+b$ system

$$\mathcal{Q} = \sum_{i=1}^{N} |\varphi_{ab,i}\rangle\langle\varphi_{ab,i}|, \qquad (2.44)$$

and the complementary projector $\mathcal{P} = 1 - \mathcal{Q}$. Then, we can identify the projected interaction $U_{ab} = \mathcal{P} V_{ab} \mathcal{P}$, which can eventually be approximated by a complex effective interaction that does not support any bound state or narrow resonance. Making use of these projectors, we rewrite the Faddeev equations as:

$$[E - \hat{T} - V_{av}]\widetilde{\Psi}^{(1)} = (V_{av} + U_{ab})\widetilde{\Psi}^{(2)} + V_{av}\widetilde{\Psi}^{(3)} \qquad (2.45\text{a})$$

$$[E - \hat{T} - V_{bv}]\widetilde{\Psi}^{(2)} = (V_{bv} + U_{ab})\widetilde{\Psi}^{(1)} + V_{bv}\widetilde{\Psi}^{(3)} \qquad (2.45\text{b})$$

$$[E - \hat{T} - V_{ab}]\widetilde{\Psi}^{(3)} = (V_{ab} - U_{ab})\widetilde{\Psi}^{(1)} + (V_{ab} - U_{ab})\widetilde{\Psi}^{(2)}. \qquad (2.45\text{c})$$

Again, one can see that, adding the three equations, one recovers the original form of the Faddeev equations. When we are interested in elastic scattering, inelastic scattering or transfer reactions, the wavefunction components $\widetilde{\Psi}^{(1,2)}$ will be important, as they appear either in the incident or the outgoing channels. However, $\widetilde{\Psi}^{(3)}$ only contributes through the couplings. Thus, insofar as the term $V_{ab} - U_{ab}$ is small, this component can be neglected. Note that this will be accurate provided that $a+b$ bound states and narrow resonances do not play a role in the reaction. Neglecting $\widetilde{\Psi}^{(3)}$, and including arbitrary potential matrices $\mathbf{U}_{Ab}$ and $\mathbf{U}_{aB}$ in both sides of Eqs. (2.45a)–(2.45c), we get

$$[E - \hat{T} - V_{av} - \mathbf{U}_{Ab}]\widetilde{\Psi}^{(1)} = (V_{av} + U_{ab} - \mathbf{U}_{aB})\widetilde{\Psi}^{(2)} \qquad (2.46\text{a})$$

$$[E - \hat{T} - V_{bv} - \mathbf{U}_{aB}]\widetilde{\Psi}^{(2)} = (V_{bv} + U_{ab} - \mathbf{U}_{Ab})\widetilde{\Psi}^{(1)}. \qquad (2.46\text{b})$$

Note that if we select the potential matrices $\mathbf{U}_{aB}$ and $\mathbf{U}_{Ab}$ as those whose matrix elements are precisely the transition potentials in the CDCC formalism, we obtain a set of equations that look very similar to the CRC equations. A detailed comparison of Faddeev and CRC is in progress, and will be published by the present authors elsewhere.

2.3 Transfer to Unbound States

So far, we have considered transfer reactions as a tool to investigate bound states of a given nucleus. However, in a rearrangement process the transferred particle can populate also unbound states of the final nucleus. This opens the possibility of studying and characterizing structures in the continuum, such as resonances or virtual states.

As in the case of transfer to bound states, the simplest formalism to analyze these processes is the DWBA method. In this case, the bound wavefunction $\varphi_{bv}(\mathbf{r}')$ appearing in the final state in Eqs. (2.8) or (2.10) should be replaced by a positive-energy wavefunction describing the state of the transferred particle v with respect to the core b. In principle, for this purpose one could use the suitable scattering state of the $v + b$ system at the appropriate relative energy. However, this procedure tends to give numerical difficulties in the evaluation of the transfer amplitude due to the oscillatory behaviour of both the final distorted wave and the wavefunction $\varphi_{bv}(\mathbf{r}')$. To circumvent this problem, several alternative methods have been used. We enumerate here some of them:

(i) The bound state approximation [36]. In the case of transfer to a resonant state, this method replaces the scattering state $\varphi_{bv}(\mathbf{r}')$ by a weakly bound wavefunction with the same quantum numbers ℓ and j. In practice, this can be achieved by starting with the potential that generates a resonance at the desired energy and increase progressively the depth of the central potential until the state becomes bound.

(ii) Huby and Mines [37] use a scattering state for $\varphi_{bv}(\mathbf{r}')$, but it is multiplied by a convergence factor $e^{-\alpha r'}$ (with α a positive real number), which artificially eliminates its contribution to the integral coming from large r' values, and then extrapolate numerically to the limit $\alpha \to 0$.

(iii) Vincent and Fortune [38] questioned the bound state approximation arguing that, in general, the bound state and resonant form factors can be very different and, even in those cases in which the fictitious form factor gives the correct shape, they can lead to very different absolute cross sections. They suggest using the actual scattering state, but choosing an integration contour along the complex plane in such a way that the oscillatory integrand is transformed into an exponential decay, thus improving the convergence and numerical stability of the calculation.

(iv) In a real transfer experiment leading to positive-energy states, one does not have access to a definite final energy, but to a certain region of the continuum. That is to say, the extracted observables, such as energy differential cross sec-

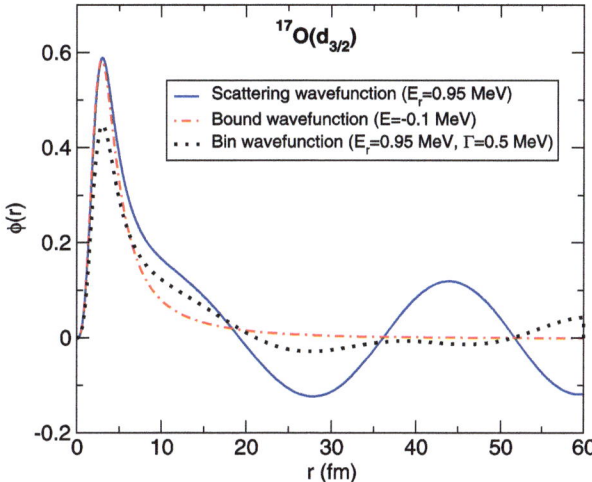

Fig. 2.3 Radial part of the $d_{3/2}$ single-particle resonance wavefunction in ^{17}O at $E_r = 0.95$ MeV compared with a slightly bound wavefunction ($E = -0.1$ MeV) and a bin wavefunction, centered at the nominal energy of the resonance and with a width of 0.5 MeV

tions, are integrated over some energy range which, at least, is of the order of the energy resolution of the experiment. This suggests a method of dealing with the unbound states consisting of discretizing the continuum states in energy bins, as in the CDCC approximation.

In Fig. 2.3, we show as an example the radial part of a $3/2^+$ resonance in ^{17}O, described in terms of a $d_{3/2}$ neutron coupled to a zero-spin ^{16}O core. The solid line is a scattering wavefunction evaluated at the nominal energy of the resonance ($E_{rel} = 0.95$ MeV). Note the oscillatory behaviour at large distances. The dotted line is a bin wavefunction, constructed by a superposition of scattering states, within the range of 0.5 MeV around the resonance energy. It is seen that, asymptotically, the oscillations are damped with respect to the original scattering states. Finally, the dot-dashed line is a bound state wavefunction, with a $1d_{3/2}$ single-particle configuration, and a separation energy of 0.1 MeV. This wavefunction is very similar to the scattering state at short distances, but decays exponentially at large distances.

An advantage of the method (iv) is that it can be equally applied to both resonant and non-resonant continuum final states. An example is shown in Fig. 2.4, which corresponds to the differential cross section, as a function of the n–^{9}Li relative energy, for the reaction ^{2}H(^{9}Li, p)^{10}Li* at 2.36 MeV/u measured at REX-ISOLDE [39]. The lines are the results of CDCC-BA calculations, including the transfer to ^{10}Li* continuum states, showing the separate contribution of the s-wave ($\ell = 0$) continuum and p-wave ($\ell = 1$) continuum. The strength of the measured cross section close to zero energy is due to the presence of a virtual state in the ^{10}Li* continuum, whereas the peak around 0.4 MeV is due to a $p_{1/2}$ resonance. This is an example of how the use of transfer reactions can provide information of the continuum structure of weakly-bound or even unbound systems.

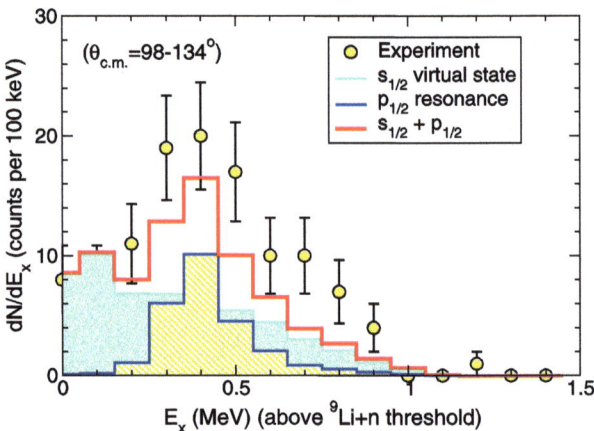

Fig. 2.4 Illustration of the transfer-to-the-continuum method, using a *binning* discretization, for the reaction $^2\text{H}(^9\text{Li}, p)^{10}\text{Li}^*$

2.3.1 Recent Applications to Weakly Bound Halo Nuclei

Many current nuclear reactions studies are done using exotic nuclei. In the light region of the nuclear chart, many of these exotic systems are weakly bound. As in the deuteron case, transfer reactions involving these nuclei must incorporate in some way the effect of the coupling to the unbound states of the weakly bound nucleus. This can be done using the ADWA or CDCC approximations discussed in the previous sections, that is, replacing the exact three-body wavefunction appearing in the transition amplitude by their adiabatic or CDCC counterparts. Although one expects that the CDCC-BA method provides more accurate results, most of these analyses have been done using the DWBA or ADWA methods, partially due to the computational difficulties inherent to the CDCC-BA method. In fact, early applications of the CDCC-BA method used invariably the zero-range approximation for the transition operator. However, there are nowadays computer codes, that permit these kind of calculations [40] using finite range transfer. An example of this kind of applications is given by the $^{14}\text{N}(^7\text{Be}, ^8\text{B})^{13}\text{C}$ reaction, which involves the weakly bound halo nucleus ^8B. This reaction was measured at Texas A&M at an energy of 84 MeV with the purpose of extracting the spectroscopic factor and the so-called astrophysical S-factor for the ^8B nucleus. The reaction was later in studied in [41, 42] using the CDCC-BA approximation in prior form.

An additional complication arises when these reactions involve deuterons in either the initial or final state, as for example in the case of (p, d) or (d, p) reactions. In this case both, the deuteron and halo nucleus continua, could play a role in the reaction dynamics. Examples of these reactions are $^7\text{Be}(d, n)^8\text{B}$ [43], $^{10}\text{Be}(d, p)^{11}\text{Be}$ [44] and $^{11}\text{Be}(p, d)^{10}\text{Be}$ [45].

To deal with these reactions one possibility is to use the ADWA and CDCC-BA formalisms discussed in previous sections. For concreteness, let us consider the $^{11}\text{Be}(p, d)^{10}\text{Be}$ case. In post form, the CDCC-BA transition amplitude reads

$$T^{(\text{post})}(p \to d) \simeq \langle \chi_{da}^{(-)}(\mathbf{R}')\varphi_{pn}(\mathbf{r}')|U_{pa} + V_{na} - U_{da}|\Psi_{na}^{\text{CDCC}}(\mathbf{R}, \mathbf{r})\rangle, \quad (2.47)$$

Fig. 2.5 CDCC-BA calculations for the ^{11}Be$(p,d)^{10}$Be transfer reaction at 38.4 MeV per nucleon. Experimental data are from Refs. [45, 48]

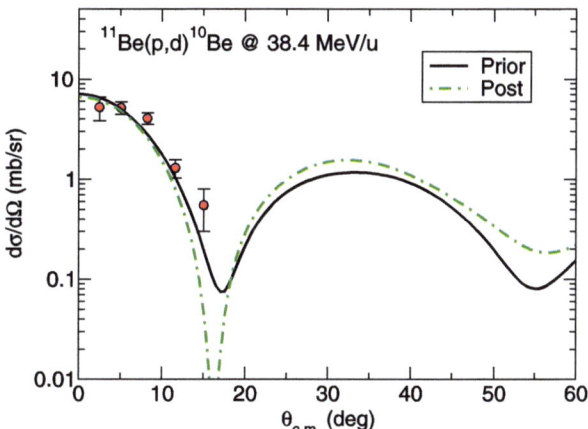

where in this case $a \equiv {}^{10}$Be. Alternatively, one can use the prior form, namely,

$$T^{(\text{prior})}(p \to d) \simeq \langle \Psi_{pn}^{\text{CDCC}}(\mathbf{R}', \mathbf{r}') | U_{pa} + V_{np} - U_{pA} | \chi_{Ap}^{(+)}(\mathbf{R})\varphi_{av}(\mathbf{r}) \rangle \quad (2.48)$$

with $A \equiv {}^{11}$Be.

Note that, in the post form, the ^{11}Be continuum is explicitly taken into account, whereas in the prior representation the deuteron continuum is considered explicitly.[1] It is not obvious to decide beforehand which of these approximations is more suitable for actual calculations. In Ref. [47], both amplitudes were compared for this reaction at a proton energy of 38.4 MeV. The angular distributions obtained from these calculations are shown in Fig. 2.5, and compared with the data from Refs. [45, 48]. As it can be seen, both representations yield very similar results. However, the convergence was found to be much faster in the prior representation, a result that can be ascribed to the shorter range of the transition operator in that case.

Another formalism specifically designed to describe (d, p) and (p, d) transfer reactions with halo nuclei is the method proposed by Timofeyuk and Johnson [17]. This approach is based on an alternative exact representation of the transfer amplitude in which the transition operator is the p–n potential and any effects due to remnant terms are included in the wave function for the initial or final channel. This is achieved by choosing the auxiliary potential appearing in the transition operator to cancel out exactly the remnant term. So, for instance, in the previous example, in the prior representation one would make the choice $U_{pA} = V_{pa}$. With this choice the transfer operator becomes simply V_{pn}. However, the initial wave function is no longer given by the factorized form $\chi_{Ab}^{(+)}(\mathbf{R})\varphi_{av}(\mathbf{r})$, but becomes a

---

[1] Strictly speaking, both two-body continua are part of the same three-body continuum, namely, $p + n + {}^{10}$Be. In principle, a complete basis of either sub-system would be sufficient to describe the three-body continuum. In practice, an accurate description of the full three-body continuum might require a very large basis and so, in actual calculations, using a truncated basis, a suitable choice of the continuum representation can be important [46].

complicated three-body wavefunction. To evaluate the resulting amplitude one may use the CDCC expansions of the initial and final states [47] or the much simpler adiabatic wave functions [17].

A similar approach was followed in Ref. [43] to study the reaction $^7\text{Be}(d,n)^8\text{B}$. In this case, the initial three-body wavefunction was approximated by the CDCC expansion in $p+n$ states, whereas the ^8B continuum was treated in the simple adiabatic prescription of Johnson and Soper [10].

2.4 Summary and Conclusions

We have presented the current status of the theoretical description of transfer reactions. We have shown, that, under certain assumptions, the complex many-body problem corresponding to a rearrangement collision can be approximated by a three-body problem in which a valence particle is transferred from one core to another. The connection between the many-body and the three-body problem is done through the introduction of spectroscopic amplitudes. The experimental determination of these spectroscopic amplitudes from the observed cross section is the key objective of the transfer reaction formalism.

The solution of the three-body problem and, in particular, the evaluation of the T-matrix element from a bound state of the valence particle with a core (initial partition), to a bound state of the valence particle with the other core (final partition), has been discussed with an increasing degree of complexity. We start from an exact formal expression, that gives the transfer transition amplitude as a matrix element of the interaction with the exact three-body wavefunction with boundary conditions on the initial partition on one side, and a two-body distorted wavefunction with boundary conditions on the final partition on the other side. In the DWBA method, the exact three-body wavefunction is approximated by a two-body distorted wavefunction, multiplied by a bound state wavefunction. This approximation, which has been the workhorse for transfer reactions in nuclear physics for years, is not accurate for weakly bound exotic nuclei, since is does not take into account the effects of break-up on the three-body wavefunction.

An improvement over the DWBA method is provided by the ADWA approximation, which relies on the adiabatic approximation of the three-body wavefunction, and assumes that the transfer interaction has a short range. The gives rise to an approximate three-body wavefunction that is accurate in the region where the valence particle and the core are very close. This leads to a formal expression of the transfer three-matrix, which has the same complexity as the DWBA, but with distorting potentials not related to the elastic scattering. The accuracy of ADWA depends on the validity of the adiabatic approximation, and the short range of the interaction. In general, ADWA is well suited for deuteron scattering. However, for high energies (about 100 MeV per nucleon), there is a significant contribution of transfer through the continuum, where excitation energy is high enough so that the adiabatic approximation may not be accurate.

The CDCC-BA approximation describes the three-body wavefunction using a finite representation for the continuum states in the initial or final partition. CDCC-BA has a wider validity compared to ADWA, as it does not rely on the adiabatic approximation, and does not require short range transition operators. However, it is much more demanding computationally and, as one goes to higher scattering energies, it would require to introduce explicitly high energy continuum states. It will, in principle, describe accurately the contribution of break-up states to transfer, provided convergence is found on the various parameters involved in continuum discretization. However, the CDCC-BA scheme describes only the three-body wavefunction on one partition, neglecting possible effects of multi-step processes involving transfer back and forth between different partitions. The CDCC-BA method should give an accurate description of transfer, including break-up effects, at energies from 10 MeV/u onwards.

The CRC approximation describes the 3-body wavefunction as a superposition of wavefunctions corresponding to different mass partitions. The solution of CRC equations involves the consideration of non-local potentials and non-orthogonality terms, and has to be carried out iteratively. CRC has proven very useful to describe elastic scattering and fusion in the presence of transfer cross sections that are comparable to the elastic cross section. This will be the case when the scattering energy is small, of a few MeV/u, and matching conditions favour transfer to bound states.

The correct formal solution of the three-body problem was formulated by Faddeev, in terms of a system of equations that involve 3 components of the wavefunction expressed in terms of the 3 possible sets of Jacobi coordinates. Each component of the wavefunction contains the short range two-body correlations for a given pair of particles, in particular the bound states. We have shown that the CRC equations can be obtained from the Faddeev equations by neglecting the component of the wavefunction containing core-core correlations. The CDCC-BA equations are obtained ignoring, for the 3-body wavefunctions, the correlations in two partitions, so that only one component of the Faddeev equations is considered. We justify the use of complex effective interactions, as the result of projecting two-body interaction in the restricted model space.

We have discussed some applications of the CDCC-BA approach to the description of transfer involving weakly bound systems, and also to unbound states. We have shown that CDCC-BA is able to include consistently the contribution of resonant and non-resonant continuum. Application of CDCC-BA allows to obtain, for example, the energy and width of a resonance, from the measurement of cross sections of transfer to the continuum as a function of the energy of the ejectile.

From our experience, we consider that the optimal energy range to measure transfer reaction of exotic nuclei would be a few tens of MeV per nucleon. At higher energies (100 MeV per nucleon), transfer cross sections diminish, and break-up is dominant. At lower energies (few MeV per nucleon), multi-step processes become important, compound nucleus effects become relevant, and the relation of cross sections with spectroscopic factors becomes more obscure.

Acknowledgements This work has been partially supported by Spanish national projects FPA2009-08848 and FPA2009-07653 and by the Consolider Ingenio 2010 Program CPAN (CSD2007-00042).

References

1. G.R. Satchler, *Direct Nuclear Reactions* (Clarendon Press, Oxford, 1983)
2. N. Austern, R.M. Drisko, E.C. Halbert, G.R. Satchler, Theory of finite-range distorted-waves calculations. Phys. Rev. **133**, B3 (1964)
3. N. Austern, *Direct Nuclear Reaction Theories* (Wiley, New York, 1970)
4. T. Tamura, Compact reformulation of distorted-wave and coupled-channel born approximations for transfer reactions between nuclei. Phys. Rep. **14**, 59 (1974)
5. N.K. Glendenning, *Direct Nuclear Reactions* (World Scientific, Singapore, 2004)
6. J. Sharpey-Schafer, The mean square radius of nuclear matter and spectroscopic factors from the DWBA. Phys. Lett. B **26**, 652 (1968)
7. J.L.C. Ford, K.S. Toth, G.R. Satchler, D.C. Hensley, L.W. Owen, R.M. DeVries, R.M. Gaedke, P.J. Riley, S.T. Thornton, Single-nucleon transfer reactions induced by ^{11}B ions on ^{208}Pb: a test of the distorted-wave Born approximation. Phys. Rev. C **10**, 1429 (1974)
8. K.S. Toth, J.L.C. Ford, G.R. Satchler, E.E. Gross, D.C. Hensley, S.T. Thornton, T.C. Schweizer, Measurements and analysis of the ^{208}Pb(^{12}C, ^{13}C), (^{12}C, ^{11}B), and (^{12}C, ^{14}C) reactions. Phys. Rev. C **14**, 1471 (1976)
9. T. Tamura, T. Udagawa, M.C. Mermaz, Direct reaction analyses of heavy-ion induced reactions leading to discrete states. Phys. Rep. **65**, 345 (1980)
10. R.C. Johnson, P.J.R. Soper, Contribution of deuteron breakup channels to deuteron stripping and elastic scattering. Phys. Rev. C **1**, 976 (1970)
11. H. Amakawa, S. Yamaji, A. Mori, K. Yazaki, Adiabatic treatment of elastic deuteron-nucleus scattering. Phys. Lett. B **82**, 13 (1979)
12. H. Amakawa, K. Yazaki, Adiabatic treatment of deuteron break-up on a nucleus. Phys. Lett. B **87**, 159 (1979)
13. R.C. Johnson, P.C. Tandy, An approximate three-body theory of deuteron stripping. Nucl. Phys. A **235**, 56 (1974)
14. J.D. Harvey, R.C. Johnson, Influence of breakup channels on the analysis of deuteron stripping reactions. Phys. Rev. C **3**, 636 (1971)
15. G.R. Satchler, Adiabatic deuteron model and the ^{208}Pb(p, d) reaction at 22 MeV. Phys. Rev. C **4**, 1485 (1971)
16. G.L. Wales, R.C. Johnson, Deuteron break-up effects in (p, d) reactions at 65 MeV. Nucl. Phys. A **274**, 168 (1976)
17. N.K. Timofeyuk, R.C. Johnson, Deuteron stripping and pick-up on halo nuclei. Phys. Rev. C **59**, 1545 (1999)
18. M. Kawai, Chapter II. Formalism of the method of coupled discretized continuum channels. Prog. Theor. Phys. Suppl. **89**(Suppl. 1), 11 (1986)
19. N. Austern, Y. Iseri, M. Kamimura, M. Kawai, G. Rawitscher, M. Yahiro, Continuum-discretized coupled-channels calculations for three-body models of deuteron-nucleus reactions. Phys. Rep. **154**, 125 (1987)
20. G.H. Rawitscher, Effect of deuteron breakup on (d, p) cross sections. Phys. Rev. C **11**, 1152 (1975)
21. Y. Iseri, M. Yahiro, M. Nakano, Investigation of adiabatic approximation of deuteron-breakup effect on (d, p) reactions. Prog. Theor. Phys. **69**, 1038 (1983)
22. H. Amakawa, N. Austern, Adiabatic-approximation survey of breakup effects in deuteron-induced reactions. Phys. Rev. C **27**, 922 (1983)
23. A. Laid, J.A. Tostevin, R.C. Johnson, Deuteron breakup effects in transfer reactions using a Weinberg state expansion method. Phys. Rev. C **48**, 1307 (1993)

24. N.B. Nguyen, F.M. Nunes, R.C. Johnson, Finite-range effects in (d, p) reactions. Phys. Rev. C **82**, 014611 (2010)
25. M. Kawai, M. Kamimura, K. Takesako, Chapter V. Coupled-channels variational method for nuclear breakup and rearrangement processes. Prog. Theor. Phys. Suppl. **89**(Suppl 1), 118 (1986)
26. T. Ohmura, B. Imanishi, M. Ichimura, M. Kawai, Study of deuteron stripping reaction by coupled channel theory. II properties of interaction kernel and method of numerical solution. Prog. Theor. Phys. **43**, 347 (1970)
27. I.J. Thompson, F.M. Nunes, in *Nuclear reactions for astrophysics, Nuclear Reactions for Astrophysics*, ed. by I.J. Thompson, F.M. Nunes (Cambridge University Press, Cambridge, 2009), p. 1
28. L.D. Faddeev, Scattering theory for a three-particle system. Zh. Eksp. Teor. Fiz. **39**, 1459 (1960)
29. N. Austern, M. Yahiro, M. Kawai, Continuum discretized coupled-channels method as a truncation of a connected-kernel formulation of three-body problems. Phys. Rev. Lett. **63**, 2649 (1989)
30. N. Austern, M. Kawai, M. Yahiro, Three-body reaction theory in a model space. Phys. Rev. C **53**, 314 (1996)
31. A. Deltuva, A.M. Moro, E. Cravo, F.M. Nunes, A.C. Fonseca, Three-body description of direct nuclear reactions: comparison with the continuum discretized coupled channels method. Phys. Rev. C **76**, 064602 (2007)
32. A. Deltuva, Spin observables in three-body direct nuclear reactions. Nucl. Phys. A **821**, 72 (2009)
33. A. Deltuva, Deuteron stripping and pickup involving the halo nuclei ^{11}Be and ^{15}C. Phys. Rev. C **79**, 054603 (2009)
34. A. Deltuva, Three-body direct nuclear reactions: nonlocal optical potential. Phys. Rev. C **79**, 021602 (2009)
35. N.J. Upadhyay, A. Deltuva, F.M. Nunes, Testing the continuum-discretized coupled channels method for deuteron-induced reactions. Phys. Rev. C **85**, 054621 (2012)
36. W.R. Coker, Gamow-state analysis of ^{54}Fe(d, n) to proton resonances in ^{55}Co. Phys. Rev. C **9**, 784 (1974)
37. R. Huby, J.R. Mines, Distorted-wave born approximation for stripping to virtual levels. Rev. Mod. Phys. **37**, 406 (1965)
38. C.M. Vincent, H.T. Fortune, New method for distorted-wave analysis of stripping to unbound states. Phys. Rev. C **2**, 782 (1970)
39. H.B. Jeppesen, A.M. Moro, U.C. Bergmann, M.J.G. Borge, J. Cederkall, L.M. Fraile, H.O.U. Fynbo, J. Gomez-Camacho, H.T. Johansson, B. Jonson, M. Meister, T. Nilsson, G. Nyman, M. Pantea, K. Riisager, A. Richter, G. Schrieder, T. Sieber, O. Tengblad, E. Tengborn, M. Turrion, F. Wenander, Study of ^{10}Li via the ^{9}Li$(^{2}$H, $p)$ reaction at REX-ISOLDE. Phys. Lett. B **642**, 449 (2006)
40. I.J. Thompson, Computer code FRESCO. Comput. Phys. Rep. **7**, 167 (1988)
41. A.M. Moro, R. Crespo, F. Nunes, I.J. Thompson, ^{8}B breakup in elastic and transfer reactions. Phys. Rev. C **66**, 024612 (2002)
42. A.M. Moro, R. Crespo, F.M. Nunes, I.J. Thompson, Breakup and core coupling in ^{14}N$(^{7}$Be, ^{8}B$)^{13}$C. Phys. Rev. C **67**, 047602 (2003)
43. K. Ogata, M. Yahiro, Y. Iseri, M. Kamimura, Determination of S_{17} from the ^{7}Be$(d, n)^{8}$B reaction. Phys. Rev. C **67**, 011602 (2003)
44. B. Zwieglinski, W. Benenson, R.G.H. Robertson, W.R. Coker, Study of the ^{10}Be$(d, p)^{11}$Be reaction at 25 MeV. Nucl. Phys. A **315**, 124 (1979)
45. S. Fortier, S. Pita, J.S. Winfield, W.N. Catford, N.A. Orr, J.V. de Wiele, Y. Blumenfeld, R. Chapman, S.P.G. Chappell, N.M. Clarke, N. Curtis, M. Freer, S. Galès, K.L. Jones, H. Langevin-Joliot, H. Laurent, I. Lhenry, J.M. Maison, P. Roussel-Chomaz, M. Shawcross, M. Smith, K. Spohr, T. Suomijarvi, A. de Vismes, Core excitation in ^{11}Be(gs) via the p (^{11}Be, ^{10}Be) d reaction. Phys. Lett. B **461**, 22 (1999)

46. A.M. Moro, F.M. Nunes, Transfer to the continuum and breakup reactions. Nucl. Phys. A **767**, 138 (2006)
47. A.M. Moro, F.M. Nunes, R.C. Johnson, Theory of (d, p) and (p, d) reactions including breakup: comparison of methods. Phys. Rev. C **80**, 064606 (2009)
48. J.S. Winfield, S. Fortier, W.N. Catford, S. Pita, N.A. Orr, J.V. de Wiele, Y. Blumenfeld, R. Chapman, S.P.G. Chappell, N.M. Clarke, N. Curtis, M. Freer, S. Galès, H. Langevin-Joliot, H. Laurent, I. Lhenry, J.M. Maison, P. Roussel-Chomaz, M. Shawcross, K. Spohr, T. Suomij, Single-neutron transfer from 11Begs via the (p, d) reaction with a radioactive beam. Nucl. Phys. A **683**, 48 (2001)

Chapter 3
What Can We Learn from Transfer, and How Is Best to Do It?

Wilton N. Catford

3.1 Motivation to Study Single-Nucleon Transfer Using Radioactive Beams

A single-nucleon transfer reaction is a powerful experimental tool to populate a certain category of interesting states in nuclei in a selective manner. These states have a structure that is given by the original nucleus as a core, with the transferred nucleon in an orbit around it. Nucleon transfer is thus an excellent way to probe the energies of shell model orbitals and to study the changes in the energies of these orbitals as we venture away from the stable nuclei. Despite a large number of detailed issues that complicate this simple picture, it remains the case that nucleon transfer reactions preferentially populate these "single particle" states in the final nucleus and also that these states are of especial interest, theoretically. Therefore, transfer reactions promise to be one of the most important sources of nuclear structure information about exotic nuclei, as more beams become available at radioactive beam facilities.

The factors that complicate the interpretation of the experiments arise primarily from the theoretical interpretation of the data. Experimentally, the selectivity of the transfer reactions is usually clear, and the states of interest—those having a large overlap with the simple core-plus-particle picture—are emphatically favoured. Often, these states will be embedded within a background of other nuclear levels. This selectivity on structural grounds is itself useful, and often allows immediate associations to be inferred between experimentally observed states and the predictions from, for example, shell model calculations. The states that are suppressed will have more complex wave functions that mix a number of configurations and are intrinsically more difficult to describe theoretically. In the first instance, it is in many ways best to focus upon the more simple states that are selected by transfer reactions,

---

W.N. Catford (✉)
Department of Physics, University of Surrey, Guildford GU2 7XH, UK
e-mail: w.catford@surrey.ac.uk

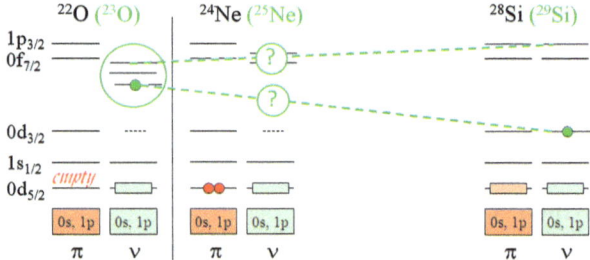

Fig. 3.1 The effective energies of the valence neutron orbitals are modified according to the number of protons present in the $0d_{5/2}$ orbital. The effect is to replace the $N = 20$ neutron shell gap by a gap at $N = 16$ when the nucleus becomes more exotic

and to use these to refine the theory. Complications begin to arise when we seek to quantify the degree to which the wave function of a particular state overlaps with the simple core-plus-particle wave function. At that level, many debates occur, regarding the quantitative interpretation of data. With suitably stated assumptions, however, quantitative analyses of experiments can be performed and confronted with theory. Thus, on a qualitative and on a quantitative level, transfer reactions provide an indispensable tool for uncovering the structures of exotic nuclei.

3.1.1 Migration of Shell Gaps and Magic Numbers, Far from Stability

Figure 3.1 shows a simplified representation of the proton and neutron shell model orbital energies and occupancies for some light nuclei. In the nuclear shell model, each nucleon is assumed to occupy an energy level (or orbital) that can be obtained by solving the Schrödinger equation for a mean field potential. This potential represents the average binding effect of all of the other nucleons. In the simplest model, the nuclear structure is obtained by filling orbitals from the lowest energies, obeying the Pauli exclusion principle. In a more sophisticated model, the interactions between valence nucleons in different orbitals (or in the same orbital) are taken into account. This allows significant mixing between different simple configurations that all have the same spin and parity and about the same (unmixed) energy. Some degree of mixing will even occur over a wide range of configuration energies. In principle the valence nucleon interaction energies, which can be represented as matrix elements in some suitable basis, can be calculated from the solutions for the mean field and an expression for the nucleon-nucleon interaction (with all of its dependence on spatial variables, spin and orbital angular momentum). In practice, the best shell model calculations in terms of agreement with experimental data are those in which the calculated matrix elements are subsequently varied by fitting them to a selection of experimental data, thus establishing an *effective interaction* that is valid in a particular model space that was used for the fitting procedure. Once we accept that valence nucleons will have an interaction potential, and hence some energy associated with the interaction, it naturally becomes possible that the valence interactions can actually change to some degree the effective energies of the orbitals themselves. Slightly more technically, the interaction potentials can be analysed in terms of a multipole expansion. It is the monopole term in the expansion that has the effect of changing the effective energies of orbitals. The energy of a single valence nucleon in a particular orbital is determined by the energy of the orbital plus the sum of the monopole components of its interaction with other active valence nucleons. A closed shell has no net effect, so it is the interactions with partially filled orbitals that needs to be considered. Whilst both the interactions between like nucleons (p–p) or (n–n) and interactions between different nucleons (p–n) are all important, the strongest effects occur when it is a proton-neutron interaction between active valence nucleons. After that, the strongest effects are between orbitals of the same number of radial nodes, and then if the angular momentum is the same this makes the effect is even stronger. This arises from the degree of spatial overlap of the wave functions. For example, the interaction between protons in an open $0d_{5/2}$ orbital and neutrons in an open $0d_{3/2}$ orbital is particularly strong.

In Fig. 3.1, the structure of the $N = 14$ isotones is shown, with the additional odd neutron for $N = 15$ being shown in an otherwise vacant $0d_{3/2}$ orbital. The $0d_{5/2}$ neutron orbital is filled, at $N = 14$. On the right hand side, we see the stable nucleus ^{28}Si, wherein the 14 protons also fill the proton $0d_{5/2}$ orbital. The $3/2^+$ state in ^{28}Si is therefore at a relatively low energy, because the sd-shell orbitals are relatively closely spaced, all lying below the $N = 20$ shell gap. As successive pairs of protons are removed from $0d_{5/2}$, the diagram indicates that the energy of the $0d_{3/2}$ orbital increases. This is actually in accord with detailed calculations and can be understood in terms of the monopole interaction [1, 2] and a version of this diagram can be found in Ref. [1]. By the time we reach the neutron rich ^{22}O, the $0d_{3/2}$ orbital has risen to such an extent that the shell gap is now below that orbital, at $N = 16$. The orbital that has moved up in energy has $j = \ell - 1/2$ and the reason for its change is that there are fewer protons in $0d_{5/2}$ (where $j = \ell + 1/2$) with which a valence $d_{3/2}$ neutron can interact. This proton-neutron interaction between $\ell + 1/2$ and $\ell - 1/2$ nucleons is attractive [1], and hence this reduction in $0d_{5/2}$ protons causes the raising in energy of the neutron $0d_{3/2}$ orbital. This is, in fact, essentially the explanation for ^{24}O being the heaviest bound oxygen isotope (with the neutrons just filling the $1s_{1/2}$ orbital). Neutrons in the $0f_{7/2}$ and $1p_{3/2}$ orbitals, with $j = \ell + 1/2$, experience a repulsive interaction with the $0d_{5/2}$ protons and hence they are lowered in energy as these protons are removed. This further confounds the previous $N = 20$ gap seen for nuclei near stability, and also tends to displace the $N = 28$ gap to a higher number ($N = 34$). In the present work, several of the example nuclei studied using transfer (^{25,27}Ne, ^{21}O) are directly of interest because of this particular migration of orbital energies. What we measure experimentally are the energies of actual states in these nuclei, and not the energies of the shell model orbitals *per se*, but there is a strong connection between the energies of the states and the orbitals in the cases that are studied here.

Fig. 3.2 Neutron structure for states in ^{21}O: (**a**) a low-lying $3/2^+$ state can be made by transferring a neutron into the vacant $0d_{3/2}$ orbital, or (**b**) by having a neutron in $1s_{1/2}$ coupled to a 2^+ ^{20}O core, where the two holes in $0d_{5/2}$ are coupled to spin 2

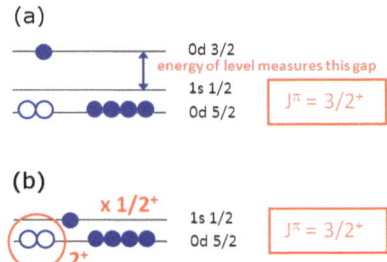

3.1.2 Coexistence of Single Particle Structure and Other Structures

Of course, it is not the case that all of the excited states in the final nucleus will have a structure that is simply explained by a neutron orbiting the original core nucleus. Such states are an important but (usually) small subset of the states in the final nucleus, and are selectively populated by transfer reactions. To measure the energy of the $0d_{3/2}$ neutron orbital, say, in ^{21}O we could imagine an experiment to add a neutron to ^{20}O and then deduce the energy of the $3/2^+$ excited state, and hence the $0d_{3/2}$ orbital energy relative to the $0d_{5/2}$ orbital of the ground state. This is shown conceptually in Fig. 3.2(a). In the lowest energy configuration, the two holes in the neutron $0d_{5/2}$ orbital are coupled to spin zero. This association of the energy of the state directly with that of the orbital is overly simplified because the state will not have a pure configuration. In Fig. 3.2(b), another relatively low energy configuration is shown, which also has spin and parity $3/2^+$. Here, the holes in $0d_{5/2}$ are coupled to spin 2, as they are in the 2^+ state of ^{20}O. A neutron in $1s_{1/2}$ can then couple with this to produce two states in ^{21}O, one of which is $3/2^+$. The residual interactions between valence nucleons will mix these two configurations and the nucleus ^{21}O will have the single particle amplitude split between the two states. Indeed, in the real nucleus, there will be even more components contributing to the wave functions with various smaller amplitudes.

3.1.3 Description of Single Particle Structure Using Spectroscopic Factors

Due to the mixing of different states with the same spin and parity, the single particle state produced by a nucleon orbiting the core of the target, in an otherwise vacant orbital, will be mixed with other nuclear states of different structures. Usually, these will be of more complex structures, or core excited structures. The contribution that this single particle amplitude makes, to the different states, will result in these states all being populated in a nucleon transfer reaction. The strength of the population of each state in the reaction will depend on the intensity of the single particle component. This intensity is essentially the quantity that is called the *spectroscopic factor*. Experimentally, it is measured by taking the cross section that is

Fig. 3.3 Spectroscopic factor versus excitation energy for $3/2^-$ (*red*) and $1/2^-$ (*blue*) states in ^{41}Ca. The strength for a given spin is split between different states in the final nucleus. This figure is due to John Schiffer [3]. Here, we have added the *inset* to indicate schematically how the weighted average of the $3/2^-$ excitation energies can be calculated, which gives a measure of the energy of the $1p_{3/2}$ single particle orbital

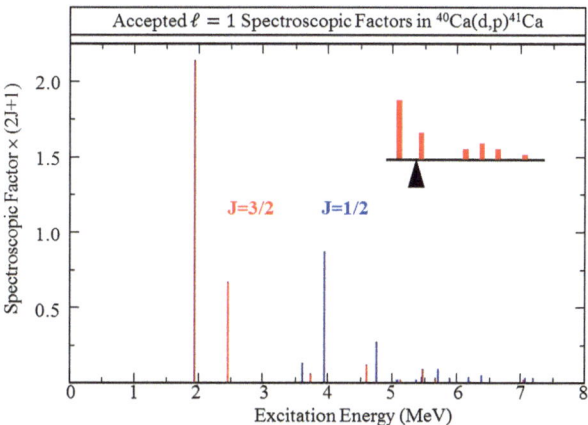

calculated for a pure single particle state and comparing it to the cross section that is measured. More specifically, this comparison is performed using differential cross sections, which are a function of the scattering angle. If the picture described here is correct, then the experimental cross section should have the same shape as the theory, and simply be multiplied by a number less than one—that is, the spectroscopic factor. To describe the sharing of intensity between states, we say that the single particle strength will be spread across a range of states in the final nucleus. This is represented in Fig. 3.3, where the single particle strength (represented as the spectroscopic factor) is plotted as a function of excitation energy. The weighted average of the excitation energies, for all states containing strength from a particular ℓj orbital, will give the energy of that orbital. Note that, for experiments with radioactive beams, the limited intensity of the beam is likely to preclude the possibility of identifying and measuring all of the spectroscopic strength, which was traditionally the aim of transfer experiments. A different approach will often be dictated by these circumstances, wherein only the strongest states are located experimentally. Then, placing more reliance on theory than was formerly done, an association can be made between the strong states experimentally and the states predicted by the theory to be the strongest. We then need to see whether the experimental data, in terms of the energies and spectroscopic factors for the strongest states, can give us enough clues about how to adapt the theoretical calculations to give an improved set of predictions. If applied consistently across a range of nuclei, using the same theory, this approach can reasonably be expected to yield good results.

3.1.4 Disclaimer: What This Article Is, and Is Not, About

This article is intended to describe briefly the general motivations for studies using (mostly) single nucleon transfer, and to provide in some detail the background, insights and perspectives relevant to designing and performing the experiments. For more details about the nuclear structure motivations in terms of nuclear structure and monopole shift the reader is referred to several excellent reviews [4–6].

This article most definitely does not seek to summarise or describe the theories that are used to interpret the data from nucleon transfer reactions, although some general features of the theoretical predictions are discussed and a justification is given for the model of choice for the examples of analysis that are described here. Detailed descriptions of the relevant reaction theory can be found in several well-known articles and books, such as those by Glendenning [7, 8] and Satchler [9, 10]. An excellent and up-to-date introduction and overview with particular reference to weakly bound and unbound states is given in this volume by Gómez Camacho and Moro [11].

With regard to the experimental results, although the main objectives of most of these measurements is to obtain the differential cross sections for individual final states, just a small number of illustrative results are shown here. In all of the discussions, the references are given for the original work, and it is to those publications that reference may be made in order to study the extent and quality achieved for the various differential cross section measurements. It is through the measurement and interpretation of these differential cross sections that the assignments of angular momentum and determinations of spectroscopic single-particle strength are made, for the nuclear states.

3.2 Choice of the Reaction and the Bombarding Energy

In this section, some features of transfer reactions as traditionally performed using stable targets and a low-mass beam (for example, the (d, p) reaction) are reviewed. Some of the differences in the case of inverse kinematics are introduced.

3.2.1 Kinematics and Measurements Using Normal Kinematics

A good way to measure (d, p) reactions when using a beam of deuterons and a stable target is to use a high resolution magnetic spectrometer to record the protons from the reaction, because this can be done with a high precision and a low background. The proton peaks observed at a particular angle will have different energies for different excited states and hence will be dispersed across the focal plane of the spectrometer. The spacings of the proton energies will be almost the same as the spacings of the energy levels in the final nucleus. An example of the kinematical variation of proton energies with laboratory angle is shown in Fig. 3.4. The lines that are almost horizontal are calculations of the proton energies from the (d, p) reaction on a ^{208}Pb target, with the uppermost being for protons populating the ground state of ^{209}Pb. The energies have little variation with laboratory angle because the very heavy recoil ^{209}Pb nucleus takes away very little kinetic energy. The uppermost line, with a much bigger slope, is for the (d, p) reaction on a much lighter target, ^{12}C.

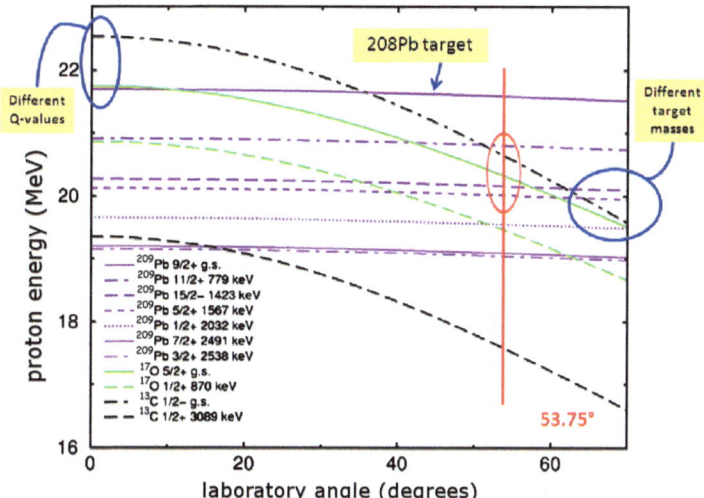

Fig. 3.4 Kinematics plots showing proton energy as a function of laboratory angle for the reaction (d, p) initiated by 20 MeV deuterons. Different curves represent the population of different excited states formed by reactions on ^{208}Pb, ^{12}C and ^{16}O (see text). The angle 53.75° is relevant to Fig. 3.5

The lines of intermediate slope are for the (d, p) reaction on a target of ^{16}O. The energy at zero degrees is different for the different targets because of the different reaction Q-values, whereas the slopes reflect the target mass. The carbon and oxygen calculations are shown, because these isotopes are typical target contaminants. In a study of ^{208}Pb$(d, p)^{209}$Pb, the contaminant reactions will give proton energies that overlap the energy region of interest for the ^{209}Pb states, but these can be identified by comparing data taken at different laboratory angles, since the contaminant peaks will shift in energy, relative to the ^{209}Pb peaks. The example of the proton energies seen in a measurement made at 53.75° is shown in Figs. 3.4 and 3.5.

The peaks corresponding to different final states in ^{209}Pb, measured at 53.75° for the (d, p) reaction [12], are shown in Fig. 3.5. The different intensities reflect both the spectroscopic strengths and the dynamical effects of different angular momentum transfers. It is apparent that different states can easily be resolved and studied. The peaks in the shaded region of Fig. 3.5 correspond to the reactions populating the ground states of ^{17}O and ^{13}C from the oxygen and carbon contamination in the target. At increasing laboratory angles, these peaks would be seen to move to the left in the spectrum, relative to the ^{209}Pb peaks.

3.2.2 Differential Cross Sections: Dependence on Beam Energy and ℓ Transfer

The principal piece of information (after excitation energy) that is measured directly, via transfer studies, is the orbital angular momentum that is transferred to the

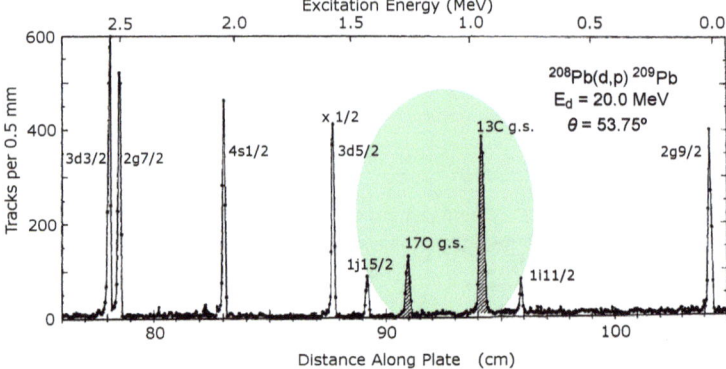

Fig. 3.5 Magnetic spectrometer data for protons from ^{208}Pb$(d,p)^{209}$Pb at a beam energy of $E_d = 20.0$ MeV and a laboratory angle of $53.75°$. Data are from Ref. [12]. Excitation energy increases from *right* to *left* and the *unshaded peaks* correspond to states in ^{209}Pb. The *shaded region* is where reactions on the ^{12}C and ^{16}O in the target produce contaminant peaks

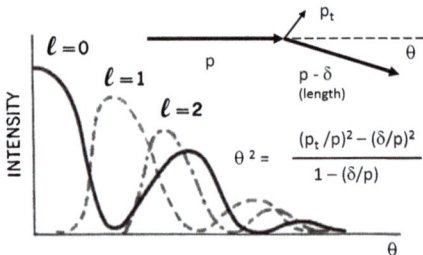

Fig. 3.6 A consideration of the conservation of linear momentum in transfer implies a relationship of the laboratory scattering angle to the transferred momentum, and therefore to the transferred orbital angular momentum, ℓ. This implies that the location of the primary maximum in the angular distribution will be approximately proportional to the transferred ℓ (see text)

target nucleus. This comes from the shape of the differential cross section. Next, the magnitude of the cross section can tell us the magnitude of the single-particle component of the wave function, or the spectroscopic factor.

The transferred angular momentum will indicate, for single-nucleon transfer, into which orbital the nucleon has been transferred. The transferred angular momentum is measured via the angular distribution of the reaction products. In this type of reaction, the differential cross section will tend to have some diffraction-like oscillatory behaviour, with the angle of the main maximum being related to the magnitude of the transferred angular momentum. We can see how the transferred angular momentum affects the angular distribution by considering a simple momentum diagram. Suppose as in the inset of Fig. 3.6 that the incident projectile has momentum of magnitude p and that the momentum transferred to the target nucleus has magnitude p_t. For a small scattering angle, θ, the beam particle will have only a small reduction in the magnitude of its momentum, as seen by construction of

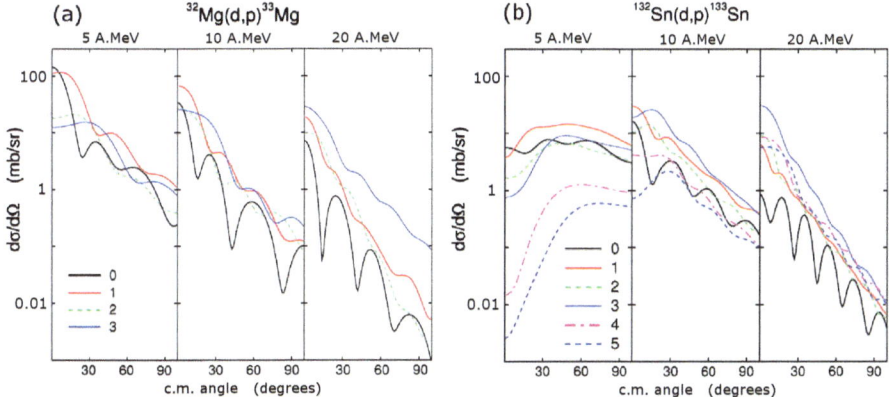

Fig. 3.7 Differential cross sections for single nucleon transfer in (**a**) ^{32}Mg$(d, p)^{33}$Mg, and (**b**) ^{132}Sn$(d, p)^{133}$Sn. The three panels in each case are for three different bombarding energies, namely 5, 10 and 20 A MeV. Each panel shows calculations for several different ℓ-transfers. The ADWA model was used (see text, Sect. 3.2.3). These plots are in terms of the centre of mass reaction angle, $\theta_{c.m.}$.

the vector diagram for momentum conservation (cf. Fig. 3.6). From the application of the cosine rule to this triangle, the formula for θ^2 as shown in the Figure can be derived, where we make use of the expansion to second order for cosine: $2(1 - \cos\theta) \approx 2(1 - [1 - \theta^2/2!]) = \theta^2$. From inspection of the diagram, the reduction δ in the length of the p vector is small compared to the magnitude of the actual transferred momentum, p_t. Hence, we can drop the terms in (δ/p) in the expression for θ^2 and we obtain $\theta^2 \approx (p_t/p)^2$. If the nucleon is transferred at the surface of the target nucleus, which has radius R, then the transferred angular momentum ℓ is given by $p_t \times R = \sqrt{\ell(\ell+1)}\hbar$. This immediately indicates that $\theta \approx$ constant $\times \ell$, and in a full quantum mechanical treatment we will not see a single angle but can expect a peak to occur in the differential cross section, at a laboratory angle that is approximately proportional to the transferred angular momentum, ℓ. This is shown schematically in Fig. 3.6, which also includes the diffractive effects in a schematic fashion. In fact, for deuterons incident at a kinetic energy of E (MeV) on a target of mass A this simple picture gives θ(degrees) $\approx 217/(\sqrt{E} \times A^{1/3}) \times \sqrt{\ell(\ell+1)}$. For 20 MeV deuterons incident on a target of mass 32, the constant term evaluates to $15°$, which of course can serve only as a guide, but is in reasonable agreement with the trend in the primary maxima observed in the middle panel of Fig. 3.7(a) which shows proper calculations for (d, p) on ^{32}Mg at 10 A MeV. The Figure is actually plotted in terms of $\theta_{c.m.}$, but would look very similar when plotted in terms of θ_{lab} for normal kinematics (which refers to the situation where the target is heavier than the projectile). The preceding discussion of the vector diagram is adapted from reference [13].

Calculations are shown in Fig. 3.7 for various ℓ-transfers at several different bombarding energies, and for two different targets. The first point to note is that the shapes of the distributions for different ℓ-transfer are distinctive, especially for

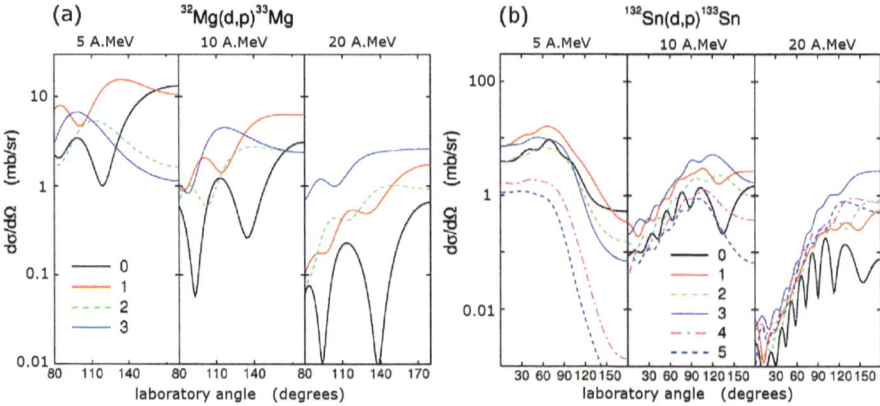

Fig. 3.8 As for Fig. 3.7, but in terms of the laboratory angle for the detected proton. In this reference frame, the extreme right of each panel corresponds to the point at the extreme left of the panels in Fig. 3.7

10 A MeV (the middle panels). For the light nucleus ^{32}Mg, the 5 A MeV distributions are also characteristic of the transferred ℓ. For the heavier ^{132}Sn target, the distributions are less distinctive due to the forward angle parts (small $\theta_{c.m.}$) being suppressed. This is due to the Coulomb repulsion between the projectile and the target, which means that the small angle scattering (especially) has a suppressed nuclear component. In addition to the above considerations, there is a general trend towards lower cross sections as the bombarding energy increases, of around half to one order of magnitude per 10 A MeV. Taken together, this information suggests that 10 A MeV is an ideal bombarding energy for this type of study, and this can be relaxed down to 5 MeV perhaps, for lighter nuclei. The remaining question is whether the existing theories are equally valid at all energies, and the ADWA model used here (see Sect. 3.2.3) should have good validity at both 5 and 10 A MeV, although probably not at energies much lower than this.

The aim of a typical nucleon transfer experiment is to measure the differential cross sections for different states in the final nucleus. From the shape of the cross section plot, the transferred angular momentum can then be deduced. The calculations shown in Fig. 3.7 are for pure single-particle states. That is, it is assumed that the structure of the final state is given perfectly by the picture of the target core with the transferred nucleon in an associated shell model orbital. Hence, another important experimental result will be the scaling factor between the theoretical calculation and the data, which will give the experimental value for the spectroscopic factor.

If experiments are performed in normal kinematics, with a deuteron beam, then the cross sections will look much like Fig. 3.7 whether we plot them in terms of the centre of mass angles or the laboratory angles. However, in reality the isotopes ^{32}Mg and ^{132}Sn used in these examples are radioactive and the experiments need to be performed in inverse kinematics: where the deuteron is the target and the heavier particle is the projectile. In Fig. 3.8, the same calculations as in Fig. 3.7 are plotted, but using the laboratory angles and assuming an inverse kinematics experiment. The same relative velocities of beam and target, i.e. the same values of the beam energy in MeV per nucleon, are employed. It can be seen that the structure characteristic of ℓ is maintained, for the cases where it was previously evident. The transformation takes zero degrees in the centre of mass frame to 180° in the laboratory frame. Now, the first peak observed relative to 180° is further from 180° as the ℓ-transfer increases. From inspection, an experimental measurement should include at least the region from 90° to 180° in order to allow an assignment of the transferred ℓ according to the observed shape of the distribution. The situation with the heavier target is more problematic, especially at the lowest energy shown here.

The transformation from the centre of mass to the laboratory reference frame, and in particular the transformation of the solid angle, is discussed further in Sect. 3.3.2.

3.2.3 Choice of a Theoretical Reaction Model: The ADWA Description

The perfect theoretical model to interpret experimental data for transfer reactions does not exist. The scattering theory is most often treated in an *optical model* approach, where the scattering potential is complex and has attractive and absorptive components. As in optical light being scattered from a cloudy crystal sphere, the loss of flux (by whatever process) is represented mathematically by the imaginary part of the potential. Most often, but not of necessity, the final state populated in a reaction such as (d, p) is represented as a core (being the original target nucleus) with the transferred nucleon in an eigenstate of the potential that arises due to the core. This implies a perfect single-particle structure for the final state, and the ratio between the experimental and theoretical cross sections is then the spectroscopic factor, as previously discussed. The simplest scattering theory, described in introductory quantum mechanics texts (for example Ref. [14]), is in terms of a *plane wave Born approximation* (PWBA). An improved model [14] replaces the plane waves by the wave solutions that are distorted by the presence of the scattering potential, giving the *distorted wave Born approximation* (DWBA). Even though transfer has been a widely used and valuable tool in nuclear spectroscopy for well over 50 years, there are still new and important developments occurring in quite fundamental aspects of the theory. One aspect of this concerns the spatial localisation of the transferred nucleon in the projectile and the final nucleus, or what is known as the *form factor*. Another important aspect, particularly for the (d, p) reaction, concerns the coupling to continuum states. Because the deuteron is weakly bound, it very easily disintegrates in the field of the target nucleus when used as a projectile. When the (d, p) reaction is applied to weakly bound exotic nuclei, the problem also occurs for the final nucleus. What is more, the coupling is not necessarily one-way: continuum states can couple back to the bound state, which can have important effects on the reaction cross section. One way to take this into account is via *coupled reaction channels* (CRC) calculations, in which all of the different contributing reaction pathways are explicitly included in the calculation. In order to include the continuum contribution, the theory usually considers hypothetical energy bins in the continuum and treats them as different states that can couple into the intermediate stages of the reaction. These are *coupled discretized continuum channels* (CDCC) calculations. The challenges of such calculations are many, including the computational power required and the choices of parameters for the various coupling strengths.

An ingenious analytical short-cut to include continuum contributions was developed by Johnson and Soper [15]. In the scattering process, the neutron and the proton inside the deuteron have complex histories, and in particular when continuum states for the neutron are included—that is, deuteron or final-nucleus breakup. The Johnson-Soper method relied on the observation that certain integrations over all spatial coordinates are dominated by the contributions wherein the neutron and proton are within a range determined by the neutron-proton interaction. Within a *distorted wave* formulation, certain energy differences are ignored, which means that the approximations of the model become less applicable at lower beam energies. However, at 5 to 10 A MeV they should remain substantially valid. The coupling to the continuum, subject to these approximations, is included exactly and to all orders by means of the simplified integrations. This theoretical method has become known as the *adiabatic distorted wave approximation*, or ADWA. A convenient feature is that the calculations are largely identical (but with different input) to those required for the DWBA, and hence the pre-existing DWBA computer codes can be adapted to perform ADWA calculations. The DWBA remains another popular choice for the analysis of transfer reactions. Descriptions can be found, for example, in the articles and books by Glendenning [7, 8] and Satchler [9, 10]. The DWBA uses imaginary potentials to take into account the loss of reaction flux from the elastic channel, which allows for deuteron breakup but not for a proper two-way coupling with the continuum. The extensions via CDCC are computationally intensive and often incomplete in terms of the contributing physics. Therefore, the ADWA has important advantages in the case of (d, p) reactions and is adopted for all such analysis in the present work. The calculations are performed using a version of the code TWOFNR [16]. The ADWA method has recently been refined to take into account the zero-point motion of the neutron and proton inside the bound deuteron [17].

3.2.4 Comparisons: Other Transfer Reactions and Knockout Reactions

In the discussion in this article the emphasis is on single-nucleon transfer, and primarily (d, p) reactions, studied in inverse kinematics with radioactive beams. In terms of physics, the aim which is emphasised is the understanding of single particle structure and the evolution of shell orbitals and shell gaps as nuclei become more exotic. There are certainly other types of transfer reaction and other ways with which to probe single particle structure. Some of those topics are briefly described here. This article aims to identify the experimental challenges and techniques of transfer reaction studies, rather than to provide a review of all such studies in the literature; some more details of other work can be found, for example, in Ref. [18] or in other papers by many of the groups cited in Sect. 3.4.

Nucleon removal reactions include (p, d) and (d, t) which are discussed in Sects. 3.4.1 and 3.4.8 respectively, and $(d, ^3\text{He})$. An alternative to the first two is $(^3\text{He}, \alpha)$ whilst an alternative to (d, p) is $(\alpha, ^3\text{He})$. The choice of which reaction to use should not be random. The helium-induced reactions will generally show a different selectivity due to the different reaction Q-value, and could be chosen to highlight higher-ℓ transfers. In terms of the discussion in Sect. 3.2.2, a more negative Q-value will reduce the kinetic energy in the exit channel so that the exiting particle takes away less orbital angular momentum than it brings in. This will tend to favour the higher ℓ-transfers. In practice, the helium-induced reactions are harder to study using radioactive beams. No simple and thin solid helium target exists, so it is necessary either to use a gas target (a windowed cell, or a differentially pumped jet) or an implanted helium-in-metal target or a cryogenic target. Each has its own challenges, but can be built and will find an increased application in the future.

Another important type of transfer reaction is when a cluster is transferred. It is always the case that when multiple particles are transferred then the process could be single-step (when the whole cluster is preformed and is transferred) or could have two or even more steps involved. Multiple-step processes are modelled theoretically using *coupled reaction channel* (CRC) extensions of the DWBA. In the case of heavy-ion transfer, they can also be modelled semiclassically, as mentioned in Sect. 3.5.2. Traditionally, anything heavier than helium is called a heavy ion, and two heavy-ion induced transfer reactions of particular importance are $(^6\text{Li}, d)$ and $(^7\text{Li}, t)$, which transfer an α-particle. Various heavy-ion transfer reactions, including α-transfer, are discussed for example in Ref. [19].

The simplest form of cluster transfer is probably the (t, p) reaction in which the transfer of two neutrons, coupled to spin and relative orbital angular momentum zero, is the dominant mechanism. These can carry various amounts of angular momentum with them as a cluster, into the final nucleus. Experimentally, it is a challenging reaction: historically, the tritium nucleus was the projectile and would pose particular problems due to its radioactivity, and with the advent of radioactive beams the tritium has to be incorporated into a compact target and then be bombarded, which potentially poses even greater problems. Nevertheless, these problems have been solved in a study of shape coexistence in ^{32}Mg via the (t, p) reaction in inverse kinematics [20]. The beam was 5×10^4 pps of ^{30}Mg at 1.8 A MeV at ISOLDE, CERN. This experiment used the T-REX array which is described in Sect. 3.4.2. The target was a foil of titanium metal (500 μg/cm^2) into which 40 μg/cm^2 of ^3H had been absorbed. There was a ratio of $\approx$1.5 of hydrogen atoms to lattice Ti atoms, giving a radioactivity of the target of 10 GBq. For stable targets, the (t, p) reaction tends to have a large positive Q-value. For the more neutron-rich radioactive isotopes, such as ^{30}Mg, the Q-value drops to be close to zero but the kinematics remain quite similar to (d, p), which is discussed in Sect. 3.3.1.

With the advent of radioactive beams at the extremes of measured nuclear existence, obtained via intermediate and high energy fragmentation reactions at laboratories such as MSU, GANIL, GSI and RIKEN, a new type of nucleon removal reaction was developed and exploited. This type of reaction is sometimes called a *knockout* reaction, but it is completely separate from true knockout reactions such as $(e, e'p)$ and $(p, p'p)$. The nucleus in the beam is incident on a light target nucleus that acts like a black disk and ideally cannot be internally excited without disintegrating—the usual choice is ^{9}Be. The experimental requirement is that the projectile survives the reaction, with just the single nucleon removed, and this automatically selects very peripheral collisions. The black disk essentially erases some part of the tail of the wave function of the removed nucleon [21]. This method, which was originally developed to study the ground states of halo nuclei, provides another way in which to study the single particle structure of nuclei. Nucleon removal from the ground state of a projectile simultaneously studies the structure of the projectile state and the structure of the final nucleus. Individual states in the final nucleus can be identified using gamma-ray spectroscopy. The angular momentum transfer and the spectroscopic factor are deduced, respectively, from the width of the longitudinal momentum distribution of the beam fragment and from the magnitude of the cross section. A very successful method of analysing these reactions was developed using high-energy Glauber approximations that were previously used to describe high energy deuteron-induced reactions (the deuteron being the archetypal halo nucleus) and this theory is outlined in Ref. [22], with a more extensive discussion of results in the review of Ref. [23]. A currently very topical result from the extensive studies using knockout reactions is the apparent quenching of single-particle spectroscopic factors relative to the predictions of large-basis shell model calculations [24]. The quenching appears to be correlated with the binding energy of the removed nucleon, which suggests some connection with higher-order correlations of nucleons, coupling to configurations outside of the shell model basis. Various different explanations have been advanced for this effect, for example those in Refs. [25–27]. The observations appear to be consistent with previously observed quenching of spectroscopic strength in stable nuclei using $(e, e'p)$. One way to investigate this for radioactive nuclei, and also to check the reaction dependence, is via $(p, p'p)$ knockout reactions such as those performed in Japan [28] and GSI [29]. Another is to compare neutron and proton knockout with results from (d, t) and $(d, {}^3\text{He})$ studies, as has been performed for the neutron deficient ^{14}O nucleus [30].

3.3 Experimental Features of Transfer Reactions in Inverse Kinematics

This section addresses some simple and rather general features of reactions such as (d, p) and (p, d) when studied in inverse kinematics. Instead of the centre of mass frame being almost at rest in the laboratory frame, as in normal kinematics experiments, the centre of mass frame moves with nearly the beam velocity. The kinematical variation of energy with angle therefore bears no resemblance to the situation for normal kinematics shown in Fig. 3.4. In a (d, p) or (p, d) reaction, the mass of the light (target) particle is substantially changed by the transfer, being

Fig. 3.9 Classical velocity addition diagram for elastic scattering in inverse kinematics, showing that the light (target) particles emerge at angles just forward of 90° for small centre of mass scattering angles

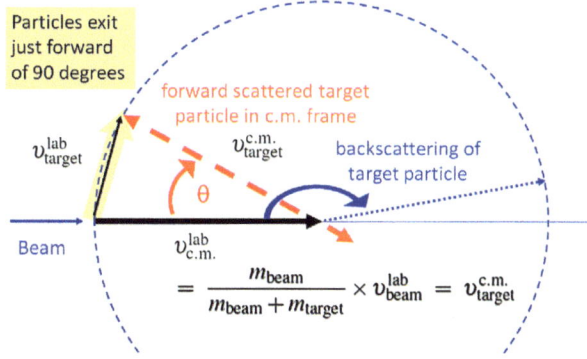

halved in (d, p) or doubled in (p, d). This in itself turns out to be a major factor in determining the two-body kinematics of the reaction. In order to illustrate this, it is convenient to use velocity addition diagrams, where we add the velocities of particles as measured in the centre of mass frame to a vector representing the velocity of the centre of mass frame in the laboratory. The resultant vectors give the velocities of the final particles in the laboratory frame, and of course this is using the Galilean transformation and thus is strictly correct only for non-relativistic situations. This is no great problem if we are working at the energies of order 10 A MeV that were suggested in Sect. 3.2.2. The discussion in the following section follows that in Ref. [31].

3.3.1 Characteristic Kinematics for Stripping, Pickup and Elastic Scattering

The vector diagram describing elastic scattering in inverse kinematics is shown in Fig. 3.9. The velocity of the centre of mass in the laboratory frame is given by a large fraction of the beam velocity, since the target is light. Measured in the centre of mass frame, taking into account conservation of momentum, we can also note that the velocities of the two particles after the collision must be in inverse proportion to their masses. Thus, the target-like particle has a velocity $v_{\text{target}}^{\text{c.m.}}$ that is much greater than that of the beam-like particle in this frame. This is shown in the Figure by the red dashed vectors. Furthermore, the target particle is initially at rest and hence the length of the target-like vector $v_{\text{target}}^{\text{c.m.}}$ is equal to the length of the centre of mass velocity as measured in the laboratory frame, $v_{\text{c.m.}}^{\text{lab}}$. The scattering angle as measured in the centre of mass frame is given by the angle enclosed between $v_{\text{c.m.}}^{\text{lab}}$ and $v_{\text{target}}^{\text{c.m.}}$, indicated by θ in Fig. 3.9. For a scattering angle of zero in the centre of mass frame, the light particle in the final state is stationary. For small scattering angles (where the cross section is highest, for elastic scattering) the light particles emerge just forward of 90° and with a velocity (energy) that increases approximately linearly (quadratically) with centre of mass angle. Also, the centre of mass angle is

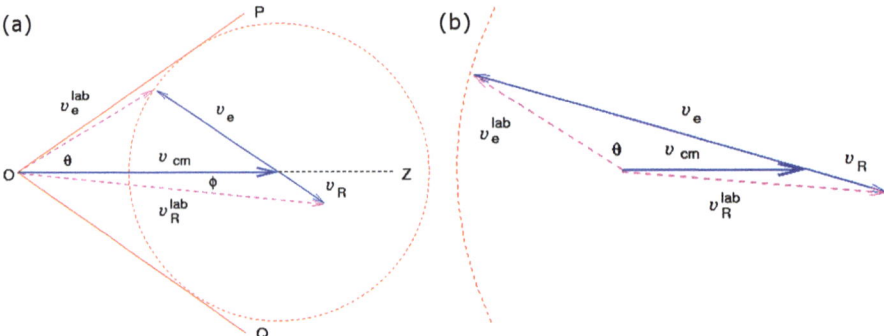

Fig. 3.10 Velocity addition diagrams (**a**) for a typical *pickup* reaction such as (p,d) or (d,t), and (**b**) for a typical *stripping* reaction such as (d,p). Certain assumptions about the beam energy and the reaction Q-value are described in the text

simply twice the difference between the laboratory angle and 90° in this classical approximation, since the velocity addition triangle is isosceles. The beam particle continues in the forward direction with little change in either energy or direction. In the case of backscattering in the centre of mass frame, the light particles travel rapidly in the direction of the incoming beam, and the beam particle also continues in that direction, being just slightly slowed down. An important result here, of experimental significance, is that elastically scattered target particles will be detected just forward of 90° and their energies will increase rapidly with angle. In general, they will require a thick detector for them to be stopped and their energy measured precisely.

The vector diagrams describing reactions in which there is *pickup* of a nucleon by the light particle, or *stripping* of a nucleon from the light particle are shown in Fig. 3.10 (adapted from Ref. [31]). The lengths of the vectors in these diagrams are given in terms of the masses involved, and the reaction Q-value, by formulae included in Refs. [31, 32]. As shown by those formulae, the diagrams shown here implicitly assume a small reaction Q-value, or at least that the Q-value in units of MeV is small compared to the energy of the beam as expressed in MeV per nucleon. Especially for reactions involving exotic neutron rich projectiles, the Q-values for neutron addition or removal will typically be small, and similarly for a reaction such as $(d, {}^3\text{He})$ on the proton-rich side of the nuclear chart.

In the case of a reaction such as (p,d), corresponding to Fig. 3.10(a), it is easy to obtain a rough estimate of the length of the light particle vector in the centre of mass, labelled v_e in the figure. Firstly, the heavy particle is going to continue with little change in velocity or direction, much as in the case of elastic scattering. Now, the centre of mass vector in elastic scattering was required to be the same length as the centre of mass velocity vector in the laboratory frame, denoted by v_{cm} in Fig. 3.10. In the case of (p,d), the mass of the light particle is doubled relative to the elastic scattering situation, but the momentum that this particle must carry in the centre of mass frame is about the same as in the elastic case, which follows from the remark about the velocity of the beam particle. Thus, this vector v_e is about half the length of v_{cm}. The precise value depends upon the reaction Q-value of course, but the basic form of the vector diagram is always the same, subject to the assumptions mentioned above. The result is that the light reaction products are forward focussed into a cone of angles or around 40° relative to the beam direction. There will be two energy solutions for each angle, within this cone, wherein the lower energy corresponds to the smaller centre of mass reaction angle and hence (typically) the higher cross section. The low energy solution may be very low indeed, in energy.

In the case of a reaction such as (d, p), corresponding to Fig. 3.10(b), the mass of the light particle is halved in the reaction and hence its velocity vector in the centre of mass frame is approximately doubled in length, in the approximate picture. The small centre of mass angles (and typically the higher cross section) will correspond to light particles that emerge travelling opposite to the beam direction. They will have energies that may be quite low, and will increase in energy all the way to zero degrees in the laboratory frame, which corresponds to a centre of mass angle of 180°. For reactions that populate an excited state in the final nucleus, there will be less energy available in the final state than for the ground state, and hence the vectors in the centre of mass are shorter and the laboratory energies of the light particle will be lower than for the ground state, at all laboratory angles.

When planning an experiment in inverse kinematics, it can be useful to construct a velocity addition diagram such as those in Fig. 3.10. It allows an intuition about the reaction kinematics to be gained, easily. The form of the diagram depends only on the ratio of the length of v_e to that of v_{cm}. This ratio is given [31] by

$$\frac{v_e}{v_{cm}} = \left(qf \frac{M_R}{M_P}\right)^{1/2} \approx \sqrt{qf} \quad \text{if } M_R \approx M_P$$

where the masses of the projectile and recoil are denoted by M_P and M_R. The quantity f is related to the change in mass of the target particle, $f = M_T/M_e$ where M_T and M_e are the masses of the target and light ejectile respectively. The quantity q is of order unity but has a Q-value dependence and typically varies between 1 and 1.5. Specifically, $q = 1 + Q/E_{cm}$ where $Q = Q_{g.s.} - E_x$ for an excited state and E_{cm} is the kinetic energy in the centre of mass frame. Given that the target is much lighter than the projectile, most of the kinetic energy in the centre of mass frame is carried by the target particle, so $E_{cm} \approx M_T(E/A)_{beam}$. Then $q \approx 1 + Q/2(E/A)_{beam}$ and q is closer to unity for small Q-values or as the E/A for the beam is increased. In the limit that $q = 1$ then for a pickup reaction such as (p, d) or (d, t) the size of the cone around the beam direction that contains all of the events is given by $\theta_{max} = \sin^{-1}\sqrt{f}$ where $f = 1/2$ for (p, d) and $f = 2/3$ for (d, t). This gives, as a first approximation, a cone of about 50° degrees half-angle in each case. Similarly, it is possible to estimate that in (d, p) the laboratory angle corresponding to 30° in the centre of mass frame is about 110°, so a (d, p) experiment will typically need to measure at least the angular range from 110° to near 180° in the laboratory.

Another interesting feature of the velocity addition diagram is how it scales with the E/A of the beam [33]. Whilst the relative lengths of the vectors are determined largely by the masses of the various particles, with some residual dependence on the Q-value and the beam energy, the length scale (as given in Ref. [32]) is

$\sqrt{2q(M_R + M_e)}$ which, with the assumption again that $M_P \gg M_T$ is approximately proportional to $\sqrt{(E/A)_{beam}/M_P}$. Now, with the assumption that $M_P \approx M_R$ (because the transfer hardly changes the mass), the lengths of the vectors such as v_e and v_{cm} scales as $\sqrt{M_P}$. Thus, the whole diagram scales as the product of these lengths and the length scale itself, and the $\sqrt{M_P}$ factor cancels. The diagram therefore scales roughly as $\sqrt{(E/A)_{beam}}$ and the energies will scale roughly as $(E/A)_{beam}$. The approximation is better, the closer the Q-value is to zero, but the expression at least gives a guide to the behaviour that can be expected: the detected energy scales with the beam energy. For elastic scattering, the Q-value is zero, so the result is accurate: the rate of increase of the energy of the scattered particle with angle, for angles moving forward of 90°, scales with the beam energy.

The results of proper (relativistically correct) kinematics calculations for two very different beams and energies are shown in Fig. 3.11. In Fig. 3.11(a), the results are for a beam of ^{16}C at 35 A MeV such as might be produced by a fragmentation beam facility. The central solid line near 90° shows the energy of elastically scattered deuterons, rising steeply as the centre of mass angle increases and the laboratory angle slightly decreases. On the right hand side of Fig. 3.11(a) are the results for the protons from the (d, p) reaction populating the ground state in ^{17}C (upper curve) and a hypothetical excited state at 4 MeV excitation energy. The pair of curves with the lowest energies at zero degrees are for the (d, t) reaction. The faint dotted line near 90° shows the energies of elastically scattered protons, if there were to be any ^{1}H in the target along with the ^{2}H (a situation commonly encountered experimentally). The curve with the highest energy at zero degrees is for tritons from the (p, t) reaction populating the ground state of ^{14}C. The remaining curves at the intermediate energies at zero degrees are for the reactions $(d, {}^3\text{He})$ and (p, d) initiated by the different isotopes in the target. Lines connecting all of the curves show the points corresponding to the indicated centre of mass angles. Note that the energies of the particles from (d, p) and (d, t) are less than or equal to 5 MeV over the most interesting range of relatively small centre of mass angles, where the differential cross section will be largest and most structured. Also, the maximum energies reached over the interesting range are all less than about 30 MeV. Figure 3.11(b) is for ^{74}Kr at 8.16 A MeV. The curves on the right are for (d, p) to the ground state of ^{75}Kr and a hypothetical state at 5 MeV excitation. At forward angles, the two lower curves are for $(d, {}^3\text{He})$ from this neutron-deficient nucleus. The next two curves are for (d, t). In each of these cases, the calculations are for the ground state and a 5 MeV state. The final kinematic curve in Fig. 3.11(b), intersecting at 15 MeV at 0°, is for the (p, d) reaction to the ground state of ^{73}Kr. Once again, the particles of principal interest are generally of about 5 MeV or less, and the energies of interest range up to about 30 MeV. This consistency of the relevant kinematic energy-angle domains has important implications for the design of particle detection systems aimed at studying transfer in inverse kinematics. It indicates that a static array could be optimised to such measurements and would be applicable to a wide range of reaction studies.

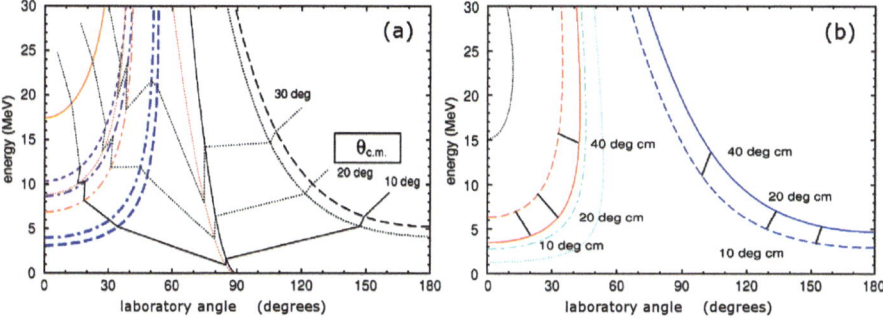

Fig. 3.11 Two-body relativistic kinematics calculations for two very different beams in terms of mass and energy, including results for elastic scattering and several different single-nucleon transfer reactions: (**a**) ^{16}C at 35 A MeV, (**b**) ^{74}Kr at 8.16 A MeV

3.3.2 Laboratory to Centre of Mass Transformation

It is common to transform results for the measurement of differential cross sections from the laboratory frame into the centre of mass frame, for comparison with the results of reaction theory calculations. The theory is of course naturally calculated in the centre of mass frame. In the days when the experiments were performed in normal kinematics, the shape of the cross section plot would be similar in both the laboratory and the centre of mass reference frame, because the target was typically much heavier than the incident deuteron. In the case of inverse kinematics, this is no longer the case, as shown by comparing Figs. 3.7 and 3.8. It is important to note that it is not simply a transformation from one angle to another that changes the differential cross sections between the two reference frames, but the solid angle is also transformed. The ratio of differential factors that describes this transformation is known as the Jacobian. Inspection of Fig. 3.10(b), which describes the (d, p) reaction, shows that for backward laboratory angles (as illustrated) the laboratory angle (for v_e^{lab} measured relative to v_{cm}) varies much more rapidly than the centre of mass angle (enclosed between v_{cm} and v_e). In the diagram there is a factor of about two, between the rates of change. This means that a small solid angle in the centre of mass frame is spread out over a rather large solid angle in the laboratory frame. Thus, while $d\sigma/d\Omega_{\text{c.m.}}$ is largest at small $\theta_{\text{c.m.}}$ or near 180° in the laboratory frame, the effect of the Jacobian is that $d\sigma/d\Omega_{\text{lab}}$ is reduced relative to less backward angles. That is, while the very backward laboratory angles are still important in (d, p) measurements, for determining the shape of the differential cross section, there are very few counts there.

The transformation from centre of mass to laboratory angles, as just mentioned, has the effect of spreading out the counts from (d, p) at small centre of mass angles, so that they are spread over a wider solid angle. This reduces the yield of counts observed near 180° in the laboratory frame. A completely separate effect to also remember is the "$\sin\theta$" effect. This will further emphasise the importance of detectors close to 90° compared to those near 180°, assuming that the charged particle detection is cylindrically complete, or approaching this. Then, since the solid angle in a range $d\theta$ at angle θ is given by $2\pi \sin\theta d\theta$, the cross section that measures the number of counts in an experiment is not $d\sigma/d\Omega$ but

$$\frac{d\sigma}{d\theta} = 2\pi \sin\theta \frac{d\sigma}{d\Omega}.$$

Thus, if a coincidence experiment is considered, for example measuring gamma-rays in a $(d, p\gamma)$ experiment, many of the gamma-rays will come in association with particles detected towards 90°.

The transformation between the centre of mass and laboratory reference frames, for the differential cross section, is given for example by Schiff in his classic *Quantum Mechanics* text [14]:

$$\frac{d\sigma}{d\Omega}\bigg|_{\text{lab}} = \frac{(1+\gamma^2+2\gamma\cos\theta_{\text{c.m.}})^{3/2}}{|1+\gamma\cos\theta_{\text{c.m.}}|} \times \frac{d\sigma}{d\Omega}\bigg|_{\text{c.m.}}$$

where $\gamma = v_{\text{c.m.}}/v_e$. This complicated transformation, as noted above, changes the shape of the curves significantly. Therefore, a plot in the centre of mass frame of experimental data for the differential cross section will retain very little information about any experimental constraints or impacts of the laboratory angles. For example, the physical gaps between detectors, or the differing thicknesses of target through which the particles must exit: these often have important implications for the data but the information is lost in the transformation to the centre of mass frame. For this reason, some workers choose to plot experimentally measured cross sections in the laboratory frame, for inverse kinematics experiments, following the ethos of presenting the data in a form as close as possible to what is actually measured experimentally.

3.3.3 Strategies to Combat Limitations in Excitation Energy Resolution

In trying to do experiments using radioactive beams, there are two properties of the beams that tend to influence the experimental design more than any others. Firstly, the beams are radioactive. That means that care must be taken regarding the eventual dumping of the beam and also, quite often, to deal with the angular scattering of the beam in the target [34]. Secondly, the beams are generally weak, maybe up to a million times weaker than a typical stable beam that one might have used for an equivalent normal kinematics experiment with a stable target. This means that, in practice, there will be a minimum sensible value for the target thickness in order to perform the experiment in a reasonable time. In turn, this will affect the energies and angles measured for the particles produced in the reaction. As discussed above, the particles of interest are often of rather low energy. The energy that is measured may depend quite significantly on where the reaction takes place—at the front or at the back surface of the target, or somewhere in between. Also, for the lowest energy particles, the direction may be affected by multiple low-angle scattering of the charged particle as it leaves the target material.

In an experiment to identify and study the unknown excited states of an exotic nucleus, the kinematical formulae used to produce a plot such as Fig. 3.11 will be inverted so that the measured energy and angle of a particle are used to calculate the excitation energy of the final state. Any process that modifies the measured energy and angle from the actual reaction values will lead to a limitation on the achieved resolution for excitation energy, even if the best possible computed corrections are applied. All of these factors were included in a detailed analysis of the resolution that could be expected from transfer reactions, under different experimental conditions [35]. The two basic categories of experiment were as follows:

I *rely on detecting the beam-like ejectile in a magnetic spectrometer*
II *rely on detecting the light, target-like ejectile in a silicon detector*

with a third option arising which is

III *detect decay gamma-rays in addition to the charged particles.*

A magnetic spectrometer or a recoil separator is essential in the first case, in order to separate the reaction products from the direct beam and to measure the ejectile properties with sufficient accuracy. Operated at zero degrees, it will need to be instrumented to allow the accurate measurement of the scattering angles for the very forward-focussed beam particles. The degree of forward focussing, and hence the requirements placed upon the resolution of the angular measurements, become more and more demanding as the mass of the projectile increases. For heavier beams, it becomes impractical for existing detectors. Furthermore, any spread in the beam energy translates directly to a spread in the measured excitation energies, and any nucleon transfer reactions on contaminant material in the hydrogen targets (usually plastic) will contribute to the observed yield.

If the second method is employed, then we know from the discussion of the kinematics that the particles of interest are spread over a significant range of angles. In order to detect particles over this range, and with good resolution in both energy and angle, the most obvious choice is an array of semiconductor detectors, and silicon is by far the most versatile material at present. This method is less sensitive than the first, to a spread in the beam energy, but is limited as discussed above by the effects of the target thickness on the measured energies and angles. In practice, it is hard to imagine resolutions of better than 100–200 keV or so, for excitation energy, if the experiment demands targets of 0.5 mg/cm^2 or more [35]. (This assumes $(CD_2)_n$ deuterated polythene targets,[1] and with a thickness determined by beam intensities that may be as low as 10^4 pps.) In some experiments, thinner targets could be used and hence better resolution achieved, if the beam intensity were to allow it. In any

---

[1]For convenience, the $(\ldots)_n$ part of $(CD_2)_n$ will subsequently be omitted, and similarly for the non-deuterated $(CH_2)_n$ targets.

case, to achieve the best resolution, the detector array for the light particles should achieve good measurement of the particle angles. In the case of a silicon strip detector array, this requires a high degree of segmentation, or in some cases a resistive strip readout is possible.

A variant of the second method, which avoids the need for an extensive and highly segmented silicon array, is to use a magnetic solenoid aligned with the beam axis to collect and focus the charged particles onto a more modest array of silicon detectors located around the axis of the solenoid. This is the *HELIOS* concept, named after the first device of this type to be implemented [36]. The elegant feature of HELIOS is that it removes the kinematic compression of energies. Considering the kinematics of a typical (d, p) reaction as shown for example in Fig. 3.11, the lines for a difference of 1 MeV in excitation energy are separated in terms of proton energy by less than 1 MeV at a particular laboratory angle. In HELIOS, when the protons are focussed back to the axis of the solenoid, they are dispersed in distance according to a linear dependence on excitation energy. For a detector located at a particular distance along the axis, it measures particles emitted at different angles for different excitation energies. The net result is to disperse excited states in energy in such a way that any limitation due to the intrinsic resolution of the silicon detector becomes significantly less important. However, if the limitation lies in the target thickness and the ensuing deleterious effects on the energy and angle of the particles leaving the target, then there is little benefit to be obtained from simply using a different method of measurement. Ultimately, if the experiment demands a relatively thick target, the resolution will be as estimated in Ref. [35]. The helical detector concept is discussed again in Sect. 3.4.3.

It may be that the limits to resolution imposed by a reasonable target thickness are not acceptable for a good measurement to be performed. This is likely to happen in the case of heavier nuclei where levels are more closely spaced than the light nuclei, or it can occur in any odd-odd nucleus for any mass. In this situation, which can be expected to be common, the third solution—measuring decay gamma-rays—becomes attractive.

The higher energy resolution that can be achieved with gamma-ray detection then gives a much better energy resolution for excited states. This of course works only for bound states in the final nucleus. In addition to the improved energy resolution, another feature of possibly comparable importance is that the gamma-ray decay pathway for a particular final state may help to identify the state more precisely. From the particle transfer measurement, it is only possible to infer the ℓ-transfer, which leaves an uncertainty according to whether the spin is $(\ell + 1/2)$ or $(\ell - 1/2)$ since the transferred nucleon has spin 1/2. The gamma-ray decay branches may resolve this ambiguity. It should be noted that there is an experimental challenge in detecting the gamma-rays with a high enough efficiency and with the ability to apply a sufficiently good Doppler correction. For the typical beam energies discussed above, the final gamma-ray emitting nuclei will have velocities of the order of $0.10c$ in the laboratory reference frame, always aligned almost exactly along the beam direction (c is the speed of light). In order to correct for the substantial Doppler shift that this implies, the gamma-ray detectors will also need to measure the angle of emission for the gamma-ray, relative to the incident beam. Doppler shift corrections are discussed further in Sect. 3.4.4.

3.4 Examples of Light Ion Transfer Experiments with Radioactive Beams

Having described the various approaches to designing an experiment in the previous section, these possibilities are now illustrated by means of specific examples. Mostly, the examples are early experiments which helped in developing the techniques and/or (for convenience) experiments by the author with collaborators.

3.4.1 Using a Spectrometer to Detect the Beam-like Fragment

An example of an experiment in which the beam-like particle is measured, and used to extract all of the experimental information, is provided by the early experiment performed by the Orsay and Surrey groups at GANIL [37, 38] and illustrated in Fig. 3.12. The aim was to study the (p, d) reaction with ^{11}Be in order to study the parentage of the ^{11}Be halo ground state. Because the projectile is relatively light, then it is a reasonable approach to measure the beam-like particle (method I of Sect. 3.3.3). A magnetic spectrometer was used, for two reasons. Firstly, the beam was produced by secondary fragmentation and therefore has significant spreads in both energy and angle. In order to resolve final states in ^{10}Be that were separated by less energy than the spread in the beam, a dispersion matched spectrometer was required. This experiment used the *spectromètre à perte d'energie du GANIL*, SPEG [39]. Secondly, in order to measure the scattering angle of the ^{10}Be it was necessary to separate the ^{10}Be from the beam and track its trajectory to a precision that required a spectrometer. In order to recover the scattering angle, it was also necessary to track the incident beam particles, which required detectors placed before the first (dispersive) dipole element of SPEG. The beam intensity was 3×10^4 particles per second (pps) and the mean beam energy was 35.3 A MeV. The measured ^{10}Be particles, at the focal plane, were dominated by the yield from carbon in the polythene CH_2 target. Reactions on just the protons in the target were isolated by recording the deuterons from the reaction in an array of ten large area silicon detectors. This experiment was successful in measuring the parentage of the ^{11}Be ground state, which has a neutron halo, in terms of the s-wave and d-wave components (the latter with an excited ^{10}Be core). In addition to the innovative experimental techniques, the experiment also highlighted some important complexities in the theory and made innovations in the theoretical interpretation. Specifically, it was necessary to go beyond the normal simplification of modelling the transferred nucleon in a potential well that is due to the core. It was necessary to use a dynamic picture of ^{11}Be in terms of a particle-vibration coupling model, in order to calculate the overlap functions in the transfer amplitude directly from the nuclear structure model.

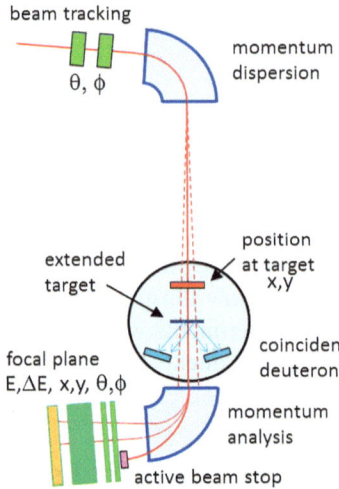

Fig. 3.12 In this (p,d) study using a secondary ^{11}Be beam [37, 38], the beam had a large energy spread, so a dispersion matched spectrometer was used. This, together with the limited spatial focussing of the beam required beam tracking detectors at the target and in the beam line. Coincident deuteron detection allowed background from the carbon in the CH_2 target to be removed in the analysis, but the ^{10}Be measurement in the spectrometer gave all of the critical energy and angle information. The active beam stop comprised a plastic scintillator that allowed the intensity of the beam to be monitored

3.4.2 Using a Silicon Array to Detect the Light (Target-like) Ejectile

The first high resolution example of this kind of experiment, aimed at measuring spectroscopic quantities using a radioactive beam, was an experiment employing a previously-prepared source of radioactive ^{56}Ni in order to measure the reaction (d, p) in inverse kinematics [40]. Useful and astrophysically relevant results were obtained. The experiment used silicon strip detectors arranged in the backward hemisphere with a solid target of CD_2 deuterated polythene and a recoil separator device—in this case, the fragment mass analyser (FMA) at Argonne [41]. The beam was produced in the normal way for a tandem accelerator using a source of radioactive nickel material, and had a typical intensity on target of 2.5×10^4 pps at an energy of 4.46 A MeV. An additional challenge was the isobaric impurity of ^{56}Co which was a factor of seven more intense than the ^{56}Ni and was separated using differential stopping foils within the FMA.

A particular silicon array that was developed specifically for experiments with radioactive beams is MUST [42], which uses large area highly segmented silicon strip detectors with CsI detectors in a telescope configuration. MUST led the way in developing electronics that could cope with the many channels required for highly segmented detectors. Excellent particle identification is achieved. MUST has been used to study a range of reactions including inelastic scattering of a range of nuclei,

3 What Can We Learn from Transfer, and How Is Best to Do It? 91

and with regard to transfer it was very often targeted at experiments to study the structure of very light and even unbound exotic nuclei, for example ^{7,8}He [43, 44]. Another major silicon array is HiRA which was developed initially for experiments using radioactive beams produced by fragmentation at MSU [45]. The MUST array was combined with SPEG spectrometer in a study of neutron-rich argon isotopes with a pure reaccelerated beam of 2×10^4 pps of ^{46}Ar at 10.7 A MeV from SPIRAL at GANIL, incident on a CD_2 target of 0.4 mg/cm^2 [46]. Good resolution in excitation energy was achieved, in part by exploiting the special optics of the SPEG beamline. The detection of argon ions in SPEG was useful in helping to identify and eliminate background from carbon in the target, and also allowed the identification of bound and unbound states in ^{47}Ar according to whether ^{47}Ar or ^{46}Ar was detected in SPEG, although the spectrometer acceptance was limited and prevented a full coincidence experiment. Another interesting experiment that used a silicon array by itself was the study of (d, p) using a beam of ^{132}Sn at 4.8 A MeV from the Oak Ridge radioactive beam facility [47]. As seen from the calculated cross sections in Figs. 3.7 and 3.8, this was not really the ideal energy for such a heavy beam, but it was the maximum possible. The resolution achieved for excitation energy was limited, for this heavy beam, not by the silicon array but by the target thickness of 0.16 mg/cm^2. As also suggested by the cross section plots in Fig. 3.8, the silicon detectors were optimised by mounting them in a range of angles around 90° in the laboratory.

The TIARA array [48] was the first purpose-built array to combine silicon charged particle detectors with gamma-ray detectors for transfer work and was first employed with a radioactive ^{24}Ne beam at the SPIRAL facility at the GANIL laboratory [49]. Initial tests and benchmarking were performed with a stable beam and a reaction that was previously studied in normal kinematics [48]. TIARA was designed, taking into account the experience gained from using a high intensity radioactive beam of nearly 10^9 pps of ^{19}Ne in the TaLL experiment at Louvain-la-Neuve [34, 50, 51]. This led to a design in which radioactive beam particles that are scattered at significant angles by the reaction target will be carried away from the immediate vicinity of the target, and hence away from the field of view of the gamma-ray array [34].

TIARA is shown schematically in Fig. 3.13. It is designed to be operated with four HPGe clover gamma-ray detectors from the EXOGAM array [34, 52] mounted at 90° and at a distance of only 50–55 mm from the centre of the target. The space available in the forward hemisphere was also severely restricted due to the design requirement of coupling to the VAMOS spectrometer [53]. The spectrometer allows reaction products to be measured with high precision and to be identified according to Z and A. The exceptionally large angular acceptance of VAMOS (up to 10°) also allows the efficient detection of recoils from the decay of unbound states via neutron emission. Examples of the gamma-ray and spectrometer performance are given in Sects. 3.4.4, 3.4.5 and 3.4.7.

Figure 3.14 shows in detail the geometry of the central barrel in TIARA relative to the segmented HPGe clover detectors of EXOGAM. The front faces of the clovers are mounted 54 mm from the centre of the target in this configuration with two layers

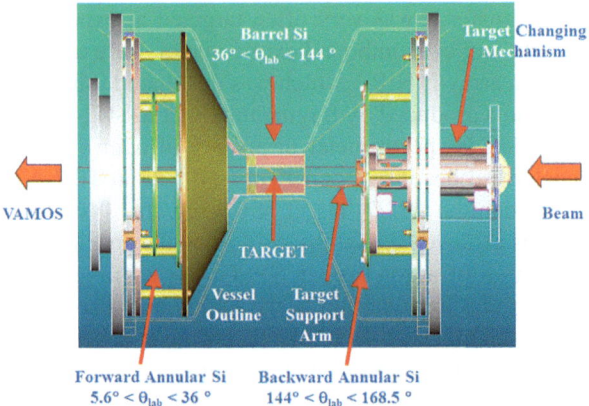

Fig. 3.13 The TIARA array was designed specifically to measure nucleon transfer reactions in inverse kinematics with radioactive beams. It has an octagonal barrel of position-sensitive silicon detectors, with annular silicon arrays at forward and backward angles. In total, approximately 90 % of 4π is exposed to active silicon. The vacuum vessel is designed so that EXOGAM gamma-ray detectors can be placed very close to the target, achieving a gamma-ray peak efficiency of order 15 % at 1 MeV. A robotic target changing mechanism allows different targets to be placed at the centre of the barrel

of silicon in the barrel. The inner layer of silicon is position sensitive parallel to the beam direction, which is the most important direction in defining the scattering angle of any detected particle. Each of the 8 inner detectors has four resistive strips and is 400 µm thick. The second layer of silicon is 1 mm thick but non-resistive. The 4 strips per detector align behind the strips on the inner barrel. The primary purpose of the second layer of the barrel is to indicate when particles punch through the inner layer. The target is placed at the geometric centre of the barrel. The targets are typically 0.5 mg/cm^2 self-supporting foils of CD_2 mounted on thin holders with holes of diameter 40 mm, where the large hole diameter is chosen so as to minimize the shadowing of the barrel by the target frame.

Subsequent developments of the TIARA approach are represented by T-REX [56] and SHARC [57], which are shown in Fig. 3.15. Another key development, with a barrel design similar to TIARA, is ORRUBA [60] (and its non-resistive strip version super-ORRUBA) which was developed at Oak Ridge. The most obvious feature of these arrays, relative to TIARA, is that they are designed to fit inside a more conventional gamma-ray array. To some extent, this is equivalent to accepting a limitation on the beam intensity that can be used—certainly at an intensity of 10^9 pps as envisaged in the TIARA design, an enormous amount of radioactivity would be deposited inside the gamma-ray array by the elastic scattering of beam particles from a typical CD_2 target. However, at beam intensities of up to a about 10^8 pps, the radioactivity deposited inside the array will be tolerable and there will be a real benefit in having the silicon array inside a more extensive array of gamma-ray detectors. The advantages lie in the energy resolution achievable with improved Doppler correction, and in simply having a wider range of gamma-ray angles included in the

3 What Can We Learn from Transfer, and How Is Best to Do It?

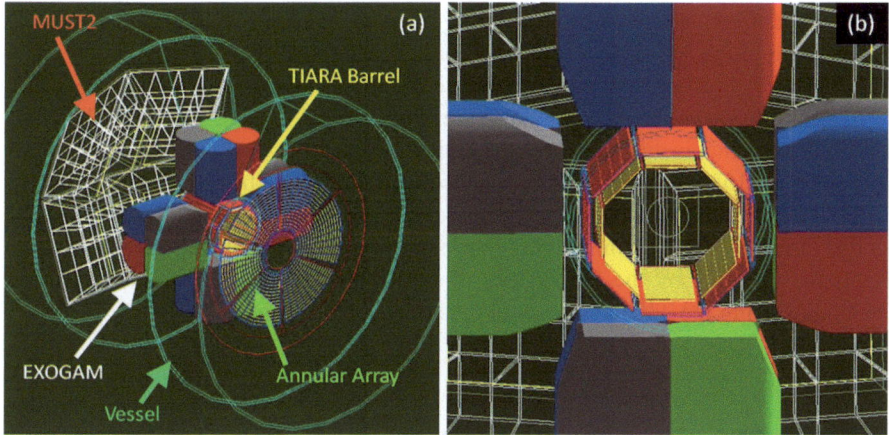

Fig. 3.14 The TIARA setup as modelled in *géant4* [54]: (**a**) overview, including MUST2 [55] and the EXOGAM clover HPGe gamma-ray detectors [52]. The four leaves of each of the 4 are shown; (**b**) the central silicon array comprises two concentric octagonal barrels and the clover front faces are 54 mm from the beam axis. The view is looking with the beam from just in front of the annular array. Beyond the barrel, the detectors of MUST2 are glimpsed. The circular target is mounted at the centre of the barrel

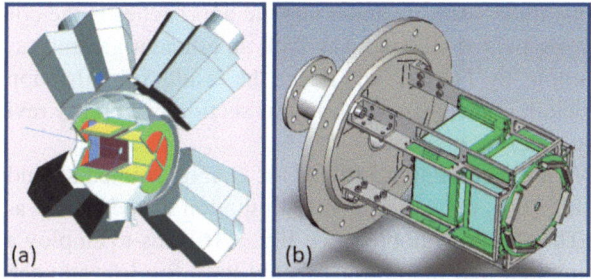

Fig. 3.15 Two post-TIARA silicon arrays developed for use completely inside a large gamma-ray array: (**a**) T-REX [56], which is operated inside the MINIBALL array of HPGe cluster detectors at ISOLDE [58], and (**b**) SHARC [57] which is operated inside the TIGRESS array of segmented HPGe clover detectors [59]. Both include silicon boxes situated forward and backward from the target

measurements. A wide range of gamma-ray angles may open up additional physics possibilities in the interpretation of the data. The planned deployment (GODDESS) of ORRUBA inside Gammasphere [61] with around 100 gamma-ray detectors is perhaps the pinnacle of this approach. The two arrays T-REX and SHARC, coincidentally, have extremely similar geometries. The choice of rectangular boxes allows the silicon detector designs to be relatively simple and hence economical, and the ends of the array are completed with compact annular detectors of a pre-existing design. T-REX (as in the case of ORRUBA, and the original TIARA) includes resistive strips, which helps to keep the number of electronics channels manageable using conventional electronics. However, the price that is paid for using resistive strips is quite high, in terms of performance. Firstly, such detectors typically have higher energy thresholds than non-resistive strips, because they have an electronic noise contribution related to the resistance of the strip [62, 63]. Secondly the position resolution that is achieved is dependent on the energy deposited, being proportional to $1/E$ [64]. SHARC is the first dedicated compact transfer array to utilise double-sided (non-resistive) silicon strip detectors completely, resulting in superior energy thresholds and a consistency in position resolution. This choice of detector was made possible by the availability of up to 1000 channels of high resolution electronics using the TIGRESS digital data acquisition system [59].

3.4.3 Choosing the Right Experimental Approach to Match the Experimental Requirements

As will be apparent from the examples already discussed, a variety of experimental approaches are chosen by different experimenters, for transfer experiments. Largely, these are driven by specific experimental requirements, of which two of the most important are: beam intensity limitations, and the required resolution in excitation energy. One of the most versatile and complete approaches is the combination of a compact, highly segmented silicon array with an efficient gamma-ray detection (as adopted, for example, by TIARA) and hence the results from that approach are presented in some detail, in this document. In this section, we briefly review alternative choices made by experimenters.

In the case of an experiment at SPIRAL at GANIL, aimed at studying ^{27}Ne via the (d, p) reaction [65], the experimental limitation at the time was the available beam intensity. The solution adopted (see Fig. 3.16) was to employ a much thicker target, but this implied that the protons would have too low an energy to exit and be detected. Therefore the experiment was focussed on using the heavier beam-like particle, as in the ^{11}Be experiment discussed in Sect. 3.4.1. The final nucleus had a reasonably complex structure, and hence gamma-ray detection was considered vital and would possibly offer additional information on spin, since the proton differential cross sections could not be observed. The EXOGAM array of segmented Ge gamma-ray detectors was employed [52]. The required target thickness, in order to achieve sufficient gamma-ray detection, was then achieved by using a solid cryogenic pure D_2 target of 17 mg/cm^2. In terms of an equivalent CD_2 thickness of deuterons, the energy loss in the cryogenic target is reduced by a factor of three, so this is equivalent in energy terms to 6 mg/cm^2 of CD_2 but has three times the number of target nuclei. In addition, the absence of carbon in the target removes the problem of background reactions that was mentioned in Sect. 3.4.1. A microchannel plate detector (MCP) before the target assisted in particle identification using the VAMOS spectrometer [53]. Inside VAMOS, the particles were focussed by two quadrupole elements (Q_1, Q_2) through a dipole magnet and then detectors in the

3   What Can We Learn from Transfer, and How Is Best to Do It?

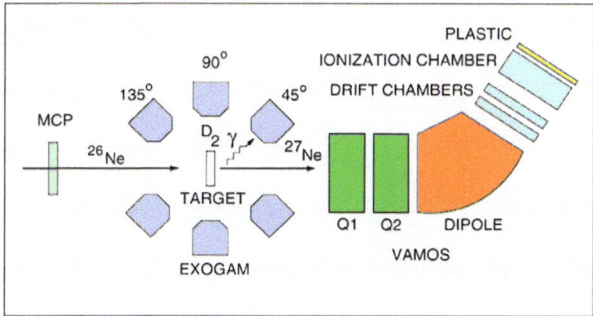

Fig. 3.16 This is a variant on the technique of extracting spectroscopic information from the beam-like particle, rather than the light target-like particle. The aim was to use a thicker target to compensate for a low beam intensity, and the background from target contaminants such as carbon was minimized by using a solid deuterium target. Gamma-ray detection allowed precise excitation energy measurements. See text for definition of other terms

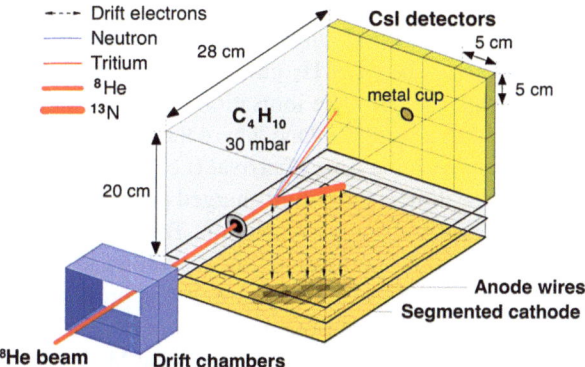

Fig. 3.17 The MAYA detector [66] is an *active target* in the sense that the gas that fills MAYA acts both as the target for the nuclear reactions and also as the fill gas of a time projection chamber. Ionisation paths in the gas are drifted to readout planes, and using the drift time it is possible to reconstruct every individual nuclear reaction in three dimensions (and with particle identification). The diagram shows a reaction on the ^{12}C in the C_4H_{10} gas, but reactions on the hydrogen, or other fill gases, can also be studied

focal plane region recorded the particles' positions, angles and energies. An example of the particle identification that can be achieved in VAMOS is included in Sect. 3.4.5.

Most experimental methods discussed here are limited in resolution by the energy loss effects in thick targets. However, this problem is largely removed if it is possible to determine the precise point of interaction within the target. By turning the target into an active detector, designs such as MAYA [66] (shown in Fig. 3.17) achieve this objective and hence can be used with the lowest beam intensities. In fact, for higher beam intensities it is usually necessary in this type of detector to place an electrostatic screen around the path of the beam itself. The classic model

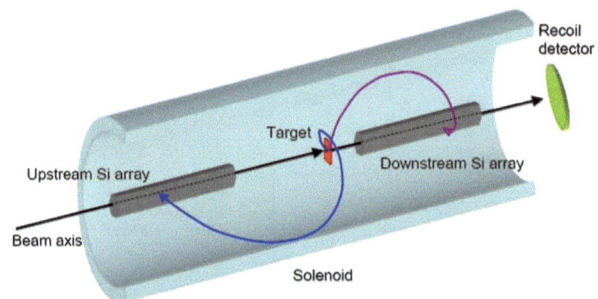

Fig. 3.18 The HELIOS device [36, 69] collects particles magnetically at all angles and focusses them to compact detectors along the axis. The angular information is reconstructed from the measured energy and the distance from the target to its point of return to the axis, and is generally more accurate than can be obtained by direct angle measurements. The way in which the spectrometer operates has the effect of reducing the limitations arising from the detector energy resolutions themselves

for this type of detector is IKAR, which was produced for high energy beams and operates with multiple atmospheres of H_2 gas [67, 68]. In MAYA, the reaction can occur at any point through the gas. The ionisation by all the particles in the gas is drifted in an electric field to a readout plane where the position and amount of ionisation are recorded, along with the time of arrival (i.e. the drift time). This allows a full reconstruction in three dimensions of all charged particle trajectories, subject to various limitations in spatial and energy resolution. The measurement of the ionisation along the whole path of the particles in the gas allows the particle types to be identified. In order for proper drifting of the charge and proper readout, the choice of gas pressure is subject to some restrictions, and hence some particles might easily penetrate beyond the confines of the gas volume. The MAYA detector includes a forward wall of CsI detectors, to deal with these penetrating particles.

A novel approach to achieving 4π detection efficiency is the HELIOS concept that has been developed by the Argonne group and collaborators [36]. Particles emerging at almost all angles from the target are focussed in a large-volume solenoidal field and are brought back to a position-sensitive silicon array aligned along the solenoid axis. This is shown schematically in Fig. 3.18, which is adapted from Ref. [36]. The targets are typically CD_2 foils, but a gas cell target has also been constructed to allow the study of ^{3,4}He-induced reactions. The ideal design parameters for the solenoid are remarkably similar to those for medical MRI scanners and indeed the original HELIOS is a decommissioned MRI device [69]. The energy limitations arise not only from the field strength and radius, but also the length along the axis. It is shown in Ref. [36] that the limitations are much more significant for a typical 0.5 m long device (or a 1 m long device with the target at the centre) than they are for a 1.5 m long device (Fig. 8 of Ref. [36]). The detection limits as a function of angle, for a device between the quoted lengths, are well matched to the kinematics of (d, p) in inverse kinematics. As shown in Ref. [36], the Q-value (excitation energy) is calculated directly from the measured energy and distance along the axis for each detected particle (Eq. (5) [36]). So also is the centre of mass angle (Eq. (7) [36]). At an intermediate point in the calculation, the measured time of flight is used to measure the charge to mass ratio A/q for the particle (Eq. (2) [36]) but once the particle identification is made, the exact value is substituted in further calculations. Thus, apart from the measured energy and position, the calculations rely only upon the precise value/stability and the homogeneity of the magnetic field. The particle identification (apart from one A/q ambiguity between deuterons and ^{4}He^{++}) is a significant bonus, although it does have some implications for the time structure of pulsed beams. As mentioned in Sect. 3.3.3, any impact on the excitation energy resolution arising from the detector energy resolution is significantly reduced in the HELIOS method, because particles are compared at the same z (distance along the axis) rather than at the same θ_{lab} (angle of emission in the laboratory frame). This turns out to have the effect of removing the *kinematic compression* observed in Fig. 3.11, wherein (particularly at backward angles in the laboratory) the kinematic lines are closer together in proton energy than in excitation energy.

An example of the use of HELIOS with an online produced radioactive beam is the study of ^{16}C via the (d, p) reaction in inverse kinematics with a thin CD_2 target of 0.11 mg/cm^2 and a beam of 10^6 pps of ^{15}C [70]. Interestingly, the ^{15}C secondary beam was itself produced using the (d, p) reaction in inverse kinematics with a ^{14}C primary beam. Good resolution was achieved, but one key doublet of states at 3.986/4.088 MeV in ^{16}C could not be resolved. Each of these states gamma-decays to the 1.766 MeV level, and the 100 keV difference in the energies of these 2.2 MeV gamma-rays would be easily resolvable with a modern Ge gamma-ray array. It is a considerable challenge to combine the HELIOS technique with state-of-the-art gamma-ray detection. One very appealing future direction of development would be to combine the MAYA and HELIOS concepts, so that particles could be completely tracked in three dimensions but with the focussing and collection advantages of the magnetic field.

3.4.4 Using (d, p) with Gamma-Rays, to Study Bound States

Typical data for the energies of the measured particles, as a function of their deduced laboratory angle, are shown in Fig. 3.19 for an experiment using a silicon array with a large angular coverage [71]. This experiment was performed with a beam of 3×10^7 pps of ^{25}Na at 5 A MeV, using the SHARC array [57] at TRIUMF. Provided that calibrations have been performed in advance, this type of spectrum can be created online, during data acquisition. Once the kinematic lines are seen, the first hurdle is crossed, and the experiment is seen to be working correctly. Then, the discussion can turn to the specifics of the physics to be measured and the statistics that are required. The most intense lines will typically be those due to elastic scattering. In the figure, the data show lines that are recognisable as coming from the elastic scattering of both deuterons and protons in the 0.5 mg/cm^2 CD_2 target. It is typical that any deuterated target will have some fraction of non-deuterated

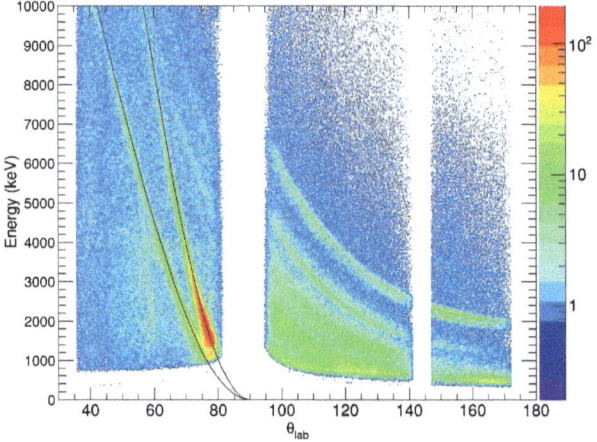

Fig. 3.19 Raw data for a typical experiment [71] using a silicon array to detect the light particles, to be compared with Fig. 3.11. Kinematic lines are overlaid over the deuterons (higher energies) and protons from elastic scattering. At larger angles, the loci are clearly seen for protons from (d, p) reactions. The apparent angular dependence of the lower energy threshold is due to corrections that are applied to compensate for energy losses in the target

molecules. The intensity falls away, generally, as the energy increases and the centre of mass scattering angle also increases. The angular distribution may show oscillations, but the fall in intensity is the general tendency. In this particular experiment, there is a gap in the data near 90° due to a physical gap in the array, related to the target mounting and changing mechanism. A further gap exists in the backward angle region due to the silicon detector support structure. In the region backward of 90°, the kinematic lines arising from (d, p) reactions are evident. In this angular range, there are no other deuteron-induced reactions (apart form (d, p)) that can contribute to the charged particle yield. From that perspective, no particle identification is needed for the backward angles. In fact, because of the low energies, no $\Delta E - E$ identification technique would be appropriate, but time-of-flight or silicon pulse-shape techniques would be feasible. The reason that particle identification could indeed be useful is that not all reactions will be induced on the deuterons in the target. Assuming a CD_2 target as in the experiment shown here, the reactions induced on carbon nuclei can produce charged particles at any angle. Typically, the compound nuclear reactions that arise from the carbon will produce both protons and alpha-particles (and possibly other species) by evaporation from the excited compound nucleus. Standard codes exist, to estimate the evaporation channels that will be important for a particular beam and energy combination (e.g. LISE++ [72], which includes the fusion-evaporation code PACE4 [73]). These evaporated particles will not have a specific angle-energy relationship because several particles will be evaporated. Also, alpha-particles can deposit much more energy than protons in a given thickness of silicon because of their shorter range. Thus, the kinematic lines from (d, p) and elastic scattering will in general appear on a smooth background arising from evaporated charged particles from compound nuclear reactions. This is evident to some extent in Fig. 3.19, even though some experimental techniques have been applied so as to reduce the compound nuclear contribution (see below).

Figure 3.20 summarises a range of experimental results from a $(d, p\gamma)$ study using a radioactive beam of 2×10^4 pps of ^{24}Ne at 10 A MeV [49]. The energy

3 What Can We Learn from Transfer, and How Is Best to Do It? 99

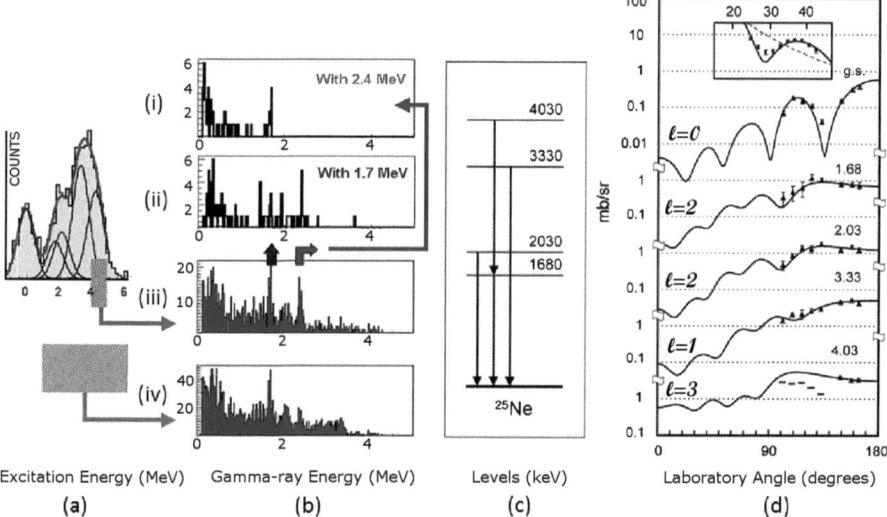

Fig. 3.20 Results from a (d, p) study of ^{25}Ne using a beam of ^{24}Ne at 10 A MeV [49]: (**a**) example excitation energy spectrum reconstructed from the measured proton energies and angles and showing gating regions used to extract coincident gamma-ray spectra, (**b**) gamma-ray energy spectra (iii), (iv) from p–γ coincidences for highlighted regions of excitation energy in (**a**), spectra (i), (ii) from p–γ–γ data with the events and γ-ray gates indicated in (iii), (**c**) summary of the level and decay scheme deduced from this experiment, (**d**) differential cross sections for the indicated ℓ transfers to states in ^{25}Ne. Elastic scattering data are inset (see text)

and angle information as shown in the previous figure can be combined to calculate the excitation energy in the final nucleus, assuming that the reaction was (d, p) initiated by the beam. Angular regions where other reactions dominate can be removed in the analysis. Figure 3.20(a) shows an excitation energy spectrum for ^{25}Ne calculated from the kinematic formulae, for one particular angle bin. The fit to the various excited state peaks in this spectrum was informed and constrained by the observed gamma-ray energies. The gamma-ray energy spectrum observed with specific limitations on the excitation energy are shown in part (b) of the figure, where parts (iii) and (iv) correspond to the indicated excitation energy limits in ^{25}Ne. For the events included in Fig. 3.20b(iii), the results of gating on particular gamma-ray peaks are shown in parts (i) and (ii). The p–γ–γ triple coincidence statistics in these two spectra are sufficient (just) to deduce that the two observed gamma-ray transitions are in coincidence. (Actually, the experiment in Ref. [49] also measured the heavy (^{25}Ne) particle after the reaction, so the data in Fig. 3.20b(i)–(ii) actually represent quadruple coincidence data.) Taking into account the excitation energies at which the nucleus is fed by the (d, p) reaction, and the observed gamma-ray cascade, the level scheme in Fig. 3.20(c) was inferred. The angular distributions shown in Fig. 3.20(d) were used to deduce the transferred angular momentum carried by the neutron, according to the best-fit shape. The calculations that are shown were performed using the ADWA method. Different angular momenta were deduced for the various states. For example, the ground state has a clear $\ell = 0$ distribution. The scaling of the theory to the experimental data gave the measured spectroscopic factors. In the case of the 4.03 MeV state, it was only possible to set a lower limit on the cross section at certain angles. This was related to the energy thresholds of the silicon detectors used for the proton detection. As shown in the kinematics diagrams in Fig. 3.11, and illustrated in the data of Fig. 3.19, the observed particles from (d, p) are lower in energy for states with higher excitation energy and hence the higher states are subject to this type of threshold effect. Raising the beam energy will give access to higher excitation energies. The observed lower limits on the cross section for the 4.03 MeV state were nevertheless sufficient to rule out the alternative angular momentum assignments and $\ell = 3$ could be assigned. Finally, an inset in Fig. 3.20(d) shows the differential cross section for deuteron elastic scattering, measured as a function of the centre-of-mass scattering angle. This was derived from the rapidly rising locus of data points observed in the data, similar to that for the elastics shown in Fig. 3.19. This will be discussed further, in Sect. 3.4.6.

The gamma-ray energy spectra of Fig. 3.20 include a correction, applied event-by-event, for a very significant Doppler shift. At the recommended beam energies of 5–10 A MeV, the projectiles have a velocity of approximately $0.10c$. Actually, the velocity is sufficient for the Doppler shift at 90° due to the second-order terms to be easily measured. Hence, the full relativistically correct formula should be used, to apply Doppler corrections to the measured gamma-ray energies so that they accurately reflect the emission energies in the rest frame of the nucleus. The Doppler-corrected energy E_{corr} is given by

$$E_{\text{corr}} = \gamma(1 - \beta \cos\theta_{\text{lab}}) E_{\text{lab}}$$

where $\gamma = 1/\sqrt{1-\beta^2}$ and $\beta = v/c$ where v is the velocity of the emitting nucleus. The angle θ_{lab} is measured for the gamma-ray detector relative to the direction of motion of the nucleus. In practice, and taking into account the accuracy with which the gamma-ray angle can be determined, it is usually sufficient to assume that the emitting nucleus is travelling along the beam direction in these inverse kinematics experiments (although it is also easy to calculate it's angle from the measured light-particle angle). It will be relevant later, to note that another relativistic effect related to gamma-rays is significant at these beam energies. The angle of emission relative to the beam direction, as measured in the frame of the emitting nucleus, is different from the angle measured in the laboratory frame of reference. This consequence of relativistic aberration means that the gamma-rays emitted by a moving nucleus are concentrated conically towards its direction of motion, which is known as relativistic beaming or as the relativistic headlight effect. For isotropic centre of mass emission at $\beta = 0.1$, the fraction of gamma-rays emitted forward of 90° in the laboratory will be about 55 %. The yield of gamma-rays observed at 10° in the laboratory will be larger than the yield at 170° by a factor of $1.22/0.82 = 1.49$. The relativistic aberration formula is given by

$$\cos\theta_{\text{lab}} = \frac{\cos\theta_{\text{c.m.}} - \beta}{1 - \beta\cos\theta_{\text{c.m.}}}$$

3 What Can We Learn from Transfer, and How Is Best to Do It?

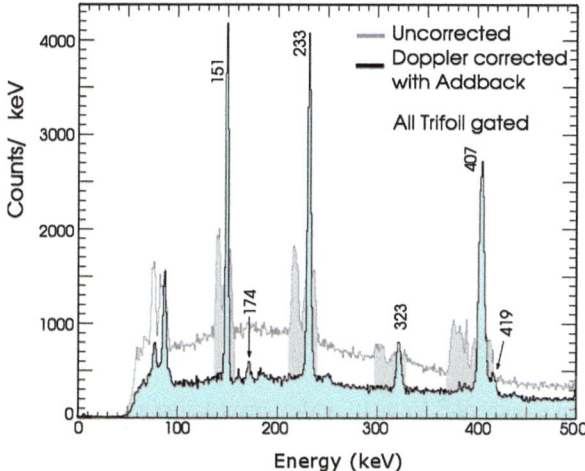

Fig. 3.21 Results of the Doppler shift correction procedure applied to ^{26}Na gamma-rays produced in the reaction of 5 A MeV ^{25}Na with deuterons [71]. The upper spectrum (outlined and *partly shaded in light grey*) is uncorrected, with the *shaded parts* indicating the spread of counts contributing to four of the strongest peaks in the lower spectrum. The lower spectrum (*darker shading*) is corrected for the Doppler shift. In addition to the Doppler correction, an add-back procedure has been applied to account for Compton scattering (see text). This lowers the continuum background. All of these data are "Trifoil gated" to remove or minimize events of a compound nuclear origin, as explained in Sect. 3.4.5

where $\theta_{c.m.}$ is measured in the rest frame of the nucleus and other terms are as defined above.

The relativistic Doppler shift correction was already performed for the gamma-rays in Fig. 3.20 and is shown in more detail for a different experiment, in Fig. 3.21. In the case of Fig. 3.20, the gamma-ray angle could be determined only according to which leaf (crystal) of the clover detector recorded the initial interaction. The resolution at 1 MeV was 65 keV FWHM (full width at half maximum) after correction [49], limited by the high value of $\beta = 0.1$, the close proximity of the detectors to the target (50 mm) and the lack of any further gamma-ray angle information. This is reduced by a third to just under 45 keV (FWHM) at 1 MeV in the TIARA configuration if the clover segmentation information is used [48]. In the experiments [71] with SHARC, using TIGRESS, the distance to the front face of the gamma-ray detectors was nearly three times larger than TIARA, at 145 mm. The gamma-ray clover detectors were centred at either 90° or 135° and each leaf of the clover was four-fold segmented electronically. An add-back procedure was applied, to account for Compton scattering between different leaves of the same clover. This involved adding the energies together and then adopting the segment with the highest energy as indicating the angle of the initial interaction (a criterion that is justified by simulations [48]). For a $(d, p\gamma)$ gamma-ray at 1806 keV, the observed resolution after Doppler correction was 23 keV (FWHM) or 18 keV (FWHM) for detectors at 90° and 135° respectively (reflecting the Doppler broadening, as opposed to shift, that

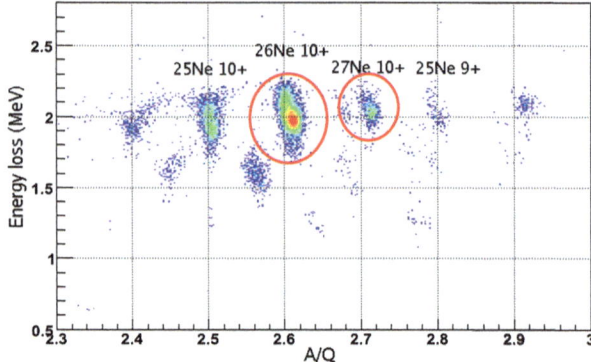

Fig. 3.22 Data from a study of ^{26}Ne at 10 A MeV bombarding a CD_2 target [75, 76]. Particles were detected in the wide-acceptance spectrometer VAMOS centred at zero degrees and were identified using the parameters measured at the focal plane. This determined the reaction channel and effectively eliminated any contribution from carbon in the target

contributes at 90°). Scaling this to the previously quoted energy of 1 MeV gives a resolution of 10–12 keV (FWHM). This resolution is a factor of 10–50 better than the resolution in excitation energy obtained from using the measured energy and angle of the proton. Thus, the resolution in excitation energies for states populated in (d, p) reactions can be improved by a similar factor.

3.4.5 The Use of a Zero-Degree Detector in (d, p) and Related Experiments

The ability to detect the beam-like particle, as well as the light particle, from transfer reactions in inverse kinematics is a big advantage for several reasons. It was therefore a fundamental design constraint, for TIARA [48, 74], that it should be coupled to the magnetic spectrometer VAMOS. The advantages are partly evident from inspection of Fig. 3.22. The different particle types observed at angles around zero degrees, following the bombardment of a CD_2 target with a ^{26}Ne beam, are clearly identified. The beam in this case was 2500 pps at 10 A MeV and the target thickness was 1.20 mg/cm^2. Two further features make this zero degree detection even more useful. Firstly, the silicon array will record the coincident particles only for the reactions induced on the hydrogen in the target; the recoil carbon nuclei for this constrained kinematics will essentially all stop in the target. Secondly, the spectrometer gives not only the particle identification but also the angle of emission for the heavy particle, which can be exploited, for example as in Sect. 3.4.7. In the example shown here, the reaction products could be simultaneously collected and identified for (d, p) to bound states of ^{27}Ne, (d, p) to unbound ^{27}Ne that decays back to ^{26}Ne, and (d, t) to bound states of ^{25}Ne.

3 What Can We Learn from Transfer, and How Is Best to Do It?

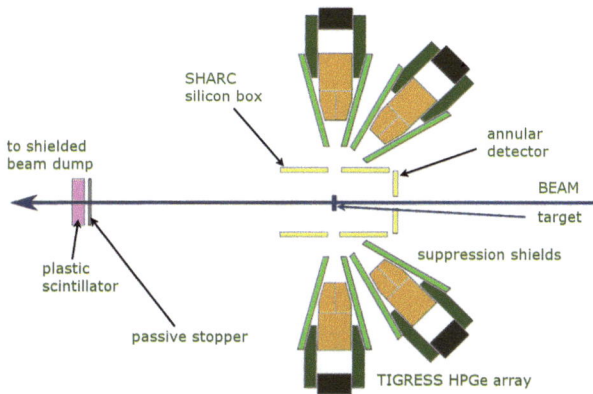

Fig. 3.23 Schematic of the experimental setup for experiments combining the SHARC Si array with the TIGRESS gamma-ray array [71]. A plastic scintillator detector was introduced at zero degrees, 400 mm beyond the target, to help in identifying and eliminating events arising from reactions on the carbon component of the CD_2 target. The performance of this *trifoil* detector [78] is discussed in the text

In experiments currently performed at TRIUMF, there is no access to a spectrometer such as VAMOS, and hence a less elaborate solution was implemented, and is described here. Note that, in the longer term, the purpose-built fragment mass separator EMMA [77] will become available at TRIUMF. In the meantime, a detector developed at LPC Caen and called the *trifoil* was adapted [78] from its original purpose, which was to provide a timing signal for secondary beams produced via projectile fragmentation at intermediate energies. The experimental layout for the first experiment [71] using the trifoil in this fashion is shown in Fig. 3.23. In this implementation, the plastic scintillator in the trifoil will record signals arising from unreacted beam particles or transfer and similar reactions in the target, i.e. where the beam-like particle is not slowed down. If the reaction in the CD_2 target was induced by the carbon, then it could be either a transfer reaction (if peripheral) or a compound nuclear reaction. In the former case, no particle would be observable in the silicon array SHARC. In the second case, the evaporated charged particles could be observed, but also the product at zero degrees would be slower moving and would have a higher Z than for a transfer reaction induced by the hydrogen in the target. The compound nuclear products are then stopped by a passive layer of aluminium, whilst still leaving the direct reaction products with sufficient energy to be recorded in the trifoil and then pass through to a remote beam dump. The present trifoil detector is big enough to span the cone of recoil beamlike particles corresponding to protons from (d, p) collected over a wide range of angles. Compound nuclear events are completely prevented from producing a valid trifoil signal, by means of the passive stopper, but depending on the beam rate there may be random coincidences with other beam particles arriving in the same bunch of the pulsed beam. (Ideally, the detector would be insensitive to unreacted beam particles, and this was achieved to some extent.)

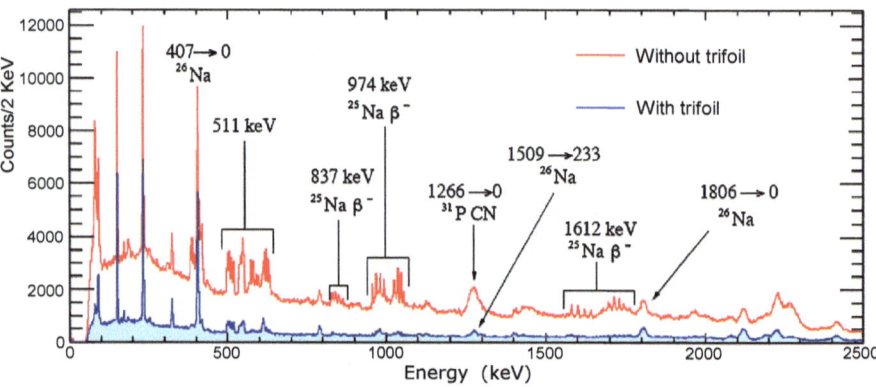

Fig. 3.24 Gamma-ray energy spectrum acquired for a beam of ^{25}Na at 5 A MeV incident on a CD_2 target, using the full TIGRESS array (see text). The requirement of a trifoil signal eliminates a large fraction of the smooth background, and largely removes the peaks due to scattered radioactivity and compound nuclear reactions. The radioactivity peaks are dispersed by the Doppler correction

The effect of the zero degree trifoil detector in reducing the background in the gamma-ray energy spectra is illustrated in Fig. 3.24. This spectrum was acquired for a beam of ^{25}Na at 5 A MeV incident on a CD_2 target with an average intensity of 3×10^7 pps. The spectrum includes data from the full TIGRESS array, comprising 8 detectors with 4 placed at 90° and 4 at 135° in this experiment [71]. The spectrum is Doppler corrected as described above, and hence the gamma-rays produced by a source at rest (such as the 511 keV annihilation gamma-ray and those originating from the radioactive decay of scattered and then stopped ^{26}Na projectiles) have been transformed into multiple peaks depending on their angle of detection relative to the target. Escape suppression has also been applied, using the signals from the scintillator shields for each clover detector. The first thing to note is that the smooth background, arising from unsuppressed Compton scattering events due to higher energy gamma-rays, is massively reduced by applying the trifoil requirement. This is quantified below. Secondly, with regard to the peaks, it can be seen for example that the 1806 keV peak arising from the (d, p) product ^{26}Na is retained in the trifoil-gated spectrum with high efficiency whereas the 1266 keV peak arising from the compound nuclear product ^{31}P is mostly eliminated. In fact, the elimination of the ^{31}P peak reveals an underlying ^{26}Na peak at 1276 keV.

In order to quantify the improvement in peak:background ratio that was achieved by using the trifoil, spectra such as those in Fig. 3.25 were produced. The gamma-ray energy spectrum in Fig. 3.25(a) is for a single clover at a single laboratory angle. The data were analysed in this way, in order to be sure to separate as much as possible the gamma-rays arising from transfer and compound nuclear reactions. The optimal value of the velocity β for the Doppler correction is of course different for these two different categories of reaction, so the correction procedure produces relative movement in energy between counts from transfer and compound reactions depending upon the angle of the gamma-ray detection. The proton energy data in Fig. 3.25(b) are for a thin slice in a spectrum of energy versus angle such as that

3 What Can We Learn from Transfer, and How Is Best to Do It?   105

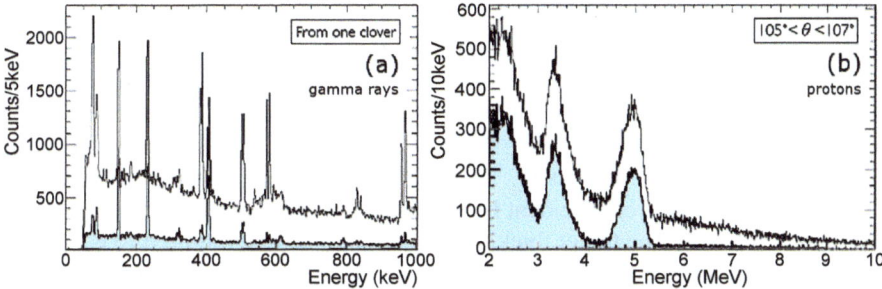

Fig. 3.25 Energy spectra accumulated for a beam of ^{25}Na at 5 A MeV incident on a CD_2 target, showing the rejection of background using the trifoil detector as discussed in the text: (**a**) expanded view of the low energy *gamma-ray* spectrum, for a single clover crystal at 82° to the beam direction, (**b**) example *proton* energy spectrum for measured proton angles between 105° and 107° in the laboratory frame, i.e. a vertical slice in Fig. 3.19

shown in Fig. 3.19. Already, in Fig. 3.19, the trifoil requirement was applied and this reduced a smooth background arising from compound nuclear events. The extent of this background reduction can be measured using Fig. 3.25(b). In this particular experiment, the average efficiency for successfully tagging a genuine proton or a genuine gamma-ray (i.e. one arising from a transfer or other direct reaction) was about 80 %. The shortfall relative to 100 % was due to the intrinsic efficiency properties of this particular trifoil detector. The average probability for incorrectly tagging a charged particle or gamma-ray of compound nuclear origin was about 15 %, or for a gamma-ray from radioactive decay it was about 10 %. The origin of this unwanted probability lay in the high beam intensity and the chance of recording an unreacted beam particle in the same nanosecond sized beam bunch as a compound reaction. Taken overall, the peak:background ratio in each of the proton energy spectrum and the gamma-ray energy spectrum was improved by nearly an order of magnitude. The two reductions of the background are not independent. For a particular gamma-ray peak, an enhancement in the peak:background ratio of typically a factor of 40 was observed, and there is scope for improvement upon this as noted above.

3.4.6 Simultaneous Measurements of Elastic Scattering Distributions

In the experiments with TIARA [49, 75, 84] and SHARC [71], the absolute normalisation was provided by a simultaneous measurement of the elastic scattering cross section. An example of the data obtained for the cross section, plotted as a function of the centre of mass scattering angle, is shown as the inset in Fig. 3.20. This technique works well, so long as the elastic scattering can be measured sufficiently close to 90° in the laboratory that it includes the small values of the centre of mass angle where the elastic cross section can be calculated reliably. The method relies upon being able to evaluate the cross section theoretically using an optical model calculation. At small centre of mass angles, the deviation from Rutherford scattering will be small and the cross section will be reliable. Assuming that the measurements can be made, there are significant advantages in using this technique. The three main advantages concern (a) the beam integration, (b) the target thickness and (c) the dead time in the data acquisition system. The beam integration would normally require the direct counting of every incident beam particle, with a detector of a known and consistent efficiency. The target thickness would normally be required to be known precisely. However, the measurement of the yield for elastic scattering allows the product of these two quantities ((a) and (b)) to be measured, including any necessary correction for the dead time (c) of the acquisition. In the experiments described, the trigger for the acquisition was for a particle to be detected in the silicon array. The elastic scattering and (d, p) reaction events were then subject to the same dead time constraints. It is still necessary to have a reasonable measurement of the target thickness, so that corrections can be applied for the energy lost by the incident beam and by charged particles as they leave the target.

3.4.7 Extending (d, p) Studies to Unbound States

The extension of (d, p) studies to include transfer to states in the continuum of the final nucleus is relatively straightforward experimentally compared to the theoretical interpretation. In fact, this issue highlights situations in the development of the reaction theory that have remained unresolved, or partially unresolved, from the days when (d, p) reactions in normal kinematics were a major topic of research.

An experimental example that is relatively simple to treat, both experimentally and theoretically, is provided by a study of the lowest $7/2^-$ state in ^{27}Ne, populated via (d, p) with a ^{26}Ne beam [75]. This state is observed as an unbound resonance at an excitation energy of 1.74 MeV in ^{27}Ne, compared to the neutron separation energy of 1.43 MeV. For reasons of both the relatively small energy above threshold and the relatively large neutron angular momentum of $\ell = 3$, this unbound state is quite narrow. In fact, the experiment implies the natural width to be 3–4 keV (but in the data it is observed with a peak width of 950 keV due primarily to target thickness effects). In the case of a relatively narrow resonance, meaning a resonance with a natural width that is small compared to its energy above threshold, it is possible to carry out a theoretical analysis with relatively small modifications to the theory. One method is to make the approximation that the state is bound, say by 10 keV, in order to calculate the form factor (i.e. overlap integral) for the neutron in the transfer; this can satisfactorily describe the wave function in the region of radii where the transfer takes place. An improved approach is to use a resonance form factor, following the method of Vincent and Fortune [79]. In this theory, the magnitude of the differential cross section scales in proportion to the width of the resonance. If a barrier penetrability calculation is used, to estimate the width for a pure single particle state, then the cross section can again be interpreted in terms of a spectroscopic factor. The Vincent and Fortune method has been implemented [80] in the Comfort extension [81] of the widely used DWBA code DWUCK4 [82]. For these narrow, almost bound resonances, the structure of the differential cross section retains its characteristic shape, determined by the transferred angular momentum.

It has long been known [83] that the oscillatory features of the differential cross sections, which allow the transferred angular momentum to be inferred from experimental data, are less prominent or even absent when the final state is unbound and broad in energy. The method of Vincent and Fortune also ceases to be applicable, for these broad resonances. Because of the lack of structure, it becomes difficult to interpret the experimental data so as to determine the spins of final states. An experimental example is provided by the study of unbound states in ^{21}O via the (d, p) reaction with a beam of ^{20}O ions [84]. The analysis in Ref. [84] included calculations using the CDCC model mentioned in Sect. 3.2.3, wherein the continuum in ^{21}O was considered to be divided into discrete energy bins with particular properties.

It may be possible to recover some sensitivity to the transferred angular momentum by observing the sequential decay of the resonance states. The observed angular distribution should reflect the angular momentum of the decay of the resonance, with a dependence on the magnetic substate populations for the resonant state in the transfer reaction. An attempt to exploit this effect was made in the study of $d(^{26}\text{Ne}, ^{27}\text{Ne})p$ mentioned above [75, 76]. The ^{26}Ne products were identified in a magnetic spectrometer as shown in Fig. 3.22. By a process of ray tracing [53] it was also possible to reconstruct the magnitude and direction (θ, ϕ) of the ^{26}Ne momentum. Combining this with the momentum of the incident beam and the light particle detected in TIARA, it was possible to reconstruct the missing momentum [76]. It was assumed that the light particle was a proton, arising from (d, p). The primary aim of this particular analysis was to be able to separate the events arising from (d, p) from those arising from (d, d) or (p, p) in the part of the TIARA array forward of 90°. In this sense, it was very successful, as shown by the separation of the main elastic peak from the sequential decay peak in Fig. 3.26. A threshold of 40 MeV/c effectively discriminates between these two reaction channels. Unfortunately, the resolution in terms of the reconstructed angle (rather than the magnitude) of the unobserved neutron momentum was inadequate to take this further. No useful angular correlation could be extracted, for the sequential $^{27}\text{Ne}^* \rightarrow ^{26}\text{Ne} + n$ decays.

3.4.8 Simultaneous Measurement of Other Reactions Such as (d, t)

Radioactive beams are so difficult to produce that an experiment should make the best possible use of the beam delivered to the target. The compact silicon arrays such as TIARA were designed to cover the whole range of laboratory angles with particle detectors that would assist in this aim. The detectors in the forward hemisphere can record the particles from reactions such as (d, t) or $(d, ^3\text{He})$, at the same time as those just forward of 90° record elastic scattering and those in (predominantly) the backward hemisphere record the (d, p) reaction products. Indeed, the experiment

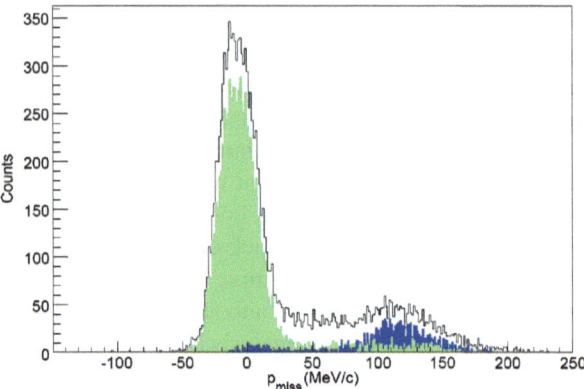

Fig. 3.26 Reconstructed magnitude for the momentum of any missing particle when ^{26}Ne and *a light charged particle* (assumed to be a proton) are detected from the reaction of a ^{26}Ne beam on a CD_2 target [76]. The *upper histogram* is for all data where a ^{26}Ne was positively identified. The *green shaded area* giving mainly a peak near zero is for data selected to highlight elastic scattering, which in fact is $d(^{26}\text{Ne}, ^{26}\text{Ne})d$. The *dark blue shaded area* with fewer counts is selected to highlight the reaction $d(^{26}\text{Ne}, ^{27}\text{Ne}^* \rightarrow {}^{26}\text{Ne} + n)p$ where the neutron was undetected

using TIARA to study ^{21}O via (d, p) with a beam of ^{20}O [84] was also designed to measure the (d, t) reaction to ^{19}O at the same time. The (d, t) measurements [85, 86] employed the telescopes of MUST2 [55] which were mounted at the angles forward of the TIARA barrel (cf. Fig. 3.14). The gamma-ray coincidence measurements with EXOGAM allowed new spin assignments as well as the spectroscopic factor measurements for ^{19}O states [85, 86]. Any studies with (d, t) are immediately useful for comparing to the sorts of knockout studies described in Sect. 3.2.4. The work of Ref. [84] was able to take the additional step of combining the spectroscopic factors measured for (d, p) and (d, t) from ^{20}O. In an analysis based on sum rules and the formalism of [87] and [88], it was possible to derive experimental numbers for the single particle energies for this nucleus. The values were in good agreement [84] with the effective single particle energies of the USDA and USDB shell model interactions for the sd-shell obtained in Ref. [89]. The previously discussed experiment using a beam of ^{26}Ne [75, 76] used the same TIARA + MUST2 experimental setup as the ^{20}O experiment. The data for the (d, t) reaction from ^{26}Ne are still under analysis [90] but an interesting feature here is that the (p, d) reaction was also able to be measured at the same time. The separation experimentally of the d and t products of the (p, d) and (d, t) was possible in MUST2 with a suitable combination of time-of-flight identification and kinematical separation.

3.4.9 Taking into Account Gamma-Ray Angular Correlations in (d, p)

It is well known that gamma-ray angular correlations will be observed for gamma-rays de-exciting states that are populated in nuclear reactions. These correlations have been widely exploited to reveal information about transition multipolarities and mixing, and hence to deduce spin assignments. For a nucleus produced in a reaction, and having some spin J, the angular distribution of gamma-rays measured relative to some z-axis (such as the beam direction) will depend on the population distribution for the magnetic substates $m_j = -J$ up to $+J$. If $J = 0$, the gamma-rays will necessarily be isotropic. However, for other J-values the population of substates will be determined by the reaction mechanism and other details of the reaction. Thus, in (d, p) reactions for example, the gamma-ray angular distribution can depend on details such as the angle of detection of the proton. Certain simplifications can be made. For example, if $J = 1/2$ then, for an unpolarised incident beam, and for protons detected symmetrically around zero degrees (with respect to the beam) the gamma-rays will necessarily be isotropic. Historically, experiments performed with stable beams and targets were designed to restrict the detection parameters in such a way as to simplify the angular momentum algebra, so as to remove any need to understand the magnetic substate populations, and hence the reaction mechanism, in detail. One of the most widely used classifications for angular correlation experiments are the Methods I and II of Litherland and Ferguson [91]. These methods are discussed in some detail in various texts, for example Ref. [92]. A simple and relevant example of the application of Method II is a study of the ^{26}Mg$(d, p\gamma)^{27}$Mg reaction, in which the spins of the first three states in ^{27}Mg were deduced from the measured gamma-ray angular correlations [93]. Method I of Litherland and Ferguson involves measuring a γ–γ angular correlation relative to a particular fixed angle for the first gamma-ray. The quantisation axis is defined by the direction of the incident beam. Method II, the more relevant here, is to measure a particle-γ angular correlation where the outgoing particle from the reaction is measured at either 0° or 180°. This limits the orbital magnetic quantum numbers of the projectile and ejectile to be $m_\ell = 0$ and the consequences of this eliminate the need for any detailed knowledge of the reaction in order to know the substate populations for the final nucleus.

In the present work, we consider a more general situation where we retain one major simplification, namely the cylindrical symmetry of the particle detection, around the beam axis. The discussion is based around the previously discussed experiment using a ^{25}Na beam to study (d, p) reactions populating states in ^{26}Na [71]. The SHARC experimental setup (cf. Fig. 3.23) gives essentially cylindrically symmetrical detection of the protons. The simplification that is produced by this symmetry in the angular description of the angular correlation is dramatic and is described in Sects. III.E and III.F of the article by Rose and Brink [94]. Rose and Brink define an *alignment condition* which means that $w(-M_1) = w(M_1)$ for all values of the magnetic substate quantum number M_1 of the emitting nucleus with spin J_1. Here, w is the weight (i.e. population probability) for a given magnetic substate and is subject to the normalisation

$$\sum_{M_1} w(M_1) = 1.$$

As described in their method 2 of Sect. III.F, entitled *the alignment is achieved by a particle-particle reaction*, the alignment condition will be achieved if the outgoing particle is detected with cylindrical symmetry (assuming that the beam and target particles are unpolarised). Method II of Litherland and Ferguson is simply a very restricted instance of this stipulation. The results used here to describe angular correlations are taken from Rose and Brink's article [94], and they have also been summarised and discussed in the book by Gill [95].

Suppose we have an experiment where the outgoing particle (for example, the proton in a (d, p) reaction) is detected in a cylindrically symmetric fashion at some particular angle with respect to the beam direction. Let the spin of the excited state be J_1. Suppose also, for simplicity, that the gamma-ray transition by which the excited state decays is a pure transition of a particular mulitpole L (the more general cases of mixed multipolarity transitions with a mixing ratio δ are discussed in Refs. [94, 95]). If a gamma-ray detector with a fixed solid angle were then to be moved sequentially to various angles θ with respect to the beam direction, then the angular distribution observed for the gamma-rays would be given by Eq. (3.38) of Ref. [94],

$$W_{\exp}(\theta) = \sum_K a_K P_K(\cos\theta)$$

where it can be shown that K runs from 0 to $2L$ and is even, the P_K are the Legendre polynomials and the a_K can be calculated (as described below) provided that we know the magnetic substate populations of the initial state J_1 and the spin of the final state J_2. Outside of the summation, there will also be an additional factor, usually denoted A_0, to normalise W to the data. The definition of $W(\theta)$ is chosen so that isotropic emission corresponds to $W(\theta) = 1$. Note that this implies that the constant term in the expansion is always $a_0 = 1$. The number of gamma-rays in total that are emitted at an angle θ into an angular range $d\theta$ is given by $W(\theta) \times 2\pi \sin\theta d\theta$. In the case of a transition with pure multipolarity (i.e. with a mixing parameter of $\delta = 0$) Eq. (3.47) of Ref. [94] states that the theoretical form for the angular distribution is given by

$$W(\theta) = \sum_K B_K(J_1) \times R_K(LLJ_1J_2) \times P_K(\cos\theta)$$

where the R_K are independent of the reaction mechanism and basically contain coefficients to describe the angular momentum coupling. The expression for R_K is given by Eq. (3.36) of Ref. [94],

$$R_K(LL'J_1J_2) = (-)^{1+J_1-J_2+L'-L-K} \times \sqrt{(2J_1+1)(2L+1)(2L'+1)}$$
$$\times (LL'1-1 \mid K0) \times W(J_1J_1LL'; KJ_2)$$

where the final two terms are the Clebsch-Gordon coefficient and the Racah W-coefficient describing the indicated angular momentum couplings. These coefficients may be obtained from tables or recursion formulae or from a suitable computer code such as Ref. [96]. In the present case, for a pure multipolarity, we have

$L' = L$. It is the B_K coefficients that contain the information from the reaction mechanism, via the magnetic substate population parameters, $w(M_1)$. The expression for B_K is given by Eq. (3.62) of Ref. [94],

$$B_K(J_1) = \sum_{M_1 = 0 \text{ or } 1/2}^{M_1 = J_1} w(M_1) \times \rho_K(J_1 M_1)$$

where the statistical tensor coefficients ρ_K are given by

$$\rho_K(J_1 M_1) = (2 - \delta_{M_1, 0}) \times (-)^{J_1 - M_1} \times \sqrt{2J_1 + 1} \times (J_1 J_1 M_1 - M_1 \mid K 0)$$

and the final term is again a Clebsch-Gordon coefficient. For most normally-arising cases, the values of ρ_K and R_K are tabulated in the appendix of Ref. [94]. The above description has followed exclusively the formulation of Rose and Brink [94]. Other authors have also presented formulae to describe these angular correlations, but it should be remembered that the different authors often adopt different phase conventions, etc., and hence the tables of symbols appropriate to one description can not be assumed to be appropriate for a different description: one particular formulation must be used consistently. Also, in Ref. [94] the formalism is extended to the case where a gamma-ray cascade occurs, and the second (or subsequent) gamma-ray is the one that is observed. In this case, as given by Eq. (3.46) of Ref. [94], the R_K coefficient in the expression for $W(\theta)$ is replaced by a product of coefficients $U_K R_K$ where U_K depends on J_1 and J_2 for the initial gamma-ray transition and R_K depends on J_2 and J_3 for the second gamma-ray transition. The extension to a longer gamma-ray cascade is straightforward.

Thus, if the spins of the states are known, it is possible to calculate the a_K coefficients, $a_2, a_4, \ldots$, of the Legendre polynomials in the gamma-ray angular distribution $W(\theta)$ provided that the magnetic substate weights $w(M_1)$ are known—at least, for pure multipolarity transitions. These expressions all rely on the particle detection being cylindrically symmetric at some angle (or range of angles) with respect to the beam direction. This ensures that $w(-M_1) = w(M_1)$ for all M_1.

The values of the population parameters $w(M_1)$ depend on the reaction mechanism and, in general, on the angle of the particle detection. An ADWA calculation for a (d, p) reaction can be used to calculate the population parameters $w(M_1)$ and their evolution with the detection angle of the proton. Examples of this are shown in Fig. 3.27, for the (d, p) study discussed above, using a beam of ^{25}Na at 5 A MeV [71]. The different panels correspond to different assumptions about the final orbital for the transferred neutron, and also for the final spin in ^{26}Na. The different panels are for ℓ transfers of $\ell = 0, 1, 2$ and 3. The different lines are for different values of M_1 from 0 to J_1. The main point to note is that in general the populations change dramatically, for different angles of observation. The obvious counter example is the panel for $s_{1/2}$ transfer. The symmetry imposed by s-wave transfer forces all five substates, from $M_1 = -2$ to $+2$ to have equal weights of 0.2 at every observation angle and the gamma-ray emission will always be isotropic in this case.

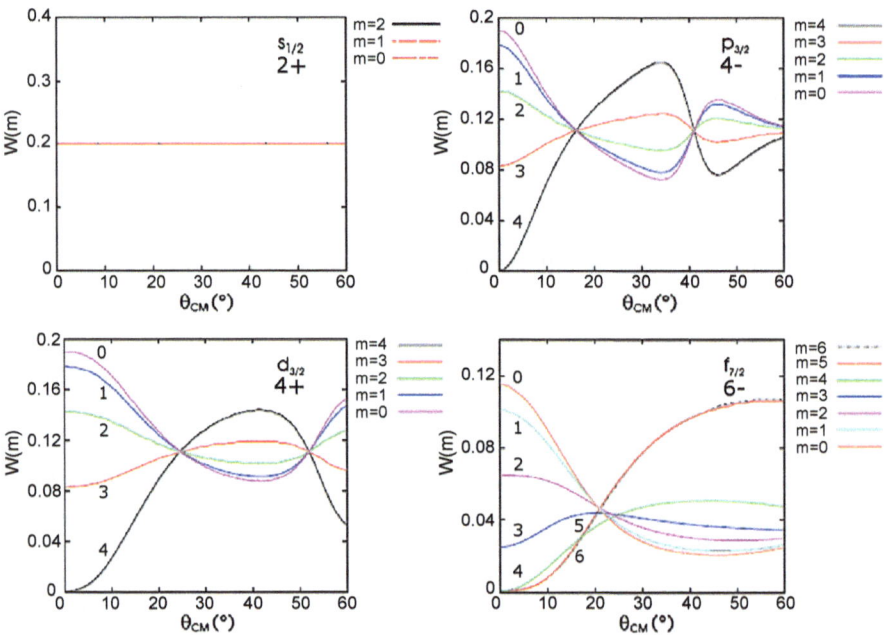

Fig. 3.27 Calculations of magnetic substate population parameters as a function of centre of mass angle, performed using the ADWA model with the code TWOFNR [16]. The calculations all suppose a final state at 2.2 MeV excitation, formed in the (d, p) reaction with ^{25}Na to make ^{26}Na. The orbital into which the neutron is transferred is indicated, along with the assumed final state spin. It can be seen that, in general, the populations vary dramatically. In the experiment, centre of mass angles out to approximately 30° were studied

In Fig. 3.28 the gamma-ray angular distributions determined by the substate populations are plotted, for the upper right hand case in Fig. 3.27, namely $1p_{3/2}$ transfer populating a hypothetical 4^- state at 2.2 MeV excitation energy in ^{26}Na. The gamma-ray decay is assumed to be a pure dipole decay to the 3^+ ground state. Since the multipolarity of this decay is $L = 1$, the maximum value of K for the a_K coefficients is 2. In the centre of mass frame (rest frame) of the emitting nucleus, the gamma-ray angular distribution with respect to the beam axis is given by a constant term plus a term proportional to $a_2 P_2(\cos\theta)$, and the value of a_2 depends on the detection angle of the proton. It is assumed that, for a given proton angle θ(proton) with respect to the beam direction, the protons are detected with cylindrical symmetry at all polar angles, ϕ. For the centre of mass gamma-ray angular distributions, the functions are necessarily symmetric around 90°. The three curves intersecting the axis higher up at $\theta = 0$ are plotted with the horizontal axis representing the gamma-ray angle as measured in the laboratory frame. There is a focussing of the gamma-rays towards zero degrees, due to the relativistic headlight effect as discussed in Sect. 3.4.4.

3 What Can We Learn from Transfer, and How Is Best to Do It?

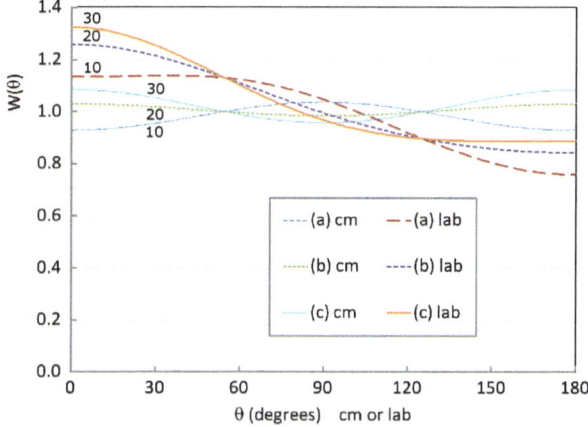

Fig. 3.28 Gamma-ray angular distributions for different detection angles θ_{cm} (proton) for the proton from (d, p). Calculated for ^{25}Na incident on deuterons at 5 A MeV, with $1p_{3/2}$ transfer populating a hypothetical 4^- state at 2.2 MeV excitation energy. For the three symmetric curves, the horizontal axis shows the gamma-ray angle in the centre of mass frame of the emitting nucleus. For the other three curves, the horizontal angle is the gamma-ray angle measured in the laboratory, with respect to the beam direction. The proton centre of mass angles are (**a**) 10°, (**b**) 20°, (**c**) 30°

In Fig. 3.29, the differential cross sections in the laboratory frame are shown, for the population of states in ^{26}Na via the (d, p) reaction in inverse kinematics. The curves for $\ell = 0, 1, 2$ and 3 show the expected movement of the main peak progressively further away from 180° as ℓ increases. The parallel curves with the lower cross sections are actually the computed curves, assuming a gamma-ray coincidence requirement. The angular distributions for a gamma-decay to the ground state were computed using TWOFNR and the ADWA model, for each proton laboratory angle. The gamma-ray angular distributions were then integrated over the appropriate range of angles, corresponding to the laboratory angles spanned by the TIGRESS detectors in the experiment [71]. The relativistic aberration effect was also taken into account. The important point here is that the curves, whilst not perfectly parallel, are very little modified in shape from the ungated curves, i.e. those that have no coincidence requirement. This means that the experimental data can simply be corrected for the measured efficiency of the gamma-ray array and then compared with the unmodified ADWA calculations. This simplification was achieved in this experiment by the large angular range spanned by the gamma-ray array, which meant that the various changes in the angular distributions of the gamma-rays had little net effect after integration. The slight distortions that do occur are negligible (in this case) compared to the statistical errors in the data points and to the inevitable discrepancies that typically occur, between the theoretical and experimental shapes of the differential cross sections. The results from this experiment [71] are currently being prepared for publication.

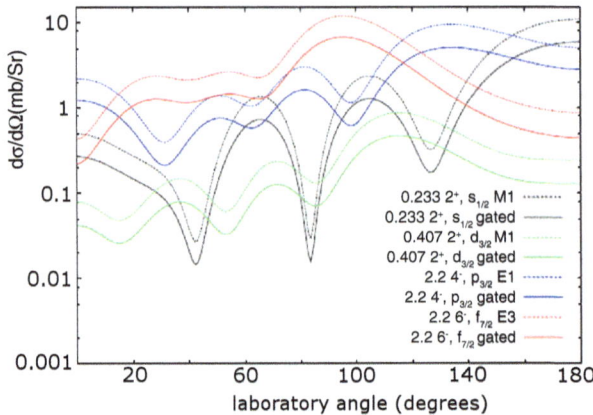

Fig. 3.29 Differential cross sections in the laboratory frame, calculated for the (d, p) reaction leading to four different states in ^{26}Na for an experiment at 5 A MeV, in inverse kinematics. Pairs of almost parallel curves are shown for (**a**) $1s_{1/2}$ transfer to a 2^+ state at 0.233 MeV, (**b**) $1p_{3/2}$ transfer to a hypothetical 4^- state at 2.2 MeV, (**c**) $0d_{3/2}$ transfer to a 2^+ state at 0.407 MeV, (**d**) $0f_{7/2}$ transfer to a hypothetical 6^- state at 2.2 MeV in ^{26}Na. In each case, the *upper curve* is the ADWA calculation and the *lower curve* is the calculated curve for a gamma-ray coincidence requirement (see text)

3.4.10 Summary

Section 3.4 was headed *examples of light ion transfer experiments with radioactive beams* and in this section a range of different experimental approaches have been reviewed. With a relatively light projectile such as ^{11}Be it was possible to make all of the detailed spectroscopic measurements using the beam-like particle. For the alternative approach using a silicon array for the light (target-like) particle, the TIARA array and subsequent developments such as T-REX and SHARC were described. Gamma-ray detection was shown to be useful, or in many cases essential, in order to resolve different excited states and to identify them on the basis of their gamma-ray decay pathways. Hence, the related issues of Doppler correction and angular correlations were discussed. The use of a detector centred at zero degrees for the beam-like reaction products was shown to be a great advantage. Whilst a large-acceptance spectrometer such as VAMOS gives superior performance including full particle identification, it was shown that even a simple detector such as the *trifoil* can substantially assist in the reduction of background. The background arises from compound nuclear reactions induced by the beam on contaminant materials in the target, such as carbon. A common target choice is to use normal $(CH_2)_n$ or deuterated $(CD_2)_n$ polythene self-supporting foils. The option of using a helical orbit (solenoidal) spectrometer instead of a conventional silicon array, for the light particle detection, was described. An example of the use of a cryogenic target of deuterium was included: in the example described, the target was thick and largely absorbed the low energy target-like particles, but it is worth noting that there is research aimed at producing much thinner cryogenic targets that could be used with light particle detection. Finally, an important different approach was described, wherein the target thickness is essentially removed as a limitation because the target becomes the detector itself. This is sometimes called an *active target*. With a time projection chamber (TPC) such as MAYA, the fill-gas of the detector includes within its molecules the target nuclei, and the measurements make it possible to reconstruct the full kinematics of the nuclear reaction in three dimensions. This makes an active target the ideal choice for very low intensity beams, where a thick target is indispensable. With more development to improve the resolution and dynamic range, this type of detector could eventually have the widest applicability of all experimental approaches.

3.5 Heavy Ion Transfer Reactions

For the transfer of a nucleon between two heavy ions, there is an important selectivity in favour of certain final states which allows the spins of the final states to be deduced. This is known as $j_>/j_<$ selectivity because it can tell us whether the final orbital for the transferred nucleon has $j = \ell + 1/2$ or $j = \ell - 1/2$. The origin of the effect is two-fold [97]. Firstly, a heavy ion at the appropriate energies will have a small de Broglie wavelength because of its large mass, and hence its path can be reasonably described as a classical trajectory. Secondly, the transfer must take place in a peripheral encounter between projectile and target because a smaller impact parameter will result in a strongly absorbed compound nuclear process and a larger impact parameter will keep the nuclei from interacting except through the large repulsive coulomb interaction. Therefore, we can consider classical trajectories for peripheral transfer and take into account quantum mechanical factors in a semiclassical fashion. Of course, a full quantum mechanical treatment using the normal reaction theories is possible. The advantage of the semiclassical model is that it allows the origins of the particular selectivity in heavy ion transfer to be understood more readily.

3.5.1 Selectivity According to $j_>$ and $j_<$ in a Semi-classical Model

The semiclassical model for nucleon transfer between heavy ions has been described by Brink [98] and is represented in Fig. 3.30. At the moment of transfer, the mass m has some linear momentum in the beam direction due to the beam velocity v and also due to the rotational motion of m around M_1. Just after the transfer, it is orbiting M_2 which is at rest, and all of the linear momentum is due to the orbital motion. The initial and final linear momenta should be approximately equal by conservation of momentum. Quantum mechanically, they need not be exactly equal because of the uncertainty in momentum introduced by the spatial uncertainty in the precise point of transfer as measured in the beam direction (which can be estimated).

Fig. 3.30 Sketch of the transfer of a mass m from the projectile M_1 to the target M_2 in a heavy ion collision, showing the variables used to derive the Brink matching conditions [98] (see text)

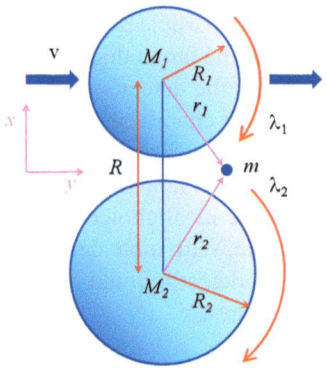

A similar condition can be formulated for the angular momentum of the transferred mass m. Before the transfer, this has contributions from the relative motion between the two colliding heavy ions and from the internal orbital angular momentum of the transferred nucleon. These are the only parts that change: the former due to the adjustments in mass and possibly charge, and the second due to the change of orbital. Once again, the initial and final values should be almost equal.

The two kinematical conditions given by Brink [98] are:

$$\Delta k = k_0 - \lambda_1/R_1 - \lambda_2/R_2 \approx 0$$

$$\Delta L = \lambda_2 - \lambda_1 + \frac{1}{2}k_0(R_1 - R_2) + Q_{\text{eff}}R/\hbar v \approx 0$$

where the orbital angular momentum and projection on the z-axis for the transferred particle are given by (ℓ, λ) with subscripts 1 and 2 for before and after the transfer, respectively. The quantity Q_{eff} is equal to the reaction Q-value in the case of neutron transfer, but otherwise has an adjustment due to changes in Coulomb repulsion: $Q_{\text{eff}} = Q - \Delta(Z_1 Z_2 e^2/R)$. The beam direction is y and the z direction is chosen perpendicular to the reaction plane. A further pair of conditions arise from the requirement that the transfer should take place in the reaction plane, where the two nuclei meet, and hence the spherical harmonic functions $Y_{\ell m}$ should not be zero in that plane:

$$\ell_1 + \lambda_1 = \text{even}$$

$$\ell_2 + \lambda_2 = \text{even}.$$

The two kinematical conditions arising from linear momentum and angular momentum conservation will each, separately, imply a particular *well matched* angular momentum value, for a given reaction, bombarding energy and final state energy (Q-value). Alternatively, for a given ℓ-transfer they will each imply a particular excitation energy at which the matching is optimal. If the values implied by the two equations are equal, then the reaction to produce a state of the given spin and excitation energy will have a large cross section (if such a state exists, with the correct structure in the final nucleus). If the two values are not equal, then the cross section will be reduced by an amount that depends on the degree of mismatch.

3.5.2 Examples of Selectivity Observed in Experiments

A detailed inspection of the Brink matching conditions for Δk and ΔL, given above, implies that a reaction with a large negative Q-value will favour final states with high spin, or more specifically a large value of λ_2 in the notation of Fig. 3.30. This arises because the conservation of linear momentum favour a high value of $\lambda_2 + \lambda_1$ and the conservation of angular momentum implies a large value for $\lambda_2 - \lambda_1$. This selectivity, which occurs for heavy ion transfer with a negative Q-value, is discussed in detail by Bond [99] with a derivation in terms of DWBA formalism. As further noted by Bond [97] the large negative Q-value will imply that the projectile has reduced kinetic energy after the collision and hence is slowed down, which implies a significant transfer of angular momentum. Being heavy ions, the angular momentum of relative motion is large, and hence a relatively small reduction corresponds to transfer into a relatively high spin orbital. In Fig. 3.31 for the (^{16}O, ^{15}O) reaction, which has a large negative Q-value, the neutron is transferred from the $0p_{1/2}$ orbital. For the best matching, there will be a maximum ℓ-transfer which implies that the nucleon will change λ, i.e. the projection of the angular momentum in the direction perpendicular to the reaction plane, as much as possible. For example, from a $0p_{1/2}$ orbital (with orbital angular momentum $\ell = 1$) and an initial projection of $m_\ell = -1$ (which implies also that $m_s = +1/2$) the transfer will favour $m_\ell = +\ell$ for a high-ℓ orbital in the final nucleus. It is reasonable to assume that there is no interaction in the transfer to change the direction (projection) of the intrinsic spin of the nucleon. Therefore the relative directions of orbital and spin angular momentum for the nucleon become swapped in the transfer process. The preferred transfer in this case is from $\ell - 1/2$ (denoted as $j_<$) to $\ell + 1/2$ (denoted as $j_>$). In general, if the Q-value is negative, the transfer from an orbital with $j_<$ ($j_>$) will favour the population of orbitals with $j_>$ ($j_<$) in order to achieve the largest change in λ for the transferred nucleon. Therefore, in Fig. 3.31, the reaction (^{12}C, ^{11}C) shows the opposite selectivity to (^{16}O, ^{15}O). In the upper panel we see a favouring of the ($j_>$) $7/2^-$ state corresponding to the $1f_{7/2}$ orbital, and a relative suppression of the ($j_<$) $9/2^-$ state corresponding to the $0h_{9/2}$ orbital. This selectivity is reversed in the lower panel, and we also see that the ($j_>$) $13/2^+$ state (corresponding to the $0i_{13/2}$ orbital) follows the ($j_>$) $7/2^-$ in becoming weaker relative to the favoured ($j_<$) $9/2^-$ state.

The discussion for single nucleon transfer can be simply extended to include cluster transfer [98]. A further step is to describe reactions in which nucleons are transferred in both directions, to and from the projectile, or in two independent transfers in the same direction. In the work of Ref. [100], the ideas developed by Brink [98] and described by Anyas-Weiss et al. [19] are extended to describe the reactions (^{18}O, ^{17}F) and (^{18}O, ^{15}O) where one of the two steps is the transfer of a dineutron cluster. The trajectories of the transferred particles between the two heavy

Fig. 3.31 Illustration of the $j_>/j_<$ selectivity exhibited in heavy ion transfer when the Q-value is large and negative. The data are for the reactions (^{16}O,^{15}O) and (^{12}C,^{11}C) on a ^{148}Sm target with the same beam velocity, defined by a beam energy of 7.5 A MeV. The selectivity is reversed due to the parent orbitals of the transferred neutron being $0p_{1/2}$ ($j_<$) and $0p_{3/2}$ ($j_>$) respectively. Therefore the *upper panel* favours $j_>$ states and the *lower panel* favours $j_<$ states. The *two highlighted peaks* correspond to populating the $0h_{9/2}$ ($j_<$) and $1f_{7/2}$ ($j_>$) orbitals. The *biggest peak (unshaded)* corresponds to the $0i_{13/2}$ ($j_>$) orbital. Figure adapted from Ref. [97]

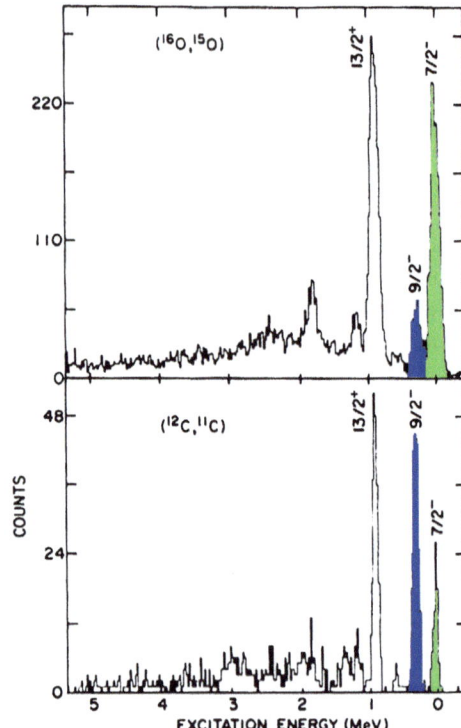

ions are represented in Fig. 3.32 for the favoured (well matched) and unfavoured trajectories. The proton is required in each case to make a transition from $j_<$ to $j_>$ in a stretched trajectory as shown, so as to form the $5/2^+$ ground state in ^{17}O, which was observed in the experiment [100]. Figure 3.32(a) shows that the favoured final states in ^{19}N will have a total spin where 1/2 from the $0p_{1/2}$ proton is added collinearly with the orbital angular momentum transferred by the dineutron cluster. This type of selectivity was observed in the experiment and was used to interpret the states populated in ^{19}N and ^{21}O. In the case of the ^{21}O there has been independent verification of the interpretation via the previously-mentioned study of the (d, p) reaction with a beam of ^{20}O using TIARA [84].

3.6 Perspectives

It is always dangerous to speculate about the future directions for the development of instrumentation or experimental techniques. The experimental devices described here are all likely to deliver a range of new results in nucleon transfer, as new facilities and more beams at the appropriate energies become available. It is, however, perhaps worth taking note of some of the new developments that might be expected. These developments will in part be enabled by an increased capability to deal with

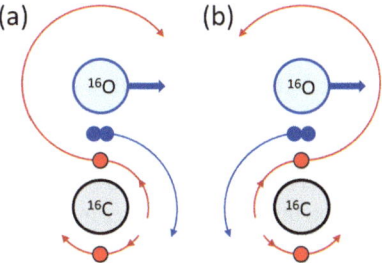

Fig. 3.32 The semiclassical model of Brink [19, 98] can be extended to two-step transfer reactions, such as this (^{18}O, ^{17}F) reaction on a target of ^{18}O (figure adapted from Ref. [100]). The reaction is modelled as a dineutron transfer from the ^{18}O projectile and the pickup of a proton from the ^{18}O target: (**a**) the strongly favoured senses for the two transfers, (**b**) the less favoured transfer directions

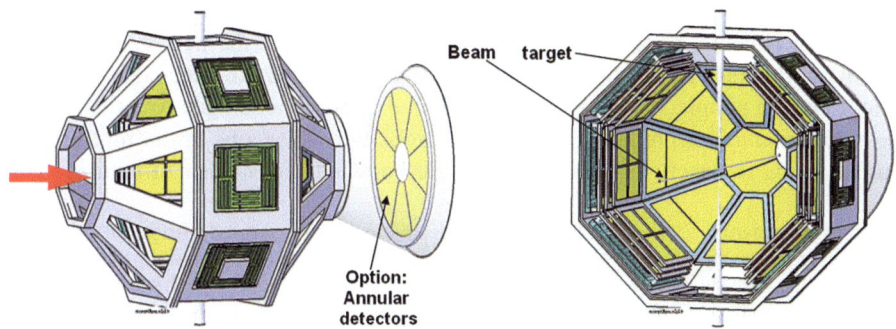

Fig. 3.33 Preliminary design for a new array GASPARD [101] which would represent a new generation of device for the approach using a compact particle array with coincident gamma-ray detection. Multilayer highly segmented particle detectors with enhanced particle identification properties, plus the ability to use cryogenic targets, are amongst its advantages

large numbers of electronics channels, due to innovations in electronics design. One development is to take the simple idea of a highly efficient silicon array (i.e. with a large geometrical coverage) mounted inside a highly efficient gamma-ray array (as adopted by TIARA, SHARC, T-REX, ORRUBA, ...) and improve it. This is the aim of GASPARD [101] which is an international initiative based originally around the new SPIRAL2 Phase 2 facility but also able to be deployed potentially at HIE-ISOLDE. A preliminary design is shown in Fig. 3.33. The particle detection is based on one to three layers of silicon, depending on angle. The segmentation of the silicon is sub-millimetre, but with the detectors still able to supply particle identification information based on the pulse shape. The array is sufficiently compact to fit inside newly developed gamma-ray arrays such as AGATA or PARIS. The geometry is chosen to allow innovative target design, and in particular to have operation with the thin solid hydrogen target CHyMENE, currently under development at Saclay. Another current development is the AT-TPC detector at MSU [102] which aims to combine the advantages of the active target MAYA and the helical spectrometer HELIOS. As noted previously, a key advantage of an active target is that it can, in principle, remove the limitations on energy resolution (or, indeed the limitations on even being able to the detect reaction products) that arise from target thickness. An alternative approach to minimising the target thickness effect is to use an extremely thin target but to compensate by passing the beam through it many times, say 10^6 times. Under certain circumstances, a beam of energy 5–10 A MeV as suitable for transfer could be maintained and recirculated in a storage ring for this many revolutions. A thin gas jet target would allow transfer reactions to be studied in inverse kinematics. The ring could be periodically refilled and the beam cooled, in a procedure that was synchronised with the time structure of the beam production. This is one of the ideas behind the proposed operation of the TSR storage ring with reaccelerated ISOL beams at ISOLDE [103].

In summary, there are some very powerful experimental devices already available and able to exploit the existing and newly developed radioactive beams. In addition, there are challenging and exciting developments underway, that will create even better experimental possibilities to exploit the beams from the next generation of facilities. Because of their unique selectivity, and because the states that are populated have a simple structure that should be especially amenable to a theoretical description and interpretation, transfer reactions will always be at the forefront of studies using radioactive beams to extend our knowledge of nuclear structure.

Acknowledgements Thank you to all of my colleagues who have assisted with putting together, performing and analysing the results from the experiments described in this work: these include N.A. Orr, M. Labiche, R.C. Lemmon, C.N. Timis, B. Fernandez-Dominguez, J.S. Thomas, S.M. Brown, G.L. Wilson, C. Aa. Diget, the VAMOS group at GANIL and the TIGRESS group at TRIUMF, the nuclear theory group at Surrey, plus all of the other colleagues from the TIARA, MUST2, TIGRESS and Charissa collaborations: *we together weathered many a storm* [104].

References

1. T. Otsuka et al., Phys. Rev. Lett. **87**, 082502 (2001)
2. T. Otsuka et al., Phys. Rev. Lett. **95**, 232502 (2005)
3. J. Schiffer, Talk at transfer reaction workshop, Oak Ridge, TN, USA, 21–22 June 2002; available online at http://www.phy.ornl.gov/workshops/transfer/docs/jps_transfer.pdf
4. O. Sorlin, M.-G. Porquet, Prog. Part. Nucl. Phys. **61**, 602 (2008)
5. O. Sorlin, M.-G. Porquet, Phys. Scr. T **152**, 014003 (2013)
6. T. Otsuka, Phys. Scr. T **152**, 014007 (2013)
7. N.K. Glendenning, One- and two-nucleon transfer reactions, in *Nuclear Spectroscopy and Reactions, Part D*, ed. by J. Cerny (Academic Press, London, 1975). ISBN 0121652041
8. N.K. Glendenning, *Direct Nuclear Reactions* (World Scientific, Singapore, 2004). (Originally published, Academic Press, 1983)
9. G.R. Satchler, *Introduction to Nuclear Reactions*, 2nd edn. (Macmillan, London, 1990). ISBN 033351484-X
10. G.R. Satchler, *Direct Nuclear Reactions* (Oxford University Press, Oxford, 1983). ISBN 0198512694
11. J. Gómez Camacho, A.M. Moro, Chapter 2 in this volume
12. D.G. Kovar et al., Nucl. Phys. A **231**, 266 (1974)

13. B.L. Cohen, *Concepts of Nuclear Physics* (Tata McGraw-Hill, New York, 1971/2004). ISBN 9780070992498
14. L.I. Schiff, *Quantum Mechanics*, 3rd edn. (McGraw-Hill, New York, 1968). http://archive.org/details/QuantumMechanics_500
15. R.C. Johnson, P.J. Soper, Phys. Rev. C **1**, 976 (1970)
16. J.A. Tostevin, M. Toyama, M. Igarashi, N. Kishida, University of Surrey version of the code TWOFNR. http://www.nucleartheory.net/NPG/code.htm
17. N.K. Timofeyuk, R.C. Johnson, Phys. Rev. Lett. **110**, 112501 (2013)
18. K.L. Jones, Phys. Scr. T **152**, 014020 (2013)
19. N. Anyas-Weiss et al., Phys. Rep. **12**, 201 (1974)
20. K. Wimmer et al., Phys. Rev. Lett. **105**, 252501 (2010)
21. P.G. Hansen, Phys. Rev. Lett. **77**, 1016 (1996)
22. J.A. Tostevin, J. Phys. G, Nucl. Part. Phys. **25**, 735 (1999)
23. P.G. Hansen, J.A. Tostevin, Annu. Rev. Nucl. Part. Sci. **53**, 219 (2003)
24. A. Gade et al., Phys. Rev. C **77**, 044306 (2008)
25. N.K. Timofeyuk, Phys. Rev. C **84**, 054313 (2011)
26. C. Barbieri, Phys. Rev. Lett. **103**, 202502 (2009)
27. C. Barbieri et al., Nucl. Phys. A **834**, 788c (2010)
28. T. Kobayashi et al., Nucl. Phys. A **805**, 431c (2008)
29. V. Panin et al. GSI Scientific Report PHN-ENNA-EXP-27 (2012). http://repository.gsi.de/record/52060
30. F. Flavigny et al., Phys. Rev. Lett. **110**, 122503 (2013)
31. W.N. Catford, Nucl. Phys. A **701**, 1 (2002)
32. W.N. Catford et al., Nucl. Instrum. Methods A **247**, 367 (1986)
33. W.N. Catford, Eur. Phys. J. A **25**, 245 (2005). doi:10.1140/epjad/i2005-06-171-4
34. W.N. Catford et al., J. Phys. G, Nucl. Part. Phys. **24**, 1377 (1998)
35. J.S. Winfield, W.N. Catford, N.A. Orr, Nucl. Instrum. Methods A **396**, 147 (1997)
36. A.H. Wuosmaa et al., Nucl. Instrum. Methods A **580**, 1290 (2007)
37. S. Fortier et al., Phys. Lett. B **461**, 22 (1999)
38. J.S. Winfield et al., Nucl. Phys. A **683**, 48 (2001)
39. L. Bianchi et al., Nucl. Phys. A **276**, 509 (1989)
40. K.E. Rehm et al., Phys. Rev. Lett. **80**, 676 (1998)
41. C.N. Davids et al., Nucl. Instrum. Methods B **70**, 358 (1992)
42. Y. Blumenfeld et al., Nucl. Instrum. Methods A **421**, 471 (1999)
43. F. Skaza et al., Phys. Rev. C **73**, 044301 (2006)
44. N. Keeley et al., Phys. Lett. B **646**, 222 (2007)
45. M.S. Wallace et al., Nucl. Instrum. Methods A **583**, 302 (2007)
46. L. Gaudefroy et al., Phys. Rev. Lett. **97**, 092501 (2006)
47. K.L. Jones et al., Nature **465**, 454 (2010)
48. M. Labiche et al., Nucl. Instrum. Methods A **614**, 439 (2010)
49. W.N. Catford et al., Phys. Rev. Lett. **104**, 192501 (2010)
50. W.N. Catford et al., Nucl. Instrum. Methods A **371**, 449 (1996)
51. W.N. Catford et al., Nucl. Phys. A **616**, 303c (1997)
52. J. Simpson et al., Acta Phys. Hung. **11**, 159 (2000)
53. S. Pullanhiotan et al., Nucl. Instrum. Methods A **593**, 343 (2008)
54. S. Agostinelli et al., Nucl. Instrum. Methods A **506**, 250 (2003)
55. E. Pollacco et al., Eur. Phys. J. A **25**, s01, 287 (2005)
56. V. Bildstein et al., Eur. Phys. J. A **48**, 85 (2012)
57. C.Aa. Diget et al., J. Inst. **6**, P02005 (2011)
58. J. Eberth, Prog. Part. Nucl. Phys. **46**, 389 (2001)
59. M.A. Schumaker, C.E. Svensson, Nucl. Instrum. Methods A **575**, 421 (2007)
60. S.D. Pain et al., Nucl. Instrum. Methods B **261**, 1122 (2007)
61. M. Devlin et al., Nucl. Instrum. Methods A **383**, 506 (1996)
62. T. Yanagimachi et al., Nucl. Instrum. Methods A **275**, 307 (1989)

63. V. Radeka, IEEE Trans. Nucl. Sci. **21**, 51 (1974)
64. R.B. Owen, M.L. Awcock, IEEE Trans. Nucl. Sci. **15**, 290 (1968)
65. A. Obertelli et al., Phys. Lett. B **633**, 33 (2006)
66. C.E. Demonchy et al., Nucl. Instrum. Methods A **573**, 145 (2007)
67. A.A. Vorobyov et al., Nucl. Instrum. Methods A **119**, 509 (1974)
68. A.A. Vorobyov et al., Nucl. Instrum. Methods A **270**, 419 (1988)
69. J.C. Lighthall et al., Nucl. Instrum. Methods A **622**, 97 (2010)
70. A.H. Wuosmaa et al., Phys. Rev. Lett. **105**, 132501 (2010)
71. G.L. Wilson, Ph.D. thesis, University of Surrey (2012). Available online at http://epubs.surrey.ac.uk/775380/
72. O.G. Tarasov et al., Nucl. Instrum. Methods B **266**, 4657 (2008)
73. A. Gavron, in *Computational Nuclear Physics 2*, ed. by K. Langanke et al. (Springer, New York, 1993), p. 108
74. W.N. Catford et al., Proc. CAARI 2002, AIP Conf. Proc. **680**, 329 (2003). http://personal.ph.surrey.ac.uk/~phs1wc/report
75. S.M. Brown et al., Phys. Rev. C **85**, 011302(R) (2012)
76. S.M. Brown, Ph.D. thesis, University of Surrey (2010). Available online at http://epubs.surrey.ac.uk/2829/
77. B. Davids, C.N. Davids, Nucl. Instrum. Methods A **544**, 565 (2005)
78. G.L. Wilson et al., J. Phys. Conf. Ser. **381**, 012097 (2012)
79. C.M. Vincent, H.T. Fortune, Phys. Rev. C **2**, 782 (1970)
80. S.G. Cooper, R. Huby, J.R. Mines, J. Phys. G, Nucl. Part. Phys. **8**, 559 (1982)
81. J. Comfort, Extended version of DWUCK4, University of Pittsburg (1979, unpublished)
82. P.D. Kunz, Code DWUCK4, University of Colorado reports COO-535-606, COO-535-613 (1978, unpublished)
83. E.I. Dolinsky, P.O. Dzhamalov, A.M. Mukhamedzhanov, Nucl. Phys. A **202**, 97 (1973)
84. B. Fernández-Domínguez et al., Phys. Rev. C **84**, 011301(R) (2011)
85. A. Ramus et al., Int. J. Mod. Phys. E **18**, 1 (2009)
86. A. Ramus, Ph.D. thesis, Université de Paris XI, 2009. See http://tel.archives-ouvertes.fr/docs/00/45/75/09/PDF/TheseRamus.pdf
87. M. Baranger, Nucl. Phys. A **149**, 225 (1970)
88. A. Signoracci, B.A. Brown, Phys. Rev. Lett. **99**, 099201 (2007)
89. B.A. Brown, W.A. Richter, Phys. Rev. C **74**, 034315 (2006)
90. J.S. Thomas, W.N. Catford et al., to be published
91. A.E. Litherland, A.J. Ferguson, Can. J. Phys. **39**, 788 (1961)
92. J.B.A. England, *Techniques in Nuclear Structure Physics, Part 2* (Macmillan, London, 1974). ISBN 0333174763
93. M.A. Eswaran, M. Ismail, N.L. Ragoowansi, Phys. Rev. **185**, 1458 (1969)
94. H.J. Rose, D.M. Brink, Rev. Mod. Phys. **39**, 306 (1967)
95. R.D. Gill, *Gamma-Ray Angular Correlations* (Academic Press, London, 1975). ISBN 0122838505
96. P.D. Stevenson, Clebsch-o-matic on-line calculator, http://personal.ph.surrey.ac.uk/~phs3ps/cleb.html
97. P.D. Bond, Comments Nucl. Part. Phys. **11**(5), 231 (1983). http://personal.ph.surrey.ac.uk/~phs1wc/report
98. D.M. Brink, Phys. Lett. B **40**, 37 (1972)
99. P.D. Bond, Phys. Rev. C **22**, 1539 (1980)
100. W.N. Catford et al., Nucl. Phys. A **503**, 263 (1989)
101. W.N. Catford et al. Letter of Intent for GASPARD at HIE-ISOLDE, http://indico.cern.ch/getFile.py/access?contribId=35&sessionId=0&resId=0&materialId=0&confId=96297
102. W. Mittig et al., http://fribusers.org/4_GATHERINGS/4_ARCHIVE/02_10/presentations/mittig.pdf
103. M. Greiser et al., Eur. Phys. J. Spec. Top. **207**, 1 (2012)
104. R.A. Zimmerman, Bob Dylan's dream, in *The Freewheelin' Bob Dylan*, Columbia (1963)

Chapter 4
Effective Field Theories of Loosely Bound Nuclei

U. van Kolck

4.1 Introduction

Exotic nuclei challenge models constructed for ordinary nuclei, which differ in their predictions for the positions of the driplines. Near these borders of the nuclear chart, a nucleus has one or more loosely bound nucleons that can easily be separated from the rest. In the simplest "halo" or "cluster" configurations, one or more clusters of tightly bound nucleons ("cores") are surrounded by a few nucleons at relatively large distances, which exceed the range of the strong force, $r_0 \sim \hbar/m_\pi c \simeq 1.4$ fm where $m_\pi \simeq 140 \, \text{MeV}/c^2$ is the pion mass. Since in classical mechanics the orbital distance is given by the range of the force, these loosely bound systems are intrinsically quantum mechanical and can display a variety of peculiar phenomena. For example, in a "Borromean" halo, the system is bound even though its subsystems are not.

Nevertheless, loosely bound systems are theoretically simpler than their more deeply bound counterparts. The reason is a fundamental "decoupling" principle according to which physics at a given distance scale is insensitive to the *details* of dynamics at much shorter distances. The short-distance dynamics can be captured instead by a finite number of parameters, whose number depends on the precision we want to achieve at the scale of interest. For the large distances characteristic of loosely bound systems, we can take the potential among constituents to be, in a first approximation, Dirac delta functions. Such a simplification means that systems with different constituents, say nucleons or atoms, can have very similar dynamics, differing only in the strength of the delta functions and the relative importance of the various possible interactions. This "universality" means that one can explain phenomena across subfields of physics using the same theoretical concepts and tools.

U. van Kolck (✉)
Institut de Physique Nucléaire, CNRS/IN2P3, Université Paris Sud, 91406 Orsay, France
e-mail: vankolck@ipno.in2p3.fr

U. van Kolck
Department of Physics, University of Arizona, Tucson, AZ 85721, USA
e-mail: vankolck@physics.arizona.edu

The decoupling principle has underlaid physics research from its beginning, and for the last thirty years or so has been formalized in the concept of effective field theories (EFTs). EFTs provide a systematic method to account for short-range dynamics even when the latter is unknown. As such, this principle can be, and of course has been, applied much more widely than the exotic nuclei of interest in this school. Even in nuclear physics, its first applications, started some twenty years ago, have been to distances scales of order r_0 [1–10]. We know that the theory of strong interactions at distances small compared to $\hbar/M_{QCD}c$, where $M_{QCD} \sim 1$ GeV/c^2 is the hadronic mass scale, is given by quantum chromodynamics (QCD), a gauge theory of quarks and gluons. At larger distances QCD is non-perturbative in its coupling constant and more easily described in terms of hadrons. An EFT—"Chiral" (or "Pionful") EFT—can be constructed at distances comparable to r_0 which includes pions and correctly incorporates the approximate symmetries of QCD such as chiral symmetry. This EFT forms the basis for a description of all nuclei, and is now the main input to the rapidly developing "*ab initio*" methods for the derivation of nuclear structure and reactions.

For distances much larger than r_0, pion exchange can be regarded as a short-range effect, and nuclear interactions reduce to delta functions and their derivatives. This "Contact" (or "Pionless") EFT is relevant for light nuclei because the two-nucleon scattering lengths—that is, the two-nucleon scattering amplitude at zero energy—are much larger than r_0, for reasons that are not well understood but we will return to. Large scattering lengths signal loosely bound states, and indeed the low-energy behavior of the lightest nuclei can be described systematically in this EFT. Universality means that with relatively small modifications this EFT can be applied to atomic systems with scattering lengths that are large compared to the Van der Waals length scale. It also means that a similar EFT—"Halo/Cluster" EFT—can be constructed for larger loosely bound nuclei, where cores are treated on the same footing as valence nucleons.

These lectures are an introduction to both the general ideas behind EFTs and the specific applications to nuclear physics. The first lecture presents the ingredients of an EFT, articulates the view of the world afforded by EFTs, and gives both classical and quantum-mechanical examples. The second lecture introduces the Chiral EFT relevant for ordinary nuclei, and describes some of its features with an emphasis on the crucial, singular character of pion exchange. In the final lecture I come to the EFTs most relevant for loosely bound systems, Pionless and Halo/Cluster EFTs. These lectures are not meant as a comprehensive review of the field, for which I refer you to, for example, Refs. [11–16]. Instead, they stress basic ideas and some of the conceptual subtleties and open problems, which are often shoved under the technical rug weaved by the many successful applications of nuclear EFTs.

4.2 Nuclear Physics Scales and Effective Field Theories

Nuclear physics has a long and venerable history. A large amount of nuclear data can be described within a picture that emerged in large part before QCD:

- nuclei are essentially made out of non-relativistic nucleons with two isospin states (protons and neutrons) of nearly equal mass $m_N \simeq 940$ MeV, which interact via a potential;
- the potential is mostly two-body, with an important one-pion-exchange component, but there is evidence for smaller three-body forces;
- isospin is a good symmetry, except for electromagnetic interactions, a sizable breaking in the two-nucleon scattering lengths, and other, smaller effects—for example, the neutron-proton mass difference is just $m_n - m_p \simeq 1.3$ MeV;
- external probes, such as photons, interact mainly with each nucleon separately, although there is evidence for smaller few-nucleon currents.

In contrast, QCD with the lightest quarks has almost opposite features:

- up and down quarks have an average ("current") mass $\bar{m} = (m_u + m_d)/2$ that is relatively small, so that they can easily be relativistic, and interact via (relativistic) gluon exchange;
- the interaction is a multi-gluon, and thus multi-quark, effect;
- isospin symmetry is not obvious since the relative mass splitting $\varepsilon = (m_d - m_u)/2\bar{m} \sim 1/3$ is not particularly small;
- external probes can interact with the collective of quarks called a hadron.

This situation automatically begs a question that is now central to the field: how does nuclear structure emerge from QCD? This is a contemporary version of a problem that has defied legions of researchers for decades: what holds the nucleus together?

In these lectures we will see how EFTs help us answer this question. The key to start tackling this problem lies on its multi-scale character. If you go through the tables of the Particle Data Book [17] you will see that hadron masses cluster—with a few notable exceptions to which we will come back in the next lecture—in the few-GeV region. This observation suggests that QCD has an intrinsic mass scale $M_{QCD} \sim 1000$ MeV/c^2. On the other hand, when you put A nucleons together to form nuclei, you find, very roughly, binding energies per nucleon $B/A \sim 10$ MeV and charge radii $\langle r^2 \rangle_{ch}^{1/2}/A^{1/3} \sim 1$ fm. This is consistent with a non-relativistic dispersion relation where the typical binding momentum is $M_{nuc}c \sim 100$ MeV/c. Thus, we face three energy scales,

$$M_{QCD}c^2 \sim m_N c^2 \simeq 1000 \text{ MeV}, \qquad M_{nuc}c^2 \sim 100 \text{ MeV},$$
$$M_{nuc}^2 c^2/M_{QCD} \sim 10 \text{ MeV}. \tag{4.1}$$

There is, of course, a very familiar multi-scale problem: the H atom, or more generally, a two-body Coulombic state with reduced mass μ. The Hamiltonian can be written in the center-of-mass frame with relative coordinate r and momentum p as

$$H = \left(\frac{p^2}{2\mu} - \frac{\alpha \hbar c}{r}\right)\left[1 + \mathcal{O}\left(\alpha; \frac{p^2}{\mu^2 c^2}; \frac{\hbar^2}{\mu^2 c^2 r^2}\right)\right], \tag{4.2}$$

where $\alpha \equiv e^2/4\pi\hbar c \simeq 1/137 \ll 1$ is the small fine-structure constant. A quick and dirty way to uncover the scales of the resulting quantum-mechanical states is to say that they are characterized by a size $r \sim r_{at}$ and a momentum $p \sim p_{at} \sim \hbar/r_{at}$, so that the energy $E \sim \hbar^2/(2\mu r_{at}^2) - \alpha \hbar c/r_{at}$, has a minimum at $r_{at} = \hbar/(\alpha \mu c)$. There are thus three energy scales, which for the H atom are

$$\mu c^2 \simeq m_e c^2 \simeq 0.5 \text{ MeV}, \qquad p_{at} c \sim \alpha \mu c^2 \simeq 3.6 \text{ keV},$$
$$p_{at}^2/(2\mu) \sim \alpha^2 \mu c^2/2 \sim 13.6 \text{ eV},$$
(4.3)

where m_e is the electron mass. As this very simple analysis shows, the three separate scales arise from the smallness of α, which also allows a controlled exploration of the effects of the corrections in Eq. (4.2).

In low-energy QCD, however, there is no obvious small coupling constant. In contrast to α, which increases (albeit slowly) as the energy scale increases, the analogous QCD fine-structure constant α_s increases as the energy scale decreases, and it becomes of $\mathcal{O}(1)$ around $M_{QCD}c^2$. Hadronic models of low-energy data have also failed to unveil any small coupling constant. So, for a controlled approach to nuclear physics we need a method to deal with multi-scale problems that does not rely on small coupling constants. Such a method is EFT.

4.2.1 Basic Ideas

An EFT is from the outset designed to address physics at the desired resolution scale, and it substitutes a small ratio of physical scales for a coupling constant as an expansion parameter. The framework evolved from the work of Weinberg, Wilson, and many others in the 60s and 70s, and was clearly articulated at the end of the 70s [18]. For a sample of introductions to EFT, see Refs. [19–22].

An EFT puts together in a single formal framework four basic ingredients, a couple of which are frequently used separately in model building:

1. **Relevant degrees of freedom.** The degrees of freedom one should use depend on the resolution we aim for. Although physics is independent of the specific choice of coordinates, some choices simplify the theoretical description significantly. Take, for example, a painting by the French Neo-Impressionist master, Georges-Pierre Seurat. Looking closely, you see that it is made of colorful blobs of paint. Yet, at the resolution scale relevant for viewing in, say, a museum, the blobs fuse together in larger-scale images, *e.g.* a face. Although we can describe any part of the painting by giving a list of blob coordinates, the same part of the painting might be more efficiently described by concepts appropriate to the scale of the larger image, *e.g.* a mustache. In other words, *choose the degrees of freedom that best fit your problem.*
2. **All possible interactions.** The effective degrees of freedom interact in all possible ways. Take another example: a system consisting of a satellite around the Earth, and the nearby Moon. Certainly, there are gravitational interactions between any two bodies of this three-body system. But there is also an indirect

effect of the Moon on the satellite motion because of the tidal deformation of the Earth. On the large distance scale set by the Earth-Moon separation ($\simeq 384$ Mm), the radii of Earth ($\simeq 6.4$ Mm) and Moon ($\simeq 1.7$ Mm) are small and they can be well approximated by points where the whole mass of the body is concentrated. In the point-like picture, the tide-mediated interaction is represented by a three-body force, which is an effect among three bodies that disappears when any of the bodies is removed. More generally, *whatever is not forbidden is compulsory*, and exists at some level of precision.

3. **Symmetries.** Symmetries play a fundamental role in physics, which is stressed in a traditional joke. There are several versions, but they all involve a cow with deficient milk production, two other scientists (typically a biologist and a chemist) who view the problem as complex, and a physicist who thinks it simple. The physicist's solution starts with "First, consider a spherical cow...". The reason this joke is well known is that it does capture what we do. In this case, if you consider two vectors (**u**, **v**) used in the description of the cow, spherical symmetry will ensure that a bilinear combination of them appears in scalar quantities in the form of the scalar product (**u** · **v**) rather than the most general combination of components (*e.g.* $u_1 v_2$). The lesson is that, thanks to symmetries, *not everything is allowed*. Of course, most of the time symmetries are only approximate, and indeed the joke often continues, "Next, we treat the head in perturbation theory...". That is, the other bilinear combinations do appear, but as long as the cow is pretty healthy, they are preceded by dimensionless parameters which are small compared to 1, and thus amenable to perturbation theory.

4. **Naturalness.** This is perhaps the most distinguishing feature of EFTs, adapted from a principle proposed by 't Hooft [23]. After relevant scales have been identified, the remaining, dimensionless parameters are $\mathcal{O}(1)$, unless suppressed by a symmetry. (Recall the cow non-sphericity.) The justification is Occam's razor: this is the simplest assumption one can make about an infinite number of interaction strengths. It is crucial for an EFT, because in the absence of large quantities, we can expect observables to be amenable to *an expansion in the small ratio between the scale characteristic of short-range physics left out of the theory and the distances of interest*—or, alternatively, the inverse of the corresponding momentum or energy scales. This assumption is, of course, to be revised when necessary. If interactions strengths are found to deviate systematically from it, it is possible that a particular scale has been left out or misidentified; after we account for it, naturalness is expected once again. Of course we might expect deviations for a finite number of interaction strengths anyway, in which case we speak of *fine-tuning* and incorporate it in a case-by-case basis.

Let me consider a simple classical example: a light object of mass m near the surface of a very large body that produces a gravitational acceleration of magnitude g. Simple experiments at energies $E \sim mgh$, where h is the height of the object, suggest that the important degree of freedom is the position, and that there is an (approximate) translation symmetry, so that the effective potential $V_{\it eff}$ is a function of h only. This description will break down at some energy, which I will call $E_{und} \equiv$

mgR. For $E \ll E_{und}$, or equivalently $h \ll R$, the most general effective potential can be written as an expansion

$$V_{eff}(h) = m \sum_{i=0}^{\infty} g_i h^i = \text{const} + mgh(1 + \eta h + \cdots), \qquad (4.4)$$

where g_i are parameters ($g_1 \equiv g$, $g_2 \equiv \eta g$) and I am neglecting quantum corrections that could give rise to non-analytic h dependence. Naturalness means

$$\frac{mg_{i+1}h^{i+1}}{mg_i h^i} = \frac{E}{E_{und}} \times \mathcal{O}(1) = \frac{h}{R} \times \mathcal{O}(1), \qquad (4.5)$$

which in turn implies

$$g_{i+1} = \mathcal{O}\left(\frac{g}{R^i}\right). \qquad (4.6)$$

If we carry out very precise experiments we can obtain values not only for g but also for η and other parameters, testing Eq. (4.6). In fact, given a desired accuracy, we need only a few terms in this expansion—in daily life only the linear one. Of course this example is somewhat artificial because we already know, thanks to Newton's apple, that, if the large object has mass M and is approximately spherical with a radius R, it produces a gravitational potential energy

$$V = -m\frac{GM}{R+h} = m\frac{GM}{R^2} \sum_{i=0}^{\infty} \left(\frac{-1}{R}\right)^{i-1} h^i. \qquad (4.7)$$

In this case

$$g_{i+1} = (-1)^i \frac{g}{R^i}, \quad g = \frac{GM}{R^2}, \qquad (4.8)$$

so that Eq. (4.6) is indeed fulfilled, and we can identify the breakdown scale R of the effective theory with the radius of the large body. Still, this simple example illustrates why naturalness is such a, well, natural assumption: it is not so easy to come up with situations where Eq. (4.6) would fail. If, say, we had misidentified the scale R by a factor of $\mathcal{O}(1)$, Eqs. (4.4) and (4.6) would still apply. Of course, in a more realistic case, the lack of exact spherical symmetry of the large body (say a mountain nearby) manifests itself in a breaking of translation symmetry. This breaking leads to a dependence of V_{eff} on the other two spatial coordinates, but the corresponding terms are relatively suppressed by powers of the ratio between a parameter encoding spherical asymmetry and the large body size—the cow raises its head again.

It is in the quantum context, however, that EFTs come to full force, because virtual processes explore all possibilities allowed by symmetries. In this context it is perhaps easiest to think in terms of path integrals. When there are various possibilities for a process, the probability of the outcome is the square of the sum of the

4 Effective Field Theories of Loosely Bound Nuclei

amplitudes for each possibility, each amplitude being proportional to the exponential of $(i/\hbar)$ times the Hamilton action. After integrating over momenta, the total amplitude is expressed in terms of trajectories $q(t)$ in coordinate space as

$$A = \int \mathscr{D}q \exp\left(\frac{i}{\hbar} \int dt \mathscr{L}(q(t))\right), \qquad (4.9)$$

where $\mathscr{L}$ is the Lagrangian (not necessarily the classical one...). The measure $\mathscr{D}q$ stands for the sum over all possible trajectories. To make it well defined, we divide the time interval for the process in a set of discrete values t_i, $i = 0, \ldots, N$, with $t_{i+1} - t_i \equiv \hbar/(\Lambda c)$. (More complicated slicings are certainly also possible.) The action becomes a sum over each time slice and the measure is the well-defined product of integrals,

$$\int \mathscr{D}q = \prod_{i=1}^{N-1} \int dq(t_i). \qquad (4.10)$$

This procedure is called regularization and the (momentum) scale Λ is referred to as the ultraviolet (UV) regulator parameter or cutoff. The classical path arises as the one that extremizes the action.

But what do we do with Λ? It is something I introduced by hand: psychology if you will, not physics. Physics is obtained from the S matrix, which is related to the scattering amplitude or the T matrix. Taking elastic two-particle scattering for simplicity, in the prototypical experiment two asymptotically free particles approach with a relative (on-shell) momentum $\mathbf{p}$ and scatter into an asymptotically free state of relative (on-shell) momentum $\mathbf{p}'$, with $|\mathbf{p}'| = |\mathbf{p}| \equiv k$ from energy conservation. The probability for the final state depends on k and the angle θ of $\mathbf{p}'$ with respect to $\mathbf{p}$. It is usually convenient to expand the θ dependence in Legendre polynomials corresponding to angular momenta $l = 0, 1, \ldots$. The coefficients $T_l(k)$ are the partial-wave amplitudes, which are usually parametrized in terms of phase shifts $\delta_l(k)$, from which the cross section can be obtained. The poles of $T_l(k)$ in the complex momentum plane can be of two types: (i) a pole in the imaginary axis, $k = i\kappa_B$, corresponding to a real ($\kappa_B > 0$) or virtual ($\kappa_B < 0$) bound state of energy $E = -B = -\kappa_B^2/2\mu + \cdots < 0$; (ii) a pair of poles elsewhere in the lower half-plane, $k = \pm\kappa_R - i\kappa_I$, $\kappa_{R,I} > 0$, representing a resonance of complex energy.

The procedure of ensuring observable quantities are independent of the regularization is called renormalization. Depending on the interactions, the most relevant paths will have structure at a certain time scale, let me call it $\hbar/Mc^2$. Certainly we want $\Lambda \gg Mc$ in order to capture the structure at this scale. But let me suppose we are interested in dynamics over a larger time scale $\hbar/mc^2$ in terms of a mass scale $m \ll M$. In this case we might be satisfied with a smaller cutoff, as long as $\Lambda \gg mc$. In the coarse-graining procedure of reducing the cutoff—called renormalization-group (RG) running—we induce errors by approximating trajectories with fine structure by a coarse set of points. We can mitigate these errors by keeping in the Lagrangian not only the positions at each slice $q(t_i)$, but also the first derivative $(dq/dt)(t_i)$ and higher derivatives, for progressively better accuracy. In general, as discussed earlier, we also might want to change our coordinates to a new, more efficient set which is some function of the older one, $\tilde{q}(t_i) = f(q(t_i))$. This is accomplished by introducing

$$1 = \prod_i \int d\tilde{q}(t_i)\delta\big(\tilde{q}(t_i) - f\big(q(t_i)\big)\big) \equiv \int \mathscr{D}\tilde{q}\,\delta\big[\tilde{q}(t) - f_\Lambda\big(q(t)\big)\big] \qquad (4.11)$$

in Eq. (4.9). Inverting the order of integrals we arrive at

$$A = \int \mathscr{D}\tilde{q}\,\exp\left(\frac{i}{\hbar}\int dt\,\mathscr{L}_{\it eff}\big(\tilde{q}(t)\big)\right), \qquad (4.12)$$

where, schematically, the effective Lagrangian is

$$\mathscr{L}_{\it eff}(\tilde{q}) = \sum_{n,d,m} c_{ndm}(M,\Lambda) O\big(\tilde{q}^n, \big(d^d\tilde{q}/dt^d\big)^m\big). \qquad (4.13)$$

Here O represents a combination of various powers of the new coordinate and its derivatives at the same instant, since we cannot resolve time intervals $\lesssim \hbar/(\Lambda c)$ because of the uncertainty principle. The respective coefficient, c_{ndm}, is called a low-energy constant (LEC), and depends in general not only on the underlying dynamics at scale M but also on the regulator Λ.

In principle the effective Lagrangian can be obtained from the integral over the original coordinates, but regardless of our ability to do so, the path integral (4.12) forms the basis for the effective theory. From it we can obtain the T matrices for various low-energy processes, the goal of effective theory being to write each in a controlled expansion. Again taking a simple two-body elastic scattering for illustration, we want that, for $k \sim mc$,

$$T_l(k) = \sum_{\nu \geq \nu_{min}}^\infty \tilde{c}_\nu(M, \Lambda)\left(\frac{k}{Mc}\right)^\nu F_{l;\nu}\left(\frac{k}{mc}; \frac{k}{\Lambda}\right), \qquad (4.14)$$

where ν is a counting index with a minimum value ν_{min}, the new coefficients $\tilde{c}_\nu$ are related to the c_{ndm} appearing in the Lagrangian, and the $F_{l;\nu}$ are calculable functions of the light scales and the cutoff, which are obtained by solving the dynamical equations (like the Schrödinger or Lippmann-Schwinger equations) of the theory. We refer to the terms with $\nu = \nu_{min}$ as leading order (LO), $\nu = \nu_{min} + 1$ as next-to-leading order (NLO), and so on. Since Λ is arbitrary, the coefficients $\tilde{c}_\nu$ (and thus the c_{ndm}) have to be such as to ensure "RG invariance",

$$\frac{dT_l(k)}{d\Lambda} = 0. \qquad (4.15)$$

The relation between ν and the labels (n, d, m) of the expansion (4.13) is called power counting. Frequently the least trivial aspect of an effective theory, power counting is necessary for any predictive power. A truncation of Eq. (4.14) guarantees that only a finite number of LECs appear, but in principle introduces regularization

4 Effective Field Theories of Loosely Bound Nuclei

errors. These errors will be relatively small as long as they scale as inverse powers of k/Λ,

$$T_l(k) = T_l^{(\bar{\nu})}(k)\left[1 + \mathscr{O}\left(\frac{k}{Mc}, \frac{k}{\Lambda}\right)\right], \qquad \frac{\Lambda}{T_l^{(\bar{\nu})}(k)}\frac{dT_l^{(\bar{\nu})}(k)}{d\Lambda} = \mathscr{O}\left(\frac{k}{\Lambda}\right), \quad (4.16)$$

with $\bar{\nu}$ denoting the chosen truncation. For this to happen, there need to be enough LECs at each order to remove non-negative powers of Λ, otherwise the power counting is not consistent (with the RG). Once we have ensured that errors scale appropriately, we want at least $\Lambda \gg mc$ and, optimally, $\Lambda \gtrsim Mc$, as in the latter case the regularization errors are no larger than the errors coming from the incomplete accounting of short-range physics. In fact, as long as regulator errors come in the form (4.16) one can use a variation of Λ from Mc to very large values as an estimate of the full truncation error.

So far I have presented the ideas of EFT without invoking the notion of "field" directly, although historically they were first formalized in terms of relativistic quantum fields. The path $q(t)$ can be thought as a field over time, but closer contact with field theory arises if we "second quantize" the system, by elevating the wavefunction ψ to an operator. In the path integral formulation above one replaces

$$q(t) \to \psi(\mathbf{r},t), \psi^*(\mathbf{r},t), \qquad t \to \mathbf{r},t, \qquad dt \to d\mathbf{r}dt, \quad (4.17)$$

so ψ is now a non-relativistic field over the four spacetime coordinates $\mathbf{r},t$. Relativity can be introduced by enforcing $SO(3,1)$ invariance, which is most easily accomplished by employing a field with definite transformation properties under the Lorentz group. We will return to the connection between relativistic and non-relativistic field theories in the next subsection. From now on I use units where $\hbar = 1$ and $c = 1$, so the dimensions of mass, energy, momentum, inverse position, and inverse time are all the same.

The EFT Lagrangian has a form similar to Eq. (4.13), which includes terms with an arbitrary number N of fields and their derivatives. It is usually convenient to discuss quantum field theory starting from an expansion around the free theory and, in non-perturbative situations, "resum" parts of this expansion. The free terms define the "canonical (mass) dimension" of the fields. With our choice of units, the action is dimensionless so each term in the Lagrangian has mass dimension 4. If an operator O has canonical dimension D, the LEC has mass dimension $4 - D$. The EFT involves interactions with arbitrary D. The terms in the expansion of the exponential of the action have a correspondence to more intuitive Feynman diagrams, which can be thought as representing the propagation of particles interspaced with their interactions. The former is represented by lines ("propagators") carrying a four-momentum and the latter by line intersections ("vertices") where four-momentum is conserved, each associated with a specific factor. A closed loop implies a free four-momentum that is integrated over. The corresponding integrals are well-defined in general only after regularization, which removes the contributions from high momenta at the cost of the arbitrary UV regulator parameter Λ. Details can be supplied by a good book on quantum field theory, such as Ref. [24].

In these terms, the recipe for an EFT is essentially:

1. identify the relevant degrees of freedom (represented by fields) and symmetries (groups of discrete or continuous, global or local transformations);
2. construct the most general Lagrangian with these ingredients;
3. postulate a power counting to truncate physical amplitudes;
4. run the methods of field theory to calculate amplitudes, that is, compute Feynman diagrams for momenta $Q < \Lambda$ (regularization) and relate the LECs to observables so that the latter are independent of Λ (renormalization);
5. if this is achieved and the now well-defined expansions are well behaved, declare victory; otherwise, return to step 3, or earlier if necessary.

Renormalization in step 4 is crucial. Short-range physics appears in both the high-momentum components of loops and in the LECs, and changes in Λ merely shuffle it from one to the other. RG invariance effectively guarantees that the relevant momenta in diagrams are set by the external momenta, so that, as long as the external momenta are relatively small, successive terms in the expansions of the various amplitudes will be smaller and smaller.

The observables used as input in step 4 can be experimental data or the result of a calculation in the underlying theory, if the latter is known and can be solved in the low-energy domain of the EFT. In this case we speak of matching the EFT to the underlying theory. When the EFT shares the symmetries of the underlying theory this matching must be possible, if one accepts Weinberg's "theorem" [18]:

> The quantum field theory generated by the most general Lagrangian with some assumed symmetries produces the most general S matrix incorporating quantum mechanics, Lorentz invariance, unitarity, cluster decomposition and those symmetries, with no further physical content.

This "theorem" has not been proved in general but to my knowledge no counterexamples are known. In specific examples, such as the one in the next subsection, one can verify that this "theorem" works.

Weinberg's "theorem" embodies the most important difference between an EFT and simple models. Models can be very useful in guiding step 1 (and sometimes 3), but they fail to be fully consistent with all the low-energy consequences of the underlying theory, which is only guaranteed with step 2. Sometimes models do contain interactions with arbitrarily many derivatives in the form of *ah hoc* "form factors" attached to vertices with otherwise no (or few) derivatives. These form factors, involving a finite number of parameters (usually one), play the dual role of representing a physical effect (the coordinate profile of a particle) and of regulating integrals. In EFT the physics of particle structure is represented by the higher-derivative interactions, each with its own LEC, while the infinities in integrals are avoided with an unphysical regulator. (Without step 4, the regulator would become indistinguishable from an universal form factor and observables would depend on Λ in addition to the infinite number of LECs that parametrize the most general Lagrangian.) A model of the underlying theory, if it includes the right symmetries, can also be represented at low energies by the EFT, with specific relations between the LECs given by the limited number of parameters of the model (including form-factor parameters, if any).

But, as long as RG invariance has been achieved, the arbitrarily higher-derivative interactions ensure that the EFT represents *all* models with the right symmetries. When the underlying theory is not known, or cannot be solved, EFT provides a model-independent approach to data.

Even when the matching of the EFT to the underlying theory can be done, there are advantages in using the EFT at low energies, because it is inefficient to keep explicit in the theory very massive degrees of freedom instead of emergent, low-mass states. No progress in nuclear theory will make it preferable to calculate atomic and molecular properties directly from a collection of protons and neutrons, rather than from a point nucleus with a few relevant parameters (which in turn can be calculated from a collection of protons and neutrons). The heavy degrees of freedom are "integrated out", their contribution subsumed in the LECs.

If the EFT is natural, one expects no cancellations between the bare LECs and the high-momentum components of loops, at least for changes in Λ of $\mathcal{O}(1)$. Now, loops typically come with factors of 4π. For a LEC c of an operator of canonical dimension D involving N fields, it is convenient to define a "reduced LEC" that is dimensionless and includes some factors of 4π,

$$c_R \equiv M^{D-4} c / (4\pi)^{N-2}. \tag{4.18}$$

It can be inferred [25, 26] from examples based on perturbative matching that c_R is usually of the order of the product of reduced couplings of the underlying theory that generate it. Examples of this "naive dimensional analysis" (NDA) will be given later, starting in the next subsection. There is no guarantee that it will always work, but it is usually at least a good first guess on which to base a power counting.

For an effective field theorist, nature has the onion-like structure of a sequence of EFTs ordered according to energy (or inverse distance) scale. (Whether this sequence ends at some high energy is metaphysics.) Thus, EFT provides a framework for both reductionism and emergence in physical theories. This and other philosophical implications of EFT are lucidly discussed in Refs. [27, 28].

In the remaining lectures I will focus on the EFT at a few GeV and its strong-interaction sector as a starting point for traditional nuclear physics. It will prove useful, however, to spend some time in the next subsection with a simpler EFT, which provides an archetype for the approach we will follow in nuclear physics.

4.2.2 An Example: NRQED

Let me now return to atoms. At a momentum scale comparable to the electron mass, the relevant degrees of freedom are electrons and nuclei interacting through the exchange of photons, as given by quantum electrodynamics (QED). As we have seen above, atomic bound states exist at a much smaller momentum scale $p_{at} \sim \alpha m_e$. Now I discuss in qualitative terms an EFT tailored to this scale [29], termed Non-Relativistic QED (NRQED) because even electrons move slowly. I will stress the features that find similar expression in the Chiral EFT of nuclei in the next lecture.

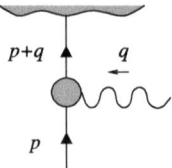

Fig. 4.1 Part of a Feynman diagram where an on-shell fermion of momentum p first interacts with a real or virtual photon of incoming momentum q, then propagates with momentum $p+q$. A fermion (photon) is denoted by a *solid* (*wavy*) *line*. The *shaded areas* represent parts of the diagram whose details are not important now

For simplicity, let me take at the higher scale a single spin-1/2 fermion represented by a Dirac field ψ of mass m and charge $Q_\psi e$, interacting with a spin-1 boson A_μ, subjected to Lorentz, parity, time-reversal, and $U(1)$ gauge invariance. As it is well known, the latter is most easily enforced using gauge-covariant derivatives:

$$D_\mu \psi = (\partial_\mu + ieQ_\psi A_\mu)\psi, \qquad F_{\mu\nu} = \partial_\mu A_\nu - \partial_\nu A_\mu. \tag{4.19}$$

The underlying Lagrangian is thus

$$\mathscr{L} = -\frac{1}{4} F_{\mu\nu} F^{\mu\nu} + \bar{\psi}(i\slashed{D} - m)\psi + \cdots, \tag{4.20}$$

where $\bar{\psi} = \psi^\dagger \gamma_0$ and $\slashed{D} = \gamma_\mu D^\mu$ in terms of the Dirac matrices γ_μ. From the terms shown explicitly in this Lagrangian we can read off the fermion and gauge-boson propagators and an interaction vertex, as discussed in textbooks [24]. More-derivative interactions, represented by the "...", give contributions suppressed by powers of momentum over the mass scale of the physics we have integrated out (such as heavier fermions and weak-gauge bosons).

If we are only interested in processes with external momenta $Q \ll m$, we can consider an additional expansion in Q/m. Take the fermion propagator after the fermion with momentum p is kicked by a photon of momentum q, see Fig. 4.1. If $|\mathbf{p}| \sim |\mathbf{q}| = \mathcal{O}(Q)$, then $q^0 \sim |\mathbf{q}| = \mathcal{O}(Q)$ but $p^0 = \sqrt{|\mathbf{p}|^2 + m^2} = m + \mathcal{O}(Q^2/m)$. The propagator can then be written

$$\frac{i}{\slashed{p} + \slashed{q} - m + i\varepsilon} = \frac{i(p^0 \gamma^0 + m - \mathbf{p}\cdot\boldsymbol{\gamma} + \slashed{q})}{2p^0 q^0 + q^{02} - 2\mathbf{p}\cdot\mathbf{q} - \mathbf{q}^2 + i\varepsilon} = \frac{i}{q^0 + i\varepsilon} P_+ + \cdots, \tag{4.21}$$

where in the last line I introduced one of the projectors onto positive/negative energy states,

$$P_\pm \equiv \frac{1 \pm \gamma^0}{2}, \qquad P_\pm P_\pm = P_\pm, \qquad P_\pm P_\mp = 0. \tag{4.22}$$

This represents, in a first approximation, a static *two*-component fermion propagating forward in time, as can be seen from a Fourier transformation. To capture the importance of this limited set of degrees of freedom it is convenient to split the field into two two-component "heavy fermion" fields [30]

$$\Psi_\pm \equiv e^{imt} P_\pm \psi, \qquad \psi = e^{-imt}(\Psi_+ + \Psi_-). \tag{4.23}$$

The effective Lagrangian then employs the relevant degrees of freedom embodied in Ψ_+ instead of the full ψ. It can be obtained [31] by substituting Eq. (4.23) into Eq. (4.20), integrating over the Ψ_- field in the path integral as well as the high-momentum components of Ψ_+ and A_μ, and expanding in powers of $1/m$. With an appropriate redefinition of A_μ to keep its bilinear form unchanged, and rewriting Ψ_+ as a Pauli spinor Ψ (with $\overline{\Psi} \equiv \Psi^\dagger$), the result is

$$\mathscr{L}_{NRQED} = -\frac{1}{4} F_{\mu\nu} F^{\mu\nu} + \frac{\beta_0}{m^4} (F_{\mu\nu} F^{\mu\nu})^2 + \frac{\beta_1}{m^4} (F_{\mu\nu} \tilde{F}^{\mu\nu})^2 + \cdots$$
$$+ \overline{\Psi} i D_0 \Psi + \frac{1}{2m} \overline{\Psi} \mathbf{D}^2 \Psi + \frac{e}{2m}(Q_\psi + \kappa) \overline{\Psi} \sigma_i \Psi \tilde{F}^{0i} + \cdots$$
$$+ \mathscr{L}_{f \geq 2} \tag{4.24}$$

where $\tilde{F}_{\mu\nu} = \varepsilon_{\mu\nu\alpha\beta} F^{\alpha\beta}/2$, σ_i are the Pauli spin matrices, $\beta_{0,1}$ and κ are dimensionless LECs that can be obtained from the parameters appearing in Eq. (4.20), and $\mathscr{L}_{f \geq 2}$ involves four or more fermion fields, as discussed below. As one would expect, this is simply the most general Lagrangian built out of Ψ_+ and A_μ and subjected to the assumed symmetries. Invoking Weinberg's "theorem" and our experience that for $Q \ll m$ the appropriate fermion degree of freedom is a bi-spinor, one can write this Lagrangian directly, without explicitly performing a path integration. In this case, the LECs are obtained by the matching of physical amplitudes.

Even though the Lagrangian is more complicated, the structure of the EFT is much simpler than that of the underlying theory. The absence of Ψ_- implies that there is no explicit pair creation in the EFT, which is an effect of range $\sim 1/(2m)$ that is absorbed in the LECs. Fermion lines just go through Feynman diagrams of the EFT. As a consequence, operators with $2f$ fermion fields do not contribute to processes that involve less than f incoming fermions. We can tackle the various sectors of the theory successively by increasing f, which is particularly important for the treatment of non-perturbative physics for $f \geq 2$. In the relativistic theory the few-body problem is instead a many-body problem.

The $f = 0$ sector of Eq. (4.24) is the Euler-Heisenberg Lagrangian, one of the earliest examples of EFT ideas, which is the basis for studies of low-energy processes involving photon fields alone, such as light-by-light scattering. A remarkable feature of Eq. (4.24) is that it contains photon self-interactions already at tree level. The most important such interactions have four derivatives and thus by NDA are expected to be inversely proportional to m^{-4}. The LECs $\beta_{0,1}$ originate in fermion-loop diagrams of the underlying theory and would be $\mathscr{O}(1)$ if α were $\mathscr{O}(1)$. Since $\alpha \ll 1$, we can match the EFT to the underlying theory in perturbation theory, obtaining $\beta_{0,1}$ in an expansion in powers of α. The most important diagram has a single fermion loop with four photon lines attached. Since each vertex is $\propto e$ and the loop typically brings in a $(4\pi)^{-2}$, we expect $\beta_{0,1} = \mathcal{O}(\alpha^2)$. Alternatively we can use Eq. (4.18). The reduced charge is $e_R = e/(4\pi)$ and for $c = \beta_{0,1}/m^4$ we expect $c_R = m^4 c/(4\pi)^2 = \mathcal{O}(e_R^4)$, from which, again, $\beta_{0,1} = \mathcal{O}(\alpha^2)$. One can similarly write and calculate the LECs of higher-derivative terms. They may contain further powers of α, but are additionally suppressed by at least $(Q/m)^2$ in processes with typical external momentum Q. These LECs are of course needed to ensure renormalizability, Eq. (4.16), in light-by-light scattering at loop level. An explicit comparison between EFT and QED amplitudes can be found, for example, in Ref. [32].

In the $f = 1$ sector, we see that the exponential in Eq. (4.23) removes from the evolution of the fermion field the relatively large and inert mass m, leaving a dispersion relation of the non-relativistic, "residual" form $p^0 = \mathbf{p}^2/(2m) + \cdots$. The kinetic part has a static piece $\overline{\Psi} i D_0 \Psi$, a recoil correction, and, with more derivatives, relativistic corrections. The strengths of these terms are not arbitrary, but fixed by the mass. If we neglected the "..." in Eq. (4.20) and loop corrections, the same would be true for the $\overline{\Psi} \sigma_i \Psi \tilde{F}^{0i}$ interaction, or "Pauli term", which represents the interaction of the magnetic dipole moment with a magnetic field. The existence of further physics and of radiative corrections in the underlying theory leads to an "anomalous" magnetic moment κ, which again would be expected to be of $\mathcal{O}(1)$ if α were $\mathcal{O}(1)$. The contributions from loops to κ can be estimated again using Eq. (4.18): for $c = e\kappa/m$, $c_R = mc/(4\pi) = \mathcal{O}(e_R^3) = \mathcal{O}(e^3/(4\pi)^3)$, which gives $\kappa = \mathcal{O}(\alpha/(4\pi))$, in agreement with the classic Schwinger result $\kappa = \alpha/(2\pi) + \cdots$ for the electron.

The fact that some coefficients are entirely determined by m applies also to terms contained in the "..." of Eq. (4.24), and is merely a consequence of Lorentz invariance. Despite its non-relativistic appearance, Eq. (4.24) does respect Lorentz invariance, but in a Q/m expansion. If we start directly with Eq. (4.24), rather than integrating degrees of freedom out of a manifestly Lorentz-invariant Lagrangian, Lorentz invariance can be implemented by demanding "reparametrization invariance" [33], namely that the Lagrangian should be invariant under small changes in the velocity of the heavy fermion. This method has the advantage of applying also when the relativistic version of the theory does not have a well-defined expansion and/or when the explicit integration cannot be performed explicitly. A recent discussion of the NRQED Lagrangian can be found in Ref. [34].

With the $f = 0, 1$ sectors of the Lagrangian we can study low-energy processes in which a number of photons scatter from a slow-moving fermion. As we have seen, for probe energies comparable to the fermion three-momentum, the fermion is, in a first approximation, static. We are close to the limit $m \to \infty$, where the spin operator does not appear in the Lagrangian (4.24) and there is an $SU(2)$ symmetry of rotations in spin space. This is an example of an "accidental" symmetry, a symmetry that emerges at LO in the EFT.

The simplest process is Compton scattering. In the Coulomb gauge $A_0 = 0$, the most important contribution comes from the "seagull" diagram stemming from the two $\mathbf{A}$ fields contained in $\overline{\Psi} \mathbf{D}^2 \Psi$; the corresponding contribution to the amplitude, $\propto 1/m$, is nothing but the spin-independent "Thomson amplitude". In the underlying theory, this term arises from the transition to negative-energy states. One can go on

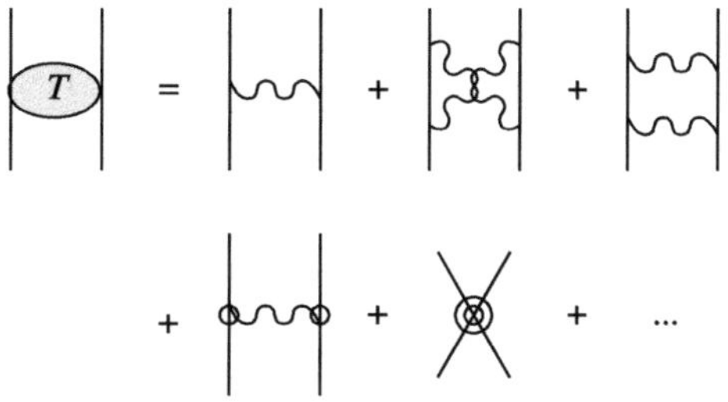

Fig. 4.2 Diagrams representing the T matrix for elastic scattering of two heavy fermions in NRQED. A *circle* at the vertex denotes an inverse power of the heavy mass. Other notation as in Fig. 4.1

and calculate higher-order terms in the combined Q/m and α expansions using the interactions in Eq. (4.24). For details at a pedagogical level, see Ref. [35].

Let me now turn to the $f \geq 2$ sectors, more germane to my goal of tackling nuclear bound states later. It is here that EFT becomes particularly useful, because the formulation of the non-perturbative problem is exceedingly complicated in the underlying theory, where we retain contributions of momentum comparable to the fermion mass. But, as we have seen, such momenta are also not very relevant for electromagnetic bound states! Here I will emphasize qualitative aspects of the problem, omitting for example a discussion of infrared divergences. For a fuller account in the case of positronium, for example, see Ref. [36].

So I consider the T matrix for the elastic scattering of two of our heavy fermions in the center-of-mass frame, where both initial and final relative three-momenta are, respectively, $|\mathbf{p}| \sim |\mathbf{p}'| = \mathcal{O}(Q)$. I will denote the transferred momentum by $\mathbf{q} \equiv \mathbf{p} - \mathbf{p}'$. See Fig. 4.2.

The simplest diagram one can draw represents the exchange of a single photon, where the photon-fermion interaction comes from the $\bar{\Psi} i D_0 \Psi$ term in Eq. (4.24). It gives

$$T_{1\gamma} = \frac{Q_\psi^2 e^2}{(p^0 - p'^0)^2 - (\mathbf{p} - \mathbf{p}')^2 + i\varepsilon} \simeq -\frac{Q_\psi^2 e^2}{\mathbf{q}^2 - i\varepsilon}\left(1 + \mathcal{O}(Q^2/m^2)\right). \quad (4.25)$$

The dominant term above, which has magnitude $\mathcal{O}(4\pi\alpha Q_\psi^2/Q^2)$, is, as we are going to see shortly, just a fancy way to generate the Coulomb potential. Of course there are other one-photon-exchange diagrams with derivatives at the vertices, which start at relative order $\mathcal{O}(Q^2/m^2)$ just like the non-static corrections in Eq. (4.25). An example comes from magnetic interactions at both vertices, which gives a dipole-dipole interaction of contact form because momenta from the vertices and propagator cancel out.

In quantum field theory, nothing forbids the exchange of more than one photon. For example, each fermion can emit a photon which is subsequently absorbed by the other fermion, forming a one-loop "crossed-box" diagram. According to the Feynman rules, the corresponding contribution is

$T_{2\gamma \times}$

$$= -iQ_\psi^4 e^4 \int \frac{d^4l}{(2\pi)^4} \frac{1}{l^0 + p^0 - (\mathbf{l}+\mathbf{p})^2/2m + i\varepsilon} \frac{1}{l^0 + p'^0 - (\mathbf{l}-\mathbf{p}')^2/2m + i\varepsilon}$$

$$\times \frac{1}{(p^0 - p'^0 + l^0)^2 - (\mathbf{p} - \mathbf{p}' + \mathbf{l})^2 + i\varepsilon} \frac{1}{l^{02} - \mathbf{l}^2 + i\varepsilon}$$

$$= Q_\psi^4 e^4 \int \frac{d^3l}{(2\pi)^3} \frac{1}{|\mathbf{l}| - p^0 + (\mathbf{l} + \mathbf{p})^2/2m - i\varepsilon} \frac{1}{|\mathbf{l}| - p'^0 + (\mathbf{l} - \mathbf{p}')^2/2m - i\varepsilon}$$

$$\times \frac{1}{(p^0 - p'^0 - |\mathbf{l}|)^2 - (\mathbf{p} - \mathbf{p}' + \mathbf{l})^2 + i\varepsilon} \frac{1}{2|\mathbf{l}| - i\varepsilon} + \cdots. \qquad (4.26)$$

Here I integrated over the zeroth component of the loop momentum using contour integration. Closing the contour on the upper plane, we get two contributions from the poles in the photon propagators, only one of which I show explicitly—the other, of a similar form, is in the "...". Because the three-momentum scale is Q, these poles lie typically a distance Q from the origin in the l^0 complex plane. As a consequence, the most important terms in the denominators after the first integration involve $|\mathbf{l}|$ and $|\mathbf{q} + \mathbf{l}|$, which, in particular, implies static fermion propagators as for $f = 1$. The propagators and integration measure typically contribute, respectively, $\mathcal{O}(Q^{-5})$ and $\mathcal{O}(Q^3/(4\pi)^2)$ to the final result, which is then $\mathcal{O}((Q_\psi^2 \alpha/4\pi)(4\pi\alpha Q_\psi^2/Q^2))$. Just as it could have been expected from experience with $f = 0, 1$ diagrams, the result is smaller than one-photon exchange by $\mathcal{O}(Q_\psi^2 \alpha/4\pi)$.

One might further expect that the smallness of α implies that all other diagrams are small and amenable to perturbation theory. However, if that were true there would be no electromagnetic bound states. So, how can a bound state arise in a weakly coupled theory? More generally, what makes the problem non-perturbative? (Since I am considering explicitly only one type of fermion, there is obviously no electromagnetic bound state because the Coulomb interaction is repulsive, but still perturbation theory fails at low energies.)

The reason is apparent already when we consider the other two-photon exchange diagram, the "box", in which the photons are exchanged sequentially. In this case there are subtle differences compared to the crossed-box diagram due to a different routing of the loop momenta in one of the fermion propagators:

$$T_{2\gamma\square}$$

$$= -i Q_\psi^4 e^4 \int \frac{d^4l}{(2\pi)^4} \frac{1}{l^0 + p^0 - (\mathbf{l} + \mathbf{p})^2/2m + i\varepsilon} \frac{1}{-l^0 + p^0 - (\mathbf{l} + \mathbf{p})^2/2m + i\varepsilon}$$

$$\times \frac{1}{(p^0 - p'^0 + l^0)^2 - (\mathbf{p} - \mathbf{p}' + \mathbf{l})^2 + i\varepsilon} \frac{1}{l^{02} - \mathbf{l}^2 + i\varepsilon}$$

$$= Q_\psi^4 e^4 \int \frac{d^3l}{(2\pi)^3} \frac{1}{-2p^0 + (\mathbf{l}+\mathbf{p})^2/m - i\varepsilon} \frac{1}{p^0 - (\mathbf{l}+\mathbf{p})^2/2m - \mathbf{l}^2 + i\varepsilon}$$

$$\times \frac{1}{(2p^0 - p'^0 - (\mathbf{l}+\mathbf{p})^2/2m)^2 - (\mathbf{p} - \mathbf{p}' + \mathbf{l})^2 + i\varepsilon} + \cdots. \qquad (4.27)$$

Because of the different signs, closing the contour on the upper half-plane now involves a third pole, which stems from one of the fermion propagators and lies only a distance Q^2/m from the origin. The contribution from this pole is the one displayed explicitly above, the other two being relegated to the "...". You should convince yourself that these other two contributions are similar to Eq. (4.26), and thus also small by $\mathcal{O}(Q_\psi^2 \alpha/4\pi)$ compared to one-photon exchange. The contribution from the fermion-propagator pole, on the other hand, is larger because the remaining fermion propagator contains only the difference between small fermion kinetic energies, and is thus $\mathcal{O}(Q^2/m)$. We refer to this as an infrared enhancement, as it becomes more pronounced as Q decreases. Note that the static approximation is no longer good, although of course relativistic corrections remain small. In the photon propagators, on the other hand, the kinetic energies can still be neglected in a first approximation, so that as before each photon denominator is $\mathcal{O}(Q)$. What we have here are simply two sequential Coulomb photon exchanges separated by the usual non-relativistic two-fermion propagation: it is the iteration of one-photon exchange. Such a non-relativistic integral over the three-momentum typically contributes $\mathcal{O}(Q^3/(4\pi))$, an extra 4π compared to integrals not originating in the heavy fermion propagator poles. The size of this term is then $\mathcal{O}(Q_\psi^2 \alpha m/Q)$ compared to one-photon exchange.

This argument can be generalized to diagrams with more photon exchanges, and even more fermions. It is convenient to introduce "old-fashioned" time-ordered perturbation-theory diagrams, which represent contributions to amplitudes *after* integration over the zeroth-component of loop momenta, as in the last lines in Eqs. (4.26) and (4.27). In this case vertices are drawn in a specific time order, intermediate states are associated with energy differences, and loops represent three-momentum integrations. The one-photon-exchange Feynman diagram becomes two time-ordered diagrams depending on which fermion first emits the photon. The crossed-box Feynman diagram becomes four time-ordered diagrams, all of the same crossed-box type. The box Feynman diagram becomes six time-ordered diagrams: two "stretched-box" diagrams where at any time there is a photon "in the air", representing the "..." in Eq. (4.27), and four "true-box" diagrams where there is an intermediate state without photons, representing once-iterated one-photon exchange. In this language, we define "reducible" diagrams as those that contain intermediate states with fermions only, and are thus infrared enhanced. These intermediate states contribute $\mathcal{O}(mQ/(4\pi))$ to the amplitude. The potential is defined as (minus) the sum of irreducible diagrams.

We expect the potential to have a simple perturbative expansion, which in the case considered here starts with the one-photon exchange of $\mathcal{O}(4\pi\alpha Q_\psi^2/Q^2)$. Fourier transforming Eq. (4.25), we obtain, indeed, the Coulomb potential in coordinate space. Each extra iteration of this potential adds a factor $\mathcal{O}(Q_\psi^2 \alpha m/Q)$, as seen above. When $Q \sim Q_\psi^2 \alpha m$, all iterations are equally important, and the amplitude is given by an integral equation, the Lippmann-Schwinger equation—which, in turn, can be shown to be equivalent to the Schrödinger equation. Schematically the LO amplitude is

$$T^{(0)} \sim \frac{4\pi\alpha Q_\psi^2}{Q^2}\left[1 - \mathscr{O}\left(\frac{Q_\psi^2 \alpha m}{Q}\right)\right]^{-1}, \qquad (4.28)$$

which can have a pole at $Q \sim |Q_\psi^2|\alpha m$, corresponding to energies $|E| \sim Q_\psi^4 \alpha^2 m$. If we were considering two fermions with opposite charge, say $Q_\psi^2 \to -1$, we would expect a bound state with binding momentum $p_{at} \sim \alpha m$ and energy $\sim p_{at}^2/m \sim \alpha^2/m$—just the back-of-the-envelope estimate given at the beginning of this lecture.

But NRQED gives also a way to systematically go beyond LO. First, it allows us to calculate corrections to the Coulomb potential, such as the Q/m ($\sim Q_\psi^2 \alpha$ in the bound state) corrections to one-photon exchange and the irreducible two-photon exchange we discussed above. Second, it tells us that these corrections to the potential are also corrections in the scattering amplitude, which should be calculated in perturbation theory on top of the LO ("distorted-wave perturbation theory").

Note that because each iteration of one-photon exchange scales with a negative power of Q, the LO amplitude (4.28) satisfies the RG condition (4.16) without invoking contact interactions in $\mathscr{L}_{f \geq 2}$. However, such four- and more-fermion terms must exist because they are allowed by the symmetries. And indeed, as the order increases and the corrections to the potential become more singular (since they have more powers of momentum or, alternatively, inverse distance), loop diagrams will bring in new cutoff dependence, which can only be compensated by the LECs. As for $f = 0, 1$, the LECs with $f \geq 2$ can in principle be obtained from a fully perturbative matching calculation in the window $m \gg Q \gg Q_\psi^2 \alpha m$.

The total spin of two fermions can be $s = 0, 1$ depending on the individual spins being antiparallel or parallel. It is useful to think of the projectors onto spin s,

$$P_s \equiv \frac{1}{4}\left[2s + 1 + (2s - 1)\boldsymbol{\sigma}_1 \cdot \boldsymbol{\sigma}_2\right], \qquad P_s P_{s'} = \delta_{ss'} P_s, \qquad (4.29)$$

where $\boldsymbol{\sigma}_i/2$ is the spin of fermion i. Because I am considering a single two-state fermion, the Pauli principle ensures that an S-wave two-fermion contact interaction can only contribute when the spins are antiparallel. In other words, there is only one four-fermion interaction without derivatives, corresponding to P_0. The many-fermion Lagrangian has thus the form

$$\mathscr{L}_{f \geq 2} = \frac{\gamma}{4m^2}(\overline{\Psi}\Psi\overline{\Psi}\Psi - \overline{\Psi}\boldsymbol{\sigma}\Psi \cdot \overline{\Psi}\boldsymbol{\sigma}\Psi) + \cdots, \qquad (4.30)$$

where γ is a LEC. NDA for $c = \gamma/m^2$ gives, assuming it accounts for photon exchange at momenta $\gtrsim m$, $c_R = m^2 c/(4\pi)^2 = \mathscr{O}(e_R^2)$ or $\gamma = \mathscr{O}(4\pi\alpha)$, suggesting a relative $\mathscr{O}(Q^2/m^2)$ effect in the two-fermion scattering amplitude. The "..." in Eq. (4.30) include not only more-derivative four-fermion interactions but also six- and more-fermion interactions, which because of their larger canonical dimensions are expected to lead to contributions suppressed by powers of Q/m. Again thanks to the Pauli principle, for a two-state fermion there are no six- or more-fermion interactions without derivatives, which means three- and more-body forces are very small.

One can now consider arbitrary processes involving two or more fermions—in particular, in a bound state—and external low-momentum photons, for example *Bremsstrahlung*. NRQED provides the framework to describe atomic and molecular physics, and indeed state-of-the-art calculations are carried out within this framework (Ref. [37] is but one example). I will now set up an analogous framework for nuclear physics.

4.2.3 Summary

Nuclear systems involve multiple scales but no obvious small coupling constant. EFT is a *general* framework to deal with multi-scale problems using small ratios of scales as expansion parameters. Applied to low-energy QED, EFT reproduces some well-known results but also provides a systematic expansion for scattering amplitudes.

4.3 QCD at Low Energies

For the rest of these lectures I apply the EFT framework to nuclear systems. I start with the Standard Model (SM) of particle physics at a few GeV and construct the EFT relevant for momenta $\mathcal{O}(M_{nuc})$, which should form the basis for a description of typical nuclei. This Chiral EFT has indeed become the starting point for the rapidly developing "*ab initio*" methods that harness ever-growing computational power for the solution of the nuclear dynamics from interactions determined in few-nucleon systems. However, as we are going to see in some detail, some basic issues in this EFT, related to the impact of the renormalization of singular potentials on power counting, are still not fully understood.

4.3.1 Building Blocks

The SM has been repeatedly validated, particularly at scales above a few GeV where most processes can be studied with relatively little reliance on non-perturbative physics. The successes of the SM can be understood if it is considered as an EFT at energies around 100 GeV. It is constructed (in its minimal version) out of quarks, leptons, gauge bosons and a Higgs boson subjected to an $SU(3)_c \times SU(2)_L \times U(1)_Y$ gauge symmetry. All interactions of canonical dimension up to four have now been established directly or indirectly, and the only allowed dimension-five operator, which violates lepton number, is a candidate to explain the small neutrino masses. Operators of dimension six and higher can cause smaller effects still, for example baryon-number violation.

As the energy scale is lowered, we can integrate out the Higgs and weak-gauge bosons, which leaves as only gauge symmetries the $SU(3)_c$ of color and the electromagnetic $U(1)_{em}$, for which the force carriers are, respectively, eight gluons $G_\mu \equiv G_\mu^a \lambda^a$, with λ^a, $a = 1, \ldots, 8$, the Gell-Mann $SU(3)$ matrices and sum over a implied, and the photon A_μ. Likewise, we can integrate out the heaviest quarks (top, bottom, charm). Although not irrelevant for nuclear physics, strange quark effects are not dominant for ordinary nuclei because, as we are going to see, in a hadronic theory the quark masses come together with the relatively large scale M_{QCD}. If the strange quark is kept explicit in the theory, the relatively heavy strange hadrons pose significant difficulties to the convergence of the low-energy EFT. In contrast, if the strange quark is integrated out, as I will do, its effects are suppressed by the strange hadron masses. The relevant quark fields are then conveniently written as an isospin doublet,

$$q = \begin{pmatrix} u \\ d \end{pmatrix}. \tag{4.31}$$

Here u and d are Dirac spinors for the up and down quarks, taken to be mass eigenstates with real masses $m_u = \bar{m}(1-\varepsilon)$ and $m_d = \bar{m}(1+\varepsilon)$. Matrices in isospin space can be expressed in terms of the unit matrix, which I will not write explicitly, and the isospin Pauli matrices τ_a, $a = 1, 2, 3$. Quarks transform under $U(1)_{em}$ according to the charge matrix

$$Q_q = \begin{pmatrix} 2/3 & 0 \\ 0 & -1/3 \end{pmatrix} = \frac{1}{6}(1 + 3\tau_3) \tag{4.32}$$

and under $SU(3)_c$ with a universal strength g, which also governs gluon self-interactions. If we define the covariant derivatives

$$D_\mu q = (\partial_\mu + ieQ_q A_\mu - igG_\mu)q, \qquad G_{\mu\nu} = \partial_\mu G_\nu - \partial_\nu G_\mu + ig[G_\mu, G_\nu], \tag{4.33}$$

the most general Lagrangian with Lorentz invariance and these gauge symmetries is (see, *e.g.*, Ref. [38])

$$\mathscr{L}_{QCD} = -\frac{1}{4}F_{\mu\nu}F^{\mu\nu} - \frac{1}{2}\text{Tr}\{G_{\mu\nu}G^{\mu\nu}\} + \bar{q}i\slashed{D}q$$

$$- \bar{m}\,\bar{q}\left[1 - \varepsilon\tau_3 + \frac{1-\varepsilon^2}{2}\bar{\theta}i\gamma_5\right]q + \cdots \tag{4.34}$$

where $\bar{\theta}$ is the so-called QCD vacuum angle (assumed to be small on phenomenological grounds) and "..." represent higher-dimensional operators. For simplicity I have dumped into the "..." also leptonic terms, whose virtual importance in purely hadronic process is small. They are of course needed when considering processes with external leptons, which I will not do explicitly in these lectures.

In order to understand some of the low-energy consequences of the theory based on the Lagrangian (4.34), I will first neglect the "...", in which case I am left with five parameters: g, $\bar{m}$, ε, e, and $\bar{\theta}$. I will argue based on *a posteriori* agreement with phenomenology that the last four parameters can, in some sense, be considered small. I will start with the "chiral limit" in which they are all set to zero, and then consider the changes their non-vanishing values cause. Some of this material (and original references) can be found in a good advanced textbook, such as Ref. [39].

4.3.1.1 Chiral Limit

In the chiral limit QCD has a single *dimensionless* parameter, g, and the action is invariant under scale transformations ($x \to \lambda^{-1}x$, $q \to \lambda^{3/2}q$, $G_\mu \to \lambda G_\mu$, $A_\mu \to \lambda A_\mu$, with λ a real parameter). However, in the path integral obtained from $\mathscr{L}_{QCD}$, scale invariance is "anomalously" broken by the inevitable presence of a *dimensionful* regulator. For renormalization, the strong constant $\alpha_s \equiv g^2/4\pi$ "runs" with the energy scale: it decreases as the energy increases—"asymptotic freedom"—and conversely increases as the energy decreases, so that $\alpha_s(1\,\text{GeV}) \sim 1$. Assuming "confinement", that is, that only colorless states ("hadrons") are asymptotic, and naturalness, the fact alluded to in the previous lecture that *almost* all hadrons have masses $\mathscr{O}(1\,\text{GeV})$ indicates that QCD has a characteristic scale $M_{QCD} \sim 1\,\text{GeV}$.

I seek here the EFTs of QCD for momenta $Q \ll M_{QCD}$. The first clue comes from the observation that the three pions—the lightest mesons—form a nearly degenerate isospin triplet,

$$\vec{\pi} = \begin{pmatrix} (\pi^+ + \pi^-)/\sqrt{2} \\ -i(\pi^+ - \pi^-)/\sqrt{2} \\ \pi^0 \end{pmatrix}, \qquad (4.35)$$

of pseudoscalar mesons with an approximate common mass $m_\pi \simeq 140\,\text{MeV} \ll M_{QCD}$. Would this be an indication of a breakdown in the naturalness assumption? No! The smallness and near degeneracy of pion masses can be naturally explained if we assume the "spontaneous" breaking of "chiral symmetry".

If I define the projectors

$$P_{L,R} \equiv \frac{1 \mp \gamma_5}{2}, \qquad P_{L,R}P_{L,R} = P_{L,R}, \qquad P_{L,R}P_{R,L} = 0, \qquad (4.36)$$

I can split the quark field into two components

$$q_{L,R} \equiv P_{L,R}q, \qquad q = q_L + q_R. \qquad (4.37)$$

In the limit I am considering, the quarks are massless and can have a definite left (L) or right (R) chirality according to its spin being against or in the direction of the momentum. Moreover, in this limit the quark kinetic term in $\mathscr{L}_{QCD}$ splits into two terms, which involve q_L and q_R separately. Each term is invariant under a separate $SU(2)$ transformation in isospin space, so $\mathscr{L}_{QCD}$ has a global $SU(2)_L \times SU(2)_R$ chiral symmetry. Since $SU(2)_L \times SU(2)_R$ is the covering group of $SO(4)$, it is sometimes sufficient to consider the latter, more intuitive group.

However, chiral symmetry is certainly not realized in the low-mass hadronic spectrum, where the lightest scalar meson and the lowest negative-parity spin-1/2 baryon have masses that are several hundreds of MeV above pions and nucleons, respectively. Still, the spectrum can be qualitatively understood if I assume that the solution of the path integral has, instead, a smaller, global symmetry of isospin given by the diagonal subgroup $SU(2)_{L+R}$, when again it is sometimes easier to talk instead of $SO(3)$. Goldstone's theorem tells us that there are massless (pseudo)scalars in the coset space $SO(4)/SO(3)$. The phenomenon here is completely analogous to the spontaneous breaking of $SO(3)$ rotational invariance down to $SO(2)$ that gives rise to spontaneous magnetization and spin waves in a ferromagnet.

An intuitive picture of this effect comes from considering the effective potential of QCD in the mesonic sector, as function of the four components of the $SO(4)$ vector

$$S = \begin{pmatrix} -i\bar{q}\gamma_5 \vec{\tau} q \\ \bar{q}q \end{pmatrix}. \tag{4.38}$$

$SO(4)$ symmetry of the potential means that it is invariant under rotations of S. If the potential had a minimum at the origin, $SO(4)$ would be manifest in the spectrum. Non-perturbative physics in QCD must be such as to make the potential have instead a "Mexican hat" shape: a degenerate set of absolute minima—a four-dimensional "chiral circle"—a distance away from the origin, which defines the pion decay constant f_π. If we take the "true" minimum to be in the $\bar{q}q$ direction, there are three massless excitations in the $\bar{q}i\gamma_5\tau_a q$ directions, which we can identify as the pions. In contrast, in the $\bar{q}q$ direction there is curvature in the potential, which we could expect to be characterized by M_{QCD} so that the corresponding scalar "sigma" meson has a mass $m_\sigma \sim M_{QCD}$.

The massless pions, but not excitations with mass $\mathcal{O}(M_{QCD})$, need to be accounted for explicitly in the EFT, since they give rise to long-distance interactions. In the limit we are considering, $SO(4)$ is an exact symmetry of the dynamics, so the EFT Lagrangian has to be invariant under small $SO(4)$ rotations, which are pion-field translations $\vec{\pi} \to \vec{\pi} + \vec{\varepsilon}$, with ε_i three constants. The simplest way to implement the symmetry is to choose pion fields such that all their interactions involve $\partial_\mu \vec{\pi}$, although, the manifold being a circle, a derivative is always accompanied by a factor of $(1 - \vec{\pi}^2/4f_\pi^2 + \cdots)$. Because there are only three pions, $SO(4)$ cannot be realized linearly, but there is a well-developed technology to incorporate chiral symmetry in the Lagrangian, the theory of non-linear realizations of symmetries [39]. Pions and fermion fields transform non-linearly, but covariant derivatives can be defined which transform in a simple way:

$$D_\mu \vec{\pi} = \left(1 - \frac{\vec{\pi}^2}{4f_\pi^2} + \cdots\right)\partial_\mu \vec{\pi},$$
$$\mathcal{D}_\mu \psi = \left(\partial_\mu + \frac{i}{2f_\pi^2}\vec{I}_\psi \cdot \vec{\pi} \times D_\mu \vec{\pi}\right)\psi,$$
(4.39)

where $\vec{I}_\psi$ is the generator of isospin in the representation of the fermion ψ (e.g. $\vec{I}_N = \vec{\tau}/2$ for the nucleon field N). Similarly one can define covariant derivatives of the covariant derivatives, and so on. Under the full chiral group these derivatives transform as under the unbroken isospin subgroup, but with a pion-field-, and thus position-, dependent parameter. The consequence is that an interaction built of these covariant ingredients to be isospin symmetric is automatically chiral invariant.

Each chiral-invariant effective interaction will have its LEC. Since the QCD dynamics that generate them is non-perturbative, they should contain arbitrary powers of g. As far as NDA goes, that means reduced couplings with arbitrary dependence on $g_R \equiv g/4\pi$. Thus consistency requires that we take $g_R \sim 1$, and thus a LEC of a chiral-invariant operator of canonical dimension D involving N fields is expected to be $c = \mathcal{O}((4\pi)^{N-2}/M_{QCD}^{D-4})$. Even though these LECs might not be particularly small, chiral-invariant interactions become weak at sufficiently low energies because each derivative brings in a power of momentum Q. And, for interactions with fixed N, NDA suggests that an extra derivative gives a relative factor Q/M_{QCD}. As we are going to see, this is a good guide for the perturbative sector of the EFT, but it is not always true in the nuclear sector where the EFT is non-perturbative.

4.3.1.2 Away from the Chiral Limit

Pions are, however, not massless. Let me take a second step and consider $\bar{m} \neq 0 \ll M_{QCD}$, but still $\varepsilon = 0$, $e = 0$, and $\bar{\theta} = 0$. Now there is an explicit breaking of chiral symmetry in Eq. (4.34), since $\bar{q}q$ is one of the components of the $SO(4)$ vector S in Eq. (4.38). The Lagrangian is still invariant under the $SO(3)$ isospin subgroup.

In the Mexican-hat picture, the effect of this term is to lower (raise) the potential in the direction of positive (negative) $\bar{q}q$, breaking the degeneracy of the now-deformed chiral circle. We can still talk of a tilted, approximately circular bottom of the hat, with a slightly different radius $f_\pi \simeq 92$ MeV. The distorted potential shape should not greatly affect the mass of the sigma or any other non-Goldstone state. But because the bottom is no longer flat, pions acquire a mass, and we speak of them as pseudo-Goldstone bosons. Note that it is the explicit breaking of chiral symmetry that justifies our choice of true vacuum in the chiral limit. Spontaneous breaking of a continuous symmetry only exists in a well-defined limit of the explicitly broken case. In a ferromagnet, this "vacuum alignment" is manifest in a spontaneous magnetization in the direction of a previously applied magnetic field.

In the EFT with pions, the Lagrangian no longer has $SO(4)$ invariance; there are now terms that break chiral symmetry in the $\bar{q}q$ direction, that is, the fourth component of S. Such terms do not necessarily contain derivatives but are proportional to powers of $\bar{m}$—actually $\bar{m}_R \equiv \bar{m}/M_{QCD}$ if we invoke NDA. One example is the pion mass term, for which we expect $m_\pi^2 = \mathcal{O}(M_{QCD}\bar{m})$, since from Eq. (4.18) $(m_\pi^2)_R \equiv m_\pi^2/M_{QCD}^2 = \mathcal{O}(\bar{m}_R)$. This roughly gives an average quark mass in the ballpark of 10 MeV. Another example is the so-called nucleon sigma term, the leading change Δm_N in the nucleon mass away from the chiral limit: by NDA $(\Delta m_N)_R \equiv \Delta m_N/M_{QCD} = \mathcal{O}(\bar{m}_R)$, or $\Delta m_N = \mathcal{O}(\bar{m}) = \mathcal{O}(m_\pi^2/M_{QCD})$. Terms with higher powers in $\bar{m}$ come from the fourth components of tensor products of S. In this way we are able to produce an S matrix with the correct chiral-symmetry breaking, not restricted to first-order in the breaking parameters. For details of how to construct these operators, see Refs. [38, 39].

The low-energy effects of the remaining terms in Eq. (4.34) can be analyzed in similar fashion [38]. When we allow $\varepsilon \neq 0$, even the $SO(3)$ group of isospin is explicitly broken, since the corresponding term in Eq. (4.34) transforms as the third component of another $SO(4)$ vector,

$$P = \begin{pmatrix} \bar{q}\vec{\tau}q \\ i\bar{q}\gamma_5 q \end{pmatrix}. \tag{4.40}$$

In the EFT Lagrangian, this means there is going to be another class of terms that break isospin like $\bar{q}\tau_3 q$, which are proportional to powers of $\varepsilon\bar{m}$—in fact $(\varepsilon\bar{m})_R \equiv \varepsilon\bar{m}/M_{QCD}$ according to NDA. An example is the quark-mass contribution to the neutron-proton mass difference: from NDA, $(\delta m_N)_R \equiv \delta m_N/M_{QCD} = \mathcal{O}((\varepsilon\bar{m})_R)$ or $\delta m_N = \mathcal{O}(\varepsilon\bar{m}) = \mathcal{O}(\varepsilon m_\pi^2/M_{QCD})$. Among terms of higher order in the symmetry-breaking parameters one finds a contribution to the pion mass splitting, which by NDA is expected to be $\delta m_\pi^2 = \mathcal{O}((\delta m_N)^2)$.

Allowing for $e \neq 0$ introduces further isospin breaking since the photon couples differently to up and down quarks. There result two types of interactions in the EFT. First, "soft" photons interact in the standard NRQED fashion with other low-energy degrees of freedom: either through $U(1)$ covariant derivatives, with

$$\partial_\mu \pi_a \to (\delta_{ab}\partial_\mu + e\varepsilon_{3ab}A_\mu)\pi_b, \qquad \partial_\mu \psi \to (\partial_\mu + ieQ_\psi A_\mu)\psi \tag{4.41}$$

in Eq. (4.39), or through $F_{\mu\nu}$. In either case, the interactions are proportional to e. Second, "hard" photons, which are integrated out, give rise to purely hadronic interactions with strengths proportional to $e_R^2 = \alpha/4\pi$. In this case chiral symmetry is broken as the 34-component of the $SO(4)$ antisymmetric tensor

$$T_\mu = \begin{pmatrix} \varepsilon_{abc}\bar{q}\gamma_\mu\gamma_5\tau_c q & \bar{q}\gamma_\mu\tau_b q \\ -\bar{q}\gamma_\mu\tau_a q & 0 \end{pmatrix}. \tag{4.42}$$

Among the effects from hard photons one finds contributions to both the pion mass splitting and the neutron-proton mass difference. From NDA, since both are expected to require at least one photon exchange, for the former $\bar{\delta}m_\pi^2 = \mathcal{O}(\alpha M_{QCD}^2/(4\pi))$ and for the latter $\bar{\delta}m_N = \mathcal{O}(\bar{\delta}m_\pi^2/M_{QCD})$.

In order to isolate the quark mass difference encoded in ε, one should look at quantities that depend linearly on ε, such as the neutron-proton mass difference.

It is useful to focus on a discrete subgroup of $SO(3)$, called charge symmetry, of rotations that interchange up and down quark (up to a sign). Observables that are charge-symmetry breaking are linear in ε, because the ε term in Eq. (4.34) changes when up and down quark are interchanged. Electromagnetism also breaks charge symmetry, so its effects have to be estimated. When this is done, one finds $\varepsilon \sim 1/3$ [40]. Observables that are isospin but not charge-symmetry breaking—sometimes called "charge-independence" breaking—are proportional to ε^2 at best, so are usually dominated by electromagnetic effects. NDA suggests that this is the case, for example, for the pion mass splitting.

In the next step we allow $\bar{\theta} \neq 0$. The corresponding term in Eq. (4.34) now breaks both parity (P) and time-reversal (T) invariance. To arrive at this term I have followed Ref. [41] and performed an anomalous chiral rotation to eliminate a term of the type $\text{Tr}\{G_{\mu\nu}\tilde{G}^{\mu\nu}\}$ that leads to T violation through non-perturbative effects. There is an infinite number of rotations that perform this task, but I have selected the one for which the vacuum is stable. The form in Eq. (4.34) is useful because it shows that the $\bar{\theta}$ term leads also to chiral-symmetry breaking, as the fourth component of the *same* $SO(4)$ vector P that appears in quark-mass isospin violation, Eq. (4.40). In the EFT, thus, there is a intimate relationship between the strengths of T-violating and those of T-conserving, charge-symmetry-breaking interactions [38]. Exploiting this connection and assuming naturalness in Chiral EFT, we can convert the tight limit on the neutron electric dipole moment [42] to a bound $\bar{\theta} \lesssim 10^{-10}$. Why $\bar{\theta}$ is so much smaller than 1 is an open naturalness problem in the SM, known as the strong CP problem. (Assuming CPT is a good symmetry, T violation implies CP violation and *vice-versa*.)

Finally, one can go on and construct the low-energy interactions coming from the higher-dimensional operators in the "..." in Eq. (4.34), such as P- [43, 44] (and T- [45]) violating interactions stemming from dimension-four (and -six) interactions in the SM.

All the EFT interactions that originate beyond the basic quark-gluon interaction are proportional to powers of relatively small parameters: $\bar{m}/M_{QCD}$, $\varepsilon\bar{m}/M_{QCD}$, e, $\alpha/(4\pi)$, etc. Thus, these interactions also tend to be weak, so that the full EFT allows for a controlled expansion of hadronic amplitudes. But obviously I am not being precise here. We have already seen, for example, how inverse powers of Q can compensate for the smallness of α at low energies to give rise to electromagnetic bound states. We are going to witness a failure of a simple perturbative expansion in nuclei as well. In both cases, however, an expansion still exists on top of a non-perturbative LO.

4.3.2 Chiral EFT

I am now in position to formulate Chiral EFT. The goal is to describe nuclear processes with typical external momenta $Q \sim M_{nuc} \ll M_{QCD}$ consistently with QCD. I start by building the chiral Lagrangian and then discuss at a qualitative level its implications, as we did for NRQED.

4.3.2.1 Chiral Lagrangian

At a minimum, we need to include the lightest baryons, the proton and neutron, which are (at least nearly) stable in the time scale of strong-interaction processes. Just as in NRQED, we can include only their two non-relativistic components, since pair creation requires $\gtrsim 2m_N \sim M_{QCD}$ in energy. At $Q \sim m_\pi$ we are probing distances at which pion-exchange effects can be resolved, so the three pions are also relevant degrees of freedom. In contrast, all other mesons can be integrated out because they have masses $\mathcal{O}(M_{QCD})$. If we are interested in going a bit further in energy, or increase the convergence of the theory at low energies, we should also include nucleon excitations. The four charge states of the isospin-3/2, spin-3/2 Delta isobar have approximately equal masses, with $m_\Delta - m_N \sim 3 f_\pi$, which numerically is about $2m_\pi$. One should thus expect considerable effects from the Delta, which are best reproduced if the Delta is not integrated out but instead kept explicitly as a heavy fermion field, from which we remove the same phase as for the nucleon field. (It is not difficult to generalize the heavy fermion formalism to spin higher than 1/2.) It is convenient to include also an explicit field for the next excitation, the Roper, which has the same quantum numbers of the nucleon, even though with $m_R - m_\Delta \sim 2 f_\pi$ we are getting close to M_{QCD}. For simplicity, here I omit the Roper and other nucleon resonances. My baryon fields are thus

$$N = \begin{pmatrix} p \\ n \end{pmatrix}, \qquad \Delta = \begin{pmatrix} \Delta^{++} \\ \Delta^{+} \\ \Delta^{0} \\ \Delta^{-} \end{pmatrix}. \qquad (4.43)$$

In order to couple the nucleon to the Delta, one introduces [10] 2×4 spin and isospin transition operators, respectively $\mathbf{S}$ and $\vec{T}$, normalized so that $S_i S_j^\dagger = (2\delta_{ij} - i\varepsilon_{ijk}\sigma_k)/3$ and analogously for $\vec{T}$.

The next step is to write the most general Lagrangian with the same symmetry structure as QCD: Lorentz, $SU(3)_c$, $U(1)_{em}$, and approximate, spontaneously broken $SU(2)_L \times SU(2)_R$. I will consider explicitly here only the terms shown in Eq. (4.34) with $\bar{\theta} = 0$, in which case P and T are also symmetries. This is sufficient for describing the essence of nuclear physics, but of course one can later add interactions reflecting the neglected terms, which give rise to smaller (but sometimes important!) effects from weak interactions and physics beyond the SM. Color gauge invariance is trivial because every field is a singlet. Lorentz and electromagnetic gauge invariance can be implemented in the usual way, as we have done in NRQED. Chiral symmetry is a bit less familiar, but is implemented as described in the previous subsection. It is important to notice that in QCD chiral symmetry is broken in a specific way, which is reproduced in the EFT. In contrast, most hadronic models do not account for the correct pattern of chiral-symmetry breaking.

Each interaction in the chiral Lagrangian has a LEC, which in principle can be obtained from a direct calculation of QCD amplitudes at low energies, for example using lattice simulations. In fact, since in Chiral EFT the quark masses can be varied independently, EFT amplitudes can be matched to lattice QCD amplitudes at the somewhat larger quark masses amenable to today's computers. Chiral EFT provides an extrapolation tool to smaller quark masses, and can be used to pre- or post-dict observables at the long distances not accessible directly in lattice QCD. For the time being, however, one has to resort to fitting LECs to experimental, rather than lattice, data.

In any case, we need to start with an assumption about the sizes of the LECs, so as to calculate amplitudes in the EFT before matching them to data. I assume NDA, Eq. (4.18), with $M = M_{QCD}$, and take $f_\pi = \mathcal{O}(M_{QCD}/4\pi)$ and $m_\Delta - m_N$ as low-energy scales. In addition to an expansion in powers of momenta and $m_\Delta - m_N$, there are separate expansions in the chiral-breaking parameters $\bar{m}/M_{QCD}$, $\varepsilon\bar{m}/M_{QCD}$, e, $\alpha/(4\pi)$, etc. How we combine them with the expansion of chiral-invariant operators is to some extent a matter of choice. The first parameter can be converted into m_π^2/M_{QCD}^2 and paired with the momentum expansion for $Q \sim m_\pi$. The second parameter, now $\varepsilon m_\pi^2/M_{QCD}^2$, depends on the dimensionless number $\varepsilon \sim 1/3$. Although one can certainly entertain a counting where ε is taken as comparable to $\mathcal{O}(m_\pi/M_{QCD})$, I prefer to err on the side of overestimating rather than underestimating isospin breaking and count it as $\mathcal{O}(1)$. Similar choices affect the electromagnetic interactions. Since e appears explicitly in covariant derivatives, it is natural to count it as Q. On the other hand, $\alpha/4\pi$ is numerically not very far from $\varepsilon m_\pi^3/M_{QCD}^3$, so having chosen $\varepsilon = \mathcal{O}(1)$ leaves $\alpha/4\pi$ as giving suppression comparable to three powers of the expansion parameter. (Anything less would make the pion mass splitting appear in LO, clearly too much of overestimate.)

With these choices, it is convenient to write the chiral Lagrangian as

$$\mathscr{L}_{ChEFT} = \sum_{\Delta=0}^{\infty} \mathscr{L}_{f=0,1}^{(\Delta)} + \mathscr{L}_{f\geq 2}, \quad (4.44)$$

where I introduced the "chiral index" [2, 18]

$$\Delta = d + f - 2 \geq 0 \quad (4.45)$$

of an interaction with d derivatives and powers of m_π or $m_\Delta - m_N$ or e or $(\alpha/4\pi)^{1/3}$, and $2f$ fermion fields. This index counts inverse powers of the high scale M_{QCD}—even for hard-photon operators, as long as we count $\alpha/4\pi \sim \varepsilon m_\pi^3/M_{QCD}^3$. It is bounded from below because chiral symmetry guarantees that terms with $f = 0$ have at least two derivatives or powers of m_π and thus $d = 2$, while terms with $f = 1$ have at least one derivative and thus $d = 1$:

$$\mathscr{L}_{f=0,1}^{(0)} = \frac{1}{2}(D_\mu \vec{\pi})^2 - \frac{m_\pi^2}{2}\vec{\pi}^2\left(1 - \frac{\vec{\pi}^2}{4f_\pi^2} + \cdots\right) - \frac{1}{4}F_{\mu\nu}F^{\mu\nu}$$

$$+ \bar{N}i\mathscr{D}_0 N + \frac{g_A}{2f_\pi}\bar{N}\vec{\tau}\sigma N \cdot \mathbf{D}\vec{\pi}$$

$$+ \bar{\Delta}[i\mathcal{D}_0 - (m_\Delta - m_N)]\Delta$$
$$+ \frac{h_A}{2f_\pi}[\bar{N}\vec{T}\mathbf{S}\Delta + \text{H.c.}] \cdot \mathbf{D}\vec{\pi} + \cdots \quad (4.46)$$

where g_A and h_A are LECs, and "…" contain terms with more pion and/or Delta fields. Increasing the chiral index we find

$$\mathcal{L}^{(1)}_{f=0,1} = -\frac{\bar{\delta}m_\pi^2}{2}(\vec{\pi}^2 - \pi_3^2)\left(1 - \frac{\vec{\pi}^2}{2f_\pi^2} + \cdots\right)$$
$$+ \bar{N}\left[\frac{\mathcal{D}^2}{2m_N} + \Delta m_N\left(1 - \frac{\vec{\pi}^2}{4f_\pi^2} + \cdots\right)\right.$$
$$\left. - \frac{\delta m_N}{2}\left(\tau_3 - \frac{1}{2f_\pi^2}\pi_3\vec{\pi}\cdot\vec{\tau} + \cdots\right)\right]N$$
$$+ \frac{1}{f_\pi^2}\bar{N}\left[b_1(D_0\vec{\pi})^2 - b_2(\mathbf{D}\vec{\pi})^2 + ib_3\varepsilon_{ijk}\varepsilon_{abc}(D_i\pi_a)(D_j\pi_b)\sigma_k\tau_c\right]N$$
$$- \frac{g_A}{4m_N f_\pi}[i\bar{N}\vec{\tau}\sigma\cdot\mathcal{D}N + \text{H.c.}]\cdot D_0\vec{\pi}$$
$$- \frac{h_A}{4m_N f_\pi}[\bar{N}\vec{T}\mathbf{S}\cdot\mathcal{D}\Delta + \text{H.c.}]\cdot D_0\vec{\pi}$$
$$+ \frac{e}{4m_N}\bar{N}\left\{1 + \kappa_0\right.$$
$$\left. + (1+\kappa_1)\left[\tau_3 - \frac{1}{2f_\pi^2}(\vec{\pi}^2\tau_3 - \pi_3\vec{\pi}\cdot\vec{\tau}) + \cdots\right]\right\}\sigma_i N\tilde{F}^{0i}$$
$$+ \cdots, \quad (4.47)$$

where $\kappa_{0,1}$ and $b_{1,2,3}$, are new LECs, and so on.

Equations (4.46) and (4.47) are sufficient to illustrate some of the characteristics of Chiral EFT. First, notice that the electromagnetic interactions are similar to those in NRQED (4.24) once we account for the isospin of the nucleon. For example, the anomalous magnetic moment of the proton (neutron) is $\kappa_{p(n)} = [\kappa_0 + (-)\kappa_1]/2$. Terms like the $\beta_{0,1}$ in Eq. (4.24) appear only at higher orders. Second, the pions play a role in Chiral EFT similar to the photon in NRQED, in the sense that they are relativistic bosons that self-interact, dress and can be exchanged among fermions. Third, the fermions themselves are non-relativistic and therefore the theory can likewise be split into sectors of different fermion numbers. But, as stressed before, the theory *is* Lorentz invariant, with invariance implemented in a Q/m_N expansion, as revealed by the terms in Eq. (4.47) whose coefficients are determined by the LECs from Eq. (4.46) and m_N.

4.3.2.2 Chiral Perturbation Theory

Just like NRQED, Chiral EFT is perturbative in the sectors of $f = 0, 1$, where it is referred to as Chiral Perturbation Theory (ChPT). For a more complete introduction and references see, for example, Ref. [46].

Let us consider an arbitrary scattering process involving $A = 0, 1$ nucleon and one or more incoming and outgoing pions and photons, all with external three-momenta $Q = \mathcal{O}(M_{nuc})$. In this case, after renormalization all internal three-momenta in loops can also be taken as comparable to Q. Important for perturbation theory is the energy difference between an intermediate state and the initial state, and that is also of $\mathcal{O}(Q)$. Using topological identities about Feynman diagrams, one can show that the amplitude can be put in the schematic form of Eq. (4.14) with [2]

$$\nu = 2 - A + 2L + \sum_i V_i \Delta_i \geq 2 - A \equiv \nu_{min}, \quad (4.48)$$

where L is the number of loops, V_i is the number of vertices of type i, which have chiral index Δ_i, and the sum is over all types of vertices. The lower bound, stemming from the lower bound on the chiral index (thus ultimately from chiral symmetry) means that LO consists of all tree diagrams that can be constructed out of $\mathscr{L}^{(0)}$, NLO from tree diagrams with one insertion of a $\mathscr{L}^{(1)}$ element, *etc.* Loops start contributing at N²LO. The tree results are equivalent to old current algebra, but ChPT allows a systematic exploration of quantum corrections. One should of course keep in mind that this power counting is meant as a general guide; for a specific process at specific kinematics there might be a reorganization of the ordering that better reflects the relative importance of interactions.

The ChPT expansion in Q/M_{QCD} actually comprises: (i) a non-relativistic expansion for fermions in Q/m_N; (ii) a multipole expansion of the "heavy" meson cloud in $Q/m_\sigma, \ldots$; and (iii) a pion-loop expansion in even powers of $Q/(4\pi f_\pi)$. Resumming any of these three expansions would be great, but would not affect the overall error of a truncation unless the other two expansions can be resummed at the same time. As noted, since m_N is not smaller than the other high scales, it does not increase the error to include Lorentz invariance only approximately. Conversely, the error is not decreased in covariant versions of ChPT. Although we are not solving QCD at short distances, which might well be represented by a very dense cloud of heavy mesons, we are exploiting the fact that this "inner" cloud is short-ranged at the resolution scale $1/Q$. At this scale the pion cloud is not short-ranged and cannot be treated in a multipole expansion, but the large factor $4\pi f_\pi$ associated with loops means that the "outer" pion cloud is sparse, the probability of finding pions "in the air" decreasing with increasing number.

The $f = 0$ sector describes interactions of pions and photons. Already at LO pions self-interact via the kinetic and mass terms in Eq. (4.46), with strengths determined by f_π and m_π, a classic result due to Weinberg. At relative $\mathcal{O}(Q^2/M_{QCD}^2)$, there are further self-interactions with two extra derivatives, analogous to the $\beta_{0,1}$ terms in Eq. (4.24), which provide counterterms for the one-loop diagrams at the same order. When considering isospin-violating quantities one needs to account for the pion mass splitting from Eq. (4.47) as well. Nowadays calculations have reached relative $\mathcal{O}(Q^4/M_{QCD}^4)$.

In the $f = 1$ sector, a nucleon interacts with low-energy probes. As the fermion in the NRQED example, the nucleon is static at LO because, having three-momentum of $\mathcal{O}(Q)$, it has a much smaller recoil energy of $\mathcal{O}(Q^2/m_N)$. The latter is accounted for at NLO, and relativistic corrections appear two orders higher. The classic process is pion-nucleon elastic scattering, details of which can be found in Refs. [47, 48]. LO consists of tree diagrams: an S-wave pion-nucleon seagull ("Weinberg-Tomozawa term") from the covariant derivative (4.39) of the nucleon, which is determined by f_π; and P-wave interactions via nucleon and Delta "pole diagrams" stemming from the pion-nucleon and pion-nucleon-Delta couplings of LECs g_A and h_A, respectively. By NDA, $g_A, h_A = \mathcal{O}(1)$, and indeed $g_A \simeq 1.3$ and $h_A \simeq 2.9$. At NLO we find not only the Galilean corrections to these couplings but also the effects of new seagulls. Two are associated with the shifts Δm_N and δm_N in nucleon mass, which gives the possibility of determining these parameters from scattering. Three other seagulls have undetermined coefficients $b_{1,2,3} = \mathcal{O}(1/M_{QCD})$. At N^{2}LO one-loop diagrams appear, including the leading contribution to the Delta width. At threshold the P-wave interactions vanish and the scattering length is purely isovector at LO, the classic Weinberg-Tomozawa result, extended to loop orders in the early 90s [49]. Other reactions can be studied along analogous lines [49].

As long as one considers $Q \ll m_\Delta - m_N$, the Δ field can be integrated out. The Lagrangian is formally the same as above, just without this field, but the LECs are in general different, because they now include Delta contributions. For example, the coefficients $b_{2,3}$ are replaced by larger LECs $c_{2,3} = \mathcal{O}(1/(m_\Delta - m_N))$. Delta effects are thus relegated to subleading orders, suppressed by powers of $Q/(m_\Delta - m_N)$. However, as the energy increases it is more efficient to keep the Delta in, and not consider $m_\Delta - m_N$ as large. If the energy is increased further, into the Delta resonance region, one finds a "kinematic" cancellation in an s-channel Delta propagator between energy E and $m_\Delta - m_N$, thus enhancing width effects. In a window $|E - (m_\Delta - m_N)| \ll \mathcal{O}(Q^3/M_{QCD}^2)$, the power counting (4.48) has to be modified [47, 50]. Delta width effects have to be resummed, allowing us to push ChPT beyond the Delta region. The Roper then becomes important in some waves and better be included as well [48].

ChPT has met with much success—for a recent review, see Ref. [51]—which emboldens us to press on to the sectors of Chiral EFT with $A \geq 2$ nucleons.

4.3.2.3 Nuclear Physics

As first noted by Weinberg [1], for $A \geq 2$ we face the same infrared enhancement we found in NRQED, which more generally takes place for heavy particles exchanging light quanta. Although the details are different, here this enhancement again leads to a breakdown of perturbation theory. This is a good thing, given that nuclei and we would not exist otherwise....

Fig. 4.3 Diagrams representing the T matrix for the elastic scattering of two nucleons in Chiral EFT. A nucleon (Delta) is represented by a (*double*) *solid line*, a pion by a *dashed line*. A *circle* at the vertex denotes an inverse power of M_{QCD}

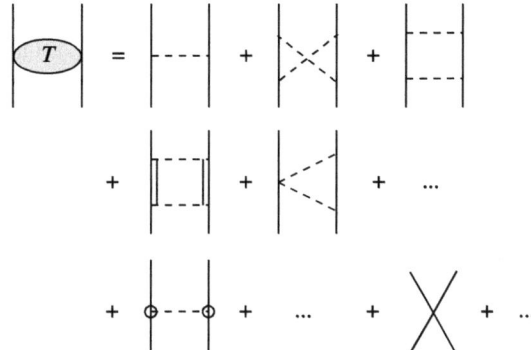

As before, I start with two-body elastic scattering. Two nucleons can exchange photons as described in NRQED, but here I focus on the strong interactions mediated at long range by pion exchange, see Fig. 4.3. For the momenta, I follow the same notation as in Sect. 4.2.2.

Analogous to Eq. (4.25), the one-pion exchange (OPE) between two nucleons is

$$T_{1\pi} \simeq \left(\frac{g_A}{2f_\pi}\right)^2 \frac{\mathbf{q}^2}{\mathbf{q}^2 + m_\pi^2 - i\varepsilon} \sigma_1 \cdot \hat{\mathbf{q}} \sigma_2 \cdot \hat{\mathbf{q}} \, \vec{\tau}_1 \cdot \vec{\tau}_2 \bigl(1 + \mathcal{O}(Q^2/m_N^2)\bigr). \quad (4.49)$$

Since $g_A = \mathcal{O}(1)$, for $Q \sim m_\pi$ this term has magnitude $\mathcal{O}(1/f_\pi^2) = \mathcal{O}(4\pi/m_N M_{NN})$, where, for reasons that will become apparent soon, I introduced the quantity $M_{NN} \equiv 4\pi f_\pi^2/m_N = \mathcal{O}(f_\pi)$. Again as in NRQED, we can, and should later, consider corrections to OPE from higher derivatives at the vertices, but they are expected to be suppressed by $\mathcal{O}(Q^2/M_{QCD}^2)$ or higher.

If we look at the crossed-box two-pion-exchange (TPE) diagrams, we obtain an expression analogous to Eq. (4.26), but with the more complicated OPE substituted for photon exchange. We can count powers of $Q \sim m_\pi$ and 4π in the same way, that is, $Q^3/(4\pi)^2$ for the integration and Q^{-5} for the propagators, plus an extra Q/f_π for each vertex, to find an overall size $\mathcal{O}(Q^2/(4\pi f_\pi^2)^2)$, or a relative $\mathcal{O}(Q^2/(4\pi f_\pi)^2) = \mathcal{O}(Q^2/M_{QCD}^2)$ with respect to OPE. The same power counting applies to the stretched-box diagrams subsumed by the box diagram, and to diagrams originating in the Weinberg-Tomozawa seagull shown in Eq. (4.46). Substituting the Delta for the nucleon in intermediate states only adds modulating factors of $Q/(m_\Delta - m_N) = \mathcal{O}(1)$ in our counting. Thus, as in NRQED, we can consider these TPE diagrams as corrections to OPE, here starting at relative $\mathcal{O}(Q^2/M_{QCD}^2)$.

The true-box diagrams representing iterated OPE, on the other hand, have the same relative $\mathcal{O}(4\pi m_N/Q)$ enhancement as seen in Eq. (4.27): all nucleon energies are small and nucleon propagators are not static. Recoil, despite appearing only in $\mathscr{L}_{f=0,1}^{(1)}$ (4.47), cannot be treated as a perturbation. (This of course has nothing to do with relativistic corrections, although sometimes one sees a confusion in the literature, where the need for Galilean corrections is mistaken for a need for relativistic resummation.) The true-box diagrams have a size $\mathcal{O}(m_N Q/(4\pi f_\pi^4))$, or a relative $\mathcal{O}(Q/M_{NN})$ with respect with OPE. More generally, iterating OPE n times gives a contribution of relative $\mathcal{O}(Q^n/M_{NN}^n)$. Qualitatively the amplitude is like a geometric series,

$$T^{(0)} \sim \frac{4\pi}{m_N M_{NN}} \left[1 - \mathcal{O}\left(\frac{Q}{M_{NN}}\right) \right]^{-1}. \tag{4.50}$$

Contrary to its NRQED counterpart (4.28), pion interactions are weak at low Q, but once $Q \sim M_{NN}$ bound states or resonances can be expected, with energies $|E| \sim M_{NN}^2/m_N$. Thus, by a reasoning entirely analogous to the one that gives the right atomic scales, we are led to identify the typical nuclear scale M_{nuc} with the low-energy scale $M_{NN} = \mathcal{O}(f_\pi)$. *Chiral-symmetry breaking*, in the form of a light pion with an interaction with strength set by $1/f_\pi$, *explains why nuclei are shallow from a QCD perspective*.

In addition to this qualitative insight, we see that in Chiral EFT the first aspect emerges of the traditional nuclear picture outlined in Sect. 4.2. Like NRQED, Chiral EFT reduces to nucleons interacting through a potential, which we define similarly and whose form we can derive. Since by construction we have no infrared enhancement in the potential, the ChPT power counting (4.48) applies to the corresponding pion-exchange diagrams. At LO, we find OPE. At NLO, or relative order $\mathcal{O}(Q/M_{QCD})$, there is nothing new because we are assuming P (and T) conservation. At N²LO, we have corrections to OPE (including isospin violation) and the TPE diagrams discussed above, and at N³LO, TPE diagrams with the seagull vertices from Eq. (4.46), in addition to isospin-breaking corrections. The isospin-symmetric potential up to this order was first derived in Refs. [3, 10], and rederived many times since (see discussion in Ref. [52]). As emphasized in Refs. [53, 54], this "chiral Van der Waals" potential has the qualitative features of heavier-meson exchange potentials, for example sigma+omega exchange in the isoscalar central channel. In Refs. [55, 56], the famous Nijmegen partial-wave analysis of two-nucleon (2N) data was redone with the chiral pion-exchange potential to N³LO without Deltas as long-range input, instead of heavier-meson exchange. A slightly better fit laid to rest the longstanding prejudice that lack of explicit heavy meson exchange was a problem for EFT. At N⁴LO, in addition to corrections to OPE and TPE, there is also three-pion exchange; the potential at this order has been derived in a *tour de force* in Ref. [57], and references therein. It seems unlikely that anything beyond this order will be needed. Note that in the literature sometimes relative $\mathcal{O}(Q^n/M_{QCD}^n)$, $n \geq 2$, is referred to as N^{n-1}LO instead of N^{n}LO as I do here.

We can go on and examine the implications of power counting for systems with $A \geq 3$ nucleons. Eq. (4.48) indeed provides the ordering of pion-exchange diagrams that generate an f-body force. For example, the leading three-nucleon (3N) force comes from tree diagrams with vertices from $\mathcal{L}^{(0)}$, and the next-to-leading component from tree diagrams with insertion of one vertex from $\mathcal{L}^{(1)}$. One can show that among the leading diagrams only TPE involving the Delta survives a cancellation against subleading terms in the OPE two-body force [8], resulting in the Fujita-Miyazawa potential as the dominant 3N force. The next-to-leading TPE diagrams in turn provide a chiral-corrected version of the Tucson-Melbourne (TM) potential,

Fig. 4.4 Sample of diagrams representing the pion-range components of the isospin-symmetric part of the nuclear potential in Chiral EFT. N^nLO stands for relative $\mathcal{O}(Q^n/M_{QCD}^n)$. (Note that in the literature sometimes terms with $n \geq 2$ are referred to as N^{n-1}LO.) Notation as in Fig. 4.3

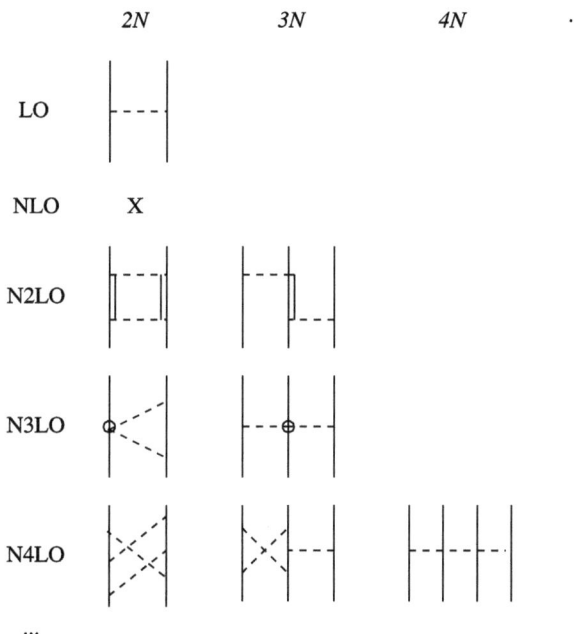

sometimes called the TM' potential [58, 59]. The leading long-range $4N$ potential has been derived as well [60, 61].

However, the relative ordering between potentials involving different numbers of nucleons is slightly more complicated. In the A-body system the A-body force connects all bodies while the $A-1$-body force, for example, leaves one of the bodies disconnected. Allowing for such disconnected diagrams in the power counting leads to an extra factor $-2C$, where C is the number of connected pieces, on the right-hand side of Eq. (4.48) [2]. We then recover the second aspect of the traditional nuclear picture: effects of the $2N$ potential are expected to be larger than those of the $3N$ potential, which in turn are expected to be larger than those of the $4N$ potential, and so on. The structure of the isospin-symmetric part of the long-range nuclear potential is shown schematically in Fig. 4.4.

Chiral EFT also provides a justification for the other two aspects of the traditional picture presented in Sect. 4.2. First, isospin is an accidental symmetry in Chiral EFT like spin symmetry in NRQED. Thus its violation is not represented by ε but at best by $\varepsilon Q/M_{QCD}$ [6]. The most important pieces of the isospin-violating potential can be obtained from the isospin-breaking terms in Eq. (4.47), and from the higher-index Lagrangians—see Ref. [62], and references therein. One finds that not only the isospin-symmetric potential tends to dominate, but also that charge independence tends to be larger than charge-symmetry breaking [9]. Second, by defining currents as the sum of diagrams with external probes that are free of infrared enhancement, we can likewise conclude that effects of one-nucleon currents are expected to be larger than those of $2N$ currents, which in turn are expected to be larger than those of $3N$ currents, and so on [5].

I have so far emphasized the long-range contributions from pions, but none of the qualitative conclusions change as long as we assume that LECs that account for short-range interactions among two or more nucleons obey NDA. In this case it is convenient to classify $\mathscr{L}_{f\geq 2}$ also according to the index (4.45),

$$\mathscr{L}_{f\geq 2} = \sum_{\Delta=0}^{\infty} \mathscr{L}_{f\geq 2}^{(\Delta)}. \tag{4.51}$$

The lowest-index terms have $f = 2$ and $d = 0$. Because of isospin, two nucleons can be found with total spin $s = 0, 1$ and we can now write two independent no-derivative interactions, corresponding to the two spin projectors (4.29):

$$\mathscr{L}_{f\geq 2}^{(0)} = -\frac{C_{0(0)}}{4}(\bar{N}N\bar{N}N - \bar{N}\boldsymbol{\sigma}N\cdot\bar{N}\boldsymbol{\sigma}N)$$
$$-\frac{C_{0(1)}}{4}(3\bar{N}N\bar{N}N + \bar{N}\boldsymbol{\sigma}N\cdot\bar{N}\boldsymbol{\sigma}N)+\cdots, \tag{4.52}$$

with two LECs $C_{0(0,1)}$ and "$\ldots$" standing for terms with Deltas. Using Fierz reordering, other isospin-symmetric forms can be written in terms of these. We can increase the index by one unit with an extra derivative or two more fermion fields,

$$\mathscr{L}_{f\geq 2}^{(1)} = \frac{D_0}{f_\pi}\bar{N}N\bar{N}\boldsymbol{\sigma}\,\vec{\tau}\,N\cdot\cdot\mathbf{D}\vec{\pi} - E_0\bar{N}N\bar{N}N\bar{N}N+\cdots, \tag{4.53}$$

where D_0 and E_0 are new LECs. Again, other forms can be reduced to these. For example, because of the Pauli principle, three nucleons at the same spacetime point can only have a total spin $s = 1/2$, so other six-nucleon operators can be rewritten in terms of E_0. Among the higher-index terms, I will also need

$$\mathscr{L}_{f\geq 2}^{(2)} = -\frac{D_{2(0)}}{4}m_\pi^2(\bar{N}N\bar{N}N - \bar{N}\boldsymbol{\sigma}N\cdot\bar{N}\boldsymbol{\sigma}N)\left(1 - \frac{\vec{\pi}^2}{4f_\pi^2}+\cdots\right)$$
$$-\frac{C_{2(0)}}{4}\left(\bar{N}N\bar{N}\mathscr{D}^2 N - \bar{N}\boldsymbol{\sigma}N\cdot\bar{N}\boldsymbol{\sigma}\mathscr{D}^2 N + \mathrm{H.c.}\right)$$
$$-\frac{C'_{2(1)}}{4}\left[3\bar{N}N(\mathscr{D}_i\bar{N})\mathscr{D}_i N + \bar{N}\boldsymbol{\sigma}N\cdot(\mathscr{D}_i\bar{N})\boldsymbol{\sigma}\mathscr{D}_i N + \mathrm{H.c.}\right]$$
$$+\cdots, \tag{4.54}$$

where $D_{2(0)}$, $C_{2(0)}$, and $C'_{2(1)}$ are further LECs. The "$\ldots$" now involve also other interactions with only nucleon fields, such as analogous terms for different spin and/or derivative combinations.

These short-range interactions are, of course, very important in quantitative applications of EFT. With the NDA assumption, we can use the index in Eq. (4.48) for the full potential, not only the pion-exchange diagrams shown in Fig. 4.4. The $C_{0(s)}$ terms appear already in the LO 2N potential, since $(C_{0(s)})_R =$

$M_{QCD}^2 C_{0(s)}/(4\pi)^2 = \mathcal{O}(1)$, or $C_{0(s)} = \mathcal{O}((4\pi)^2/M_{QCD}^2) = \mathcal{O}(4\pi/(m_N M_{NN}))$. These terms contribute to the two S-wave channels, $^1S_0$ and $^3S_1$ in the notation $^{2s+1}l_j$ where l and j are the orbital and total angular momenta, respectively. At N^{2}LO further contact interactions with two derivatives or two powers of the pion mass show up, suppressed by $\mathcal{O}((Q/M_{QCD})^2)$. For example, since by NDA $(C_{2(s)}^{(l)})_R = M_{QCD}^4 C_{2(s)}^{(l)}/(4\pi)^2 = \mathcal{O}(1)$ and $(D_{2(0)}m_\pi^2)_R = M_{QCD}^2 D_{2(0)}m_\pi^2/(4\pi)^2 = \mathcal{O}(\bar{m}_R)$, we have $C_{2(s)}^{(l)}, D_{2(0)} = \mathcal{O}(4\pi/(m_N M_{NN} \times M_{QCD}^2))$. These interactions provide (i) pion-mass- and momentum-dependent corrections in the S waves, such as $D_{2(0)}$ and $C_{2(0)}$, respectively; and (ii) short-range contributions to P waves, such as $C'_{2(1)}$. The pattern repeats at higher orders. Many of the corresponding LECs are necessary for the renormalization of the loops in the potential. The $2N$ potential then resembles some phenomenological potentials, such as AV18, where pion exchange is supplemented by a general short-range structure.

Likewise, the D_0 and E_0 terms come in the subleading $3N$ potential [8, 63]. While E_0 represents a purely short-range $3N$ effect, D_0 gives rise to a mixed-range force when the pion is attached to a third nucleon. In recent years this chiral $3N$ potential has been used in many *ab initio* nuclear calculations—see, *e.g.* Ref. [64]. For a review of chiral potentials under the assumption of NDA, including higher orders, see Ref. [16].

Following Weinberg's original suggestion [1], most calculations with these EFT-based potentials treat them in the same way as phenomenological potentials: once the form of the potential to the desired order has been derived, the appropriate dynamical equation—Lippmann-Schwinger, Schrödinger, or their few-body variants—is solved exactly (within numerical accuracy), and the unknown LECs fitted to data. After a promising start with the N^{3}LO $2N$ potential with Delta [7, 10], many years of efforts have now produced N^{4}LO $2N$ potentials without Delta that fit $2N$ data with an accuracy comparable to the best phenomenological potentials [16]. Adding the $3N$ force that appears at N^{3}LO gives a reasonable description of $A = 3, 4$ systems and beyond [64]. Still, remaining issues have led some to wait for a full N^{4}LO Deltaless potential, while others are rediscovering the Deltaful potential.

A number of processes with external probes have also been considered with Chiral EFT input, under the assumption of NDA for short-range multi-nucleon operators. They have in some cases led to results similar to earlier phenomenology, but in other cases they have given distinct, new predictions—such as, for example, the magnitude of threshold $\gamma d \to \pi^0 d$ [65] and the sign of the charge-symmetry-breaking asymmetry in $np \to d\pi^0$ [66].

But is NDA, inferred in perturbative calculations, valid for the LECs that appear in the non-perturbative, multi-nucleon sector of the EFT? And does it make sense to treat high-order corrections in the potential the same way as lowest order, instead of using perturbation theory as in NRQED? I explain next why *NO* is the right answer to both questions.

4.3.3 Renormalization of Singular Potentials and Power Counting

At the center of any EFT stands the issue of consistency, which of course is much more important than fitting data. Since EFT's model independence stems from an assumed integration over all higher-energy physics, its power counting has to yield approximate RG invariance at each order. In the case at hand, it is not obvious *a priori* that solving a dynamical equation with an NDA-based potential produces physical amplitudes free of cutoff dependence, even if the cutoff dependence of the potential has been removed. Solving a dynamical equation is just a means of accounting for reducible diagrams, which contain loops of a different type than those in the potential, but loops nevertheless. Whether such loops lead to cutoff dependence of the amplitude depends on the high-momentum, or equivalently small-distance, behavior of the potential. The problem is that an EFT-based potential gets more singular for vanishing radial distance, $r \to 0$, as the order increases.

Unfortunately, it is now known that a chiral potential based on NDA, as formulated so far, is not consistent with the RG. Despite their accuracy with respect to data, existing chiral potentials have to be replaced. Chiral potentials are just too singular, in the sense that they behave at the origin worse than r^{-2}, in both the pion-exchange and short-range components. In coordinate space, from Eq. (4.49), the LO OPE between two nucleons is, in spin-singlet and -triplet channels,

$$V_{1\pi,s=0} = \left(\frac{g_A}{2f_\pi}\right)^2 \vec{\tau}_1 \cdot \vec{\tau}_2 \left(\delta(\mathbf{r}) - \frac{m_\pi^2}{4\pi r} e^{-m_\pi r}\right) \quad (4.55)$$

$$V_{1\pi,s=1} = \left(\frac{g_A}{2f_\pi}\right)^2 \vec{\tau}_1 \cdot \vec{\tau}_2 \left[-\frac{1}{3}\left(\delta(\mathbf{r}) - \frac{m_\pi^2}{4\pi r} e^{-m_\pi r}\right)\right.$$
$$\left. + \frac{1}{4\pi r^3}\left(1 + m_\pi r + \frac{(m_\pi r)^2}{3}\right) e^{-m_\pi r} \langle S_{12}(\hat{\mathbf{r}})\rangle\right], \quad (4.56)$$

where $\langle S_{12}(\hat{\mathbf{r}})\rangle = \langle(3\sigma_1 \cdot \hat{\mathbf{r}}\sigma_2 \cdot \hat{\mathbf{r}} - \sigma_1 \cdot \sigma_2)\rangle$ is the matrix element of the tensor operator. The tensor operator mixes waves of $l = j \pm 1$, where it has one positive and one negative eigenvalue, except for $^3P_0$ where it is diagonal with a negative eigenvalue. It also acts on states with $l = j$, where it has a positive eigenvalue. Considering the matrix elements of the isospin operator, the Yukawa part of $V_{1\pi}$ is attractive in isovector (isoscalar) channels for $s = 0$ ($s = 1$). The tensor part of $V_{1\pi,s=1}$ is attractive in some uncoupled waves like $^3P_0$ and $^3D_2$, and in one of the eigenchannels of each coupled wave. OPE is thus much more singular (and complicated!) than the Coulomb potential α/r in NRQED. At LO there are minimally also the (singular) $C_{0(s)}$ delta functions from Eq. (4.52), which can be combined with the delta functions in Eqs. (4.55) and (4.56). And the potential is more singular still at higher orders.

It has been known for a long time that attractive singular potentials require additional short-range parameters, and EFT provides just the tools, via renormalization, to make the solution of the Schrödinger equation well defined. The conclusion about the bad RG behavior of NDA-based potentials is of course independent of the regularization procedure, so for illustration I will consider an intuitive regularization in coordinate space via a cutoff radius $R \equiv 1/\Lambda$. The long-range potential is unmodified for $r > R$, but for $r < R$ it is set to zero and a square well (and its derivatives) is taken instead. This short-range potential is a regularization of delta functions (and derivatives). Renormalization means that at each order the parameter(s) of the short-range potential can be found as functions of R in such a way as to keep low-energy observables R-independent, at least when R is much smaller than the distance scales of the long-range potential.

Since the complicated spin-isospin structure is not particularly relevant for renormalization, let me for simplicity consider a single uncoupled wave with a central potential

$$V(r) = -\frac{\alpha(R)}{2\mu R^2}\theta\left(1 - \frac{r}{R}\right) - \frac{\lambda}{2\mu r_0^2}\frac{f(r/r_0)}{(r/r_0)^n}, \quad (4.57)$$

where r_0 is a characteristic distance in the modulating regular function $f(r/r_0)$ such that $f(0) = 1$, λ is a dimensionless strength, n is an integer, and the dimensionless $\alpha(R)$ is a function of R. For the nuclear case with OPE, $\mu = m_N/2$, $r_0 = 1/m_\pi$, and $|\lambda| \sim m_\pi/M_{NN}$. In $s = 0$ channels, $n = 1$ and $f(x) = \exp(-x)$, while in $s = 1$ channels, $n = 3$ and $f(x)$ is slightly more complicated.

In order to solve this problem, one matches at $r = R$ the log-derivative of the solutions of the Schrödinger equation for the two regions $r < R$ and $r > R$. Let me focus first on $l = 0$, where at zero energy one obtains

$$\sqrt{\alpha(R)} \cot \sqrt{\alpha(R)} = F_n(\lambda, r_0, R), \quad (4.58)$$

with $F_n(\lambda, r_0, R)$ a complicated function obtained from the outside wavefunction. The issue is whether an $\alpha(R)$ can be found when $R \ll r_0$ such that the low-energy T matrix satisfies Eq. (4.16).

The case $n = 1$ might seem innocuous, as in this case the long-range part of the potential is *not* singular. Yet, the delta function is, and we find the first surprise here. In this case the potential is Coulombic for $R < r \ll r_0$, so the external wavefunction is given by a combination of regular and irregular Bessel functions. One finds [67]

$$F_1(\lambda, r_0, R) = -\lambda\frac{R}{r_0}\log\left(\frac{R}{R_\star}\right)[1 + \mathcal{O}(R/r_0)], \quad (4.59)$$

where $R_\star$ is a constant that determines the appropriate combination of external solutions for a given low-energy datum. The desired $\alpha(R)$ is

$$\alpha(R) = \left(k + \frac{1}{2}\right)^2 \pi^2 + 2\lambda\frac{R}{r_0}\log\left(\frac{R}{R_\star}\right) + \mathcal{O}(R^2/r_0^2), \quad (4.60)$$

where k is an integer. One can then show that the amplitude at low (not necessarily zero) energies approaches cutoff independent results.

This is quite satisfactory, *except* for the fact that $\alpha(R)$ has to have a piece linear in $\lambda/r_0 \sim m_\pi^2/M_{NN}$. The interaction that it represents in the Chiral EFT Lagrangian is the chiral-symmetry breaking term with LEC $D_{2(0)}$ in Eq. (4.54). As it is obvious, this interaction is not the same as the $C_{0(s)}$ terms in Eq. (4.52): they give rise to different pionic interactions. By NDA $D_{2(0)}$ was supposed to be N^{2}LO. Yet, we have just found that it is necessary at LO, if we take OPE as LO! Thus the magnitude of $D_{2(0)}$ is determined not by the high-energy scale M_{QCD} but instead by pion scales, and must be $D_{2(0)} = \mathcal{O}(4\pi/(m_N M_{NN}^3))$ instead. Note that this problem does not appear for $l > 0$, where one expects no delta function in LO, only the regular Yukawa potential.

Although this failure of NDA is of no particular consequence for the $2N$ problem itself, where only the combination $C_{0(0)} + D_{2(0)} m_\pi^2$ is measured, it should affect other processes like pion-nucleus scattering. Even more significantly, if NDA fails for this operator due to the non-perturbative nature of pion exchange, other operators might well suffer from the same problem. This was first pointed out in Ref. [68] and traced to the diagram where OPE is sandwiched between two contact interactions. The same authors [69] also noticed that thrice-iterated pion diagrams lead to cutoff dependence, but with two powers of momenta instead of two powers of the pion mass. Since in the NDA-based power counting two-derivative delta functions appear only at N^{2}LO, just like $D_{2(0)}$, this suggests impending disaster—it could imply the need for infinite counterterms once the whole pion ladder is considered. These authors then proposed [69] that pions be treated in perturbation theory, that is, as an expansion in $|\lambda| \sim m_\pi/M_{NN}$. If this is done, LO contains only contact interactions $C_{0(s)}$, while OPE appears first as a single insertion at NLO together with $D_{2(0)}$ and the two-derivative S-wave contacts with LECs $C_{2(s)}$. More generally, a power counting can be devised that is consistent with the RG. (This power counting is, apart from the presence of pions, the same as the one for the Pionless EFT in the next lecture.) In $s = 0$ channels it seems this expansion does converge, although in $^1S_0$ only very slowly [67]. Unfortunately, for $s = 1$, where the tensor force can be attractive, the expansion fails already at $Q \sim f_\pi$ [70].

I am thus back to trying to make sense of the renormalization of non-perturbative OPE in $s = 1$ channels, where $n = 3$. When the potential is repulsive, one can solve the Schrödinger equation in the standard manner without any subtleties. For the interesting case $\lambda > 0$, on the other hand, the two outside solutions vanish as $r \to 0$ but oscillate indefinitely on the way there. There is no way to discard one; instead, we have again a combination of Bessel functions leading for $n \geq 2$ to [71]

$$F_n(\lambda, r_0, R) = \frac{n}{4} - \frac{\sqrt{\lambda}}{(R/r_0)^{n/2-1}} \tan\left(\frac{\sqrt{\lambda}}{(n/2-1)(R/r_0)^{n/2-1}} + \phi_n\right)$$
$$\times \left[1 + \mathcal{O}\big((R/r_0)^{n/2-1}, R/r_0\big)\right], \qquad (4.61)$$

where ϕ_n is fixed by a given datum (the scattering length for $n > 3$). The corresponding $\alpha(R)$, found numerically, has a limit-cycle-like behavior: as R decreases, it decreases from $+\infty$ to $-\infty$ to start again in shorter and shorter cycles. Again, one finds that the scattering amplitude at finite energy is well behaved. We thus *can* renormalize the $l = 0$ wave for a singular potential of this type with a single counterterm, despite the wild cutoff dependence of the diagrams in the corresponding perturbative series.

This bodes well for the LO chiral potential. Indeed, for $n = 3$ $\alpha(R)$ depends on $\lambda r_0 \sim 1/M_{NN}$, which is *independent* of m_π and thus represents a chiral-invariant counterterm like $C_{0(1)}$. One can show [67] that the coupled character of the $^3S_1$-$^3D_1$ channels does not affect this conclusion. The NDA-based power counting seems to work in this case. Or does it?

In the final twist of this saga, hell breaks loose. What about channels with $l > 0$ where the singular potential is attractive? In these channels there is a repulsive centrifugal barrier $l(l+1)/r^2$, which dominates over the $-\lambda r_0/r^3$ potential at large distances. The situation gets reversed at short distances, where $-\lambda r_0/r^3$ again determines the wavefunction. Although the details of the matching change, one concludes that in *each* of these waves we need a new short-range parameter fixed by a low-energy datum. Clearly the NDA-based power counting does not provide any of these LECs at LO. If we insist in varying the cutoff in these waves, as we should, the phase shifts can be anything we want [72]. The simplest example is the $^3P_0$ wave where a bound state crosses threshold for a cutoff ~ 1 GeV. The only way to fix [72] this problem is to take as LO an interaction like the $C'_{2(1)}$ term in Eq. (4.54), which if NDA were correct would only appear at N^{2}LO. Just like $D_{2(0)}$, we must have instead an enhancement, $C'_{2(1)} = \mathcal{O}(4\pi/(m_N M_{NN}^3))$. The situation is similar in the coupled $^3P_2$-$^3F_2$ channels.

How many more LO interactions do we need? A simple estimate comes from the distance where the effective potential $-\lambda r_0/r^3 + l(l+1)/r^2$ is maximum, $r_m = 3\lambda r_0/(2l(l+1))$. When $r_m \lesssim 1/M_{QCD}$ the singular attractive potential is unimportant at large distances. The wavefunction oscillates only outside the region in which one expects the EFT to be valid. At the distances relevant to the EFT, the relatively smooth behavior of the wavefunction can be captured in perturbation theory. When one plugs in numbers, one finds that for $l \gtrsim 2$ pions are likely perturbative. This simple argument is corroborated by a more detailed calculation [73]. Thus, it seems that the correct LO consists of non-perturbative pions in S and P (and maybe D?) waves, with short-range interactions in the S and OPE-attractive P (and maybe D?) waves [72].

With the LO thus established, what about higher orders? It has been shown [74], in the context of a toy model, that as long as they are treated perturbatively (as in NRQED), the corrections in the amplitude can be renormalized with the short-range interactions given by a corrected NDA. In this corrected NDA, for a relative $(Q/M_{QCD})^n$ correction in the long-range potential, one includes short-range operators with n derivatives more than those appearing in LO. In other words, NDA applies once, but only once, we get to the perturbative corrections. For similar, but not identical, conclusions about the correct power counting from an RG-equation perspective, see Ref. [75].

A frequently asked question is, if the corrections are small enough to be perturbative, why can we not just treat them non-perturbatively, as done in existing versions of chiral potentials? The reason is simple: lack of counterterms. Take for example a two-derivative contact interaction that appears in one insertion at N²LO. In the same channel, two insertions will be in N⁴LO, giving a highly singular contribution to the T matrix. This is however no problem as there will be at the same order an equally singular four-derivative contact interaction, which will provide the necessary counterterm. Only the sum of all N⁴LO terms is cutoff independent and small. If I truncate at N²LO (one needs to truncate somewhere...) but decide to iterate both LO and N²LO, I automatically include diagrams with two (and many more!) insertions of the N²LO operator without the required four-derivative counterterm. In general, my result will now be cutoff dependent, and there are likely regions of cutoff space where the "corrections" are no longer small. Not surprisingly, variations of the cutoff for truncations at N³LO and N⁴LO in Weinberg's scheme have been shown to lead to wild variation in the phase shifts [76, 77]. Existing chiral potentials can only fit data accurately in small windows in cutoff space. Note that not everybody thinks lack of RG invariance is important in this context [78].

The first calculations of the $2N$ system based on these new power-counting ideas give encouraging results [79–83], as one might have expected from the existence of more counterterms at each order than in NDA-based potentials. We can be optimistic about the development of a chiral potential that not only fits data well, but is also consistent with the RG. However, much remains to do to gauge the impact of these discoveries in systems with more nucleons and external probes.

4.3.4 Summary

A low-energy EFT of QCD has been constructed and used as input to *ab initio* methods to describe nuclear systems. Chiral symmetry plays an important role, in particular setting the scale for nuclear bound states. Several aspects of the traditional picture of nuclear physics emerge from the chiral potential, which additionally provides consistent few-body forces and currents, and systematic treatment of loop and isospin-breaking corrections. Unfortunately, though, the simplest power counting, based on naive dimensional analysis, is inconsistent with the renormalization group. A new, consistent power counting has been formulated, but is still mostly virgin territory.

4.4 Loosely Bound Systems

Chiral EFT provides a foundation for the physics of nuclei, at least when A is not too large. However, some nuclei are loosely bound in the natural binding energy scale of $\mathcal{O}(M_{nuc}^2/M_{QCD})$. The dynamics of these nuclei mostly takes place at distances large compared to $1/M_{nuc}$. We might expect new degrees of freedom and structures to emerge and, indeed, many loosely bound states display clusterization and other phenomena like Borromean-type binding, where a system is bound even if its subsystems are not. In addition, loosely bound nuclei are important in an astrophysics context, sometimes at energies too low to achieve in the lab.

The relatively large distance scale means that fewer of the features of QCD, such as chiral symmetry, leave an imprint on the physics. On the upside, these systems share similarities with other loosely bound systems, where the underlying dynamics might be dominated by other EFTs than QCD, for example atomic systems where the underlying theory is NRQED. This universality is the overarching theme of this last lecture, where I first discuss how a low-momentum scale might arise, and then how two EFTs—Pionless and Halo/Cluster—describe nuclei at this scale.

4.4.1 Fine-Tuning

It has long been remarked that the deuteron is relatively large. From the deuteron binding energy $B_d \simeq 2.2$ MeV, an estimate for the deuteron binding momentum is $\aleph_1 \sim \sqrt{m_N B_d} \sim 45$ MeV, smaller than m_π by a factor of about 3. This means that the two nucleons are effectively at a distance three times larger than the range of the force, which prompted very early attempts by Bethe and Peierls, and others to describe deuteron physics with only schematic short-range potentials, such as a square well. The situation is even more dramatic for the $^1S_0$ virtual state, a structure in the T matrix to which we can associate a negative energy $-B_{d^*} \simeq -0.07$ MeV and thus a momentum $\aleph_0 \sim \sqrt{m_N B_{d^*}} \sim 8$ MeV, almost 20 times smaller than the pion mass.

For a generic short-range potential of range R, the two-body amplitude in the S wave can be written for $kR \ll 1$ in the form of the effective range expansion (ERE),

$$T_0(k) = \frac{2\pi}{\mu}\left[-\frac{1}{a_2} + \frac{r_2}{2}k^2 + \cdots - ik\right]^{-1}, \qquad (4.62)$$

with a_2 and r_2 the scattering length and effective range parameter, and higher ERE terms not shown explicitly. If the effective range has a natural size $r_2 \sim R$, and the same is true of other ERE parameters, a shallow bound state of binding momentum $k = i\kappa$, $\kappa \sim 1/a_2$, is possible if $|a_2| \gg R$. This is the case of the $2N$ system, given that the effective ranges and other ERE parameters have natural sizes, for example $r_{2(1)} \simeq 1.75$ fm and $r_{2(0)} \simeq 2.8$ fm, but $a_{2(1)} \simeq 5.4$ fm and $a_{2(0)} \approx 20$ fm in the $^3S_1$ and $^1S_0$ channels, respectively, are much larger than $1/m_\pi$. Thus we are close to the unitarity limit defined by $a_2 \to \infty$ with other ERE parameters vanishing.

This situation suggests that the parameters of QCD are fine-tuned. Take as a very simple example a square well in the notation of Eq. (4.57), with $\lambda = 0$ and R not a cutoff but the physical scale associated with the range of the force. In that case one can find an analytic formula for the S-wave T matrix,

$$T_0(k) = i\left[1 - e^{-2ikR}\frac{\sqrt{\alpha + (kR)^2}\cot\sqrt{\alpha + (kR)^2} + ikR}{\sqrt{\alpha + (kR)^2}\cot\sqrt{\alpha + (kR)^2} - ikR}\right], \qquad (4.63)$$

which takes the form (4.62) for $kR \ll 1$, with

$$a_2 = R\left(1 - \frac{\tan\sqrt{\alpha}}{\sqrt{\alpha}}\right), \qquad r_2 = R\left(1 - \frac{R}{\alpha a_2} - \frac{R^2}{3a_2^2}\right), \qquad \ldots \qquad (4.64)$$

For generic $\alpha = \mathcal{O}(1)$, $|a_2| \sim |r_2| \sim R$. However, if we dial α close to the critical value $\alpha_c \equiv [(2n+1)\pi/2]^2$ with n an integer, that is, if $|1 - \sqrt{\alpha/\alpha_c}| \ll 1$, then $R/|a_2| = \alpha_c |1 - \sqrt{\alpha/\alpha_c}| + \cdots \ll 1$, without changing the size of other ERE parameters significantly. By this fine-tuning a low momentum scale $\aleph \equiv |1 - \sqrt{\alpha/\alpha_c}|/R \ll 1/R$ appears in the system. Since there is a zero-energy pole in (4.63) at $\alpha = \alpha_c$, the fine-tuning means a shallow real or virtual bound state. For a real one, the wavefunction is normalizable, $\psi \propto \exp(-r/a_2)/r$, indicating a large size. This type of object is intrinsically quantum mechanical, since in classical physics bound-state sizes are limited by the range of the potential.

The details are different in the nuclear case where the LO potential consists of OPE plus contact interactions, instead of a simple square well. Still, at the physical pion mass the potential parameters must conspire to give the observed large scattering lengths. Now, all chiral-symmetric parameters are tied together by the non-perturbative QCD dynamics determined by the strong-coupling g. But the (current) quark masses, and thus the pion masses, can be considered largely independent of g in the SM. Therefore we can ask the question whether the fine-tuning can be undone by a variation in m_π. In Ref. [67] it was argued, based on a incomplete N^{2}LO analysis, that this might just be the case: the deuteron and virtual state can go unbound or bound with small variations of m_π. With some reasonable assumptions, the deuteron was found to have a more natural binding energy ~ 10 MeV in the chiral limit, and to become unbound at $m_{\pi,c} \simeq 200$ MeV, where the scattering length diverges. This analysis was later refined and compared with emerging full lattice QCD data, as described in Ref. [84]. Currently there is feverish activity in lattice QCD to calculate the binding energies in the $2N$ system (and other light nuclei) at various values of m_π (see Ref. [85] and references therein). Although there is no consensus yet, it seems that $m_{\pi,c}$ might actually be just below the physical pion mass. Either way, if this picture stands the test of lattice QCD, one can see the fine-tuning scale as $\aleph \equiv |1 - m_\pi/m_{\pi,c}| M_{nuc} \ll M_{nuc}$. The curve $a_2(m_\pi)$ [67] is in fact very similar to $a_2(B)$, where B is an external magnetic field, for atoms near a Feshbach resonance. Therefore it could very well be that we can think of QCD as near a Feshbach resonance in the quark masses.

There is no good explanation for this fine-tuning, yet. But we can exploit it by devising simpler EFTs that are valid only for momenta smaller than M_{nuc}, where even pion physics can be considered short-ranged.

4.4.2 Contact EFT

For $Q \sim \aleph \ll M_{nuc}$ (with $\aleph$ some average of $\aleph_{0,1}$), pions (and Deltas) can be integrated out: in few-nucleon systems, only nucleons are relevant degrees of freedom.

Chiral symmetry is badly broken and of no use. The most general Lagrangian with Lorentz and electromagnetic gauge invariance (and P and T symmetries) is a simplified version of what we had in the previous lecture:

$$\mathscr{L}_{piless} = \bar{N}\left(iD_0 + \frac{\mathbf{D}^2}{2m_N} - \frac{\delta m_N}{2}\tau_3\right)N - \frac{1}{4}F_{\mu\nu}F^{\mu\nu}$$

$$+ \frac{e}{4m_N}\bar{N}[1 + \kappa_0 + (1+\kappa_1)\tau_3]\sigma_i N \tilde{F}^{0i} + \cdots$$

$$+ \mathscr{L}_{f\geq 2} \qquad (4.65)$$

in the same notation as before. In the $f = 0, 1$ sectors the theory reduces to NRQED, so I will focus on $f \geq 2$,

$$\mathscr{L}_{f\geq 2} = -\frac{C_{0(0)}}{4}(\bar{N}N\bar{N}N - \bar{N}\boldsymbol{\sigma}N \cdot \bar{N}\boldsymbol{\sigma}N) - \frac{C_{0(1)}}{4}(3\bar{N}N\bar{N}N + \bar{N}\boldsymbol{\sigma}N \cdot \bar{N}\boldsymbol{\sigma}N)$$

$$- \frac{\delta C_{0(0)}}{4}\left(\bar{N}\frac{1+\tau_3}{2}N\bar{N}\frac{1+\tau_3}{2}N - \bar{N}\boldsymbol{\sigma}\frac{1+\tau_3}{2}N \cdot \bar{N}\boldsymbol{\sigma}\frac{1+\tau_3}{2}N\right)$$

$$- E_0\bar{N}N\bar{N}N\bar{N}N$$

$$- \frac{C_{2(0)}}{4}(\bar{N}N\bar{N}\mathbf{D}^2N - \bar{N}\boldsymbol{\sigma}N \cdot \bar{N}\boldsymbol{\sigma}\mathbf{D}^2N + \text{H.c.}) + \cdots, \qquad (4.66)$$

where only some representative interactions are shown, which include an isospin-breaking contact for protons with LEC $\delta C_{0(0)}$. Although I am repeating symbols for some the LECs, it should be kept in mind that they are not the same as in the Chiral EFT of the previous lecture: here they implicitly include pion physics that in Chiral EFT is kept explicit.

It has sometimes been found convenient to reformulate [86] the theory in terms not only of nucleons but also "dibaryon" (or, for atoms, "dimeron") auxiliary fields $\vec{S}$ and $\mathbf{T}$ with the quantum numbers of the two S-wave $2N$ channels, that is, spin (isospin) 0 (1) and 1 (0), respectively. In this case the Lagrangian (4.66) can be rewritten [87] as

$$\mathscr{L}_{f\geq 2} = -\Delta_0 \vec{\bar{S}} \cdot \left(1 + \frac{\delta\Delta_0}{\Delta_0}\frac{1+\tau_3}{2}\right)\vec{S} - \frac{g_0}{\sqrt{2}}[\vec{\bar{S}} \cdot \vec{P}_0 NN + \text{H.c.}]$$

$$- \Delta_1 \vec{\bar{T}} \cdot \mathbf{T} - \frac{g_1}{\sqrt{2}}[\vec{\bar{T}} \cdot P_1 NN + \text{H.c.}]$$

$$+ h\bar{N}\left\{g_0^2 \vec{\bar{S}} \cdot \vec{\tau}\vec{S} \cdot \vec{\tau} + \frac{g_0 g_1}{3}[\vec{\bar{T}} \cdot \boldsymbol{\sigma}\vec{S} \cdot \vec{\tau} + \text{H.c.}]\right.$$

$$\left. + g_1^2 \bar{N}\vec{\bar{T}} \cdot \boldsymbol{\sigma}\mathbf{T} \cdot \boldsymbol{\sigma}\right\}N + \sigma_0 \vec{\bar{S}} \cdot \left(iD_0 + \frac{\mathbf{D}^2}{4m_N}\right)\vec{S}$$

$$+ \sigma_1 \bar{\mathbf{T}} \cdot \left(iD_0 + \frac{\mathbf{D}^2}{4m_N}\right)\mathbf{T} + \cdots \qquad (4.67)$$

where $\vec{P}_0 = \sigma_2 \vec{\tau} \tau_2/\sqrt{2}$ and $\mathbf{P}_1 = \tau_2 \boldsymbol{\sigma} \sigma_2/\sqrt{2}$ are the projectors onto the S-wave $2N$ channels of spins $s=0, 1$, $\Delta_{0,1}$, $\delta\Delta_0$, $g_{0,1}$, and h are LECs—which can be thought of as the residual masses and mass splitting of the dibaryons, their couplings to two nucleons, and a dibaryon-nucleon interaction— and $\sigma_{0,1} = \pm 1$. Integrating out these auxiliary fields we regain Eq. (4.66) with certain relations between the LECs.

We can define the potential as before, so that amplitudes consist simply of iterations of the potential. But here the potential takes the particularly simple form of a sum of delta functions and their derivatives or, alternatively, of dibaryon propagation in the s channel. Either way, the potential contains no loops. All there is to do is to understand the ordering of the various terms, calculate the loops that appear in the iteration of the potential, and make sure observables are RG invariant.

4.4.2.1 The Two-Nucleon System

For a system with $|a_2| \gg |r_2| \sim R$, two-body physics at momenta $Q \sim \aleph \equiv 1/|a_2|$ can be described in a simple expansion in $R/|a_2| \ll 1$. Since effective range and other ERE terms should have natural size, in a first approximation only the non-derivative contact interactions should contribute [88].

The two-body amplitude in LO is particularly simple, being a sequence of potential insertions separated by box-like loops as in Eq. (4.27), where instead of Coulomb interactions we have a contact interaction—see Fig. 4.5. The two simplest diagrams are the single and once-iterated contact interaction, in each S-wave channel, respectively,

$$T_{1c} = -C_{0(s)} \tag{4.68}$$

and

$$T_{2c} = -iC_{0(s)}^2 \int \frac{d^4 l}{(2\pi)^4} \frac{1}{l^0 + p^0 - (\mathbf{l}+\mathbf{p})^2/2m_N + i\varepsilon}$$
$$\times \frac{1}{-l^0 + p^0 - (\mathbf{l}+\mathbf{p})^2/2m_N + i\varepsilon}$$
$$= m_N C_{0(s)}^2 \int \frac{d^3 l}{(2\pi)^3} \frac{1}{\mathbf{l}^2 - 2m_N p^0 - i\varepsilon}$$
$$= \frac{m_N}{4\pi} C_{0(s)}^2 \left[\gamma_1 \Lambda + ik + \mathcal{O}(k^2/\Lambda) \right], \tag{4.69}$$

where γ_1 is a constant that depends on the exact regulator used to make the loop integral well-defined ($\gamma_1 = 2/\pi$ for a sharp momentum cutoff, for example). The potentially dangerous dependence on a positive power of Λ can be absorbed in $C_{0(s)}(\Lambda)$ itself, while the $1/\Lambda$ terms can be taken care of by higher-derivative contacts. As in any (properly renormalized) EFT, the meaningful contribution of the loop is the term that is non-analytic in the energy $2p^0 = k^2/m_N$. We can see explicitly that an

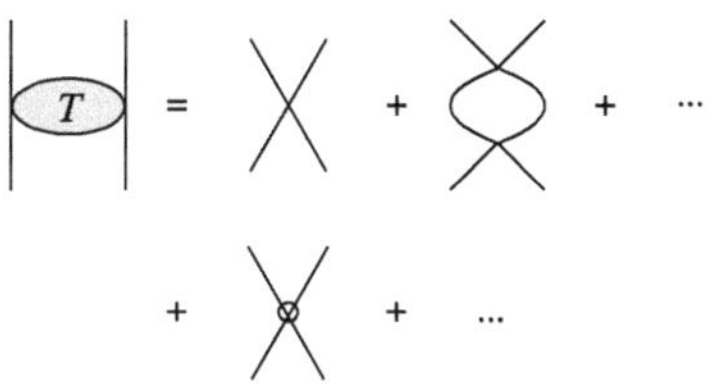

4 Effective Field Theories of Loosely Bound Nuclei

Fig. 4.5 Diagrams representing the T matrix for the elastic scattering of two nucleons in Pionless EFT. A nucleon is represented by a *solid line*. A *circle* at the vertex denotes an inverse power of M_{nuc}

intermediate state does indeed contribute $\mathcal{O}(m_N Q/(4\pi))$ to the amplitude as argued before. Each iteration therefore brings in a factor of $\mathcal{O}(m_N Q C_{0(s)}/(4\pi))$.

If $C_{0(s)} \equiv 4\pi/(m_N \aleph_s)$, then the LO two-body amplitude needs to be resummed for $Q \sim \aleph_s$, as we have done schematically before in Eqs. (4.28) and (4.50):

$$T^{(0)} \sim \frac{4\pi}{m_N \aleph_s}\left[1 - \mathcal{O}\left(\frac{Q}{\aleph_s}\right)\right]^{-1}. \tag{4.70}$$

But here we can be more explicit because, contrary to the earlier cases, the series is an exact geometric series:

$$T^{(0)} = \frac{4\pi}{m_N}\left[-\left(\frac{4\pi}{m_N C_{0(s)}} + \gamma_1 \Lambda\right) - ik\right]^{-1}\left[1 + \mathcal{O}\left(\frac{k}{\Lambda}\right)\right]. \tag{4.71}$$

Comparison with Eq. (4.62) shows that we recover the ERE in LO, with the cutoff-independent combination $C_{0(s)}^R \equiv C_{0(s)}(\Lambda)/(1 + m_N \gamma_1 \Lambda C_{0(s)}(\Lambda)/4\pi) = 4\pi a_{2(s)}/m_N$ capturing the physics of the large scattering length. The amplitude has a pole at imaginary momentum $k = i\kappa_s = i/a_{2(s)}$, which represents a real (virtual) bound state for $a_{2(s)} > 0$ (<0) with binding energy $B_{2(s)} = 1/(m_N a_{2(s)}^2)$.

Thus the fine-tuning that generates a shallow S-wave bound state can be accounted for if $C_{0(s)}$ depends on the anomalously low scale $\aleph_s$. We can consider corrections that account for natural ERE parameters if the LECs of derivative operators scale with $\aleph_s$ and M_{nuc} in a particular way. The $\aleph_s$ enhancement depends on whether the LEC contributes to the S-wave. For example, $C_{2(s)} \sim 4\pi/(m_N M_{nuc} \aleph_s^2)$ gives rise to a relative correction $\mathcal{O}(Q^2/(M_{nuc}\aleph))$ in Eq. (4.71), which incorporates physics of an effective range $r_2 \sim 1/M_{nuc}$. Similar scalings apply to higher S-wave parameters, but not in other waves, where no shallow bound states exist. Thus, for example, $C'_{2(s)} \sim 4\pi/(m_N M_{nuc}^3)$ gives the leading P-wave contribution to the amplitude at relative $\mathcal{O}(Q^3/M_{nuc}^3)$. Higher waves appear at even higher orders. For more details, see Refs. [69, 89].

Note that, as in NRQED and Chiral EFT, the corrections in Eq. (4.71) should in principle be treated in distorted-wave perturbation theory, when RG invariance can be maintained. In the particular case of the dominant correction, $C_{2(s)}$, at NLO we consider just one insertion of its vertex, and any number of $C_{2(0)}$ vertices. At N²LO we need two insertions, with one from the four-derivative interaction with

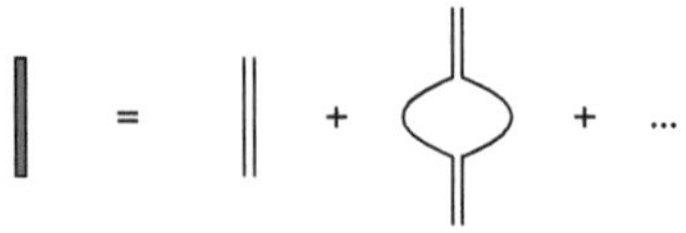

$$\Big| \;=\; \Big| \;+\; \bigcirc \;+\; \cdots$$

Fig. 4.6 Diagrams representing the reduced $\tilde{T}$ matrix for the elastic scattering of two nucleons in Pionless EFT. A nucleon is represented by a *solid line*, a dibaryon by a *double solid line*

LEC $C_{4(0)}$. Actually, the contributions from $C_{2(s)}$ can be resummed to the form (4.62) without destroying RG invariance, but only as long as $r_{2(s)} \leq 0$ [90], a form of the so-called Wigner bound. Since $r_{2(s)} > 0$ in the $2N$ case, one should indeed refrain from deviating from perturbation theory.

There is not much difficulty in adding Coulomb effects. As we have seen in NRQED, Coulomb becomes non-perturbative for $Q \lesssim \alpha m_N \sim 5$ MeV. This is not too far from the $^1S_0$ virtual bound state, that is, Coulomb effects $\mathcal{O}(4\pi\alpha/Q^2)$ become comparable to $C_{0(0)} = 4\pi/m_N \aleph_0$ for $Q^2 \lesssim \alpha m_N \aleph_0 \sim \aleph_0^2$. In the two-proton system, for such low momentum one needs to account for Coulomb in addition to a contact interaction at LO, which introduces new cutoff dependence. We therefore need to consider the additional contact interaction $\delta C_{0(0)}$, which is isospin breaking, to absorb this cutoff effect [91]. Since $\delta C_{0(0)}$ needs to be fitted to the pp scattering length, isospin breaking cannot be predicted at this order. Fortunately, at higher energies Coulomb and other electromagnetic interactions can be treated in perturbation theory.

We can recast these statements in the dibaryon formulation, where the dibaryon residual masses are taken to be fine-tuned, $\Delta_{0,1} \sim \aleph_{0,1}$, while the coupling constants are taken as natural-sized, for example $g_{0,1}^2 \sim 4\pi/m_N$. The dibaryons can be thought of as "bare" real and virtual bound-state fields, although this implies nothing about their composite nature. The S-wave $2N$ amplitudes are just the couplings of two nucleons to dibaryon propagators that are "dressed" by $2N$ loops. In LO only the dibaryon residual mass is needed in addition to the two-nucleon/dibaryon coupling. For the strong-interacting sector we can write

$$T = g_s^2 \tilde{T}, \tag{4.72}$$

with the reduced T matrix $\tilde{T}$ being the sum of successive bare dibaryon propagators depicted in Fig. 4.6,

$$\tilde{T}_{1d} + \tilde{T}_{2d} + \cdots = \frac{4\pi}{m_N g_s^2} \left[\left(\frac{4\pi \Delta_s}{m_N g_s^2}\right) - \gamma_1 \Lambda\right) - ik\right]^{-1} \times \left[1 + \mathcal{O}\left(\frac{k}{\Lambda}\right)\right], \tag{4.73}$$

from which we see that renormalizability is achieved by absorbing the Λ dependence in Δ_s/g_s^2. Coulomb can be included just like above, requiring a renormalization of $\delta\Delta_0$. Dibaryon kinetic terms generate effective ranges at NLO, but these kinetic terms have signs $\sigma_{0,1}$ given by the sign of $-r_{2(0,1)}$. Since $r_{2(s)} > 0$, $\sigma_s < 0$ and the bare dibaryons are ghosts. However, their character changes when they get dressed, and the $2N$ amplitude has no pathology.

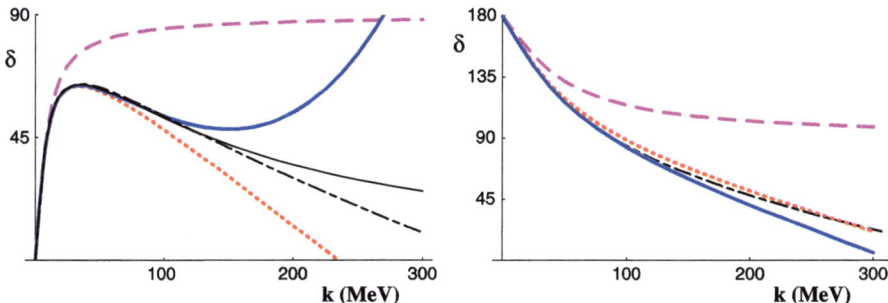

Fig. 4.7 The $^1S_0$ (*left*) and $^3S_1$ (*right*) $2N$ phase shifts (in degrees) as functions of the center-of-mass momentum (in MeV). The *dot-dashed lines* represent the Nijmegen phase-shift analysis [94]. *Left*: the *dashed*, *dotted*, and *thick solid lines* show the Pionless EFT results at LO, N^{2}LO, and N^{4}LO, respectively, while the *thin solid line* shows the ERE. *Right*: the *dashed*, *dotted*, and *thick solid lines* show the Pionless EFT results at LO, NLO, and N^{2}LO, respectively. From Refs. [12, 92], courtesy of M. Savage

This approach, in either formulation, has been shown to give a very clear path to analyze low-energy reactions involving two nucleons systematically [13]. It is a field-theoretical generalization of the ERE. The resulting $2N$ phase shifts converge to empirical values for $Q \lesssim M_{nuc}$, as shown for the S waves in Fig. 4.7 [12, 92]. Deuteron properties come out well; for example the deuteron binding energy is found to be $B_d = 1.9$ MeV in NLO, to be compared with the experimental value of 2.2 MeV. More generally, this EFT can be applied to any system with $|a_2| \gg |r_2|$, for example bosonic or fermionic atoms near a Feshbach resonance [93].

4.4.2.2 The Three-Nucleon System

The $3N$ system proves to be much more interesting, since here the EFT is not just the ERE. There is no symmetry to forbid three-body forces like the E_0 term in Eq. (4.66) or the h term in Eq. (4.67). As always in any EFT, the question is just at what order these novel effects appear. If NDA were any guide, one would expect them at relatively high orders since their canonical dimensions are high. However, we are dealing with a fine-tuned situation, and surprises are in stock.

For definiteness, let me consider neutron-deuteron (nd) scattering in the dibaryon formulation, see Fig. 4.8. The simplest diagram consists of the transfer of one nucleon from the dibaryon to the third nucleon, and it is $\mathcal{O}(m_N g_s^2/Q^2)$. A transfer back adds another $m_N g_s^2/Q^2$ multiplied by a factor $Q^3/(4\pi)$ from the loop and a $1/Q$ from the intermediate, dressed dibaryon, so it is of relative $\mathcal{O}(m_N g_s^2/4\pi) = \mathcal{O}(1)$. Thus, power counting says that in LO one has to sum all nucleon exchanges between the dimeron and the third nucleon, resulting [88] in an integral equation for the T matrix known as the Skorniakov–Ter-Martirosian equation. As always, corrections can be treated in perturbation theory.

The behavior of this T matrix at large momentum turns out to depend sensitively on the strength of the kernel of the integral equation, which in turn depends on

Fig. 4.8 Diagrams representing the T matrix for the elastic scattering of a neutron on a deuteron in Pionless EFT. Notation as in Fig. 4.6

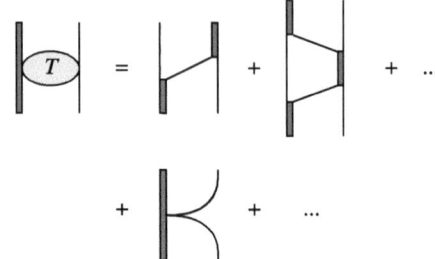

the spins of particle and dimer. For a two-state fermion instead of a nucleon, when the dimer has $s = 0$, the amplitude falls fast at large momenta and the solution of Skorniakov–Ter-Martirosian equation is RG invariant, consistent with three-body forces appearing only at high orders. The same is true for nucleons in all but the $S_{1/2}$ wave, and very accurate results for nd scattering follow from parameters fully determined in $2N$ scattering [88, 95]. For example, for the $S_{3/2}$ phase shift shown in Fig. 4.9 excellent agreement with data is achieved already at N^2LO. In particular, the scattering length is postdicted as $a_{3/2} = 6.33 \pm 0.10$ fm, to be compared to the experimental value, 6.35 ± 0.02 fm. This example shows that Pionless EFT enables nearly QED-quality nuclear physics.

On the other hand, for three bosonic particles or for nucleons in the $S_{1/2}$ wave of Nd scattering, the amplitude obtained from nucleon exchange alone has a very peculiar Λ dependence, proportional to $\cos(\ln \Lambda)$. A bound state of energy $\mathcal{O}(\Lambda^2/m_N)$ is in the spectrum, representing the well-known "Thomas collapse" of the ground state as $\Lambda \to \infty$. RG invariance can only be achieved if three-body interactions are enhanced by $\aleph^{-2}$ [96, 97]. A single non-derivative three-body interaction appears at LO, providing saturation to avoid the collapse. Higher-derivative interactions are smaller by powers of Q/M_{nuc}, and in fact to NLO there is only one parameter not fixed by $2N$ observables: the coefficient of the three-body force, h (or equivalently E_0). As a consequence, to this order three-body observables are correlated through this one parameter. This explains [87], in particular, why results obtained from $2N$ models cluster around a "Phillips line" in the plane generated by all possible values of the $S_{1/2}$ Nd scattering length $a_{1/2}$ and the triton binding energy B_t: the off-shell differences among $2N$ models are essentially captured by one parameter. Any point on the EFT line fixes the LO $3N$ parameter, which, as a function of Λ, displays an unusual, limit-cycle behavior [96]. If we use as input the $a_{1/2}$ experimental value, we find $B_t = 8.54$ MeV at NLO [98], to be compared with the experimental value of 8.48 MeV. The resulting energy dependence of $S_{1/2}$ Nd scattering also comes out very well, as shown in Fig. 4.9.

The limit cycle in the three-body force reflects a residual, approximate discrete scale invariance. At unitarity, the LO two-body T matrix is scale invariant in the limit $\Lambda \to \infty$. The only bound state is at zero energy. The regularization and renormalization of the three-body problem breaks scale invariance, except for the discrete scale invariance that survives as the cutoff takes values that are multiple of a value determined by the three-body datum. A consequence is that a geometric spectrum

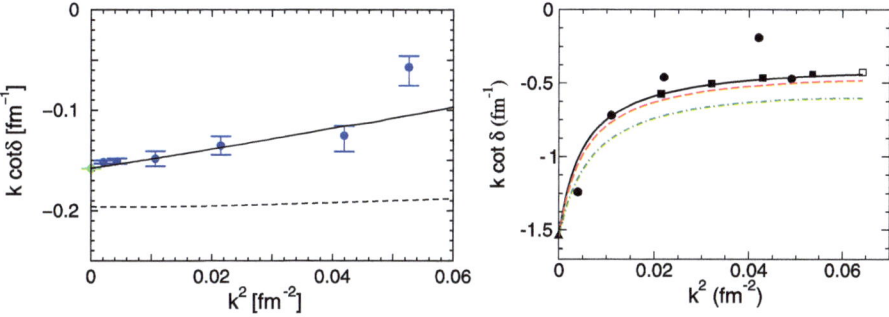

Fig. 4.9 The $S_{3/2}$ (*left*) and $S_{1/2}$ (*right*) Nd K^{-1} matrices (in fm^{-1}) as function of the square of the center-of-mass momentum (in fm^{-2}). *Dots* represent a cold-neutron measurement [99] and a phase-shift analysis [100, 101]. *Left*: *dashed* and *solid lines* show the Pionless EFT results at LO and N^{2}LO, respectively. *Right*: *dot-dashed*, *dashed* and *solid lines* show the Pionless EFT results at LO, NLO, and N^{2}LO, respectively, while the *squares* come from a phenomenological potential model [102]. From Refs. [95, 103], courtesy of H.-W. Hammer and L. Platter

of bound states appears, the famous Efimov effect. For large but finite scattering length and non-zero ERE parameters, scale invariance is only approximate to start with, and only bound states with binding energies $\lesssim 1/(2\mu R^2)$ are within the EFT.

Including Coulomb interactions in $3N$ calculations is a bit challenging, but it has been done (see Ref. [104] and references there in). Interactions with external photons, such as the triton electromagnetic form factor [105], are also beginning to be investigated. For a recent example of application to cold atoms, see Ref. [106]

4.4.2.3 The Four-Nucleon System and Beyond

I can proceed in a similar way to larger systems. Faced with the appearance of a $3N$ force at LO, an obvious question is whether other few-body forces are also leading. A hand-waving argument suggests they are not. The two-body system is made stable in Pionless EFT by a balance between kinetic repulsion and potential attraction. As we go to the three-body system, the number of pairs grows faster than the number of particles, leading to a collapse unless an effectively repulsive three-body force exists. As we add a fourth body, the number of triplets increases faster than doublets, and no instability and dramatic cutoff dependence should arise. Although a four-body force without derivatives exists, it might not be enhanced by inverse powers of $\aleph$. With four spin-isospin states, we cannot construct a five- or more-nucleon contact force without derivatives, so they are likely not to be LO either.

Since by this argument stability comes from a balance between two-body attraction and three-body repulsion, one expects properties of larger systems, such as four-body and nuclear-matter binding energies (if within Pionless EFT), to scale approximately with the LO three-body parameter. This has been shown to be true for the four-boson binding energy, at least over a limited Λ range [107]. In the $4N$

system one observes the Tjon line, which is the analog of the Phillips line, but for the alpha-particle binding energy B_α instead of the $S_{1/2}$ Nd scattering length $a_{1/2}$. This line is reproduced in Pionless EFT [108]. The LO EFT line depends a bit on which $2N$ parameters are used as input, but in any case it is close to the experimental point: at the correct B_t one finds B_α between 26.9 and 29.5 MeV, to be compared to the experimental value of 28.3 MeV. This agreement suggests that the EFT is converging for the alpha particle. An NLO calculation [109] seems to support this conclusion.

In the region $M_{nuc} \gg Q \gg \aleph$ the $2N$ T matrix has an approximate $SU(4)$ symmetry in spin-isospin space [110]. Since the LO $3N$ force is also $SU(4)$ symmetric [87], Pionless EFT provides a rationale for the emergence of Wigner's supermultiplet symmetry in nuclei.

However, a crucial question is how far we can go in A before pions can no longer be considered short-ranged. After all, binding energies per nucleon, and thus binding momenta, increase throughout the light-nuclear region as the number of nucleons increases. An answer to this question can only be provided by explicit calculation.

Because the continuum methods used so far for $A \leq 4$ tend to become unpractical quickly, one is led to introduce an explicit infrared (IR) momentum regulator or cutoff λ so as to discretize the set one-body states. One can think of $1/\lambda^3$ as providing an effective volume to which the system is confined, just as the inverse UV regulator, $1/\Lambda$, can be thought of as a minimum accessible length scale. These two regulators define the "model space" where the EFT is solved. At the end of the calculation, we need to take the limit of a large model space, $\lambda \ll Q \ll \Lambda$.

In this context, two such IR regulators have been proposed and are being actively pursued. The first [111] borrows from lattice QCD: we define the EFT at N^3 lattice points separated by a spacing $a \sim 1/\Lambda$, which make a cubic volume with sides of length $L = Na \sim 1/\lambda$. The second [112] borrows instead from an existing nuclear-structure method, the No-Core Shell Model (NCSM): the EFT is solved in a harmonic-oscillator well of frequency $\omega \sim \lambda^2/m_N$, with a maximum number of shells $2n + l \leq N_{max} \sim \Lambda^2/(m_N\omega)$—where n (l) denotes the radial (angular-momentum) quantum number—above the minimum configuration.

The limitation to an effectively finite volume poses the challenge of how to relate the LECs to observables. At LO the three LECs can be fitted to the binding energies of the lightest nuclei (deuteron, triton, alpha particle), with other binding energies being predictions (or postdictions). For example, using the NCSM method, one finds [112] for the alpha-particle excited state an excitation energy $E_{\alpha*} = 18.5$ MeV, in remarkable (for LO) agreement with the experimental value 20.2 MeV, while for the ^{6}Li ground state $B_{^6Li} = 23$ MeV to be compared to the experimental result, 32 MeV, in line with an expansion parameter $|r_2|/|a_2| \sim 1/3$. However, the growth in number of LECs with order demands the more abundant scattering data as input. Fortunately, for both methods we can related the energy levels in the model space to $2N$ scattering parameters [113, 114]. Thus, the $2N$ LECs can be fitted to $2N$ levels and, with just a few few-body inputs, predictions made for systems with larger number of particles.

Just as before, such a framework can be applied with simplifications also to cold atoms, see for example Ref. [115]. And these methods are being generalized to Chiral EFT—for reviews and more details, see Refs. [116, 117]. But these new ideas can also be used with more phenomenological input. The generic idea of using λ and Λ to extrapolate to larger model spaces, for example, is useful in calculations with phenomenological nuclear potentials [118, 119]. Conversely, other *ab initio* methods could be brought to bear on Pionless EFT. While Chiral EFT is the ultimate goal, because of its simplicity Pionless EFT plays an important role in providing a paradigm for the development of nuclear EFTs.

4.4.3 Halo/Cluster EFT

Pionless EFT simplifies the treatment of light nuclei, but its application to larger nuclei—even with a powerful *ab initio* method such as the NCSM—still faces difficult computational challenges. One would like to devise further simplifications in order to extend EFTs to even larger nuclei.

One might, in particular, wonder about the implications of the existence of the fine-tuned scale $\aleph$. While we expect the typical energy per nucleon to be $\mathcal{O}(M_{nuc}^2/m_N)$, there are some nuclear states with energies closer to $\mathcal{O}(\aleph^2/m_N)$. This is, in fact, the case of light nuclei, which, as we have just seen, seem to be within the regime of Pionless EFT. But this also happens in two more general classes of states: "halo" and "cluster" nuclei, in which one or more clusters of nucleons ("cores") with the structure of typical nuclei are surrounded by loosely bound ("halo") nucleons. Because of saturation, the radius of a typical cluster with A_c nucleons should be $R_c \sim A_c^{1/3}/M_{nuc} \equiv 1/M_c$. As long as $\aleph \ll M_c$, we can treat cores as effective degrees of freedom, thus generalizing Pionless EFT, where $A_c = 1$, to Halo/Cluster EFT [120].

These classes of systems exhibit shallow S-matrix poles, either on the imaginary axis (bound states) or in the lower half of the complex momentum plane (resonances). Many nuclei display, or are good candidates to display, halo/cluster structure with various types of cores. The simplest and perhaps most clear-cut examples involve alpha-particle cores, for which the excitation energy $E_{\text{core}} \simeq 20 \text{ MeV}$. While ^{5}He is not bound, the total cross section for neutron-alpha ($n\alpha$) scattering has a prominent bump at $E_{\text{halo}} \sim 1 \text{ MeV}$, interpreted as a shallow $P_{3/2}$ resonance. To describe scattering at such low energy, a two-body $\alpha + n$ approach should suffice. ^{6}He is bound, but the removal energy for two neutrons from ^{6}He is again $E_{\text{halo}} \simeq 1 \text{ MeV}$, making this a three-body, $\alpha + n + n$ halo nucleus. Similarly, ^{8}Be is not bound but $\alpha\alpha$ scattering shows an S_0 resonance at $E_{\text{halo}} \simeq 0.1 \text{ MeV}$. And both ^{9}Be and ^{12}C exhibit states near three-body thresholds, respectively $\alpha + \alpha + n$ and $\alpha + \alpha + \alpha$—the latter being the famous Hoyle state that plays an important role in the formation of C and O, and thus you and me, in the universe. Then we can consider also structures with protons, *e.g.* ^{5}Li as $\alpha + p$. For a compilation of data on these and some other halo/cluster states (and much more), see Ref. [121].

In Halo/Cluster EFT we thus consider explicitly a field for the core, for example a scalar field α for the α core of mass m_α:

$$\mathscr{L}_{\text{halo}} = \bar{N}\left(iD_0 + \frac{\mathbf{D}^2}{2m_N} - \frac{\delta m_N}{2}\tau_3\right)N + \bar{\alpha}\left(iD_0 + \frac{\mathbf{D}^2}{2m_\alpha}\right)\alpha$$

$$-\frac{1}{4}F_{\mu\nu}F^{\mu\nu} + \cdots + \mathscr{L}_{\geq 2}. \tag{4.74}$$

It is again extremely convenient to express $\mathscr{L}_{\geq 2}$ using dimeron fields, in this case a spin-3/2, isospin-1/2 field T_3 and a scalar, isoscalar ϕ for the ^{5}Li/^{5}He and ^{8}Be ground states, respectively:

$$\mathscr{L}_{\geq 2} = \bar{T}_3\left[\sigma_3\left(iD_0 + \frac{\mathbf{D}^2}{2(m_N + m_\alpha)}\right) - \Delta_3\left(1 + \frac{\delta\Delta_3}{\Delta_3}\frac{1+\tau_3}{2}\right)\right]T_3$$

$$+ \bar{\phi}\left[\sigma_0\left(iD_0 + \frac{\mathbf{D}^2}{4m_\alpha}\right) - \Delta_0\right]\phi + \frac{g_0}{\sqrt{2}}[\bar{\phi}\alpha\alpha + \text{H.c.}]$$

$$+ \frac{g_3}{\sqrt{2}}\left[\bar{T}_3\left(1 + \frac{\delta g_3}{g_3}\frac{1+\tau_3}{2}\right)\mathbf{S}\cdot(\alpha\mathbf{D}N + N\mathbf{D}\alpha) + \text{H.c.}\right]$$

$$+ h_3\bar{T}_3T_3\overline{(\mathbf{D}N)}\cdot\mathbf{D}N + \cdots, \tag{4.75}$$

where $\mathbf{S}$ is again the spin transition matrix, $\Delta_{0,3}$, $\delta\Delta_3$, $g_{0,3}$, δg_3, and h_3 are the most important LECs, and $\sigma_{0,3}$ are signs. This Lagrangian has a form similar to Eq. (4.67), except for the spin/isospin differences and the P-wave coupling of the ^{5}He/^{5}Li dimeron.

Similar Lagrangians can be written for other core types. If the core has a small number of low-lying excited states, they can be included as extra fields just like the Delta in Chiral EFT. The main drawback of Halo/Cluster EFT is the relatively large number of undetermined LECs, as different cores demand different LECs. As in other EFTs, one would like to eventually determine the LECs by matching low-energy amplitudes to *ab initio* calculations based on Pionless or Chiral EFTs. In the meantime they can be fitted to data.

4.4.3.1 Two-Body Systems

As with Pionless EFT, the first step is to determine the two-body LECs. Were we looking at a core-nucleon system that supports a shallow S-wave bound state, things here would be very similar to the $2N$ case. But many of the two-body systems of interest in Halo/Cluster EFT, and in particular $N\alpha$ and $\alpha\alpha$, have a shallow resonance instead of a bound state. The different pole structure requires a different power counting than in the $2N$ system. In this case the T matrix will be made out of dimeron propagators connected by two-particle bubbles just as before. But the required two-pole structure arises if both the kinetic and residual-mass terms in the dimeron propagators are of similar size.

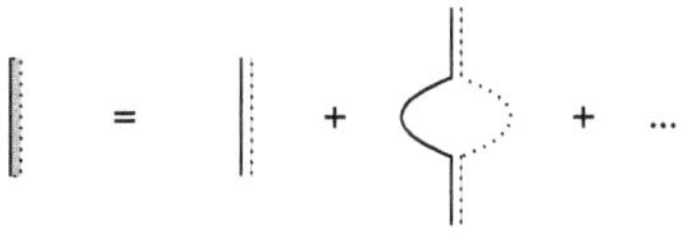

4 Effective Field Theories of Loosely Bound Nuclei

Fig. 4.10 Diagrams representing the reduced $\tilde{T}$ matrix for the elastic scattering of a nucleon on an alpha particle in Halo/Cluster EFT. A nucleon (alpha particle) is represented by a *solid (dotted) line*, and a dibaryon by a *double solid/dotted line*

The particular case of a narrow resonance, $\kappa_I \ll \kappa_R$, requires a single fine-tuning in the dibaryon mass [122]. For $N\alpha$, $\Delta_3 \sim \aleph^2/\mu_{3N}$, where $\mu_{3N} = m_N m_\alpha/(m_N + m_\alpha)$ is the $N\alpha$ reduced mass, while other parameters do not depend on $\aleph$, for example $g_3^2 \sim 2\pi/(\mu_{3N}^2 M_c)$. For simplicity, let me focus on $n\alpha$, Coulomb corrections being necessary for $p\alpha$. The T matrix in the $P_{3/2}$ channel can be written as [120]

$$T = \frac{g_3^2 k^2}{3}(2\cos\theta + i\sin\theta\boldsymbol{\sigma}\cdot\hat{\mathbf{n}})\tilde{T}, \qquad (4.76)$$

in terms of $\mathbf{n} = \mathbf{p}\times\mathbf{p}'/|\mathbf{p}\times\mathbf{p}'|$, the scattering angle θ, and the reduced $\tilde{T}$ matrix shown in Fig. 4.10. The contribution from a single dibaryon propagator is

$$\tilde{T}_{1d} = \frac{1}{\Delta_3 - \sigma_3 k^2/(2\mu_{3n})}, \qquad (4.77)$$

which is the analog of Eq. (4.68) in Pionless EFT. For $Q \sim \aleph$, this is $\mathcal{O}(\mu_{3N}/\aleph^2)$. For the once-iterated dibaryon propagator, the intermediate bubble is similar to Eq. (4.69), except for the presence of two extra momenta in the numerator inside the integral, which is therefore more sensitive to the cutoff:

$$\tilde{T}_{2d} = \frac{\mu_{3n} g_3^2}{6\pi}\tilde{T}_{1d}^2\left[\gamma_3\Lambda^3 + \gamma_1\Lambda k^2 + ik^3 + \mathcal{O}\left(\frac{k^4}{\Lambda}\right)\right], \qquad (4.78)$$

where γ_3 is another number that depends on the specific regulator choice ($\gamma_3 = \gamma_1/3$ for a sharp momentum cutoff, for example). This more severe cutoff dependence can be absorbed in a renormalization of both g_3^2 and Δ_3 and, as usual in an EFT, after renormalization the loop contributes a non-analytic term ik^3. Relative to $\tilde{T}_{1d}$, this contribution is $\mathcal{O}(\aleph/M_c)$, that is, one order down in the expansion. This means the dimeron propagator can generically be treated in perturbation theory, the LO amplitude being given by Eq. (4.77),

$$T^{(0)} \sim \frac{2\pi}{\mu_{3N} M_c}\left[1 - \mathcal{O}\left(\frac{Q^2}{\aleph^2}\right)\right]^{-1}. \qquad (4.79)$$

With appropriate signs, this amplitude generates a pair of shallow poles on the real axis at $Q \sim \pm\aleph$. The NLO unitarity correction (4.78) moves these poles below the axis, but the resulting width is relative small, meaning the resonance is narrow.

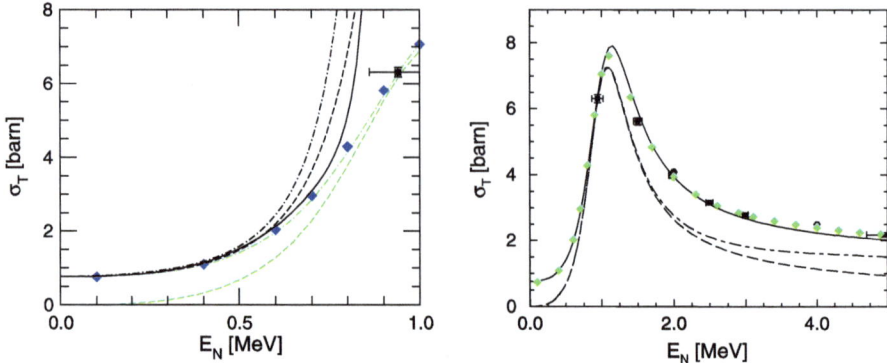

Fig. 4.11 Total cross section (in barn) for $n\alpha$ scattering as a function of the neutron kinetic energy (in MeV) in the α rest frame below (*left*) and around (*right*) the $P_{3/2}$ resonance. *Diamonds* are evaluated data [123], and *black squares* are experimental data [124, 125]. *Left*: the *dashed*, *dot–dashed* and *solid black lines* are the Halo/Cluster EFT results without resummation at LO, NLO, and N^{2}LO, and the *dashed* and *dot-dashed gray lines* the Halo/Cluster EFT results with $P_{3/2}$ resummation at LO and NLO, respectively. *Right*: the *dashed* and *solid lines* are the Halo/Cluster EFT results with $P_{3/2}$ resummation at LO and NLO, respectively. (The *dot-dashed line*, which can be ignored, shows the LO result in a modified power counting with resummation in the $P_{1/2}$ channel as well.) From Refs. [120, 122]

Now, when the external energy is in a window of $\mathcal{O}(\aleph^2/M_c)$ around the resonance the denominator in the bare propagator becomes anomalously small, requiring resummation of the propagator and bubble insertions [122] as in the S-wave bound-state case in Pionless EFT (4.73) or in the vicinity of the Delta pole in Chiral EFT:

$$\tilde{T}_{1d} + \tilde{T}_{2d} + \cdots = \frac{6\pi}{\mu_{3n} g_3^2} \left[\left(\frac{6\pi \Delta_3}{\mu_{3n} g_3^2} - \gamma_3 \Lambda^3 \right) - \left(\frac{6\pi \sigma_3}{2\mu_{3n}^2 g_3^2} - \gamma_1 \Lambda \right) k^2 - ik^3 \right]^{-1}$$
$$\times \left[1 + \mathcal{O}\left(\frac{k}{\Lambda} \right) \right], \tag{4.80}$$

which has the form of the ERE for a P wave. From this expression renormalizability is clear. It is also apparent that the scattering volume (the P-wave analog of the scattering length) is large, $\sim 1/(M_c \aleph^2)$, while the effective momentum (the P-wave analog of the effective range) is natural-sized. This leads to the characteristic bump in the cross section near the resonance energy. If the resonance is not particularly narrow a second fine-tuning is needed in the two-particle/dimeron coupling, leading to the same resummation in the whole low-energy region [120]. In this case the effective momentum is small, $\sim \aleph$. In either case, other ERE parameters come out of natural size.

A good description of $n\alpha$ scattering, displayed in Fig. 4.11, is obtained with either power counting [120, 122], the amount of fine-tuning in the ^{5}He ground state remaining unclear. When the LECs are fitted to an $n\alpha$ phase-shift analysis, one finds $M_c \sim 100$ MeV and $\aleph \sim 30$ MeV, consistent with Pionless EFT. There is in

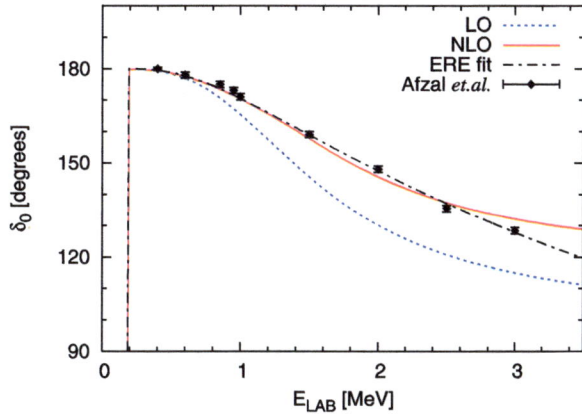

Fig. 4.12
Coulomb-corrected S-wave $\alpha\alpha$ phase shift (in degrees) as a function of the laboratory energy (in MeV). The *dotted* and *solid lines* are the Halo/Cluster EFT results at LO and NLO, respectively. *Solid circles* represent empirical phase shifts [127] and the *dash-dotted line* is an ERE fit. From Ref. [126]

principle no difficulty to include the electromagnetic interactions needed for proton halos. The extension to $p\alpha$ scattering is thus straightforward, except for a less clear separation of scales.

Coulomb is also very important for the $\alpha\alpha$ system. This system is very peculiar because the lowest resonance, with $J^\pi = 0^+$, appears at a momentum that is *small* compared to the scale Coulomb becomes important, $\alpha\mu_{\alpha\alpha}$, where $\mu_{\alpha\alpha} = m_\alpha/2$ is the $\alpha\alpha$ reduced mass. Coulomb has to be included at LO with the dimeron propagator, but it can be approximated by a $Q/\alpha\mu_{\alpha\alpha}$ expansion that makes it, in a sense, a short-range interaction. Despite its small nominal value, the resonance width is actually large in the scale set by $\aleph$. The position and width of the resonance, and the $\alpha\alpha$ phase shifts, can be well reproduced at LO and even better at NLO, as shown in Fig. 4.12 [126]. However, this is only possible only at the cost of cancellations between the short-range interaction and Coulomb at the level of factors of 100 and 10 in the scattering length and effective range, respectively. Such an additional fine-tuning is extremely surprising, and not at all understood.

External probes can be included as well. For cores for which the two-body system sustains a bound state, one can for example consider astrophysically interesting neutron-capture reactions, such as $p + {}^7\text{Be} \to {}^8\text{B} + \gamma$, which can be analyzed as was $p + n \to d + \gamma$ in Pionless EFT [128]. Halo/Cluster EFT offers the possibility of a controlled extrapolation of data to immeasurably small energies. For work in this direction, see Ref. [129] and references therein.

4.4.3.2 Three-Body Systems and Beyond

As for Pionless EFT, the real power of Halo/Cluster EFT lies in going beyond the ERE. For few-body systems, the question resurfaces of the relative size of few-body forces in the presence of fine-tuning, now with nucleon-core and core-core interactions that, as we have just seen, have quite a different power counting compared to the $2N$ system.

For α-core systems the two-body forces have been determined to NLO: the $N\alpha$ and $\alpha\alpha$ interactions from $N\alpha$ and $\alpha\alpha$ scattering, respectively, as described in Sect. 4.4.3.1, and the $2N$ interaction from $2N$ scattering, as described in Sect. 4.4.2.1. We can then test RG invariance without three-body forces for the three-body halo states in ^{6}He, ^{9}Be and ^{12}C, just as we did for the $p+n+n$ system [87] in Sect. 4.4.2.2. This issue has now been settled only in ^{6}He [130], ^{9}Be being under study [131].

^{6}He has been studied by adapting the Gamow Shell Model to the LO, energy-dependent interactions in the power counting of Ref. [120]. Without three-body forces, the three-body ground state collapses, while the simplest three-body force, shown explicitly in Eq. (4.75), was found sufficient for renormalization: its LEC can be adjusted to reproduce the experimental binding energy of ^{6}He at any cutoff. This three-body force is the EFT rendition of the phenomenological strategy of allowing the nn interaction to be empirically modified by the presence of the core. One can now proceed to calculate other ^{6}He properties, and more-neutron members of the He family, such as ^{8}He. Of course, one needs to check whether four-body forces are absent from LO as it seems to be the case for $p+p+n+n$ [108], as discussed in Sect. 4.4.2.3. Note that ^{6}He is Borromean, but its different nature compared to triton in Pionless EFT does not seem to diminish the importance of three-body forces.

The EFT is also a tool to look for Efimov-like states in halo nuclei. Several candidate nuclei with S-wave interactions have been studied in this fashion, giving some tantalizing hints of the answer (see Ref. [132] and references therein). Recently even the form factors of three-body S-wave halos have been calculated [133].

4.4.4 Summary

QCD exhibits in the two-nucleon system a certain amount of fine-tuning, which results in shallow bound states and resonances in light nuclei. These states can be described by EFTs with only contact interactions: Pionless EFT with only nucleon degrees of freedom, and Halo/Cluster EFT with additional degrees of freedom for tight clusters of nucleons. Good descriptions of data on S-shell nuclei and α-core halo/cluster states can be achieved, but in the case of the $\alpha\alpha$ system only at the cost of a baffling additional fine-tuning, this time between strong and electromagnetic interactions.

4.5 Conclusions and Outlook

I introduced the general concept of EFTs. Using the example of atoms in NRQED, I presented the EFTs of QCD for typical nuclei (Chiral EFT), for the lightest nuclei (Pionless EFT), and for larger nuclei with halo or cluster structure (Halo/Cluster EFT). Along the way, I alluded to some of the applications to cold-atom physics.

I described many of the successes of EFT approach, although my emphasis has been on the conceptual development. The main message is that EFT provides *the* framework to describe nuclear physics within the Standard Model (which itself can be viewed as an EFT): it is consistent with the symmetries, incorporates hadronic physics, and has a controlled expansion.

The frontier is to push EFTs in the direction of heavier nuclei. Is there a connection to the traditional Shell Model, perhaps generalizing Halo/Cluster EFT? For heavy, deformed nuclei, an EFT has been developed for the very low-energy rotational bands [134], and certainly other nuclear regimes await new EFTs. The EFT program is paving the road for a QCD understanding of nuclear structure and reactions, while uncovering some new, beautiful renormalization phenomena.

Acknowledgements I thank my many collaborators over the years for their help in shaping my views of EFTs, especially in the challenging nuclear context. This work was supported in part by the Université Paris Sud under the program "Attractivité 2013", and by the US DOE under grant DE-FG02-04ER41338.

References

1. S. Weinberg, Phys. Lett. B **251**, 288 (1990)
2. S. Weinberg, Nucl. Phys. B **363**, 3 (1991)
3. C. Ordóñez, U. van Kolck, Phys. Lett. B **291**, 459 (1992)
4. S. Weinberg, Phys. Lett. B **295**, 114 (1992)
5. T.-S. Park, D.-P. Min, M. Rho, Phys. Rep. **233**, 341 (1993)
6. U. van Kolck, Soft physics: applications of effective chiral Lagrangians to nuclear physics and quark models. Ph.D dissertation, University of Texas, 1993, UMI-94-01021
7. C. Ordóñez, L. Ray, U. van Kolck, Phys. Rev. Lett. **72**, 1982 (1994)
8. U. van Kolck, Phys. Rev. C **49**, 2932 (1994)
9. U. van Kolck, Few-Body Syst., Suppl. **9**, 444 (1995)
10. C. Ordóñez, L. Ray, U. van Kolck, Phys. Rev. C **53**, 2086 (1996)
11. U. van Kolck, Prog. Part. Nucl. Phys. **43**, 337 (1999)
12. S.R. Beane, P.F. Bedaque, W.C. Haxton, D.R. Phillips, M.J. Savage, nucl-th/0008064
13. P.F. Bedaque, U. van Kolck, Annu. Rev. Nucl. Part. Sci. **52**, 339 (2002)
14. E. Epelbaum, H.-W. Hammer, U.-G. Meißner, Rev. Mod. Phys. **81**, 1773 (2009)
15. H.-W. Hammer, L. Platter, Annu. Rev. Nucl. Part. Sci. **60**, 207 (2010)
16. R. Machleidt, D.R. Entem, Phys. Rep. **503**, 1 (2011)
17. J. Beringer et al. (Particle Data Group), Phys. Rev. D **86**, 010001 (2012)
18. S. Weinberg, Phys. A **96**, 327 (1979)
19. G.P. Lepage, hep-ph/0506330
20. J. Polchinski, hep-th/9210046
21. D.B. Kaplan, nucl-th/9506035
22. U. van Kolck, L.J. Abu-Raddad, D.M. Cardamone, nucl-th/0205058
23. G. 't Hooft, NATO Adv. Stud. Inst. Ser. B Phys. **59**, 135 (1980)
24. S. Weinberg, *The Quantum Theory of Fields, Vol. 1: Foundations* (Cambridge University Press, Cambridge, 1995)
25. A. Manohar, H. Georgi, Nucl. Phys. B **234**, 189 (1984)
26. H. Georgi, L. Randall, Nucl. Phys. B **276**, 241 (1986)
27. T.Y. Cao, in *Renormalization. From Lorentz to Landau (and beyond)*, ed. by L.M. Brown (Springer, Berlin, 1993), pp. 87–133

28. S. Hartmann, Stud. Hist. Philos. Mod. Phys. **32**, 267 (2001)
29. W.E. Caswell, G.P. Lepage, Phys. Lett. B **167**, 437 (1986)
30. H. Georgi, Phys. Lett. B **240**, 447 (1990)
31. T. Mannel, W. Roberts, Z. Ryzak, Nucl. Phys. B **368**, 204 (1992)
32. D.A. Dicus, C. Kao, W.W. Repko, Phys. Rev. D **57**, 2443 (1998)
33. M.E. Luke, A.V. Manohar, Phys. Lett. B **286**, 348 (1992)
34. R.J. Hill, G. Lee, G. Paz, M.P. Solon, Phys. Rev. D **87**, 053017 (2013)
35. B.R. Holstein, Am. J. Phys. **72**, 333 (2004)
36. P. Labelle, S.M. Zebarjad, C.P. Burgess, Phys. Rev. D **56**, 8053 (1997)
37. U.D. Jentschura, A. Czarnecki, K. Pachucki, Phys. Rev. A **72**, 062102 (2005)
38. E. Mereghetti, W.H. Hockings, U. van Kolck, Ann. Phys. **325**, 2363 (2010)
39. S. Weinberg, *The Quantum Theory of Fields, Vol. 2: Modern Applications* (Cambridge University Press, Cambridge, 1996)
40. S. Weinberg, Trans. N. Y. Acad. Sci. **38**, 185 (1977)
41. V. Baluni, Phys. Rev. D **19**, 2227 (1979)
42. C.A. Baker et al., Phys. Rev. Lett. **97**, 131801 (2006)
43. D.B. Kaplan, M.J. Savage, Nucl. Phys. A **556**, 653 (1993). [Erratum-ibid. A **570**, 833 (1994)] [Erratum-ibid. A **580**, 679 (1994)]
44. S.-L. Zhu, C.M. Maekawa, B.R. Holstein, M.J. Ramsey-Musolf, U. van Kolck, Nucl. Phys. A **748**, 435 (2005)
45. J. de Vries, E. Mereghetti, R.G.E. Timmermans, U. van Kolck, Ann. Phys. **338**, 50 (2013)
46. S. Scherer, M.R. Schindler, hep-ph/0505265
47. B. Long, U. van Kolck, Nucl. Phys. A **840**, 39 (2010)
48. B. Long, U. van Kolck, Nucl. Phys. A **870–871**, 72 (2011)
49. V. Bernard, N. Kaiser, U.-G. Meißner, Int. J. Mod. Phys. E **4**, 193 (1995)
50. V. Pascalutsa, D.R. Phillips, Phys. Rev. C **67**, 055202 (2003)
51. V. Bernard, Prog. Part. Nucl. Phys. **60**, 82 (2008)
52. J.L. Friar, Phys. Rev. C **60**, 034002 (1999)
53. N. Kaiser, R. Brockmann, W. Weise, Nucl. Phys. A **625**, 758 (1997)
54. N. Kaiser, S. Gerstendorfer, W. Weise, Nucl. Phys. A **637**, 395 (1998)
55. M.C.M. Rentmeester, R.G.E. Timmermans, J.L. Friar, J.J. de Swart, Phys. Rev. Lett. **82**, 4992 (1999)
56. M.C.M. Rentmeester, R.G.E. Timmermans, J.J. de Swart, Phys. Rev. C **67**, 044001 (2003)
57. N. Kaiser, Phys. Rev. C **65**, 017001 (2002)
58. J.L. Friar, D. Hüber, U. van Kolck, Phys. Rev. C **59**, 53 (1999)
59. D. Hüber, J.L. Friar, A. Nogga, H. Witała, U. van Kolck, Few-Body Syst. **30**, 95 (2001)
60. E. Epelbaum, Phys. Lett. B **639**, 456 (2006)
61. E. Epelbaum, Eur. Phys. J. A **34**, 197 (2007)
62. J.L. Friar, G.L. Payne, U. van Kolck, Phys. Rev. C **71**, 024003 (2005)
63. E. Epelbaum, A. Nogga, W. Glöckle, H. Kamada, U.-G. Meißner, H. Witała, Phys. Rev. C **66**, 064001 (2002)
64. H.-W. Hammer, A. Nogga, A. Schwenk, Rev. Mod. Phys. **85**, 197 (2013)
65. S.R. Beane, V. Bernard, T.S.H. Lee, U.-G. Meißner, U. van Kolck, Nucl. Phys. A **618**, 381 (1997)
66. U. van Kolck, J.A. Niskanen, G.A. Miller, Phys. Lett. B **493**, 65 (2000)
67. S.R. Beane, P.F. Bedaque, M.J. Savage, U. van Kolck, Nucl. Phys. A **700**, 377 (2002)
68. D.B. Kaplan, M.J. Savage, M.B. Wise, Nucl. Phys. B **478**, 629 (1996)
69. D.B. Kaplan, M.J. Savage, M.B. Wise, Nucl. Phys. B **534**, 329 (1998)
70. S. Fleming, T. Mehen, I.W. Stewart, Nucl. Phys. A **677**, 313 (2000)
71. S.R. Beane, P.F. Bedaque, L. Childress, A. Kryjevski, J. McGuire, U. van Kolck, Phys. Rev. A **64**, 042103 (2001)
72. A. Nogga, R.G.E. Timmermans, U. van Kolck, Phys. Rev. C **72**, 054006 (2005)
73. M.C. Birse, Phys. Rev. C **74**, 014003 (2006)
74. B. Long, U. van Kolck, Ann. Phys. **323**, 1304 (2008)

75. M.C. Birse, Philos. Trans. R. Soc. Lond. A **369**, 2662 (2011)
76. C.-J. Yang, C. Elster, D.R. Phillips, Phys. Rev. C **80**, 034002 (2009)
77. Ch. Zeoli, R. Machleidt, D.R. Entem, arXiv:1208.2657 [nucl-th]
78. E. Epelbaum, U.-G. Meißner, nucl-th/0609037
79. M. Pavón Valderrama, Phys. Rev. C **83**, 024003 (2011)
80. B. Long, C.-J. Yang, Phys. Rev. C **84**, 057001 (2011)
81. M. Pavón Valderrama, Phys. Rev. C **84**, 064002 (2011)
82. B. Long, C.-J. Yang, Phys. Rev. C **85**, 034002 (2012)
83. B. Long, C.-J. Yang, Phys. Rev. C **86**, 024001 (2012)
84. S.R. Beane, P.F. Bedaque, K. Orginos, M.J. Savage, Phys. Rev. Lett. **97**, 012001 (2006)
85. S.R. Beane, E. Chang et al., Phys. Rev. C **88**, 024003 (2013)
86. D.B. Kaplan, Nucl. Phys. B **494**, 471 (1997)
87. P.F. Bedaque, H.-W. Hammer, U. van Kolck, Nucl. Phys. A **676**, 357 (2000)
88. P.F. Bedaque, U. van Kolck, Phys. Lett. B **428**, 221 (1998)
89. U. van Kolck, Nucl. Phys. A **645**, 273 (1999)
90. S.R. Beane, T.D. Cohen, D.R. Phillips, Nucl. Phys. A **632**, 445 (1998)
91. X. Kong, F. Ravndal, Nucl. Phys. A **665**, 137 (2000)
92. J.-W. Chen, G. Rupak, M.J. Savage, Nucl. Phys. A **653**, 386 (1999)
93. E. Braaten, H.-W. Hammer, Phys. Rep. **428**, 259 (2006)
94. V.G.J. Stoks, R.A.M. Klomp, M.C.M. Rentmeester, J.J. de Swart, Phys. Rev. C **48**, 792 (1993)
95. P.F. Bedaque, H.-W. Hammer, U. van Kolck, Phys. Rev. C **58**, 641 (1998)
96. P.F. Bedaque, H.-W. Hammer, U. van Kolck, Phys. Rev. Lett. **82**, 463 (1999)
97. P.F. Bedaque, H.-W. Hammer, U. van Kolck, Nucl. Phys. A **646**, 444 (1999)
98. P.F. Bedaque, G. Rupak, H.W. Grießhammer, H.-W. Hammer, Nucl. Phys. A **714**, 589 (2003)
99. W. Dilg, L. Koester, W. Nistler, Phys. Lett. B **36**, 208 (1971)
100. W.T.H. van Oers, J.D. Seagrave, Phys. Lett. B **24**, 562 (1967)
101. A.C. Phillips, G. Barton, Phys. Lett. B **28**, 378 (1969)
102. A. Kievsky, S. Rosati, W. Tornow, M. Viviani, Nucl. Phys. A **607**, 402 (1996)
103. L. Platter, Phys. Rev. C **74**, 037001 (2006)
104. S. Koenig, H.-W. Hammer, Phys. Rev. C **83**, 064001 (2011)
105. L. Platter, H.-W. Hammer, Nucl. Phys. A **766**, 132 (2006)
106. C. Ji, D.R. Phillips, L. Platter, Ann. Phys. **327**, 1803 (2012)
107. L. Platter, H.-W. Hammer, U.-G. Meißner, Phys. Rev. A **70**, 052101 (2004)
108. L. Platter, H.-W. Hammer, U.-G. Meißner, Phys. Lett. B **607**, 254 (2005)
109. J. Kirscher, H.W. Grießhammer, D. Shukla, H.M. Hofmann, Eur. Phys. J. A **44**, 239 (2010)
110. T. Mehen, I.W. Stewart, M.B. Wise, Phys. Rev. Lett. **83**, 931 (1999)
111. H.M. Müller, S.E. Koonin, R. Seki, U. van Kolck, Phys. Rev. C **61**, 044320 (2000)
112. I. Stetcu, B.R. Barrett, U. van Kolck, Phys. Lett. B **653**, 358 (2007)
113. S.R. Beane, P.F. Bedaque, A. Parreño, M.J. Savage, Phys. Lett. B **585**, 106 (2004)
114. I. Stetcu, J. Rotureau, B.R. Barrett, U. van Kolck, Ann. Phys. **325**, 1644 (2010)
115. J. Rotureau, I. Stetcu, B.R. Barrett, M.C. Birse, U. van Kolck, Phys. Rev. A **82**, 032711 (2010)
116. D. Lee, Prog. Part. Nucl. Phys. **63**, 117 (2009)
117. I. Stetcu, J. Rotureau, Prog. Part. Nucl. Phys. **69**, 182 (2013)
118. S.A. Coon, M.I. Avetian, M.K.G. Kruse, U. van Kolck, P. Maris, J.P. Vary, Phys. Rev. C **86**, 054002 (2012)
119. S.N. More, A. Ekstrom, R.J. Furnstahl, G. Hagen, T. Papenbrock, Phys. Rev. C **87**, 044326 (2013)
120. C.A. Bertulani, H.-W. Hammer, U. van Kolck, Nucl. Phys. A **712**, 37 (2002)
121. TUNL Nuclear Data Evaluation Project, http://www.tunl.duke.edu/nucldata/
122. P.F. Bedaque, H.-W. Hammer, U. van Kolck, Phys. Lett. B **569**, 159 (2003)
123. National Nuclear Data Center Evaluated Nuclear Data Files, Brookhaven National Laboratory, http://www.nndc.bnl.gov/

124. B. Haesner et al., Phys. Rev. C **28**, 995 (1983)
125. M.E. Battat et al., Nucl. Phys. **12**, 291 (1959)
126. R. Higa, H.-W. Hammer, U. van Kolck, Nucl. Phys. A **809**, 171 (2008)
127. S.A. Afzal, A.A.Z. Ahmad, S. Ali, Rev. Mod. Phys. **41**, 247 (1969)
128. G. Rupak, Nucl. Phys. A **678**, 405 (2000)
129. G. Rupak, R. Higa, Phys. Rev. Lett. **106**, 222501 (2011)
130. J. Rotureau, U. van Kolck, Few-Body Syst. **54**, 725 (2013)
131. S. Ando et al., in progress
132. B. Acharya, C. Ji, D.R. Phillips, Phys. Lett. B **723**, 196 (2013)
133. P. Hagen, H.-W. Hammer, L. Platter, Eur. Phys. J. A **49**, 118 (2013)
134. T. Papenbrock, Nucl. Phys. A **852**, 36 (2011)

Chapter 5
Direct Reactions at Relativistic Energies: A New Insight into the Single-Particle Structure of Exotic Nuclei

Dolores Cortina-Gil

5.1 Introduction

Direct reactions are an excellent tool for the investigation of nuclear structure. They proceed in a single step and are very fast, $\approx 10^{-22}$ s, taking simply the time needed for the projectile to traverse a target nucleus. Usually, only a few bodies (nucleons) participate in the reaction. Moreover, only a few degrees of freedoms are involved and the momentum transfer associated is not very high. They are thus rather peripheral interactions mostly surface dominated. All these properties make the reaction mechanism easier to interpret and allow the use of a certain number of simplifications in their description. The high selectivity associated with direct reactions is also responsible for the dominance of single-particle properties over dynamical effects, opening the possibility of using them as spectroscopic tools.

The spectra of possible direct reactions is rather large and includes, among others, processes related to elastic, inelastic, transfer and knockout reactions. The use of direct reactions was vigorously extended throughout the 60–80 s with the advent of the first accelerators dedicated to the study of stable isotopes. In the 90's, the availability of exotic beams heralded a new golden age with the extension of structural studies to these new rare species. On top of the relative simplicity, the reaction cross-section is rather large, allowing their use since the first beam deployment despite the rather low intensities of these radioactive beams.

Secondary beams at different energy regimes became widespread across several experimental facilities all over the world. Elastic and inelastic reactions induced by secondary exotic projectiles took hold as a powerful tool to gain information on the radius and matter densities associated with those exotic species.

These first experiments lead to others, more specific, that addressed the study of structural properties. Low energy facilities concentrated on the study of transfer reactions, able to provide detailed information on the single-particle properties

---

D. Cortina-Gil (✉)
Universidad de Santiago de Compostela, 15782 Santiago de Compostela, Spain
e-mail: d.cortina@usc.es

of the different states connected with the reaction (see chapter by Joaquín Gómez Camacho and Antonio Moro in this volume for further information on this topic).

In addition, the exploration of removal reactions became very popular among the high energy facilities. These reactions are the result of the interaction between a fast projectile and a target at rest. The nature of the projectile-target interaction can be nuclear (light target) or Coulomb (heavy target). For large impact parameters, the reaction would be peripheral, and would result in the dissociation or breakup of the projectile into one or a few nucleons (neutrons or protons) that would be ejected and a quasi-projectile (very often called core-fragment) that would continue on its path largely unaffected by the reaction, with almost the same velocity as the incident projectile. Experimentally, the selection of the reaction channel is achieved by detecting the incident projectile and the emerging fragments (with mass A-1 in the one nucleon-removal case). The survival of the fragment is a probe of the peripheral character of the reaction.

The superposition of different removed-nucleon + core-fragment configurations resulting after breakup creates a realistic picture of the original exotic projectile wave function. Therefore, the projectile wave function can be factorised as:

$$|Projectile\rangle = \sum_i A_i \big(|core\rangle \otimes |nucleon\rangle\big)_i, \qquad (5.1)$$

where $|core\rangle$ and $|nucleon\rangle$ represent the core-fragment and removed nucleon wave function and A are the weighted probability[1] associated with each configuration. The detection of the gamma de-excitation of the fragment is used to discriminate different core configurations in the original projectile. Figure 5.1 left shows the reaction mechanism for the particular case of one-neutron knockout of a ^{23}O projectile by a light ^{12}C target.[2]

Different names have been used in the literature to refer to these kind of reactions (i.e.: nucleon removal, nucleon breakup and nucleon knockout). This last name was adopted by the NSCL scientists in their publications and is probably the most widely used (see review articles and references therein [1–3]). This term (knockout) was already used more than three decades ago to refer to Quasi-Free Scattering (QFS) reactions induced either by high energy protons and electrons, $(p, 2p)$ and $(e, e'p)$ [4, 5], and known to be very powerful spectroscopic tools. Figure 5.1 right shows the schematic representation of the quasi-free process induced by high energy exotic projectiles.[3]

QFS can be understood as a process in which a high energy particle knocks a nucleon out of a nucleus without any further significant interaction between the nucleon and the incident and the outgoing particles. After the reaction, and in the particular case of proton-knockout, the two protons (the target in our inverse kinematic example and the removed proton) emerge in the forward direction with a very strong

---

[1] Related to the spectroscopic factors.
[2] Reaction performed in inverse kinematics.
[3] Again illustrated by a reaction in inverse kinematics.

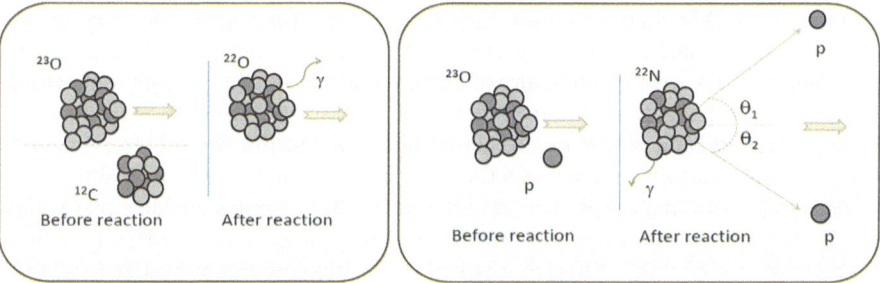

Fig. 5.1 *Left*: Schematic representation of a one-neutron knockout of an unstable ^{23}O projectile by a stable carbon target. A loosely bound neutron (not recorded) is removed during the reaction, leaving a core of ^{22}O. *Right*: Schematic representation of a quasi-free reaction induced by an unstable ^{23}O projectile on a proton target $(p, 2p)$. Two strongly correlated protons and the *A-1* fragment (^{22}N in this case) are emitted (and detected)

angular correlation.[4] The detection (momentum and direction) of both nucleons is necessary to provide kinematically complete measurements.[5]

Indeed proton-knockout using high-energy electrons $(e, e'p)$ was studied extensively in the 1980s, and was considered the only experimental method able to provide absolute spectroscopic factors in well-bound nuclei [6, 7]. QFS experiments are thus considered a quantitative tool for studying single-particle occupancies and correlation effects in the nuclei. They have been mostly exploited for reactions on stable nuclear targets and only very few QFS experiments to date involve rare species.

In this lecture, we will refer as knockout to the nucleon-removal from a projectile after reaction with a light target. The detection of the remaining fragment (with mass *A-1*) will be required but the knockouted nucleon will not be recorded. QFS will refer to reactions of a projectile with a proton (or electron). In this case, the detection of both nucleons (the removed one and the target) and eventually the *A-1* fragment is demanded and ensures fully exclusive measurements. It is our aim to provide an overview of the achievements and limitations related to the application of knockout and QFS reactions as spectroscopic tools for the particular case of exotic nuclei.

We would like to stress the complementary role of the different approaches so far discussed. Transfer reactions yield high cross sections ($\sim$1 mb) at relatively low energies (in the range from few to 10–15 MeV/nucleon). The optimum range for knockout and QFS reactions are projectile energies $\approx$100 MeV/nucleon and higher. The cross section of the nucleon knockout process can vary from well above 100 mb, for loosely bound nuclei, to $\sim$1 mb for tightly bound nuclei. The cross section for the quasi-free channel, imposes a strong kinematical condition and it, is thus smaller and also depends on the separation energy of the removed nucleon, the removal of valence nucleons translates in larger cross sections.

---

[4]The two protons are emitted back-to-back ($\Delta\phi \approx 180°$) and with an average polar opening angle $\theta \approx 90°$.

[5]In some cases the fragment is also detected providing redundant information.

In the case of knockout reactions, the strong absorption concentrates the reaction probability at the surface, allowing mainly to probe the outer part of the projectile wave-function. The peripheral nature of the reaction is also true for both transfer and Coulomb breakup reactions. On the other hand, the use of QFS reactions induced by exotic projectiles on a proton target, allows one to determine the spectral functions of protons and neutrons (in the projectile) over a wide range, from the weakly bound valence nucleons to the deeply bound core states. These studies enable a more complete investigation of the projectile wave-function, giving access to different regions of a nucleus and probing different types of correlations that could exist within the nucleons.

The experimental possibility of working with high projectile energies can also be of particular interest. In a good approximation, at beam energies above 50–70 MeV/nucleon, the internal degrees of freedom of a nucleus can be considered "frozen" during the collision [8–10]. Only the nucleons directly participating in the interaction need to be taken into account in the theoretical description, the others are considered as simple spectators, thus simplifying the reaction mechanism description. The use of a semi-classical approximation of the reaction with regard to the impact parameter of the relative motion of projectile and target is permitted.

The production of high-energy secondary beams is achieved via in-flight projectile fragmentation (or fission) [11], using inverse kinematics, which means that the projectile is heavier than the target, having interesting kinematic consequences. The fragments produced are focused in the forward direction, which contributes to increase the overall transmission of these secondary beams to the secondary reaction target. These production mechanisms originate cocktail beams, formed by different isotopes. The selection of the nuclei of interest is done by help of powerful magnetic spectrometers [11] that identify "in-flight" the nuclear species, allowing the selection of a single isotope within the spectrometer's acceptance.

5.1.1 First Experiments

The shell model of the atomic nucleus was inspired by the atomic shell model. The main assumption of the nuclear shell model consists of a description of the nuclear interaction by a central potential with the ingenious idea of a spin-orbit coupling. To first order, each nucleon in a nucleus is assumed to move independently in a mean field resulting from interactions with the rest of nucleons. At the middle of last century the basis of this model was perfectly established [12, 13]. The major evidence of the adequacy of this orbital shell configuration came from the observation of magic numbers predicted by the model in stable nuclei, the only ones available for experimentation. Many experiments contributed to this important task, including those focused on the determination of energy spectra of the low-lying states, spins, magnetic moments, the observation of polarization effects of nucleons in nuclear collisions and the internal momentum determination of nucleons in proton and electron induced QFS reactions. Since then, the nuclear shell model is considered a cornerstone to describe the structural properties of nuclei.

Experiments of $(p, 2p)$ were first undertaken in 1952 in Berkley [14, 15] with the irradiation of protons on stable nuclei and the observation of coincident proton pairs, strongly correlated, emerging from the target. The interpretation of these experiments relied on the assumption that both the incident and the knocked-out proton were free. The angular correlation of the proton pair was interpreted as a consequence of the momentum distribution of the protons in the nucleus. Additionally, the separation energy distribution, evaluated for a given projectile particle, showed different structures (peaks) that were related to the binding energies of the various nuclear shells from which the protons were ejected. A number of experiments were then performed at Chicago, Harvard, Hawell, Orsay and Uppssala [16, 17], dedicated to the study of nuclei up to ^{40}Ca. The exploration of heavier nuclei, with smaller energy differences between shells and lower cross sections, was not addressed at this early stage. The poor energy resolution achievable in these experiments was the limiting factor.

Equivalent experiments employing high energy electrons appeared as an alternative, the nuclear transparency of the electrons was considered as an advantage. The distortion of the associated momentum distributions was expected to be much smaller and offered the possibility of studying inner shells. The drawback was that, these experiments required the development of powerful electron accelerators, providing intense beams to compensate for the small electromagnetic cross-sections. Very successful experiments of $(e, e'p)$ reactions were carried out in different facilities, for nuclei ranging from ^{2}H to ^{209}Bi. Profiting from the higher experimental resolution [4, 5, 18–21], transitions to many states in the resulting nucleus could be separated and the corresponding momentum distributions accurately measured.

The 80's became again a very exciting period with the advent of the first radioactive nuclei beams. The first experiments with radioactive secondary beams were performed more than fifty years ago ([22] and references therein). They concentrated on radioactivity experiments (i.e.: decay radiation, masses and determination of ground-state properties). Subsequent technological progress made it possible to apply techniques developed for stable beams to the case of secondary beams.

The advent of fast radioactive beams produced by projectile fragmentation and the development of the in-flight identification technique for the emerging fragments was an important milestone in the systematic study of unstable nuclei. The experimental access to nuclei away from the valley of stability enabled the discovery of many interesting phenomena such as dramatic changes in the neutron density at the surface of certain nuclei producing very low density tails (i.e.: nucleon halo or skin). The observed inversions and re-arrangements in the nuclear orbitals pushed an evolution of our traditional knowledge, demanding a revision of certain aspects of the traditional shell model, and were the driving force for an intense experimental and theoretical work to better understand the single-particle properties and the role of nucleon-nucleon (NN) correlations in these newly available exotic nuclei.

It is important to mention the pioneering work of Tanihata and collaborators in the systematic investigations of matter radii of exotic nuclei [23, 24]. A few years later, nucleon-knockout [25] was used for the first time to obtain spectroscopic information on unstable nuclei. Since then, this method has been used extensively. The first knockout experiments focused on the study of neutron-halo states and were later extended to other exotic species. The first cases investigated concentrated on lighter nuclei, mainly due to the technical limitations associated with the production of secondary beams. It is important to keep in mind that even today the neutron dripline[6] has only been reached for nuclei with low Z (up to $Z = 12$). Consequently, this is where most of the knockout experiments have been performed so far.

NSCL[7]/MSU in USA, was the first laboratory to implement the nuclear-knockout technique [26]. For many years, knockout experiments at NSCL have focused on the study of n-rich nuclei at intermediate energies (50–150 MeV/nucleon) [1, 3] (and references therein). They also pioneered the application of the knockout technique to the study of heavier nuclei [27] and have more recently extended their experimental studies to the removal of two nucleons [3, 28–33]. The knockout technique was also applied on several occasions [34–36] in GANIL[8] France, working on a slightly lower energetic domain (50–80 MeV/nucleon). Scientists at RIKEN[9] in Japan, concentrated on the study of Coulomb induced breakup [37, 38]. More recently, to coincide with the commissioning of RIBF[10] and BigRIPS[11] [39, 40] they undertook the investigation of very exotic isotopes using one and two neutron knockout [41]. The German laboratory GSI,[12] working in a higher energy regime (500–1000 MeV/nucleon), has carried out investigations on both Coulomb [42–44] and nuclear [45–55] induced knockout. All this work has contributed to modify the established picture of nuclei, proving that the nuclear orbital organization far from the beta stability is different. The importance of different effects (i.e.: the tensor force, pairing interaction, three nucleon forces, coupling to the continuum, etc.), might play a prominent role in these rare species, and offer new insights on the nuclear landscape.

QFS, can be also regarded as NN scattering channels in nuclear fields and thus could give a direct access to study the modification of nucleon properties in the nuclear medium, an interesting feature of nuclear physics not much explored so far. Experiments of $(p, 2p)$ reactions on different (^{6}Li, ^{12}C, ^{16}O and ^{40}Ca) targets [56] performed at RCNP[13] in Osaka (Japan) profited from high resolution detection of the two outgoing protons. They allowed to study knockout from deeply bound nucleon states. They also allowed, in the case of ^{12}C, a first measurements of the decay particles coming from deeply bound states, offering a qualitative comparison with the shell model and microscopic cluster calculations [57]. It is also worth to mention the electron-induced QFS experiments performed at Jefferson Lab (USA) in the last

---

[6]Limit of existence as a bound nuclear state.
[7]National superconducting cyclotron laboratory.
[8]Grand accélérateur national d'ions lourds.
[9]Nishina center for accelerator-based science.
[10]Radioactive ion beam factory.
[11]Separator and zero degree spectrometer.
[12]Helmholtzzentrum für Schwerionenforschung.
[13]Research center for nuclear physics.

decade. The analysis of the reaction of ^{12}C$(e, e'p)$ reveals a large fraction of events that could be identified as proton-neutron pairs in ^{12}C. These events were interpreted as the result of the short-range component of the NN potential. The observation of a small amount of proton-proton and neutron-neutron pairs was also interpreted as a fingerprint of the tensor force [58].

Traditionally all these correlation effects and medium modifications have not been considered in the interpretation of direct reactions, but detailed studies have shown the eventual impact they could have. In the particular case of knockout reactions, this effect will not affect the shape of the momentum distribution[14] of the emerging fragment but would have an impact on the correct estimation of the cross-sections, particularly at low energies [59]. It is neither excluded, that the so-called "correlations" are simply the result of a insufficient description of the reaction mechanism.

The lecture is structured as follows. After this introduction, the main section is devoted to the analysis and interpretation of knockout reactions induced by radioactive beams, particularly in connection with their use as spectroscopic tools. We will finish with the presentation of the quasi-free channel induced by relativistic exotic nuclei on proton targets, which appears as an attractive extension of this kind of studies to probe the deeply bound states of rare species and offers the possibility of performing fully exclusive measurements.

5.2 Knockout Reactions

5.2.1 Extraction of Information in Knockout Reactions

In a knockout reaction a fast projectile with mass number A impinges on a target at rest (preferably light). The interaction projectile-target results in the removal of a single nucleon. It proceeds in a single step and is a very peripheral reaction, guaranteeing the survival of the A-1 fragment. The study of knockout reactions in the laboratory occurs at relatively high energies (typically above 100 MeV/nucleon, even though many experiments have been done at lower energies $\approx$50–80 MeV/nucleon), where the use of semi-classical descriptions of the reaction mechanism is permitted. When describing the reaction mechanism of the knockout process, different contributions have to be considered.

- Stripping refers to cases where the removed nucleon reacts with the target, the nucleon is scattered to large angles and the target is excited.
- Diffraction or elastic breakup refers to reactions where the target remains in its ground state and the removed nucleon is emitted in forward direction.

---

[14]The momentum distribution themselves have a small dependence on the binding energy of the removed nucleon, but depend on the nucleon's orbital angular momentum, which enables identification of shell occupancy.

- A third contribution, called Coulomb dissociation, corresponds to electromagnetic elastic breakup and plays a minor role in the case of light targets, as it is the case of the reactions that will be described in this paper, and thus will not be discussed in detail.

Within this scenario the cross-section can be expressed as:

$$\sigma = \sigma_{strip} + \sigma_{diff} \quad (5.2)$$

Depending on the beam energy, the relative importance of each process is different. At high beam energies the one-neutron removal cross-section is dominated by stripping whereas at lower energies, 50–60 MeV/nucleon, both contributions are similar [60]. As a general statement, the use of direct reactions as a spectroscopic tool demands a very detailed and realistic description of both the nuclear structure and the reaction mechanism.

In the case of nucleon knockout, the reaction mechanism has been often given by the (semi-classical) eikonal approximation, providing a geometrical description of the reaction in terms of the impact parameter of the relative motion of projectile and target. The strong point of the eikonal description is the relative simplicity associated with both the calculations involved and the required physical inputs. For the single step assumption to be valid, sufficiently high energies must be employed to allow for the use of the adiabatic approach, in which the internal motion of the nucleons inside the nucleus is neglected during the collision [8, 10]. The nucleons not directly involved in the reaction are considered simply as "spectators". As it will be shown in the following sections, many knockout experiments can also determine spin and parity (I^π) of the remaining fragment from the measurement of the γ de-excitation of the fragment, enabling exclusive measurements to given states (n, l, j).

There are many references describing in detail the eikonal methods [1, 8, 9, 61–63]. We will illustrate in this section the case of a system that follows a nucleon-knockout reaction, described by the subsystems after breaking: the core-target and the nucleon-target [42, 63]. The cross-sections can be calculated as follows:

$$\sigma_{diff} = \int d\mathbf{b} \left[\langle |(1 - S_c S_n)|^2 \rangle - |\langle (1 - S_c S_n) \rangle|^2 \right] \quad (5.3)$$

$$\sigma_{strip} = \int d\mathbf{b} \langle (1 - |S_n|^2)|S_c|^2 \rangle \quad (5.4)$$

The cross-sections are expressed as a function of the profile functions S [64, 65], linked in the eikonal model to the scattering matrix and evaluated from the eikonal phase-shifts. The profile functions for the core-target (S_c) and nucleon-target (S_n) depend on the impact parameter and are calculated using density distributions that reproduce measured cross-sections (see [42] for the particular case of ^{11}Be neutron-knockout). The theoretical cross-sections for each of the above mentioned processes, can thus be calculated individually for the occupancy of each subshell orbital for a given state (n, l, j), relating to the single-particle cross section for each subshell at that energy.

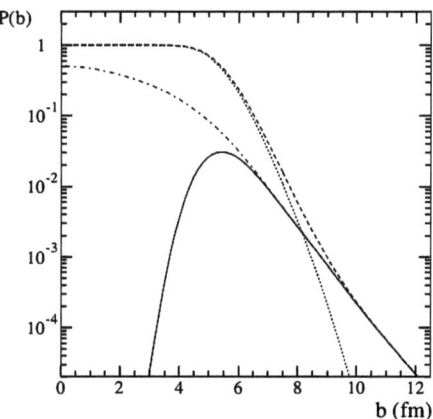

Fig. 5.2 Probability evaluation for one-neutron knockout in ^{17}O ($d_{5/2}$ neutron with $S_n = 4.1$ MeV). The *dashed curve* represents the total reaction probability. The *dash-dotted line* is the neutron reaction probability, i.e.: $\langle 1 - S_n^2 \rangle$, whereas the *dotted line* corresponds to core survival probability, i.e.: S_c^2. The *solid line* is the neutron reaction probability times the core-survival probability [68]

Other authors have worked with descriptions providing a pure quantum treatment of the reaction (the eikonal makes use of a classical description for the projectile movement). These approaches can only account for the elastic (or diffractive) term, ignoring the stripping that can be important in these reactions. Some attempts of using these approaches can be found in Refs. [66, 67].

Using the eikonal approach, we can also probe the peripheral character of this kind of reactions. If one assumes identical impact parameter for both projectile and core,[15] the core-target profile function can be taken outside the expectation value and the probability evaluation can reduce to

$$P(\mathbf{b}) = S_c^2(\mathbf{b}) \langle 1 - S_n^2(\mathbf{b}_n) \rangle = S_c^2(\mathbf{b}) \int d^3r \, |\phi_{nlj}(\mathbf{r})|^2 [1 - S_n^2(\mathbf{b}_n)], \qquad (5.5)$$

with ϕ_{nlj}, the single-particle wave function of specific states, expressed as a function of the relative core-nucleon distance. The terms involved in this expression represent the reaction probability of the nucleon with the target ($\langle 1 - S_n^2 \rangle$), and the survival of the core S_c. This calculation allows for instance to evaluate the one-nucleon knockout with a given orbital angular momentum, and is very helpful to understand the surface dominance of this kind of reaction. Figure 5.2 shows an example of the probability evaluation for the case of ^{17}O considering the knockout of a $d_{5/2}$ neutron [68]. One can observe a concentration of the neutron knockout probability at the nuclear surface.

To compare the evaluated single-particle cross-sections with the experimental ones, the former need to be normalised by the spectroscopic factor (C^2S), associated with each occupied subshell for this particular state. The result provides the theoretical cross-section (see (5.6)) for the removal of a nucleon from each of the considered subshells for this state which, when they are summed, yield the theoret-

---

[15]No recoil limit.

ical cross-section for neutron removal from this state.

$$\sigma_{theo} = \sum_j C^2 S(I_\pi, nlj)\sigma_{sp}(nlj) \qquad (5.6)$$

The sum, for all bound states, yields the complete theoretical cross section for the reaction channel under the particular conditions considered (target, beam energy, etc.). Section 5.2.3.1 includes examples of these kind of calculations applied to the ^{11}Be and ^{8}B case.

A spectroscopic factor is formally defined as the overlap of two many-body shell-model wave functions, corresponding to the initial nucleus (A) and the emerging core (A-1) in the case of nucleon-knockout. Computation of the spectroscopic factors requires a shell model calculations, the reader can find further information on the shell model in almost any nuclear physics text book. This subject has also been treated in the Euroschool notes by H. Grawe [69] and T. Otsuka [70]. Different codes are available on-line [71–73] and can be used for simple computations.

When using $C^2 S$ to weight the single-particle cross-sections, $C^2 S$ are interpreted as the intensity related to the core state (Φ_{A-1}), and thus as the pre-existing weight of a given component ($|core\rangle \otimes |nucleon\rangle$) in the incident projectile ground state wave function (Φ_A).

The comparison of the theoretical cross-section, including $C^2 S$, and the measurement would also allow to determine the factor S (also named R_S), interpreted as an experimental spectroscopic factor.

$$\sigma_{exp} = S\sigma_{theo} \qquad (5.7)$$

In a typical knockout experiment, the measurement of the remaining fragment momentum distribution, allows for the determination of the angular orbital momentum l of the removed nucleon. Taking the "adiabatic approximation" and keeping in mind the momentum conservation, the momentum of the recoil fragment after one-nucleon removal provides a measurement of the momentum of the removed nucleon. In the centre of mass system both quantities have equal modulus and opposite directions.

The shape of the momentum distribution is related to the Fourier transform of the radial wave function. It is also well known that the spatial extension of this radial wave function could be very different depending on the orbital angular momentum. In general, a lower orbital angular momentum yields a very extended spatial extension and thus a narrow momentum distribution. Figure 5.3 shows the example of radial wave functions associated with different l values and the corresponding momentum distributions applied to the ^{17}O example.

We can conclude that the shape of the residual fragment momentum distribution is dependent upon the orbital angular momentum components of the removed nucleon. The comparison of the experimental and evaluated momentum distribution thus allows the discrimination of the l of the knockout nucleon. This signature is analogue to the one extracted from the angular distributions in transfer reactions.

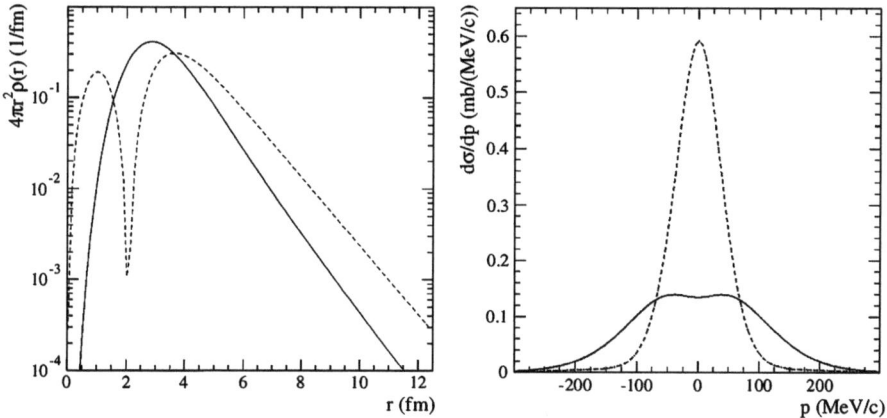

Fig. 5.3 Calculations performed for ^{17}O ($|^{16}$O$\rangle \otimes |n\rangle$). *Left*: Radial wave function for neutrons with different orbital angular momenta ($s_{1/2}$ represented by the *dashed line* and $d_{5/2}$ with *solid line*, in both cases $S_n = 4.1$ MeV). *Right*: Longitudinal momentum distribution of a remaining fragment after one-neutron knockout (^{16}O) associated with the two radial wave functions shown in the *left panel* [68]

One must remember that the independent-particle shell model relies on the assumption that each nucleon moves independently of the rest of nucleons in a nucleus. Indeed, they are not free nucleons, but subject to the action of an average potential (mean-field) induced by the neighbouring nucleons. In the nuclear shell model picture deeply-bound states are seen as fully occupied by nucleons. At the Fermi energy level (and above), configuration mixing can lead to reduced occupancies that gradually decrease to zero. The evaluation of S (R_S), and its deviation from unity, can be understood as a possible quantification of different correlation effects that are beyond the effective interactions used in the shell-model to "build" the mean-field. These kind of effects have been known for many years, observed from data of electron induced $(e, e'p)$ quasi-free scattering using stable beams with A ranging from 7 to 208, which probed the structure using knockout in both valence and deeply bound orbits (see [5] and Fig. 5.4). They show an average S value on 0.6–0.7. These data on stable nuclei can be completed with other coming from nucleon knockout of exotic projectiles (presented in detail in Sect. 5.2.3.5).

It is also worth to mention the efforts to evaluate spectroscopic factors with sophisticated calculations based on ab-initio methods which incorporate two-and three body interactions [74–78]. These calculations are only possible for the case of light nuclei.

5.2.2 Experimental Needs and Relevant Observables

The success of the nucleon knockout technique relies on two main premises:

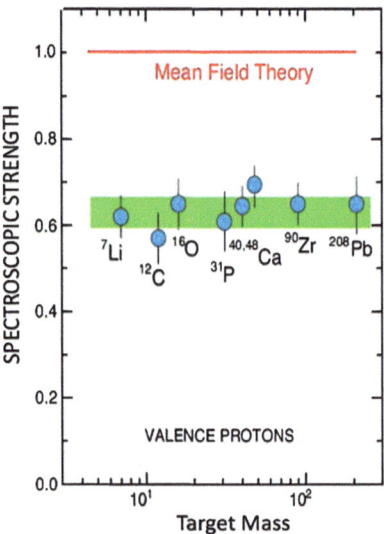

Fig. 5.4 The $(e, e'p)$ data for quasi-free scattering of valence and deeply bound orbits in nuclei gives experimental spectroscopic factors that are 60–70 % of the mean field predictions [5]. Figure extracted from [5]

(a) the use of high-energy secondary beams in conjunction with thick targets provides an efficient enhancement of the reaction yields. Reactions at high energy present an additional advantage of emitting the reaction products in a forward focused cone, offering a rather high efficiency with moderate size detectors.

(b) a detection system able to select the reaction channel and ensure a kinematically complete measurement (i.e.: identification and tracking of projectiles and core-fragments, high precision measurement of the core-fragment momentum and discrimination of the different possible core-fragment excited states).

Currently, high-energy secondary beams are produced by projectile fragmentation of stable beams that are generated by heavy ion accelerators such as synchrotrons and cyclotrons. Projectile fragmentation results in a cocktail beam composed of the various fragments produced. The intensity of the secondary beams must be sufficient to guarantee the success of the experiment: a minimum intensity of a few particles per second is necessary to perform inclusive exploratory investigations. This intensity will depend on different factors. The most important are:

- the type of accelerator used: cyclotrons generally provide higher intensities than synchrotrons. At the time of writing, the largest secondary beam intensities are achieved in RIBF [40, 41] (Japan). An important intensity upgrade is expected with the arrival of the new FAIR[16] [79] (Germany) and FRIB[17] [80] (USA) facilities in the future.
- the choice of projectile: the closer the primary projectile and the secondary beams are in A and Z, the higher the production cross section.

---

[16] Facility for antiproton and ion research.

[17] Facility for rare isotope beams.

- the target thickness: working at high energies makes it possible to use relatively thick targets, which have a larger number of atoms and thus a greater secondary particle yield.

Many knockout experiments rely on the use of very powerful magnetic spectrometers. These devices are composed of a set of electric and magnetic elements that guarantee optimum transport of secondary projectiles and emerging fragments. Electromagnetic spectrometers also act as filters allowing unambiguous identification of secondary fragments from amongst all the species produced following fragmentation of the projectile. They are equipped with various detectors to ensure identification and tracking of ions traversing the system on an event-by-event basis. Last but not least, spectrometers provide a very accurate determination of the momentum of the nuclei from the accurate measurement of the position distribution. Interested readers will find further details on these topics in Refs. [11, 81, 82].

Using knockout, the observables that provide experimental access to the exotic projectile wave function are the momentum distribution of the surviving core-fragments following nucleon knockout and the cross section associated with this reaction channel. The longitudinal momentum distribution of the core-fragment provides information about the wave function of the removed nucleon, whereas spectroscopic factors (C^2S) determined from the removal cross sections to each (n, l, j) state, represent the occupancies of the subshell orbitals in the model space considered for each state. Gamma de-excitation of the core-fragment can also be measured with gamma detection arrays. The coincident detection of core-fragment momentum distribution and removal cross-section with gamma-rays provides information on each individual contribution to the exotic projectile wave function. As a general rule, the removed nucleon is not detected. Only some experiments have concentrated on the detection of the removed nucleon allowing the experimental determination of the cross-section for each reaction mechanism (diffraction and stripping) and validating the reaction models so far used. In the rest of cases, direct discrimination of the removal mechanism is not possible and is taken into account by the reaction model.

Figure 5.5 shows a schematic view of the generic experimental setup at the FRagment Separator (FRS) at GSI [83]. The determination of all the observables introduced in this section is referred to in this example. Other devices extensively used to perform measurements of this kind are the A1900/S800 spectrograph at NSCL/MSU [84], the BigRIPS at RIBF/RIKEN [39, 40] and the SPEG energy loss spectrometer at GANIL [85].

The first section of the spectrometer is tuned to separate from the primary beam and transport the nucleus of interest to where the beam is focused at the intermediate image plane. Here a target is placed, where the knockout reaction is induced. Other fragmentation products within the spectrometer acceptance will be transported as well (as shown in the left panel of Fig. 5.6). The spectrometer section behind the knockout target is then tuned to the magnetic rigidity of the $^{A-1}$X fragments produced in the one-nucleon knockout reaction (right panel of Fig. 5.6).

It is essential to ensure an unambiguous selection of the reaction channel. This is achieved via the double identification of the exotic projectile in front of the knockout target and the remaining core-fragment following the reaction. Figure 5.6 illustrates

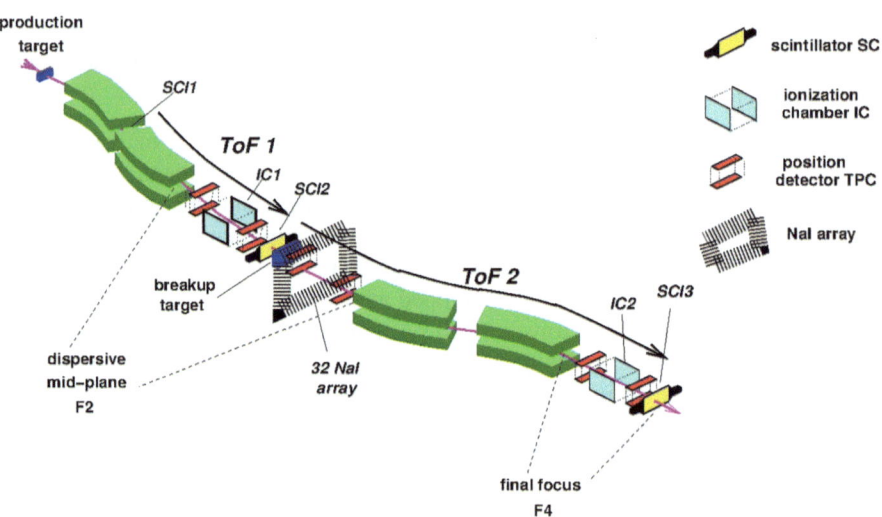

Fig. 5.5 A schematic view of the FRagment Separator (FRS) with its detection set-up. Complete identification with ToF (SC) and energy-deposition (IC) measurements is possible in both sections of the spectrometer. Several position sensitive detectors (TPC) provide projectile and fragment tracking as well as the measurement of fragment momentum distributions. A γ detector (NaI) provides the coincident measurement of fragments with prompt γ-ray deexcitation

a particular case corresponding to the ^{40}Ar fragmentation at 1 GeV/nucleon performed at the FRS. The first section of the spectrometer was tuned to select ^{20}O fragments (right panel), whereas the second section was tuned to select ^{19}O fragments emerging from one-neutron knockout reactions on a carbon target. In this particular measurement, identification was achieved by determining the A/Z ratio from Time of Flight (ToF) measurements with plastic scintillators, and energy losses ($\propto Z^2$) recorded by ionisation chambers.

As it was mentioned earlier, in the "adiabatic" approximation, the momentum of the core-fragment following one-nucleon removal provides information on the wave function of the removed nucleon. Narrow momentum distributions have been associated with a large spatial extension of the removed nucleon (associated with low angular momentum).[18] This was clearly observed in experiments with halo nuclei that will be presented in detail later in this lecture (Sect. 5.2.3.1). The core-fragment momentum is determined by measuring the velocity shift induced by the knock-out target. To determine this velocity shift, position sensitive detectors measure the position distribution of the one-nucleon removal residue at the final focal plane.

It is important to keep in mind that the experimental determination of core-fragment momentum distributions requires high-resolution measurements. The most stringent cases would correspond to the narrow momentum distributions of halo nu-

---

[18]The FWHM of the momentum distribution due to knockout is around 50 MeV/c for an s ($l=0$) neutron, and around 300 MeV/c for a d ($l=2$) neutron.

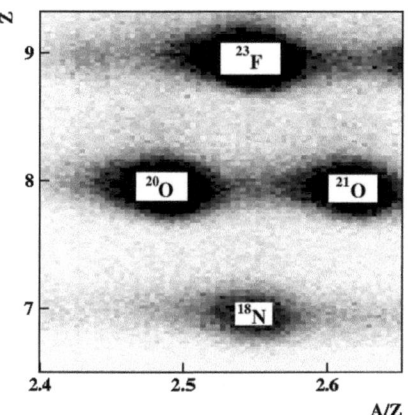 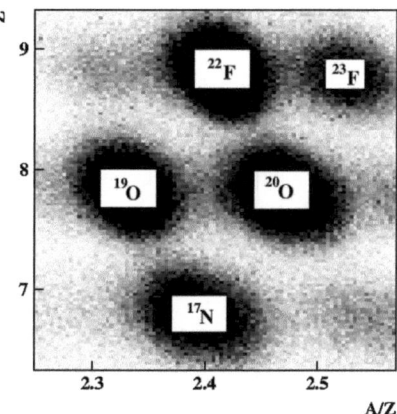

Fig. 5.6 Identification of different secondary beams (*left*) and the corresponding one-neutron removal core-fragments (*right*) emerging after one-neutron knockout on a carbon target. This cocktail beam was obtained by fragmentation of ^{40}Ar at 1 GeV/nucleon impinging on a Be target [48]

clei (see Sect. 5.2.3.1), where a single-particle hole yields a FWHM of the order of 50–80 MeV/c in the core-fragment momentum distribution.

Position distribution measurements are possible in the transversal beam directions. These quantities make it possible to determine the longitudinal (parallel) and transverse (perpendicular) contributions with respect to the beam direction of the core-fragment momentum distribution. Both projections should contain the same information but the longitudinal distribution is preferred because it is less affected by Coulomb diffraction and diffractive scattering mechanisms.

From now on, core-fragments momentum distributions will always refer to the longitudinal component. The experimental determination of the core-fragment momentum distribution is measured in the laboratory reference system and then transformed to the projectile frame using the corresponding Lorentz transformation.

The final core-fragment momentum resolution depends not only on the tracking and magnetic resolving power but also on the quality of the projectile beam (spot size and angular alignment) and the amount of matter at the mid-plane (angular energy straggling). Most of these contributions can be experimentally evaluated by measuring the momentum distribution of the projectile (without knockout target) [45, 86], which can be used for deconvolution (FWHM reported in the literature are always corrected by this value). Figure 5.7 shows these effects for the case of ^{19}C at 910 MeV/nucleon. Another possibility to get free of the effects non due to the knockout reaction on the momentum distribution would consist on the determination of the projectile incoming angle that could be subtracted on an event-by-event basis, to obtain a corrected outgoing angle for the emerging fragment.

The left panel in Fig. 5.8 shows core-fragment momentum distributions of different carbon projectiles, at almost 1 GeV/nucleon, following one-neutron knockout ranging from the bound nucleus ^{12}C with a FWHM of 220 ± 12 MeV/c to the loosely bound nucleus ^{19}C with a FWHM of 71 ± 3 MeV/c. These experimental results reveal the different initial state of the removed neutron, and indicate the dominance of

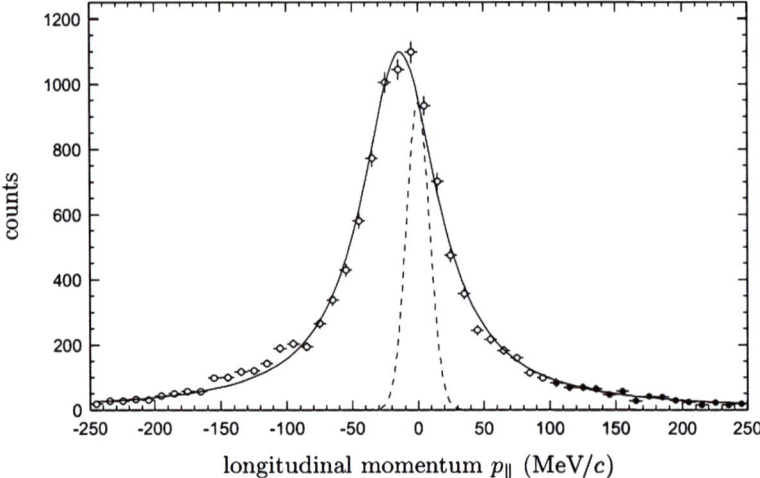

Fig. 5.7 Measured longitudinal momentum distribution of ^{18}C fragments from one-neutron knockout of ^{19}C at 910 MeV/nucleon on a carbon target (*points*). The *dashed profile* represents the measured system resolution with a width of 19.9 MeV/c [86]

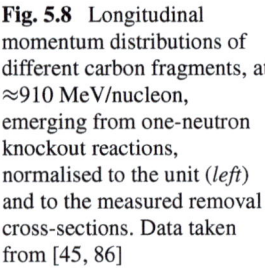

Fig. 5.8 Longitudinal momentum distributions of different carbon fragments, at ≈910 MeV/nucleon, emerging from one-neutron knockout reactions, normalised to the unit (*left*) and to the measured removal cross-sections. Data taken from [45, 86]

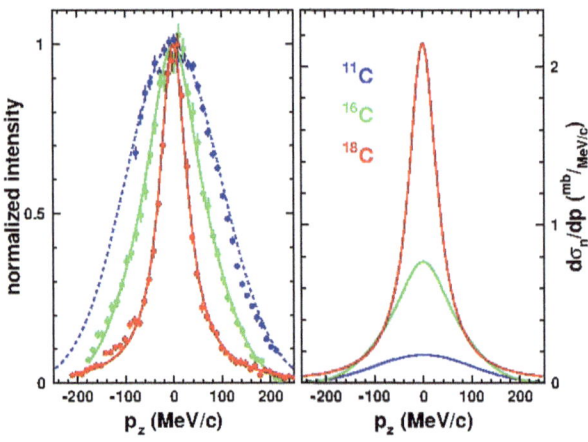

s-wave occupancy in the ground state configuration for the one-neutron halo nuclei ^{19}C [45].

The one-nucleon removal cross section is deduced from the ratio between the number of incoming exotic projectiles and the number of knockout residues. This last quantity is determined at the final focal plane of the spectrometer and must be corrected for the corresponding fragment transmission in the spectrometer region between the knockout target and the detection point. This transmission is evaluated by means of simulation programs that account for the ion-optical transport of the nuclei through the spectrometer (i.e.: LISE [87], MOCADI [88]). Other corrections, such as secondary reactions in the knockout target, detector efficiency and

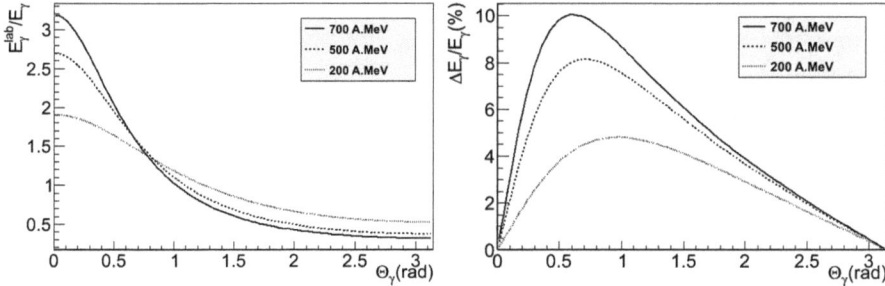

Fig. 5.9 Doppler shift and Doppler broadening for gamma-rays emitted by relativistic sources at different energies (curves generated for detectors with angular segments of about four degrees)

data acquisition dead time are also considered. The right panel of Fig. 5.8 shows the measured momentum distributions for the different C isotopes, normalized to the corresponding cross sections. These two observables, core-fragment momentum distribution and nucleon removal cross section, are determined independently.

For exclusive measurements of these observables for the bound excited states, the experimental setup must provide information on core-fragment de-excitation after knockout. The most common method for distinguishing the different core-fragment configurations contributing to the exotic nuclei wave function, requiring to differentiate between core-fragment in the ground state and in excited states, is via the coincident detection of the surviving core-fragment with a gamma ray emitted in the de-excitation process. Therefore, the gamma-ray detector must be located near the knockout target. Different gamma detector arrays are used. The first experiments were performed with scintillation-based detectors (namely NaI(Tl) and CsI(Tl)) with moderate intrinsic energy resolution. These have been gradually replaced by Ge detectors which have an excellent intrinsic energy resolution, but considerably smaller efficiency in detecting high energy gamma-rays.

The recorded gamma-ray spectra, emitted by relativistic sources would be subject to the Doppler effect (shift and broadening). The Doppler shift represents the gamma energy transformation between the laboratory and centre of mass reference systems. The left panel in Fig. 5.9 shows the evolution of the Doppler shift with the polar angle. For the forward angles ($\theta < 40$ degrees), this effect significantly increases the energy in the laboratory system, which results in lower gamma-detection efficiency (the detection efficiency of a gamma detector depends on the gamma energy, the higher the gamma energy the lower the efficiency). Doppler broadening reflects the effect of the angular aperture of the gamma detector in the final energy resolution of the system. The right panel of Fig. 5.9 shows the evolution of this effect with the polar angle (for a given detector angular aperture). The energy resolutions shown in this picture are calculated without considering the intrinsic energy resolution of the gamma detector used. We can also see (in Fig. 5.9) that the Doppler shift and broadening become more pronounced as the energy of the emitter increases. The Doppler shift can be corrected by determining the velocity of the emitter and the gamma-ray emission angle. However, the broadening effect is determined by the velocity

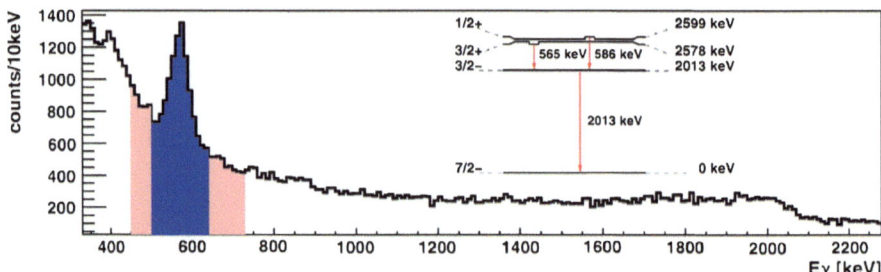

Fig. 5.10 Gamma rays recorded from the ^{47}Ca core-fragment de-excitation following the reaction ^{9}Be(^{48}Ca, ^{47}Ca $+ \gamma$)X at around 500 MeV/nucleon. Data recorded with MINIBALL Ge array in the intermediate focal plane of the FRS [90]

of the emitter and the detector angular aperture, and cannot be corrected. The final energy resolution will be dominated by this kinematic broadening. Thus, detector segmentation becomes a key factor, the finer the segmentation the better the energy resolution. These "weak" points, i.e.: small efficiency for energetic gamma-rays and finite angular segmentation, limit the performance of gamma arrays, particularly at high-energies, and constitute difficult challenges in the determination of exclusive observables.

Figure 5.10 shows an example of a gamma energy spectra recorded with the MINIBALL [89] Ge array in the intermediate focal plane of the FRS. The experiment aimed to probe the single particle properties around ^{54}Ca [90] (^{47}Ca depicted here was used as the reference case), where a shell closure effect for $N = 34$ ($Z = 20$) was predicted [91]. In this case, the emitter energy was ≈500 MeV/nucleon ($\beta \approx 0.76$). We can observe in this figure that the energy resolution achieved for the peak ≈570 keV is around a few percent. This energy resolution is considered a good result for in-beam gamma ray spectroscopy for relativistic moving sources, but it is nowhere near the intrinsic energy resolution of these Ge detectors (well below 1 %). Another example of γ-detection with segmented Ge detectors is shown in Fig. 5.17 for a moving source at 80 MeV/nucleon.

The detection of gamma-rays at very high energies (500–700 MeV/nucleon) remains a critical issue that will improve in the near future with the construction of dedicated detectors. Special mention deserves CALIFA [92], a spectrometer/calorimeter under construction for the R^{3}B/FAIR [93] experiment. CALIFA, with more than 3000 detection units, based on highly performant CsI(Tl) crystals read-out by Large Area Avalanche Photo Diodes,[19] will reduce the impact of Doppler broadening on the final energy resolution, providing $\Delta E/E \approx 5$ % at 1 MeV ($\beta \approx 0.7$) and an overall photo-peak efficiency of 40 % for γ rays up to 15 MeV (in the projectile frame).

Experiments addressing the detection of the knockout nucleons have not been so numerous. They are however very interesting to discriminate between different

---

[19]Option adopted for the backward angles of CALIFA.

reaction mechanisms. Among these exclusive experiments, some have concentrated on the detection of the diffractive component [38, 42, 94–96].

Special mention is due to the work of Bazin and collaborators [67] who reported a detailed study of the relative importance of the stripping and diffraction mechanisms involved in the one proton removal reaction, using a coincident measurement of residue and knockout proton. In typical knockout experiments, the removed nucleon is not detected and the reaction theory is used to estimate the relative weight of elastic and inelastic removal mechanism. Later these contributions are summed up and compared with the experimental result. The validation of the reaction mechanisms requires exclusive measurements detecting both momenta, one associated with the heavy residue and the other to the removed nucleon. This was successfully achieved in the one-proton knockout of ^{9}C and ^{8}B on a Be target at an incident energy of $\approx$150 A MeV [67]. The reactions took place in the scattering chamber of the A1900/S800 [84] spectrograph. This apparatus was also used to collect and identify the one-proton fragments around zero degrees, whereas the light particles emerging at large angles were measured with the HiRA [97][20] detector array (a telescope consisting in two layers of Si and CsI). The relative contribution of both reaction mechanisms was deduced from the energy sum spectra of both heavy and light fragments in coincidence (see Fig. 5.11) and showed an excellent agreement with the values predicted by the eikonal model, giving confidence for the correctness of using knockout reactions as tool to determine single-particle spectroscopic strengths in exotic nuclei.

5.2.3 Results of Knockout Measurements

Knockout experiments employing inverse kinematics started in the early 90's. The first experiments were relatively simple and consisted of the single detection of the emerging *A-1* fragment. These kind of experiments were called "inclusive" and could not distinguish neither the post-reaction fragment excitation nor the contributions due to stripping or diffraction. Soon it became evident that detailed studies were required and the coincident gamma de-excitation of the fragment began to be present for almost any knockout experiment. The discrimination on the reaction mechanism relied on the description of the reaction theory and only few experiments concentrate on the detection of the knockout nucleon. Both cases, the coincident detection of the fragment gamma de-excitation and the detection of the knocked out nucleon constitute what we know as "exclusive" measurements.

In some cases, by profiting from cocktails beams it was possible to study many different species simultaneously. Figure 5.12, shows results [98] obtained at the FRS(GSI), of inclusive momentum distributions of residual nuclei after one-neutron knockout, superimposed upon a chart of the nuclides. In this picture, the vertical axis

---

[20] HIgh resolution array.

Fig. 5.11 Energy sum spectra of one-proton knockout for ^{9}C (*upper panel*) and ^{8}B (*bottom panel*). The inelastic and elastic components of the fit are used to evaluate the exclusive contributions [67]

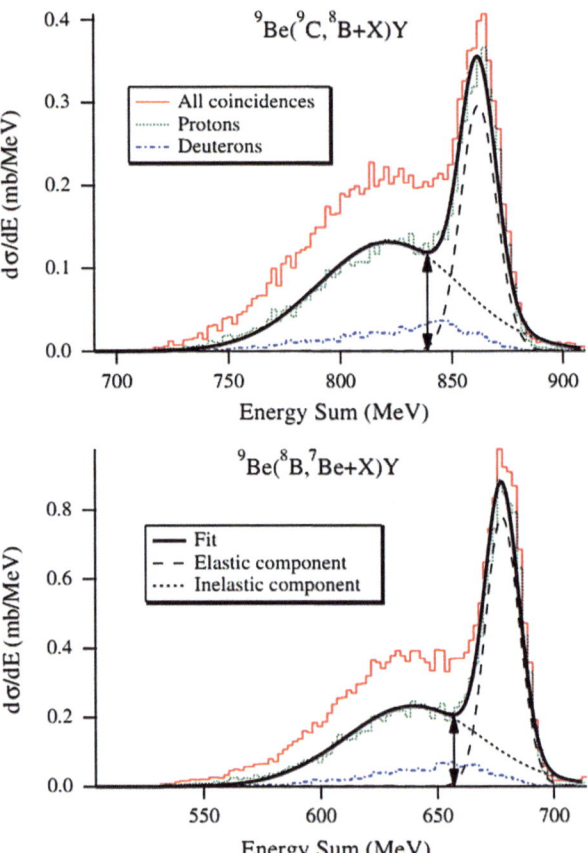

corresponds to the Z number and the horizontal axis to the N number of the exotic projectile before fragmentation. Several neutron-rich isotopes could be investigated in a single experiment where a cocktail secondary beam was produced by nuclear fragmentation of ^{40}Ar at 700 MeV/nucleon. Though qualitative, the evolution of the momentum distribution in this figure reflects the structural changes encountered by nuclei approaching the dripline. For example one can observe the narrowness of the distribution for well known "halo states" such as ^{19}C, or the $N = 14$ sub-shell effect (^{22}N, ^{23}O, ^{24}F, ^{25}Ne, ...). This figure also reveals the potential of this technique to perform exploratory investigations, as indicated by Sauvan et al. [35, 36]. The information provided by the ensemble of data obtained, in different facilities and at different energetic domains over the last 25 years is quite coherent.

In the following sections we present a selection of different experimental works that provide the reader with a picture of the progress achieved so far. They are organised in different subsections which highlight different subjects of interest. The limitations and difficulties encountered will be also commented.

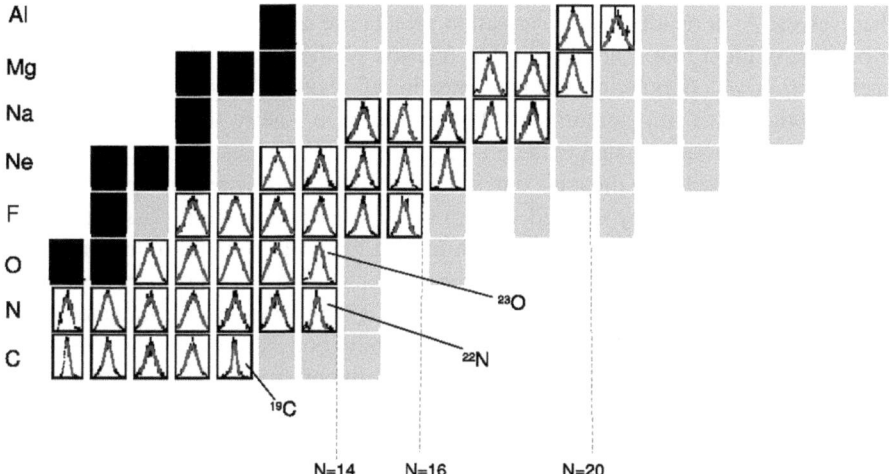

Fig. 5.12 Inclusive longitudinal momentum distributions of the *A-1* fragments after one-neutron removal from the various projectiles indicated on top of a chart of nuclides. *Black squares* correspond to stable isotopes. The measurements were performed at GSI using beam energies around 700 MeV/nucleon [98]

5.2.3.1 Study and Characterisation of Halo States

Near the neutron dripline the large neutron excess and the small neutron binding energy can lead to unexpected changes in the nuclear structure. Through the years, special attention has been given to the case of nuclear halo states. When approaching the driplines the separation energy of the last nucleon, or pair of nucleons, decreases gradually and the bound nuclear states come close to the continuum. In some cases, the combination of the short range of the nuclear force and the low separation energy of the valence nucleons results in considerable tunnelling into the classical forbidden region and a more or less pronounced halo may be formed. A halo nucleus can be visualized as an inert core surrounded by a low density halo of valence nucleon(s) [99–101]. The formation of halo states is especially characteristic for light nuclei in the dripline regions, although not all of these can form a halo.

Analysis and interpretation of knockout experiments dedicated to the study of "halos" has undoubtedly led to a better understanding of the knockout technique and its development and application as a powerful spectroscopic tool. In earlier experiments, the experimental signatures of these phenomena were the narrow momentum distribution of the emerging fragments after one-neutron knockout, reflecting the large spatial extension of the removed nucleon, and the large one-neutron removal cross sections that constitute a complementary source for structure information (see [99] for a detailed compilation). These first experiments relied uniquely on the detection of the core-fragments and are known as "inclusive" measurements. However, we should not forget that the nuclei under study are located far away from the beta-stability line and that the resulting core-fragments are exotic nuclei themselves. As a result, core-polarisation effects are quite common. Indeed, soon it became evident that a non-negligible fraction of the measured neutron-removal cross section was populating excited states in the residue. This meant that the observed longitudinal momentum distributions were in reality the superposition of broad components associated with core-fragment excited states onto the narrow distribution associated with the halo states (mainly in the core-fragment ground state).

The use of gamma-ray coincidence, mentioned earlier, made it possible to separate these different contributions which in turn made it possible to determine the partial cross-sections of the different core states. The observables extracted under these conditions are referred to as "exclusive". This exclusive experimental information, together with an adequate model describing both the structure of the nuclei involved and the reaction mechanism, allows for the experimental determination of spectroscopic factors. This coincidence technique is not exempt from experimental problems. Cases involving nuclei with complex decay schemes, and/or with many weak transitions would be associated with larger experimental errors that would make a detailed analysis difficult or impossible. However, as in the case of "halos", nuclei close to the driplines exhibit very few bound states making this determination easier.

In the following paragraphs, we will concentrate on the description of two well-known cases, ^{11}Be and ^{8}B, which correspond to one-neutron and one-proton halo states, respectively.

- One-neutron halo ^{11}Be

Our first example is ^{11}Be with only three bound excited states and a one-neutron separation energy of ∼500 keV. The ground state of ^{11}Be, considered as a $1/2^+$ intruder from the sd shell, is a well known one-neutron halo state. First experimental evidences of it came from the measurement of the half-life of the 320 keV excited state of ^{11}Be, suggesting an extremely strong E1 transition [102], and the narrow momentum distribution of ^{10}Be core-fragments resulting after one-neutron knockout of ^{11}Be [103]. The initial picture of an inert ^{10}Be core and a neutron in a $1s$ shell, soon gave way to a more complex picture where the pertinence of this inert core was questioned.

The most favourable scenario consisted of an admixture of a neutron in a $0d_{5/2}$ orbital coupled to the first excited state in ^{10}Be (2^+), but theoretical predictions ranged in their estimates from 7 % to 40 % [104, 105]. The experimental situation was ambiguous, with different results from ^{10}Be$(d, p)^{11}$Be experiments providing quite different spectroscopic factors [106–108], some of them incompatible with earlier Coulomb dissociation experiments [109, 110]. The knockout experiment performed by T. Aumann et al. [111] at NSCL (A1900/S800) shed light on this question. They produced a secondary beam of ^{11}Be at 60 MeV/nucleon via nuclear fragmentation of ^{16}O. The ^{11}Be beam impinged on a Be target producing the one-neutron knockout. A NaI(Tl) array located around the removal target recorded the gamma-rays in coincidence with the ^{10}Be fragments, which were analysed with the tracking detectors located at the end of A1900/S800 [84] (see experimental details in [111]). The resulting gamma-energy spectrum is displayed

5 Direct Reactions at Relativistic Energies: A New Insight

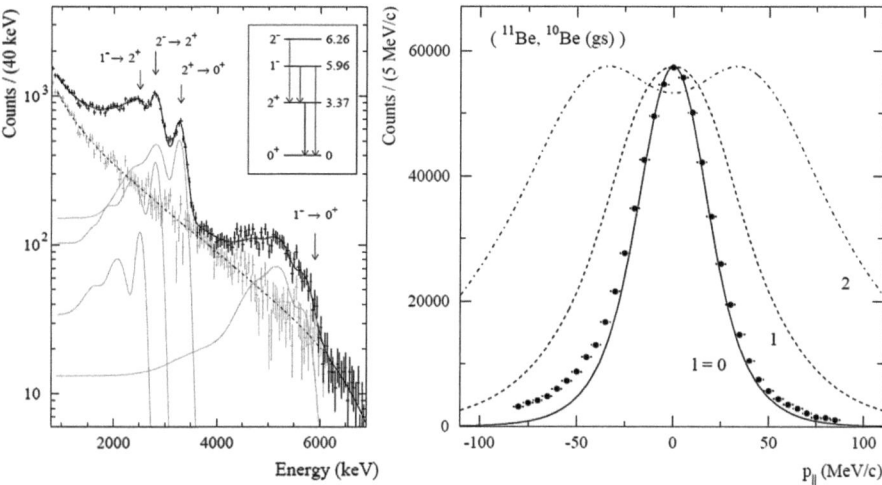

Fig. 5.13 *Left*: Doppler-corrected gamma energy spectrum measured with the NaI array in coincidence with ^{10}Be core-fragments emerging from the ^{9}Be(^{11}Be, ^{10}Be + γ)X one-neutron knockout. *Right*: Longitudinal momentum distribution of the ^{10}Be ground state fragments. The *curves* are calculations assuming a knockout reaction from s, p, and d states [111]. Pictures taken from [111]

Table 5.1 Partial cross sections (mb) to all final states I^π observed in ^{10}Be after one-neutron knockout from ^{12}Be. Different contributions of theoretical single-particle cross sections in the eikonal model are reported. The sum multiplied by the spectroscopic factor is compared with the experimental values [111]

I^π	l	C^2S	σ_{sp}^{knock}	σ_{sp}^{diff}	σ^{other}	σ^{theo}	σ^{expt}
0^+	0	0.74	125	98	10	172	203(31)
2^+	2	0.18	36	14	11	11	16(4)
1^-	1	0.69	25	9		23	17(4)
2^+	1	0.58	25	9		20	23(6)
Σ						226	259(39)

in the left panel of Fig. 5.13, where the solid line represents a fit to the experimental spectrum. The different grey lines correspond to a Monte Carlo simulation of the individual decay channels. The gamma-rays facilitated the experimental determination of partial cross-sections that are summarised in Table 5.1. The calculations shown in this table correspond to single-particle cross sections in the spectator-core eikonal three-body model. They are given separately for stripping and diffractive breakup. The theoretical cross-section for a given ^{10}Be core final state and the j value of the removed nucleon is assumed to be the product of a spectroscopic factor (C^2S) [112] and a single-particle cross section, which is the sum of the different contributions mentioned above. This theoretical cross section is then compared with the experimental values, allowing us to test the picture provided by the reaction mechanism description and the nuclear structure (spectroscopic factors calculated within the shell model). Data recorded in Table 5.1 show a quite good agreement between the experimental results and theoretical calculations. This result also corroborates a dominance of an s-wave single particle configuration for the ground-state.

With this experimental setup, it was possible to discriminate the ground-state from the other excited states for the ^{10}Be fragment momentum distribution. The results are shown in the right panel of Fig. 5.13, together with calculations assuming a knockout reaction from s, p and d neutrons. Here we see that the narrow momentum distribution (FWHM = 47.7(6) MeV/c), associated with the ^{10}Be g.s., is only compatible with the case of nucleon removal from a $1s$ state. This experiment was very successful in determining the ground state structure of ^{11}Be, quantifying the admixture of the ^{10}Be excited core to it. It provided a good understanding of the reaction mechanism and structure of the nuclei involved.

- One-proton halo ^{8}B

^{8}B with one bound state and a one-proton separation energy of 137 keV, is the only known nucleus with a proton halo structure in its ground state. Experimental evidence for it was seen in earlier measurements of a large one-proton removal cross section of 98 ± 6 mb (on carbon) and a narrow longitudinal momentum distribution of 93 ± 5 MeV/c [113–115].

Because it is an $A = 8$ nucleus, reactions involving ^{8}B are important to understand how stellar nucleosynthesis bridges the $A = 8$ mass gap. In addition, the astrophysical interest in ^{8}B stems from its key role in the production of high-energy solar neutrinos [116]. The need for accuracy in the high-energy solar neutrinos production has not diminished after the reports on neutrino oscillations [117]. Indeed, the proton capture rate of ^{7}Be strongly depends on the structure of ^{8}B.

When describing ^{8}B as a one-proton halo system, one should keep in mind that the ^{7}Be core is itself a weakly bound system, which can be considered as a two-body system (^{4}He + ^{3}He). The ^{7}Be $3/2^-$ ground state is bound by 1.587 MeV, and the only bound state below the $\alpha + ^3$He threshold is the $1/2^-$ state at 429 keV excitation energy. If ^{8}B is treated as a two-body system, there are three possible ways to couple a proton to the ^{7}Be core: the last proton in ^{8}B can be in either a $p_{3/2}$ or a $p_{1/2}$ state, and the possible ground state configurations of ^{8}B($I^\pi = 2^+$) thus are:

(a) $\psi(^7\text{Be}(3/2^-)) \otimes \psi(p(3/2^-))$
(b) $\psi(^7\text{Be}(3/2^-)) \otimes \psi(p(1/2^-))$
(c) $\psi(^7\text{Be}(1/2^-)) \otimes \psi(p(3/2^-))$

An experiment was performed at the FRS (GSI) [46, 47] with a ^{8}B beam at 936 MeV/nucleon, produced by nuclear fragmentation of a ^{12}C primary beam. This ^{8}B beam impinged on a C target located at the intermediate focal plane of the FRS. Fragment longitudinal momentum distributions after proton knockout and the cross-sections of the processes were determined in the experiment. An array of NaI(Tl) detectors, located close to the knockout target and covering the forward direction, allowed the coincident measurement of the 429 keV γ rays.

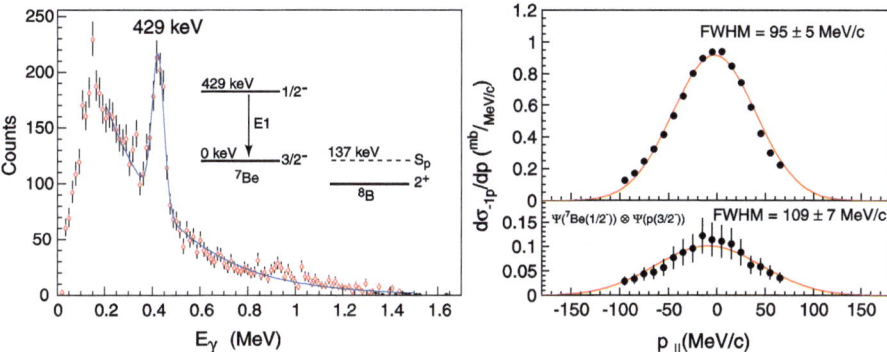

Fig. 5.14 *Left*: Energy spectrum of γ rays after Doppler correction in coincidence with ^{7}Be fragments after one-proton removal reactions of ^{8}B in a carbon target. *Right*: (*top*) Inclusive and (*bottom*) exclusive $p_\parallel$ momentum distribution of ^{7}Be core-fragments emerging from the ^{12}C(^{8}B, ^{7}Be + γ)X one-proton knockout. Exclusive data refer to contributions in coincidence with the 429 keV state in ^{7}Be. In both cases, the *full curve* represents the theoretical calculation folded with the experimental resolution and scaled to match the amplitude of the experimental spectrum [46, 47]

Table 5.2 Comparison between theoretical and experimental results for inclusive (total) and excited (in coincidence with the γ-peak at 429 keV in ^{7}Be) one-proton removal cross sections and $p_\parallel$ (FWHM) after one proton removal of ^{8}B. The theoretical widths in this table include the experimental resolution

	σ_{-1p} (mb) exp.	σ_{-1p} (mb) theo.	$p_\parallel$ (MeV/c) exp.	$p_\parallel$ (MeV/c) theo.
Total	94 ± 9	82	95 ± 5	99
Excited	12 ± 3	11.5	109 ± 7	130

Tracking detectors located at the final focal plane determined the momentum distribution of the ^{7}Be fragments after one-proton removal process of ^{8}B. This coincidence measurement provided direct information about the contribution from configuration (c) to the ^{8}B ground state wave function.

The gamma-rays recorded in the experiment are presented in the left panel of Fig. 5.14. A summary of the experimental results achieved is also shown in Table 5.2. The theoretical values were obtained using the eikonal approximation with a three body model (^{4}He + ^{3}He + t) to describe the ^{8}B wave function. The gamma coincidence also allowed to discriminate the fragment longitudinal distribution involving configuration (c) from the others. In this case the removed protons are always in a p state as shown in the right panel of Fig. 5.14, and almost no difference in the fragment momentum distribution width is observed.

The ratio of the cross section of ^{7}Be in its excited state to the total cross section is found to be 13 ± 3 % which is in excellent agreement with the theoretical value of 14 %. This indicates that the $\psi(^7\text{Be}(1/2^-)) \otimes \psi(p(3/2^-))$ component in the ^{8}B ground state wave function has a significant weight of about 16 %. The ratio

Fig. 5.15 The nuclear chart showing the region of light nuclei where different shell changes have been identified following empirical systematics of nuclear properties [120, 121]. Figure extracted from [120]

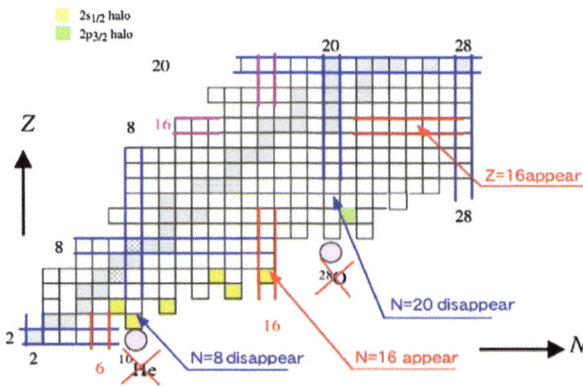

between the experimental and theoretical cross section indicates how realistic is the "prescription" used in the model. This will be explained later in Sect. 5.2.3.5.

5.2.3.2 Excursion to Nuclear Shells

Since the early work of Goeppert-Mayer and Jensen [12, 13, 118, 119], a description of the nucleus as formed by nucleons under the action of a central potential plus a spin-orbit interaction was accepted. The energy ordering of nucleons in nuclei showed important gaps at given nucleon numbers, pointing to the existence of "closed shells", where the maximum occupancy of the shell is reached. These nucleon numbers are known as "magic numbers" and provide extra stability to the nuclear system. The finding of magic numbers has been later confirmed by the presence of discontinuities associated with them, and observed in the systematic study of several nuclear properties, such as the evolution of the B(E2) strength.

This idea has been, and still is today, the robust pillar on which the shell model stands to explain the nuclear properties. The important technological improvements, achieved during the 60 years elapsed, have brought the opportunity of increasing the number of nuclei accessible for experiments.

Present nuclear studies concentrate on the investigation of species far away from the β-stability valley, characterized by the imbalanced number of protons and neutrons they own. The manifestation of new phenomena such as the nuclear halo discussed in Sect. 5.2.3.1 has been observed in some nuclei. As a general feature, unexpected structural properties result from a rearrangement of the nuclear orbitals, manifested in some of the exotic species studied.

This rearrangement has resulted in a modification of the traditional magic numbers [120] that we will address in this section and which is summarized in Fig. 5.15. A detailed review of these topics can be found in [120, 121].

To understand the importance of these modifications it is necessary to establish the connection between possible deviations from the expected picture, based on our knowledge of stable nuclei, and a more microscopical interpretation. From this point of view, a major contribution of isospin dependent terms in the nucleon-nucleon interaction could be reasonable and is today accepted [122–125] (and references therein). The need to include terms in the nuclear interaction corresponding to three-nucleon force contributions [75, 126, 127] has become obvious.

We will highlight some nucleon-knockout experiments that have contributed to the study of shell evolution close to the nuclear driplines. Again, the discussion will be restricted to the lower part of the nuclear chart.

Several experimental findings pointed to the vanishing of $N = 8$ magic number. Many experiments have addressed the study of ^{11}Li, starting with the interaction cross-section measurements [23] and followed by neutron knockout experiments [25, 128] and transfer reaction studies [129]. In a traditional shell-model picture ^{11}Li would have the valence neutron in the $1p_{1/2}$ orbital, but measurements evidenced a mixing probability of neutrons in the $1p_{1/2}$ but also in the $2s_{1/2}$ orbital. In the same direction, the study of the one-neutron knockout from ^{11}Be [111] (see Sect. 5.2.3.1) showed a major residence of neutrons in the $2s_{1/2}$ orbital. Other experiments, such as neutron transfer [130], Coulomb dissociation [42] and measurements of the magnetic moment [131] yielded similar conclusions: the breakdown of the shell gap at $N = 8$ with the presence of the intruder $2s_{1/2}$ orbital into the p-shell. The other example showed in Sect. 5.2.3.1, the one-proton knockout from ^{8}B [46], was used to explore whether a similar behaviour is present in the $Z = 8$ shell. In this case, the location of the proton halo in the usual $p_{3/2}$ orbital was confirmed and thus no shell change related.

A very intense experimental activity has concentrated on the $N = 20$ shell region. The oxygen isotopic chain ending at ^{24}O, with the non-observation of ^{28}O and ^{26}O as bound states suggested the vanishing of the $N = 20$ in this region. Many studies have addressed the identification and characterization of the so called "island of inversion" formed by nuclei whose ground states exhibit configuration mixing across the $N = 20$ shell. This weakening of the $N = 20$ shell was experimentally supported by the systematic measurements of B(E2) values for different Mg isotopes ($^{30-32}$Mg) related with the lowering of the 2^+ excitation energies. Another exhaustive work was that of the one-neutron knockout of ^{28}Ne [132] which will be described later in this lecture (see Sect. 5.2.3.3). The conclusion of this work identified the presence of fp-intruder orbitals in the ground state of ^{28}Ne. Along the same lines, one could also mention the one-neutron knockout of ^{30}Mg and ^{32}Mg [133], where the ground states show mixing configurations of $1f_{7/2}$ (dominating) and $2p_{3/2}$ orbitals and the one-neutron knockout of the very n-rich ^{33}Mg [50], which reveals in the ground state a mixing of $1f_{7/2}$, $2p_{3/2}$ and $1d_{3/2}$, with the intruder $2p_{3/2}$ contribution being quite significant.

Complementary manifestation of this shell evolution is the appearance of new shell gaps. A nice example is the observed shift of the $N = 20$ shell towards $N = 16$ for the case of light n-rich nuclei. The one-neutron knockout from the last bound oxygen isotope ^{24}O [49] was studied at the FRS. The measured inclusive momentum distribution (see Fig. 5.16) showed a clear evidence of $2s_{1/2}$ neutron occupation. The experimental determination of the associated spectroscopic factor is 1.74(19) and is compared in Table 5.3 with theoretical spectroscopic factors evaluated within the shell model using different effective interactions (SDPF-M and USDB). This large

Table 5.3 Spectroscopic factors for the $2s_{1/2}$ orbital following one-neutron removal from ^{24}O. The experimental value is compared with those determined using the SDPF-M and the USDB interactions. The experimental spectroscopic factor is obtained from a best fit of the ^{23}O momentum distribution (see Fig. 5.16)

	SDPF-M C^2S	USDB C^2S	ExpS C^2S
$1/2_{g.s.}$	1.769	1.810	1.74 (19)

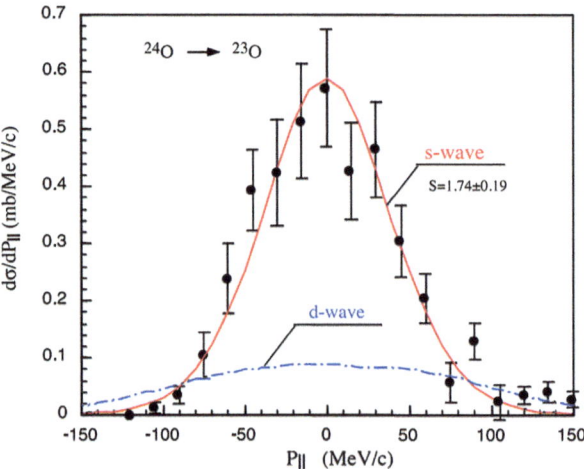

Fig. 5.16 Longitudinal momentum distribution after one-neutron removal from ^{24}O. The *solid curves* correspond to the eikonal model for neutron knockout from $2s_{1/2}$ in red, and $1d_{5/2}$ in blue [49]. Figure extracted from [49]

spectroscopic factor implies a strong concentration of the single-particle strength of the valence neutron in the $2s_{1/2}$ orbital, which indicates the existence of a large gap between the $2s_{1/2}$ and the $1d_{3/2}$ orbitals, consistent with a new shell gap at $N = 16$. This reveals ^{24}O as a new doubly closed shell nucleus.

The experimental determination of the shell gap ($\approx$4.8 MeV) has been possible from the 2^+ lowest resonance energy in ^{24}O observed in the one-proton removal from ^{26}F to ^{23}O $+ n$ [134]. This result was also confirmed by Ref. [135], measuring unbound excited states of ^{24}O via proton inelastic scattering (^{24}O$(p, p')^{23}$O $+ n$ reaction) in inverse kinematics at a beam energy of 62 MeV/nucleon at RIPS. We mention also interesting results on ^{22}C two-neutron knockout [41] performed at RIBF, that found a large spectroscopic factor of 1.403 for the $2s_{1/2}$ component in order to explain the narrow momentum distribution measured which further supports the $N = 16$ as a new magic number.

5.2.3.3 Benchmark of Nuclear Structure Models

The inherent complexity of quantifying structural properties (i.e.: the determination of C^2S) from the analysis of nucleon-removal reactions increases significantly as

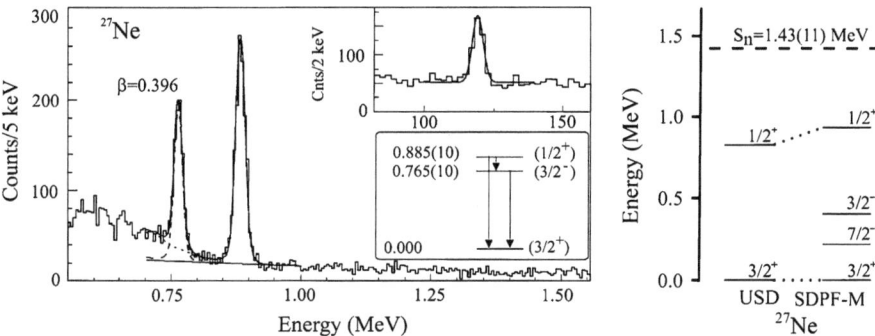

Fig. 5.17 *Left*: Doppler-reconstructed gamma-ray energy spectra for single neutron removal from ^{28}Ne. *Right*: Bound states predicted for ^{27}Ne by USD and SDPF-M calculations [132]. Figure extracted from [132]

nuclei closer to the β-stability are addressed, involving the knockout of more bound nucleons and increasing the amount of available final states.

However, it is important to stress that even in these complex cases the nucleon-knockout technique has proven to be very useful in providing structural information. An example is the neutron knockout of ^{28}Ne [132]. This experiment was performed at NSCL with a secondary beam of about 80 MeV/nucleon using SeGA [136],[21] a segmented Ge detector covering an angular range from 24° to 147°. Figure 5.17, left, shows three gamma-rays at 0.119 (inset Fig. 5.17 left), 0.765, and 0.885 MeV, recorded from the ^{27}Ne de-excitation. The 0.119 and 0.765 MeV gamma-rays were found to be in coincidence, suggesting that only two excited states were populated, as shown in the inset of this figure, and confirming the previous work [137].

The exclusive longitudinal momentum distributions were obtained by applying the coincidence method. Both ^{27}Ne excited states were associated with removed neutrons with $l = 0$ or $l = 1$ but the large experimental errors did not allow a definitive assignment. The momentum distribution of ^{27}Ne ground sate was significantly broader, but quantitative interpretation was not possible.

Figure 5.17, right, shows two different shell-model calculations for ^{27}Ne. Both interactions give very different predictions for the neutron-rich Ne isotopes, which are located in a transitional region between $N = 16$ and $N = 20$ (discussed in Sect. 5.2.3.2). Conventional calculations performed with the Universal SD (USD) interaction with a limited configuration space, do not allow for intruder configurations across the $N = 20$ gap, and thus fail to reproduce shell-breaking effects near $N = 20$. SDPF-M calculations indicate that intruder configurations are important for ^{27}Ne ($N = 17$), even in the low-energy part of the level scheme. Conversely, SDPF-M predictions in this transitional region also compare well with known level structures and electromagnetic moments.

---

[21] Segmented germanium array.

What is interesting here is that the ^{27}Ne measured gamma rays show the presence of low-lying states and are consistent with the SDPF-M shell model calculations. This contradicts the USD shell model, which predicts only one bound excited state.

Further analysis in terms of exclusive cross-sections was also carried out by the authors, who calculated single particle cross sections in the three-body reaction model [138]. The ratio of experimental and single particle cross sections made it possible to determine experimental spectroscopic factors. Spectroscopic factors from the SDPF-M model were not yet available, providing only an upper limit, which was however in quite good agreement with the experimental spectroscopic factors.

This experiment reported direct evidence of population of the $3/2^-$ intruder state in ^{27}Ne in the knockout of a single neutron from the ground state of ^{28}Ne. There are two important implications to this experimental finding; first, that this low-lying negative parity state is consistent with a narrower shell gap for exotic nuclei with $Z \ll N$ and $N \approx 20$; second, it clearly favoured Monte Carlo shell-model calculations with the modern SDPF-M interaction that successfully describe neutron-rich nuclei in the vicinity of $N = 20$, where normal and intruder configurations coexist at low excitation energy.

Another example can be found in the study of neutron-rich Ca, Ti and Cr nuclides around $N = 32$–35, where the GXPF1A interaction predicts a new doubly-magic shell closure for the $N = 34$ nucleus ^{54}Ca. This nucleus, can not be reached experimentally yet, but some nuclei in the neighbourhood, like ^{55}Ti [90], could be explored. The one-neutron knockout from ^{56}Ti using the FRS as a two-stage magnetic spectrometer and the Miniball [89] array for gamma-ray detection allowed the determination of inclusive and exclusive cross sections and longitudinal momentum distributions of ^{55}Ti, providing later the determination of the orbital angular momentum of the populated states. The measured data allowed for the first time to establish the ground state of ^{55}Ti as $1/2^-$, in agreement with shell model predictions using the GXPF1A (carefully tested in the pf-shell region [91, 139]).

The same interaction showed a reasonable agreement, reproducing the inclusive measurements of one-neutron knockout of $^{51-55}$Sc [54] for nucleons removed from p orbitals. However, an important mismatch between the experimental and theoretical cross-section was observed for the case of neutrons removed from f orbitals. The authors explained the discrepancy suggesting a migration of the spectroscopic strength of neutrons (in $f_{7/2}$) across the $N = 28$ shell gap. Indeed, this shell reduction has been recently predicted in Ca isotopes, by realistic NN interactions plus three body interactions [127], highlighting possible shortcomings in the GXPF1 interaction.

Consequently, these examples do not only demonstrate the general importance of direct reactions and particularly nucleon-knockout for the study of exotic nuclei, but can definitively be considered as excellent benchmark experiments for the predictive power of large-scale shell-model calculations and an important tool to improve them.

Fig. 5.18 Energy diagram of the neutron-rich $N = 16$ isotones (^{28}Mg, ^{27}Na, and ^{26}Ne), illustrating that direct population of the bound states of ^{26}Ne from ^{28}Mg is favoured over the two-step process of one-proton removal to excited ^{27}Na followed by proton evaporation [140]. Figure extracted from [140]

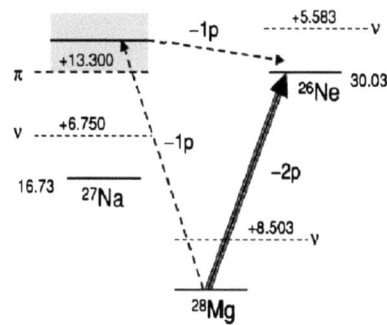

5.2.3.4 Two Nucleon Removal

The panorama of "direct reactions" induced by exotic beams was enlarged by Bazin and collaborators [28], who suggested the two-proton removal of very neutron-rich nuclei as a single step direct reaction.

It is based on the idea that the competing two-step process of a first proton-knockout followed by a proton evaporation is strongly suppressed in comparison with the neutron evaporation from the neutron-rich intermediate state, which is shown in Fig. 5.18 (extracted from [140]). This pioneering work was followed by a rather intense activity in both experimental [3, 28–33] and theoretical [140, 141] aspects (see also [1, 3]).

The inclusive cross-section of the two-nucleon knockout process is significantly smaller than the corresponding one-nucleon process, reaching maximum values of only a few mb. Even though the cross-sections are small, interesting results have been obtained. They refer, as in the case of single-nucleon knockout, to the momentum distributions of the emerging fragment and the associated cross-sections (both inclusive and exclusive).

The first experiment [28] performed at NSCL, concentrated on the study of two-proton knockout from neutron-rich nuclei (^{28,30}Mg and ^{34}Si). The $A-2$ heavy residues were detected with the A1900/S800 [84] spectrograph and the SeGA [136] array allowed the coincident detection of the γ-fragment de-excitation. The interpretation of these data pointed to the direct character of the process. This early work, which included a simplified reaction model where the two removed nucleons were uncorrelated and diffractive processes were completely neglected, provided a rather good agreement. Following experiments proved the validity of the method when applied to the two-neutron knockout of neutron-deficient species, namely ^{34}Ar, ^{30}S and ^{26}Si [29]. Similarly to the case presented above, the obtained cross-sections were consistent with direct reaction mechanisms. Cross-sections to individual excited states were measured using particle-γ coincidences with an identical experimental setup to Ref. [28]. In this work, the reaction mechanism description was improved and included ingredients of eikonal reaction theory and correlated many-body wave functions from shell-model calculations. In this model, several reaction mechanisms are considered to participate in the one-step two-nucleon removal: (a) the inelastic removal of both nucleons, (b) the elastic removal of one of them

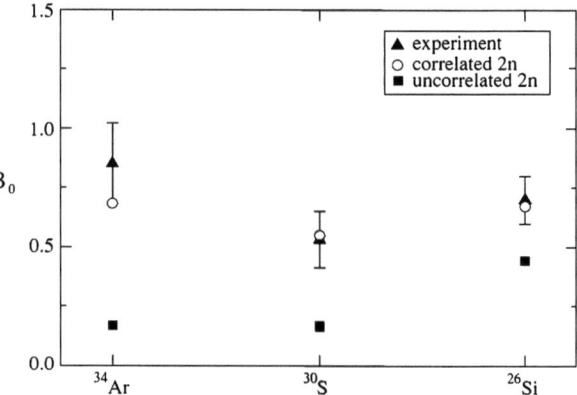

Fig. 5.19 Cross-sections for the ground state (*triangles*) in the two-neutron knockout reaction [29], compared with the corresponding theoretical cross-section [141] for correlated (*circle*) and uncorrelated (*squares*) neutrons (figure extracted from [29])

and the inelastic removal of the second and (c) the elastic removal of both nucleons. With these considerations, the cross-section of the two-nucleon removal process is expressed as (ignoring the Coulomb contribution):

$$\sigma = \sigma_{strip} + \sigma_{strip\text{-}diff} + \sigma_{diff}, \quad (5.8)$$

where σ_{strip} and σ_{diff} correspond to events where both nucleons are either inelastically or elastically scattered, and $\sigma_{strip\text{-}diff}$ corresponds to one nucleon interacting inelastically, while the other scatters elastically (details on the reaction model can be found in [140, 141]). The first two processes involve energy transfer to the target nucleus. Typically, and for the case of removal of well-bound nucleons, the diffraction mechanisms (options (b) and (c) in previous paragraph) amount for at least 40 % of the cross-section.

From the comparison of the cross-sections for the ground-state state with the theoretical estimations it was concluded that the presence of correlations between the knockout nucleons is needed in order to reproduce the experimental data (see Fig. 5.19).

Other experiments followed, extending the study to other nuclear chart regions (i.e.: $N = 20$ [31], $N = 28$ [30, 32]). The possibility of extracting structural information from the exclusive momentum distributions in a similar way as the one-nucleon knockout process was also reconsidered, and a reformulation of the theoretical calculation adapted to the two-nucleon removal case was proposed [142].

Sophistications introduced later consisted of simultaneous identification of the residual heavy fragment, the removed proton and other light charged particles after the two-proton removal reaction from ^{28}Mg projectile [143]. The experiment was again performed at NSCL, the heavy fragment was detected by the A1900/S800 [84] spectrograph and the protons and other light charged particles were detected with the HiRA [97] array. These data complement the previous work of Ref. [28], detecting the coincidence between the heavy residue and its γ de-excitation. In this new case triple coincident events were recorded, consisting of two charged particles detected in HiRA and the ^{26}Ne fragments in the spectrograph. As mentioned above, besides protons HiRA was also able to record other light charged particles like deuterons,

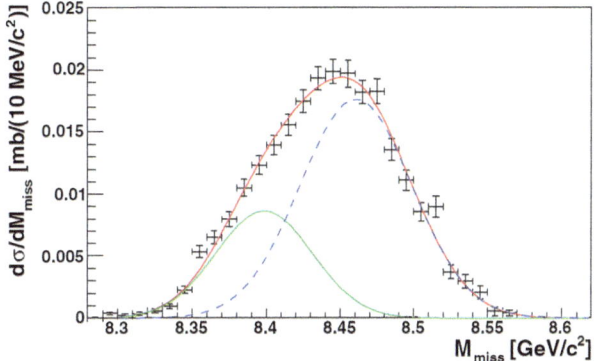

Fig. 5.20 Two-proton removal from ^{28}Mg. Missing-mass spectrum for events where two protons were detected in coincidence with the ^{26}Ne residue. The spectrum was fitted with two Gaussian peaks. The lower peak, at the target mass, is due to the diffraction mechanism (*green, solid line*), and the larger peak is attributed to events where at least one proton was removed in an inelastic collision with the target (figure extracted from [143])

tritons or α's, coming from inelastic reactions of the removed protons with target nucleons. The individual contribution of the three removal mechanisms was deduced from the missing-mass calculated for each triple-reconstructed event. Figure 5.20 shows the missing-mass spectrum for events where two protons were detected in coincidence with the ^{26}Ne residue. The experimental results obtained are consistent with the expectations of the eikonal theory described in [141].

With this scenario two-nucleon removal reactions are nowadays fully accepted as spectroscopic tools. In the case of direct knockout of "well-bound" nucleons, the associated cross-sections and A-2 fragment momentum distributions are considered as a source of information which enables to probe structural changes. They also provide information of the existence of correlations at the nuclear surface.

To date, the knockout of two-weakly bound nucleons has not been widely explored. Experimental conditions do not favour the direct process and the two step reaction mechanism is dominant in this case. Simpson and Tostevin [144] made a careful analysis of this scenario, interpreting two-neutron removal from neutron-rich carbon isotopes ($^{15-19}$C). This work determined the contribution of the single-step process to the measured cross-sections being only 10 % of the total cross-section.

Other experiments [41] have been performed at RIBF (Japan) employing unprecedented intensities for exotic secondary beams of ^{19,20,22}C at around 240 MeV/nucleon. Narrow momentum distributions were observed after one-neutron knockout of ^{19,20}C and two-neutron removal of ^{22}C, whereas the two-neutron removal of ^{20}C yielded much broader distributions. These results together with the associated cross-sections were interpreted in the case of single-nucleon knockout with help of an eikonal reaction model. The obtained results agree quite nicely with the general systematics and will be discussed in the next section.

For the interpretation of two-neutron removal cross sections, the case of one-neutron removal through unbound intermediate states followed by decay to a bound state as well as the direct two-neutron removal case were considered. A clear dominance for the first mechanism, with a direct contribution of around 8 % only and thus consistent with the early work of Simpson et al. [144] was obtained. Unfortunately the setup being used did not allow for exclusive measurements and the firm empirical confirmation of this estimation has still to wait for future experiments.

5.2.3.5 The Quest of Spectroscopic Factors

In previous sections we have discussed the ability of knockout reactions to determine the nuclear structure of rare isotopes. It has been also discussed that in order to reach this goal, it is necessary to compare the experimental observables with detailed calculations that require two main ingredients: a structure model and a proper description of the reaction mechanism (see Sect. 5.2.1).

Section 5.2.3.1 recounts in detail how knockout reactions are able to determine the nuclear structure of weakly bound halo states in which the proton (neutron) single-particle picture, the basis of the shell-model, is very realistic. But what happens when the knockout involves deeply bound states? In this case the picture provided by the shell model could be far from complete. Presently, truncated model-space effective interactions are employed and these are known to include only partially correlation effects that could be of importance in the case of deeply bound states. Indeed, the importance of these correlation effects can be evaluated from the determination of the single-particle spectroscopic strengths.

An important body of strength functions was obtained for the case of stable nuclei from electron-induced quasi-free scattering ($e, e'p$) obtained some decades ago. In these cases, it has been observed that the spectroscopic strength of valence protons (R_s)[22] exhibit a value of ≈ 0.6–0.7 [5], with R_s being the ratio between the experimental and theoretical cross-section. It will be of interest to extend the use of this technique to gain information on exotic nuclei. This could be achieved in the future with eA colliders, as the one foreseen in ELISe/FAIR [145] experiment. In the mean time, one can utilize other suitable direct reactions such as transfer and hadron knockout.

In the particular case of knockout reactions, and under the assumption of the correctness of the reaction description provided by the eikonal model, R_s (deduced by Eq. (5.6)) can be taken as a quantification of the pertinence of the pure single-particle description. The closer this value is to unity the more realistic the single-particle picture and so the effect of correlations between nucleons is less significant.

If we now come back to the first results of single-nucleon knockout of neutron and proton halo states presented in Sect. 5.2.3.1 and evaluate the corresponding spectroscopic strength we will obtain $R_s \geq 0.8$, significantly larger than the values recorded with stable nuclei. The question is now whether this technique can be applied to the study of exotic species in a general manner and what happens when the knockout of very deeply bound nucleons is addressed.

---

[22]Definition introduced in Eq. (5.7).

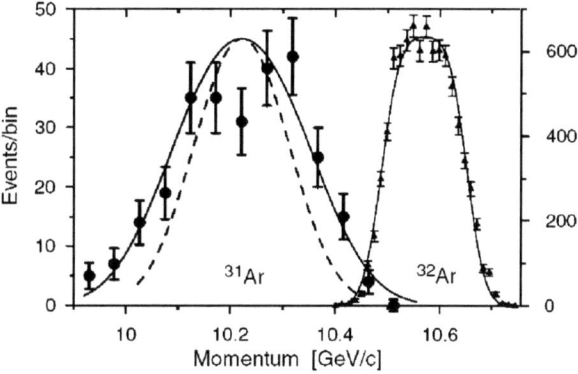

Fig. 5.21 Longitudinal momentum distribution of ^{31}Ar residues after one-neutron knockout reaction, compared with theoretical calculations assuming $l = 0$ (*dashed line*) and $l = 2$ (*solid line*) knocked out neutrons. The unreacted projectile beam ^{32}Ar fitted with a rectangular distribution folded with a Gaussian resolution function is also shown [146] (figure extracted from [146])

An instructive example will be to explore an extreme case such as the experimental study of ^{32}Ar [146] via one-neutron knockout at 61 MeV/nucleon. This nucleus is probably the most bound neutron-removal case investigated so far with this technique, which leads to the proton-dripline nucleus ^{31}Ar. This nucleus, with a $5/2^+$ ground state is peculiar because it has no bound excited states. A direct measurement of the core-fragment provides exclusive information without need of gamma coincidence.

From the experimental point of view there is an important difference between this experiment and all the others previously discussed. The S800 spectrograph was operated in "focus mode" with larger acceptance but lower intrinsic momentum resolution, as shown in Fig. 5.21. The theoretical momentum distributions depicted in this figure were calculated using the black-disk model [1] and folded with the measured response of the apparatus. This spectrum clearly associates the reaction with a $l = 2$ knockout neutron, confirming the spin and parity assignment for the ^{31}Ar ground state of $5/2^+$. Comparison of the experimental cross section 10.4(13) mb with the theoretical one yields a surprisingly low quenching factor of $R_s = 0.24(3)$. The authors of that work compared this case with ^{22}O, which has the same neutron number $N = 14$ and $Z = 8$ protons, and considered as almost doubly magic nucleus. The quenching factor, obtained in this case is 0.70(6) (using an average of two different experimental results [35, 36, 48] as experimental cross section), which is well above the result obtained for ^{32}Ar. The authors suggested that this very strong quenching is an indication of nuclear structure effects, reflecting correlations linked to the high neutron separation energies (22.0 MeV) in this very asymmetric nuclear system.

Other extremely asymmetric neutron-deficient systems were also explored (^{28}S and ^{24}Si) via one-proton and one-neutron knockout. This work presented in Ref. [147] is very interesting since it allows to explore in the same nuclei both

Table 5.4 Partial cross sections (mb) to all final states I^π after one-neutron and one-proton knockout from ^{28}S and ^{24}Si. Different contributions of theoretical single-particle cross sections in the eikonal model are reported. The sum multiplied by the spectroscopic factor is compared with the experimental values [147]

Fragment	I^π	C^2S	σ_{sp}^{knock}	σ_{sp}^{diff}	σ^{theo}	σ^{expt}	R_s
Projectile ^{24}Si							
^{23}Al$_{g.s.}$	5/2$^+$	3.42	17.56	5.18	84.68	67.3(35)	0.79(4)
^{23}Si$_{g.s.}$	5/2$^+$	1.71	10.96	2.47	25.01	9.9(10)	0.39(4)
Projectile ^{28}S							
^{27}P$_{g.s.}$	1/2$^+$	0.832	20.73	7.84	25.56	31(3)	
^{27}P$_{1.1\,MeV}$	3/2$^+$	0.82	14.61	4.40	16.75	6.8(11)	
^{27}P$_{incl.}$					42.32	38(3)	0.90(7)
^{27}S$_{g.s.}$	5/2$^+$	3.136	8.99	2.10	37.40		
^{27}S$_{e.s.}$	3/2$^+$	0.119	8.72	2.03	1.37		
^{27}S$_{incl.}$					38.77	11.9(12)	0.31(3)

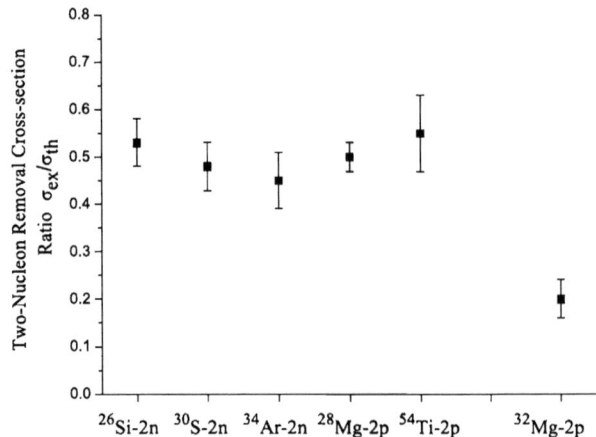

Fig. 5.22 Reduction factor R_s (defined as the ratio σ_{ex}/σ_{th}) evaluated for two-nucleon knockout cross-sections. Note that inclusive and exclusive data are included in the figure, see [31] for details

single-particle strengths associated with weakly bound nucleons (protons in the one-proton knockout case) and strongly bound nucleons (neutrons in the one-neutron case). The R_s values obtained are recorded in Table 5.4. Again, we can observe very different values of R_s for the two reaction channels studied.

These experimental results reinforce further the strong reduction observed in single particle strengths of deeply bound nucleons, whereas the reduction factor is close to unity for the case of knockout of weakly bound nucleons. A similar analysis performed for the two-nucleon removal cases discussed in Sect. 5.2.3.4, yields very similar results [31] (see Fig. 5.22), with R_s around 0.5 except for the ^{32}Mg case.

It is illustrative to analyse the systematic comparison performed by Gade and collaborators [147] (see Fig. 5.23), in which they represent an important collection

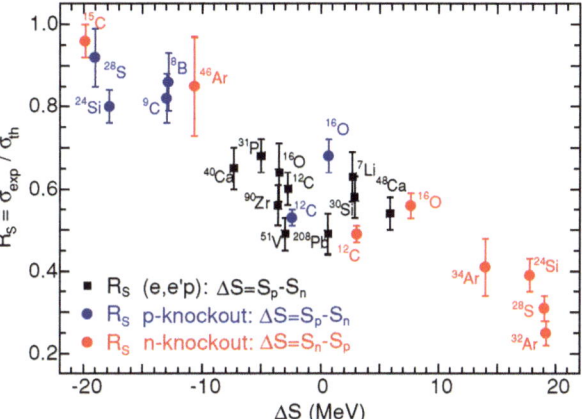

Fig. 5.23 Summary of quenching or empirical reduction factors obtained for spectroscopic factors evaluated from knockout reactions [147] (figure extracted from [147])

of the measured R_s reduction factors as a function of the difference in separation energies of the nuclei considered (ΔS). The quantity ΔS measures the asymmetry of the Fermi surface for each nuclei.[23] The following definitions $\Delta S = S_p - S_n$ and $\Delta S = S_n - S_p$ stand for the case of proton and neutron knockout respectively. Note that the figure also includes data from $(e, e'p)$ measurements mentioned earlier [5].

Is interesting to note that the physical occupancies are in general much lower than those suggested by the shell model, however this discrepancy is smaller for loosely bound nuclei (≈ 0.85 left top corner in Fig. 5.23). Cases corresponding to proton-removal from relative symmetric nuclei, with values ranging from 0.5 to 0.6, are in agreement with the findings obtained earlier with quasi-free "proton knockout" $(e, e'p)$ reactions that addressed stable nuclei and with $\Delta S \approx 0$. Knockout of deeply-bound nucleons as the ones represented by the neutron-knockout in near proton-dripline nuclei (right bottom corner in Fig. 5.23) are associated with R_s factors lower than 0.4. Using $(e, e'p)$ [148] as an analogy again, this small reduction factor is interpreted in terms of correlation effects that reveal the incomplete and simplified picture of nuclei provided by effective-interactions used nowadays. To complete the picture, note that Fig. 5.23 contains a mixture of exclusive and inclusive information. In the inclusive knockout cross section the quenching factor R_s is defined as the ratio of the experimental cross section to the sum of the theoretical cross sections to any state lying below the neutron threshold.

In spite of such significant experimental and theoretical progress, the tendency shown in Fig. 5.23 is not fully understood. The determination of absolute spectroscopic factors is a hot topic in the nuclear structure domain. The information gained with nucleon removal reactions induced by radioactive beams is undoubtedly important. The interplay between the nuclear structure and reaction input is necessary in order to extract this information, and the incomplete description of any of these ingredients could be the origin of the strong reduction observed. This interesting topic will be the subject of study in the coming years and there are plans

---

[23] Difference between the Fermi level for neutrons and protons in a given nucleus.

to extend it with the use of QFS reactions induced by radioactive beams. There is also an ongoing concerted effort to improve the reaction mechanism understanding (Refs. [152, 157]). The evaluation of spectroscopic factors in the medium-mass range coming from ab-initio calculations is expected to shed light on this exciting topic.

5.3 Quasi-Free Scattering Reactions with Rare Isotope Beams

In this last part of the lecture we will complete the introduction to the Quasi-Free Scattering (QFS) (see Sect. 5.1.1) with recent experiments addressing the study of rare isotopes produced in fragmentation facilities. We will finish reporting on the status and progress of the QFS program with rare exotic isotopes foreseen in future experimental setups such as R^3B [93] at FAIR.

As has been outlined in Sect. 5.2.3.5, further investigations on single-particle properties in rare nuclei, as well as the role of in-medium effects on the NN interactions and the existence of NN correlations, that would complement the work done so far with knockout reactions, are certainly needed.

The use of QFS reactions such as $(p, 2p)$ and (p, pn), induced by high energy proton beams offers very attractive possibilities. QFS reactions are able to excite both valence nucleons and also deep-hole states, giving access to the associated single-particle properties. Compared with the nucleon-knockout cases presented in Sect. 5.3, this reaction channel will not reduce the wave function exploration to the most external regions, thus providing a more complete picture of the structure of the nucleus. Moreover, the detection of both nucleons, the target and the removed nucleon, would provide (together with the gamma de-excitation of the *A-1* fragment) fully exclusive measurements, that in turn could help to characterize the reaction mechanism.

Up to now, we have only referred to QFS experiments performed in direct kinematics, with high energy protons impinging on the nucleus of interest [57]. It is clear that the use of inverse kinematics opens the exciting possibility of exploring nuclear structure for all kind of nuclear species.

A first attempt to this experimental approach took place few years ago at the HIMAC (Heavy Ion Medical Accelerator in Chiba) facility in NIRS (National Institute of Radiological Science in Japan) by Kobayashi and collaborators [149]. Different carbon isotopes ($^{9-16}$C) at 250 A MeV impinged on a solid-hydrogen target. These reactions provided an excellent scenario to get systematic information on weakly- to strongly bound $1p$ valence protons (lower N values) and deeply bound 1s inner shell protons (larger N values). The four-momenta of the two protons measured in a two-proton telescope located at $\pm 39°$ with respect to the beam axis allowed the determination of proton binding energy distributions (B_p)[24] of the different states

---

[24] Noted S_p in Ref. [149].

5 Direct Reactions at Relativistic Energies: A New Insight

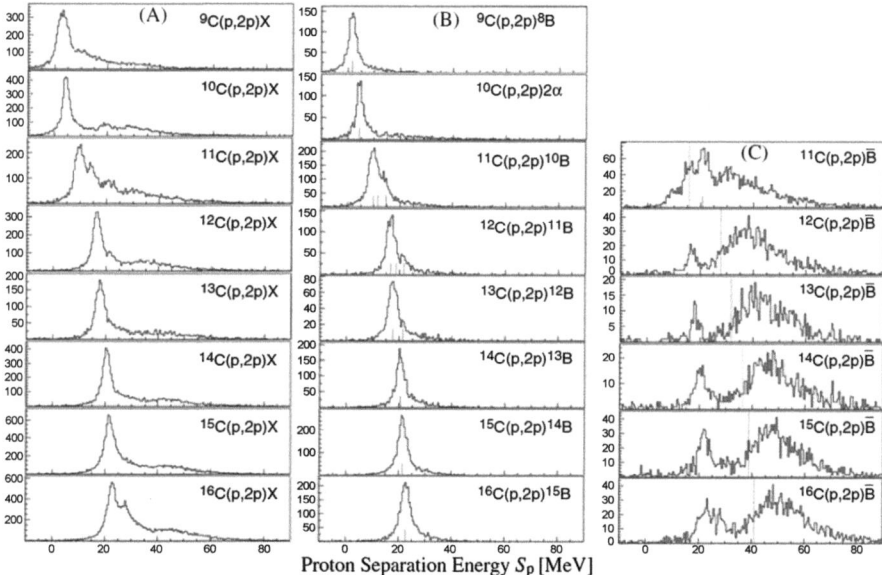

Fig. 5.24 Proton-separation energy distributions (B_p, denoted as S_p in the figure): (**A**) inclusive, (**B**) $^{A-1}$B detected in the forward spectrometer, (**C**) no boron isotopes detected in the forward spectrometer [149] (figure taken from [149])

from which the proton was removed in the projectile (see Fig. 5.24 and (5.9)).

$$B_p = S_p + E^*_{A-1},\qquad(5.9)$$

with S_p in this formula being the minimum energy to separate the last bound proton and E^*_{A-1} the excitation energy of the single particle state relative to the nucleus Fermi level (depending on how deeply bound the removed proton is this magnitude can vary from 0 MeV up to several tens of MeV).

The coincidence with the remaining fragments performed with a forward spectrometer is the responsible of the selection of the different transitions. Those to the ground state (case B in Fig. 5.24), were selected by gating on the $^{A-1}$B (boron) fragments, whereas those to the s-hole states (case C in Fig. 5.24), required the anti-coincidence with B isotopes. The s-holes states are associated with removal of deeply bound protons and are produced with high excitation energies. The high excitation energy allows to open channels decaying via charged particles emission. Those channels, in which the *A-1* fragment does not survive, provide information of the inner region of the projectile wave-function. It is important to mention the good agreement found between the ^{12}C results in this experiment and the one in direct kinematics [57].

In the same experiment, the momentum distributions of protons occupying both p- and s-orbitals could be measured and are presented in Fig. 5.25. In both cases, removal from p (A in the figure) and s-orbitals (B in the figure), one can clearly

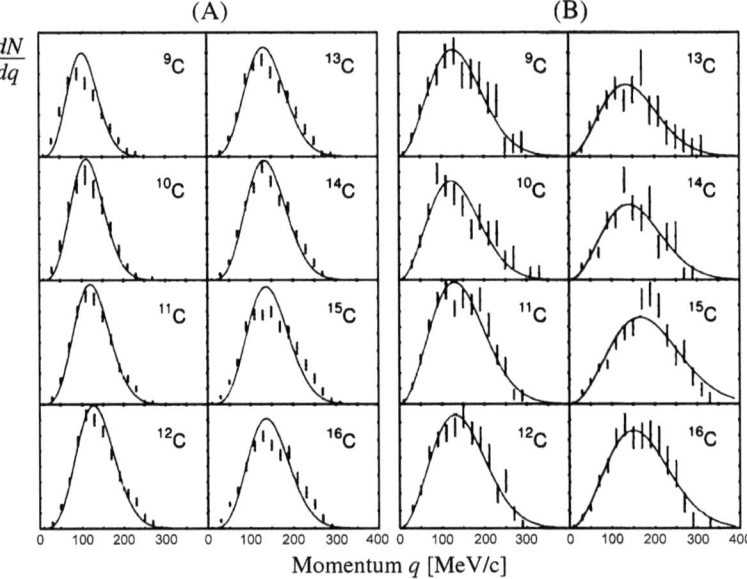

Fig. 5.25 Proton momentum distributions (**A**) for the p-hole states via $^{A}C(p, 2p)^{A-1}B_{gr}$ reaction. (**B**) for the s-hole states via $^{A}C(p, 2p)\bar{B}$ [149] (figure extracted from [149])

observe narrower distributions for lower masses (where the valence protons are more weakly bound). An accurate determination of the associated spectroscopic factors was not possible in this work but the systematic evaluation of the ratio of (p-hole)$_{yield}$/(s-hole)$_{yield}$ shows an increase for the ^{9}C case. This is also interpreted as a indication, analogous to the nucleon knockout case, of large spectroscopic factors associated with subshells for the weakly bound nucleons.

From the theoretical point of view, QFS has traditionally been interpreted in the framework of Distorted Wave Born Impulse (DWBI) approximation [150, 151]. There are new attempts to formulate a model to describe QFS observables based on the eikonal theory as it is shown in Ref. [152]. On the other hand the description of the QFS reaction mechanism on proton targets (free of stripping contributions) is also possible to be addressed with approaches such as the Continuum-Discretised Coupled Channels method (CDCC) [153–155], or even those based on complex solutions of the full Faddeev/AGS equation [156, 157].

5.3.1 Status of the QFS Program with Exotic Rare Isotopes at R^3B

The study of single particle properties of exotic nuclei is an important part of the experimental programme of several future nuclear physics instruments. First exclusive measurements of QFS reactions of rare isotopes on a proton target have been

5 Direct Reactions at Relativistic Energies: A New Insight

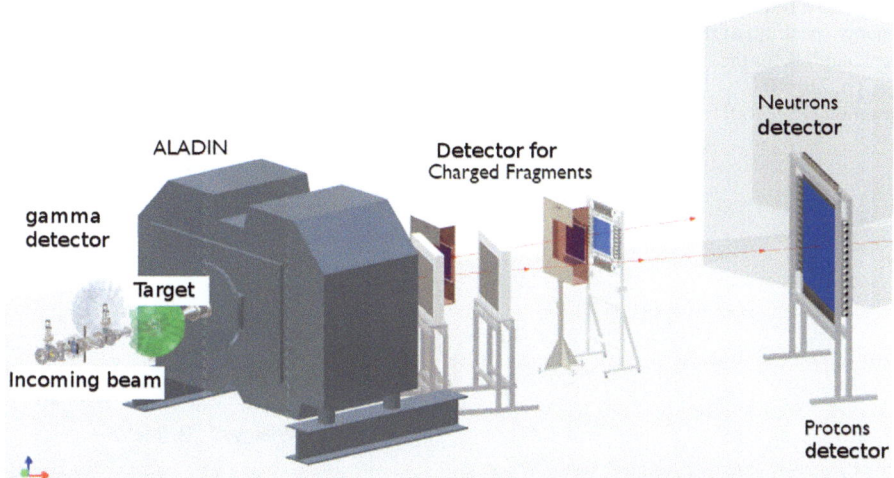

Fig. 5.26 LAND/R^{3}B setup (precursor of R^{3}B/FAIR experiment) presently installed at GSI [93]

performed recently at GSI using inverse kinematics with the LAND/R^{3}B setup (see Fig. 5.26).

The R^{3}B/NUSTAR [93] collaboration intends to apply this technique using the intense radioactive beams that will be delivered by the SuperFRS/FAIR [158] in inverse kinematics. At energies around 700 MeV/nucleon, both outgoing nucleons have energies in the range where the NN cross section is at a minimum, thus maximizing the transparency of the nucleus and minimizing final state interaction.

The large acceptance and high resolution setup proposed R^{3}B/FAIR [93] would allow background-free measurements and also for better control of final state interactions.

The well known ^{12}C$(p, 2p)^{11}$B reaction has been studied as the best validation of the experimental technique [159, 160] in the existing LAND-R^{3}B setup (see Fig. 5.26). An incident beam of ^{12}C at 400 A MeV impinged on a CH$_2$ target. A layer of silicon strip detectors surrounding this target (in a box geometry) and the 4π-calorimeter Crystal Ball. Both detectors allowed the detection of the light reaction products (in the $(p, 2p)$ case, two protons emitted back-to-back and with an average angular aperture $\approx$90°). The rest of fragments are deflected by the large acceptance magnet ALADIN and identified behind it. The neutrons, unaffected by the magnetic field of ALADIN [161], are identified with help of the LAND [162] detector.

Compared with other experiments two novelties are introduced. On the one hand, the proton detection is not limited to a co-planar geometry and on the other, the complete kinematical detection of incident beam and all the fragments emerging from the reaction allows for the first time to record redundant information on the internal nuclear momentum of the removed nucleon, either with the four-momenta determination of the two protons or with the four momenta of the fragment. Figure 5.27 shows the preliminary results [159, 160] of the total excitation energy spectrum of

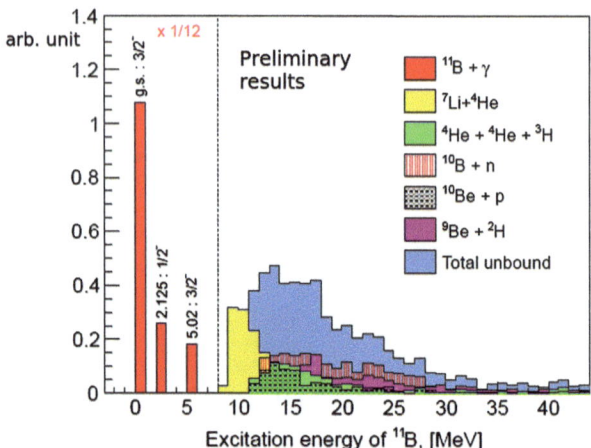

Fig. 5.27 Total excitation energy spectrum of the residual ^{11}B for the ^{12}C$(p,2p)^{11}$B reaction [159, 160]

the residual ^{11}B coming from ^{12}C$(p,2p)^{11}$B reaction that are in good agreement with other data addressing the same reaction [57, 149].

Different experiments[25] studying light neutron-rich nuclei with Z ranging from 4 to 9 produced by fragmentation of a 400 A MeV ^{40}Ar primary beam, and utilizing kinematically complete measurements of reactions at relativistic energies, were also done with the LAND/R^{3}B reaction setup. The physics topics studied comprise the measurement of astrophysical reaction rates relevant for r-process nucleosynthesis using heavy-ion induced electromagnetic excitation and quasi-free knockout reactions to study the evolution of shell and cluster structures close to and beyond the dripline. Unbound (ground and excited) states could be populated and identified in $(p,2p)$ reactions as it is shown in Ref. [163]. The quenching of single-particle strengths in neutron-proton asymmetric nuclei was also addressed by knocking out deeply bound protons and neutrons in $(p,2p)$ and (p,pn) reactions for nuclei with varying neutron-proton asymmetry. The evaluation of this interesting collection of data is presently under analysis and the first results are expected soon.

All these works open a very promising future of new investigations that will one day profit from the high-energy intense radioactive beams at FAIR. They would enable the realization of measurements on exotic nuclei, with the R^{3}B/FAIR experiment overcoming the present experimental limitations. R^{3}B/FAIR [93] will be a versatile reaction setup with high efficiency, acceptance, and resolution for kinematically complete measurements of reactions with high-energy exotic beams. The setup will be located at the focal plane of the high-energy branch of the Super-FRS. A substantial improvement is expected with respect to resolution and an extended detection scheme is foreseen. It will comprise

---

[25] S393 experiment performed in August 2010, Spokesperson T. Aumann.

- a zero-degree superconducting dipole magnet with a large vertical gap allowing an angular acceptance of ±80 mrad for neutrons and a field up to 5 T able to bend 14° charged fragments for 20 Tm beams.[26]
- a detection setup for light (target-like) recoil particles, formed by a double layer of Si strip detectors. A liquid hydrogen target could be hosted in the inner part of this Si-tracker. The recoil particle detector would provide precise tracking, vertex determination as well as energy and multiplicity measurement with high efficiency and acceptance.
- a new generation spectrometer-calorimeter CALIFA. This detector has to act as spectrometer of low to medium energy γ, being at the same time calorimeter for the target recoil (high energy gammas and light-charged particles). This will be achieved by an extremely segmented detector based on highly performant scintillation crystals (in single [92] or phoswich [164] configuration).[27]
- a dedicated tracking systems for the detection of fragments behind the dipole magnet. For the tracking of light charged particles, i.e. protons, behind the R^3B dipole magnet two identical drift chambers (DHC) are foreseen. Each DCH covers an active area of 100×80 cm^2, thus providing a large enough acceptance behind the magnet to detect decay protons. A large-area scintillating fibre detector is foreseen, for the position measurement of heavy fragments, a few meters behind the magnet. The system is completed by a ToF detector covering the full acceptance of the charged particles and ions produced in relativistic heavy-ion collisions while providing a time-of-flight resolution such that isotopes around the mass 200 can be isotopically resolved.
- a neutron detector NeuLAND featuring a high detection efficiency, a high resolution, and a large multi-neutron-hit resolving power. This is achieved by a highly granular design of plastic scintillators, avoiding insensitive converter material.
- a high-resolution fragment spectrometer that could be constructed in the second phase of the experiment and would substitute the tracking systems for the detection of fragments in those cases demanding high-resolution measurements, where $\Delta p/p$ in the order of 10^{-4} is needed. The proposed solution is based on a spectrometer placed behind the large-acceptance dipole. The spectrometer will be placed at an angle of 18°, which is the maximum bending angle of the dipole (5 Tm field integral) for a beam with magnetic rigidity of 15 Tm, and will be operated as a zero-degree spectrometer.

Lastly we mention another innovative device that will be dedicated to the study of QFS reactions. MINOS (Magic Numbers off Stability) [165], aims to investigate the properties of in-medium NN interactions through the spectroscopy of the most exotic nuclei produced at fragmentation facilities (RIKEN and GSI/FAIR). The project includes the construction of a dedicated and innovative device composed of a thick cryogenic liquid hydrogen target surrounded by a cylindrical time projection

---

[26] The maximum rigidity provided by the Super-FRS will be 20 Tm.

[27] In a phoswich detectors the energy of the particles is determined from two consecutive energy losses in consecutive detectors.

chamber devoted to determine the reaction vertex by tracking charged particles produced by $(p, 2p)$ or (p, pn) reactions with excellent resolution. MINOS has been built at IRFU, CEA Saclay[28] and will be soon used to perform the first experimental campaign in RIBF/RIKEN.

5.4 Summary and Conclusions

This lecture intended to provide a comprehensive overview of some direct reactions used at high energies to explore the nuclear structure of rare isotopes.

We have presented the interesting implications of using "knockout reactions", emphasizing the experimental and conceptual aspects associated with these kind of experiments and their possible ramifications in the analysis and interpretation of the results.

Halo nuclei, which have a relatively simple and well-known nuclear structure, were used to show the power of the method. Experiments of one-nucleon knockout from halo states provide a very satisfactory description of the reaction mechanism, making it possible to experimentally deduce the structure of the involved nuclei.

Though generally successful, application of the technique to more complex nuclei is not without experimental difficulties that complicate the extraction of quantitative information. Reactions of this kind have been used to measure the physical occupancies (C^2S) associated with the different configurations that define the bound states of the exotic nuclei under study. The experimental spectroscopic factors obtained are generally smaller than those predicted by the large-scale shell model. The existence of these quenching factors has been interpreted in terms of correlation effects pointing out the simplified picture of the nucleus provided by shell model calculations. Deficiencies in the reaction mechanism description could be also behind this effect.

Even though great progress has been made, the situation is not yet clear and the determination of absolute spectroscopic factors with nucleon removal reactions induced by radioactive beams will certainly be a hot topic in the next years. The recent extension of the use of quasi-free scattering reactions in inverse kinematics opens new experimental opportunities. The most important will be the availability of fully exclusive measurements, enabling a better understanding of the reaction mechanism. The information that one can extract about the exotic projectile wave function will not be restricted to the nuclear surface, being possible to extend this exploration to the removal of strongly bound nucleons allowing a deeper study of the role of NN correlations.

Construction and exploitation of new generation facilities (i.e.: NUSTAR/FAIR, BIGRIPS, FRIB), in the near future, will significantly increase the intensity of the available exotic beams. This, together with the development of new specific detection systems (i.e.: R^3B@NUSTAR/FAIR, MINOS ...) will open new perspectives in this field.

---

[28] Project funded by the European Research Council for the period 2010–2015.

Acknowledgements I would like to thank Dr. B. Pietras for careful reading of the manuscript, valuable comments and multiple English corrections.

References

1. P.G. Hansen, J.A. Tostevin, Direct reactions with exotic nuclei. Annu. Rev. Nucl. Part. Sci. **53**, 219 (2003)
2. T. Aumann, Reactions with fast radioactive beams of neutron-rich nuclei. Eur. Phys. J. A **26**, 441 (2005)
3. A. Gade, T. Glasmacher, In-beam nuclear spectroscopy of bound states with fast exotic ion beams. Prog. Part. Nucl. Phys. **60**, 161 (2008)
4. G. Jacob, Th.A. Maris, Nuclear structure II. Rev. Mod. Phys. **45**, 6 (1973)
5. L. Lapikás, Quasi-elastic electron scattering off nuclei. Nucl. Phys. A **553**, 297 (1993)
6. V.R. Pandharipande, I. Sick, P.K.A. de Witt Hubers, Independent particle motion and correlations in fermion systems. Rev. Mod. Phys. **69**, 981 (1997)
7. G.J. Kramer, H.P. Blok, L. Lapikas, A consistent analysis of $(e, e'p)$ and $(d, {}^3\text{He})$ experiments. Nucl. Phys. A **679**, 267 (2001)
8. M.S. Hussein, K. McVoy, Inclusive projectile fragmentation in the spectator model. Nucl. Phys. A **445**, 124 (1985)
9. C.A. Bertulani, K.W. McVoy, Momentum distributions in reactions with radioactive beams. Phys. Rev. C **46**, 2638 (1992)
10. J.S. Al-Khalili et al., Evaluation of an eikonal model for ^{11}Li-nucleus elastic scattering. Nucl. Phys. A **581**, 331 (1995)
11. D.N. Poenaru, W. Greiner, in Experimental Techniques in Nuclear Physics ed. by W. de Gruyere (Chaps. 10 and 11)
12. M.G. Mayer, On closed shells in nuclei. Phys. Rev. **74**, 235 (1948)
13. O. Haxel, J.H. Jensen, H.E. Suess, On the magic numbers in nuclear structure. Phys. Rev. **75**, 1766 (1949)
14. O. Chamberlain, E. Segrè, Proton-proton collisions within lithium nuclei. Phys. Rev. **87**, 81 (1952)
15. J.B. Cladis, W.N. Hess, B.J. Moyer, Nucleon momentum distributions in deuterium and carbon inferred from proton scattering. Phys. Rev. **87**, 425 (1952)
16. T. Berggren, H. Tryén, Quasi-free scattering. Annu. Rev. Nucl. Sci. **16**, 153 (1966)
17. G. Jacob, Th.A. Maris, Nuclear structure. Rev. Mod. Phys. **38**, 121 (1966)
18. P.K.A. de Witt Huberts, Proton spectral functions and momentum distributions in nuclei from high-resolution $(e, e'p)$ experiments. J. Phys. G **16**, 507 (1990)
19. A.E.L. Dieperink, P.K.A. de Witt Huberts, On high resolution $(e, e'p)$ reactions. Annu. Rev. Nucl. Part. Sci. **40**, 239 (1990)
20. G. van de Steehoven et al., Knockout of $1p$ protons from ^{12}C induced by the $(e, e'p)$ reaction. Nucl. Phys. A **480**, 547 (1988)
21. J. Mougey et al., Quasi-free $(e, e'p)$ scattering on ^{12}C, ^{28}Si, ^{40}Ca and ^{58}Ni. Nucl. Phys. A **262**, 461 (1976)
22. P.G. Hansen, Studies of single-particle structure at and beyond the drip lines. Nucl. Phys. A **682**, 310 (2001)
23. I. Tanihata et al., Measurements of interaction cross sections and nuclear radii in the light p-shell region. Phys. Rev. Lett. **55**, 2676 (1985)
24. I. Tanihata, Research opportunities with accelerated beams of radioactive ions. Nucl. Phys. A **693**, 1 (2001)
25. T. Kobayashi et al., Projectile fragmentation of the extremely neutron-rich nuclei ^{11}Li at 0.79 GeV/nucleon. Phys. Rev. Lett. **60**, 2599 (1988)
26. N.A. Orr et al., Momentum distributions of ^{9}Li fragment following the breakup of ^{11}Li. Phys. Rev. Lett. **69**, 2050 (1992)

27. A. Gade et al., One-neutron knockout in the vicinity of $N = 32$ sub-shell closure: ^{9}Be(^{57}Cr, ^{56}Cr $+ \gamma$)X. Phys. Rev. C **74**, 047302 (2006)
28. D. Bazin et al., New direct reaction: two-proton knockout from neutron-rich nuclei. Phys. Rev. Lett. **91**, 012501 (2003)
29. K. Yoneda et al., Two-neutron knockout from neutron-deficient ^{34}Ar, ^{30}S, and ^{26}Si. Phys. Rev. C **74**, 021303 (2006)
30. J. Fridmann et al., Shell structure at $N = 28$ near the dripline: spectroscopy of ^{42}Si, ^{43}P, and ^{44}S. Phys. Rev. C **74**, 034313 (2006)
31. P. Fallon et al., Two-proton knockout from ^{32}Mg: intruder amplitudes in ^{30}Ne and implications for the binding of ^{29,31}F. Phys. Rev. C **81**, 041302(R) (2010)
32. D. Santiago-Gonzalez et al., Triple configuration coexistence in ^{44}S. Phys. Rev. C **83**, 061305(R) (2011)
33. A. Gade et al., Cross-shell excitation in two-proton knockout: structure of ^{52}Ca. Phys. Rev. C **74**, 021302 (2006)
34. J.L. Lecouey et al., Single-proton removal reaction study of ^{16}B. Phys. Lett. B **672**, 6 (2009)
35. E. Sauvan et al., One-neutron removal reactions on neutron-rich psd-shell nuclei. Phys. Lett. B **491**, 1 (2000)
36. E. Sauvan et al., One-neutron removal reactions on light neutron-rich nuclei. Phys. Rev. C **69**, 044603 (2004)
37. T. Nakamura et al., Coulomb dissociation of ^{19}C and its halo structure. Phys. Rev. Lett. **83**, 1112 (1999)
38. N. Fukuda et al., Coulomb and nuclear breakup of a halo nucleus ^{11}Be. Phys. Rev. C **70**, 054606 (2004)
39. T. Kubo, In-flight RI beam separator BigRIPS at RIKEN and elsewhere in Japan. Nucl. Methods Phys. Res. B **204**, 97 (2003)
40. T. Ohnishi et al., Identification of new isotopes 125Pd and 126Pd produced by in-flight fission of 345 MeV/nucleon ^{238}U: first results from the RIKEN RI beam factory. J. Phys. Soc. Jpn. **77**, 083201 (2008)
41. N. Kobayashi et al., One- and two-neutron removal reactions from the most neutron-rich carbon isotopes. Phys. Rev. C **86**, 054604 (2012)
42. R. Palit et al., Exclusive measurement of breakup reactions with the one-neutron halo nucleus ^{11}Be. Phys. Rev. C **68**, 034318 (2003)
43. U. Datta-Pramanik et al., Coulomb breakup of the neutron-rich isotopes ^{15}C and ^{17}C. Phys. Lett. B **551**, 63 (2003)
44. C. Nociforo et al., Coulomb breakup of ^{23}O. Phys. Lett. B **605**, 79 (2005)
45. Th. Baumann et al., Longitudinal momentum distributions of ^{16,18}C fragments after one-neutron removal from ^{17,19}C. Phys. Lett. B **439**, 256 (1998)
46. D. Cortina-Gil et al., Experimental evidence for the ^{8}B ground state configuration. Phys. Lett. B **529**, 36 (2002)
47. D. Cortina-Gil et al., Nuclear and Coulomb breakup of ^{8}B. Nucl. Phys. A **720**, 3 (2003)
48. D. Cortina-Gil et al., Shell structure of the near-dripline nucleus ^{23}O. Phys. Rev. Lett. **93**, 062501 (2004)
49. R. Kanungo et al., One-neutron removal measurement reveals ^{24}O as a new doubly magic nucleus. Phys. Rev. Lett. **102**, 152501 (2009)
50. R. Kanungo et al., Structure of ^{33}Mg sheds new light on the $N = 20$ island of inversion. Phys. Lett. B **685**, 253 (2010)
51. C. Rodriguez-Tajes et al., One-neutron knockout from $^{24-28}$Ne isotopes. Phys. Lett. B **687**, 26 (2010)
52. C. Rodriguez-Tajes et al., One-neutron knockout from light neutron-rich nuclei at relativistic energies. Phys. Rev. C **82**, 024305 (2010)
53. C. Rodriguez-Tajes et al., Structure of ^{22}N and the $N = 14$ subshell. Phys. Rev. C **83**, 064313 (2011)
54. S. Schwertel et al., One-neutron knockout from $^{51-55}$Sc. Eur. Phys. J. A **48**, 191 (2012)

55. C. Nociforo et al., One-neutron removal reactions on Al isotopes around the $N = 20$ shell closure. Phys. Rev. C **85**, 044312 (2012)
56. T. Noro et al., A study of nucleon properties in nuclei through $(p, 2p)$ reactions. Nucl. Phys. A **629**, 324 (1998)
57. M. Yosoi et al., Structure of the s-hole state in 11B studied via de 12C$(p, 2p)$11B* reaction. Phys. Lett. B **551**, 255 (2003)
58. R. Subedi et al., Probing cold dense nuclear matter. Science **320**, 1476 (2008)
59. M. Karakoc, A. Banu, C. Bertulani, L. Trache, Coulomb distortion and medium corrections in nucleon-removal reactions. Phys. Rev. C **87**, 024607 (2013)
60. K. Henken, G. Bertsch, H. Esbensen, Breakup reactions of the halo nuclei ^{11}Be and ^{8}B. Phys. Rev. C **54**, 3043 (1996)
61. C.A. Bertulani, P. Danielewicz, Introduction to nuclear reactions (Institute of Physics Publishing, 2004)
62. http://www.nuclear.theory.net/NPE/Talentmaterial/index2.htlm
63. G.F. Bertsch, K. Henken, H. Esbensen, Nuclear breakup of Borromean nuclei. Phys. Rev. C **57**, 1366 (1998)
64. J. Tostevin, J. Al-Khalili, Sizes of the He isotopes deduced from proton elastic scattering measurements. Nucl. Phys. A **616**, 418c (1997)
65. J. Al-Khalili, J. Tostevin, I. Thompson, Radii of halo nuclei from cross section measurements. Phys. Rev. C **54**, 1843 (1996)
66. J. Tostevin et al., Single-neutron removal reactions from ^{15}C and ^{11}Be: deviations from the eikonal approximation. Phys. Rev. C **66**, 024607 (2002)
67. D. Bazin et al., Mechanisms in knockout reactions. Phys. Rev. Lett. **102**, 232501 (2009)
68. T. Aumann, Private communication (2013)
69. H. Grawe, Shell Model from a Practitioner's Point of View, in *The Euroschool Lectures on Physics with Exotic Beams*, vol. I (2004), pp. 33–76
70. T. Otsuka, Shell Structure of Exotic Nuclei, in *The Euroschool Lectures on Physics with Exotic Beams*, vol. III (2009), pp. 1–25
71. Shell model code ANTOINE, http://sbgat194.in2p3.fr/~theory/antoine/intro.html
72. Shell model code NuShell, http://www.garsington.eclipse.co.uk
73. Shell model OXBASH, http://arxiv.org/abs/nucl-th/9406020
74. R. Roth et al., Ab initio calculations of medium-mass nuclei with normal-ordered chiral $NN + 3N$ interactions. Phys. Rev. Lett. **109**, 052501 (2012)
75. G. Hagen et al., Continuum effects and three-nucleon forces in neutron-rich oxygen isotopes. Phys. Rev. Lett. **108**, 242501 (2012)
76. H. Hergert et al., Ab initio calculations of even oxygen isotopes with chiral two-plus-three-nucleon interactions. Phys. Rev. Lett. **110**, 242501 (2013)
77. Ch. Forssén et al., Systematics of 2^+ states in C isotopes from the ab initio no-core shell model. J. Phys. G, Nucl. Part. Phys. **40**, 055105 (2013)
78. V. Somà et al., Ab initio Gorkov-Green's function calculations of open-shell nuclei. Phys. Rev. C **87**, 011303(R) (2013)
79. http://www.fair-center.eu/
80. http://www.frib.msu.edu/
81. W. Mittig, Spectromètres magnetiques et electriques comme detecteurs de haute resolution et comme filtres selectifs. Ecole Joliot-Curie (1994)
82. D.J. Morrisey, B. Sherrill, *In-Flight Separation of Projectile Fragments*. Euroschool on Exotic Beams, vol. 1 (Springer, Berlin, 2004), ISBN 3-540-22255-3
83. H. Geissel et al., A versatile magnetic system for relativistic heavy ions. Nucl. Inst. Methods Phys. Res. B **70**, 286 (1992)
84. D.J. Morrissey et al., A new high-resolution separator for high intensity secondary beams. Nucl. Instrum. Methods Phys. Res. B **126**, 316 (1997)
85. D. Bianchi et al., SPEG: an energy loss spectrometer for GANIL. Nucl. Instr. Methods Phys. Res. A **276**, 509 (1989)

86. Th. Baumann, Longitudinal momentum distributions of ^{8}B and ^{19}C: signatures for one-proton and one-neutron halos. Ph.D. Dissertation, University of Giessen (1999)
87. http://lise.nscl.msu.edu/documentation.html
88. http://web-docs.gsi.de/~weick/mocadi/
89. J. Eberth et al., MINIBALL A Ge detector array for radioactive ion beam facilities. Prog. Part. Nucl. Phys. **46**, 389 (2001)
90. P. Maierbeck et al., Structure of ^{55}Ti from relativistic one-neutron knockout. Phys. Lett. B **675**, 22 (2009)
91. M. Honma et al., Effective interaction for pf-shell nuclei. Phys. Rev. C **65**, 061301 (2002)
92. Technical design report of the CALIFA Barrel Detector, http://www.fair-center.eu/en/for-users/publications/experiment-collaboration-publications.html
93. https://www.gsi.de/en/work/research/nustarenna/kernreaktionen/activities/r3b.htm
94. T. Nakamura et al., Observation of strong low-lying E1 strength in the two-neutron halo nucleus ^{11}Li. Phys. Rev. Lett. **96**, 252502 (2006)
95. H. Simon et al., Systematic investigation of the drip-line nuclei ^{11}Li and ^{14}Be and their unbound subsystems ^{10}Li and ^{13}Be. Nucl. Phys. A **791**, 267 (2007)
96. M. Zinser et al., Invariant-mass spectroscopy of ^{10}Li and ^{11}Li. Nucl. Phys. A **619**, 151 (1997)
97. M.S. Wallace et al., The high resolution array (HiRA) for rare isotope beam experiments. Nucl. Instrum. Methods Phys. Res. A **583**, 302 (2007)
98. C. Rodriguez, Ph.D. Thesis Universidad de Santiago de Compostela, http://www.usc.es/genp
99. B. Jonson, Light dripline nuclei. Phys. Rep. **289**, 1 (2004)
100. J. Al-Kahalili, An introduction to Halo Nuclei, in *The Euroschool Lectures on Physics with Exotic Beams*, vol. I (2004), pp. 77–112
101. K. Riisager, Nuclear Halos and experiment to probe them, in *The Euroschool Lectures on Physics with Exotic Beams*, vol. II (2006), pp. 1–36
102. D.J. Millener et al., Strong E1 transitions in ^{9}Be, ^{11}Be, and ^{13}C. Phys. Rev. C **28**, 497 (1983)
103. J.H. Kelley et al., Parallel momentum distributions as a probe of halo wave functions. Phys. Rev. Lett. **74**, 30 (1995)
104. F.M. Nunes, I.J. Thompson, R.C. Johnson, Core excitation in one neutron halo systems. Nucl. Phys. A **596**, 171 (1996)
105. T. Suzuki, T. Otsuka, A. Muta, Magnetic moment of ^{11}Be. Phys. Lett. B **364**, 69 (1995)
106. D.L. Auton, Direct reactions on ^{10}Be. Nucl. Phys. A **157**, 305 (1970)
107. B. Zwieglinski, W. Benenson, R.G.H. Robertson, Study of the ^{10}Be$(d, p)^{11}$Be reaction at 25 MeV. Nucl. Phys. A **315**, 124 (1979)
108. N.K. Timofeyuk, R.C. Johnson, Deuteron stripping and pick-up on halo nuclei. Phys. Rev. C **59**, 15 (1999)
109. R. Anne et al., Exclusive and restricted-inclusive reactions involving the ^{11}Be one-neutron halo. Nucl. Phys. A **575**, 125 (1994)
110. T. Nakamura et al., Exclusive and restricted-inclusive reactions involving the ^{11}Be one-neutron halo. Phys. Lett. B **331**, 296 (1994)
111. T. Aumann et al., One-neutron knockout from individual single-particle states of ^{11}Be. Phys. Rev. Lett. **84**, 35. 45 (2000)
112. E.K. Warburton, B.A. Brown, Effective interactions for the 0p1s0d nuclear shell-model space. Phys. Rev. C **46**, 923 (1992)
113. W. Schwab et al., Obsevation of a proton-halo in ^{8}B. Z. Phys. A **350**, 283 (1995)
114. M.H. Smedberg et al., New results on the halo structure of ^{8}B. Phys. Lett. B **452**, 1 (1999)
115. D. Cortina-Gil et al., One-nucleon removal cross-sections for ^{17,19}C and ^{8,10}B. Eur. Phys. J. A **10**, 49 (2001)
116. J.N. Bahcall et al., How uncertain are solar neutrino predictions? Phys. Lett. B **433**, 1 (1998)
117. Q.R. Ahmad et al., Measurement of the rate of $v_e + d \rightarrow p + p + e-$ interactions produced by ^{8}B solar neutrinos at the sudbury neutrino observatory. Phys. Rev. Lett. **87**, 071301 (2001)
118. M. Goeppert Mayer, *The Shell Model*. Nobel Lecture (1963)
119. J.H. Jensen, *Glimpses at the History of the Nuclear Structure Theory*. Nobel Lecture (1963)

120. R. Kanungo, A new view of nuclear shells. Phys. Scr. T **152**, 014002 (2013)
121. I. Tanihata, H. Savajols, R. Kanungo, Recent experimental progress in nuclear halo structure studies. Prog. Part. Nucl. Phys. **68**, 215 (2013)
122. T. Otsuka et al., Magic numbers in exotic nuclei and spin-isospin properties of the NN interaction. Phys. Rev. Lett. **87**, 082502 (2001)
123. T. Otsuka et al., Evolution of nuclear shells due to the tensor force. Phys. Rev. Lett. **95**, 232502 (2005)
124. T. Otsuka et al., Three-body forces and the limit of oxygen isotopes. Phys. Rev. Lett. **105**, 232501 (2010)
125. T. Otsuka, Exotic nuclei and nuclear forces. Phys. Scr. T **152**, 014007 (2013)
126. S.C. Pieper, R.B. Wiringa, J. Carlson, Quantum Monte Carlo calculations of excited states in $A = 6$–8 nuclei. Phys. Rev. C **70**, 054325 (2004)
127. J. Holt et al., Three-body forces and shell structure in calcium isotopes. J. Phys. G, Nucl. Part. Phys. **39**, 085111 (2012)
128. H. Simon et al., Direct experimental evidence for strong admixture of different parity states in ^{11}Li. Phys. Rev. Lett. **83**, 496 (1999)
129. I. Tanihata et al., Measurement of two-halo neutron transfer reaction $p(^{11}\text{Li},^{9}\text{Li})t$ at 3 A MeV. Phys. Rev. Lett. **100**, 192502 (2008)
130. S. Fortier et al., Core excitation in $^{11}\text{Be}_{gs}$ via the $p(^{11}\text{Be},^{10}\text{Be})d$ reaction. Phys. Lett. B **461**, 22 (1999)
131. W. Geithner et al., Measurement of the Magnetic Moment of the One-Neutron Halo Nucleus ^{11}Be. Phys. Rev. Let. **83**, 3792 (1990)
132. J.R. Terry et al., Direct evidence for the onset of intruder configurations in neutron-rich Ne isotopes. Phys. Lett. B **640**, 86 (2006)
133. J.R. Terry et al., Single-neutron knockout from intermediate energy beams of ^{30,32}Mg: mapping the transition into the island of inversion. Phys. Rev. C **77**, 014316 (2008)
134. C.R. Hofman et al., Evidence for a doubly magic ^{24}O. Phys. Lett. B **672**, 17 (2009)
135. K. Tshoo et al., The $N = 16$ spherical shell closure in ^{24}O. http://arxiv.org/pdf/1205.5657.pdf
136. W.F. Mueller et al., Nucl. Instrum. Methods Phys. Res. A **466**, 492 (2001)
137. A. Obertelli et al., $N = 16$ subshell closure from stability to the neutron drip line. Phys. Rev. C **71**, 024304 (2005)
138. J. Tostevin, Core excitation in halo nucleus break-up. J. Phys. G **25**, 735 (1999)
139. M. Honma, T. Otsuka, B.A. Brown, T. Mizusaki, Shell model description of neutron-rich pf-shell nuclei with a new effective interaction GXPF1. Eur. Phys. J. A **25**, 499 (2005)
140. J. Tostevin, G. Podolyak, B.A. Brown, P.G. Hansen, Correlated two-nucleon stripping reactions. Phys. Rev. C **70**, 064602 (2004)
141. J.A. Tostevin, B.A. Brown, Diffraction dissociation contributions to two-nucleon knockout reactions and the suppression of shell-model strength. Phys. Rev. C **74**, 064604 (2006)
142. E.C. Simpson et al., Two-nucleon knockout spectroscopy at the limits of nuclear stability. Phys. Rev. Lett. **102**, 132502 (2009)
143. K. Wimmer et al., Probing elastic and inelastic breakup contributions to intermediate-energy two-proton removal reactions. Phys. Rev. C **85**, 051603(R) (2012)
144. E.C. Simpson, J.A. Tostevin, One-and two-neutron removal from the neutron-rich carbon isotopes. Phys. Rev. C **79**, 024616 (2009)
145. http://www.gsi.de/work/forschung/nustarenna/kernreaktionen/activities/elise.htm
146. A. Gade et al., Reduced occupancy of the deeply bound $0d5 = 2$ neutron state in ^{32}Ar. Phys. Rev. Lett. **93**, 042501 (2004)
147. A. Gade et al., Reduction of spectroscopic strength: weakly-bound and strongly-bound single-particle states studied using one-nucleon knockout reactions. Phys. Rev. C **77**, 044306 (2008)
148. W.H. Dickhoff, C. Barbieri, Self-consistent Green's function method for nuclei and nuclear matter. Prog. Part. Nucl. Phys. **52**, 377 (2004)
149. T. Kobayashi et al., $(p, 2p)$ reactions on $^{9-16}$C at 250 MeV/A. Nucl. Phys. A **805**, 431 (2008)

150. N.S. Chant, P.G. Roos, Distorted-wave impulse-approximation calculations for quasifree cluster knockout reactions. Phys. Rev. C **15**, 57 (1977)
151. N.S. Chant, P.G. Roos, Spin orbit effects in quasifree knockout reactions. Phys. Rev. C **27**, 1060 (1983)
152. T. Aumann, C.A. Bertulani, J. Ryckebusch, Quasi-free $(p, 2p)$ and (p, pn) reactions with unstable nuclei, to be published. arXiv:1311.6734v1 [nucl-theo]
153. N. Austern et al., Continuum-discretized coupled-channels calculations for three-body models of deuteron-nucleus reactions. Phys. Rep. **154**, 125 (1987)
154. F. Nunes, Continuum-discretised coupled channels methods, http://www.scholarpedia.org/article/Continuum-Discretised-Coupled-Channels-methods
155. A. Moro, F. Nunes, Transfer to the continuum and breakup reactions. Nucl. Phys. A **767**, 138 (2006)
156. E. Alt et al., Reduction of the three-particle collision problem to multi-channel two-particle Lippmann-Schwinger equations. Nucl. Phys. B **2**, 167 (1967)
157. R. Crespo et al., Spectroscopy of unbound states under quasifree scattering conditions: one-neutron knockout reaction of ^{14}Be. Phys. Rev. C **79**, 014609 (2009)
158. H. Geissel et al., The super-FRS project at GSI. Nucl. Instrum. Methods Phys. Res. B **204**, 71 (2003)
159. J.T. Taylor, Proton induced quasi-free scattering with inverse kinematics. Ph.D. Thesis, University of Liverpool (2011)
160. V. Panin, Fully exclusive measurements of quasi-free single-nucleon knockout reactions in inverse kinematics. Ph.D. Thesis Technical, University of Darmstadt (2012)
161. www-aladin.gsi.de/www/kp3/aladinhome.html
162. T. Blaich et al., A large area detector for high-energy neutrons. Nucl. Instr. Methods Phys. Res. A **314**, 132 (1992)
163. Ch. Caesar et al., Beyond the neutron drip line: the unbound oxygen isotopes ^{25}O and ^{26}O. Phys. Rev. C **88**, 034313 (2013)
164. O. Tengblad et al., LaBr$_3$(Ce):LaCl$_3$(Ce) phoswich with pulse shape analysis for high energy gamma-ray and proton identification. Nucl. Instrum. Methods Phys. Res. A **704**, 19 (2013)
165. A. Obertelli, *Magic numbers off stability*, ERC Starting Grant (2010–2015) 258567

Chapter 6
Nuclear Charge Radii of Light Elements and Recent Developments in Collinear Laser Spectroscopy

Wilfried Nörtershäuser and Christopher Geppert

6.1 Introduction

It is now known for nearly a century that atomic spectra are a fabulous tool to study nuclear ground-state properties through the hyperfine structure of atomic transitions. The magnetic hyperfine splitting of spectral lines was already observed—even though not interpreted as such—by A. Michelson [1], the influence of the mass of the nucleus lead to the discovery of deuterium by Harold Urrey [2]. Schüler and Schmidt observed the electrical hyperfine structure in ^{151,153}Eu in 1935 [3], which was interpreted by H.B.G. Casimir [4] as arising from a deformation of the nuclear charge distribution. While these first investigations were performed on stable isotopes, it was soon recognized that this technique can also be applied to radioactive short-lived isotopes. In the early years, atomic beam magnetic resonance (ABMR) measurements and radiation-detected optical pumping (RADOP) were the work horses in this field since the radio-frequency transitions provided sufficient resolution to study the atomic hyperfine structure [5, 6]. With the invention of the laser, the study of optical transitions became possible with higher resolution. Resonance laser excitation was combined with the ABMR technique to study the isotopes of sodium [7]. But shortly thereafter a new technique was proposed [8, 9] and realized on-line first at the TRIGA reactor in Mainz [10, 11] and then later at ISOLDE [12]: collinear laser spectroscopy (CLS) is widely applicable and a large part of the data for radioactive isotopes available today has been harvested with this technique. As a second workhorse, resonance ionization spectroscopy (RIS) was established on-line and is now commonly used for both, the generation of isobarically pure (or at least enriched) beams at on-line mass separators and studies of hyperfine structure and isotope shifts in the ion source (hot cavity) or in gas cells (for a recent review see [13]). While the CLS technique offers high resolution and can be applied

---

W. Nörtershäuser (✉) · C. Geppert
Institut für Kernphysik, Technische Universität Darmstadt, 64289 Darmstadt, Germany
e-mail: wnoertershaeuser@ikp.tu-darmstadt.de

all over the nuclear chart, RIS inside hot cavities and gas cells is often more sensitive but has limited resolution due to the Doppler width or pressure broadening of the transition in the corresponding environment. Pulsed lasers are used in these cases, which have bandwidths matched to the width of the broadened transition line. However, based on a principle demonstrated earlier [14], a new setup has been established at ISOLDE that combines the advantages of both techniques, applying resonance ionization in collinear laser spectroscopy (CRIS, collinear resonance ionization spectroscopy) [15].

In a previous volume of this series, an overview on the determination of nuclear moments has been given by Neugart and Neyens [16]. Since then, new techniques have been established for the study of charge radii, some of them are based on collinear laser spectroscopy, others employ ion and atom traps. This lecture is intended to be an extension to [16] with its focus on high-resolution laser spectroscopy for on-line charge radius measurements. In parallel there was also progress in medium–to low-resolution work performed with pulsed lasers in hot-cavity ion sources and gas cells, which has been part of a recent review about resonance laser ionization for nuclear physics [13] and will not be discussed here. This note does also not provide a general or complete review of this field, but rather a guide for newcomers into some of the latest technical developments. Therefore, we will focus in the first section on the way to extract information about nuclear charge radii from the optical spectra and in the second part on experimental aspects. A few examples of extracted nuclear data will be given but we do not discuss the results in terms of nuclear structure in depths. Therefore we refer to the original literature. More general recent reviews can be found in [17, 18].

6.2 Atomic Theory: Isotope Shift and Charge Radii

The finite mass and size of the atomic nucleus has a small but distinct influence on the optical spectrum. If one compares the wavelength (or transition frequency ν) of an electronic transition along a chain of isotopes of a certain element, a small shift between the lines can be observed. This frequency difference

$$\delta\nu^{AA'} = \nu^{A'} - \nu^{A} \tag{6.1}$$

between the isotopes with mass numbers A and A' is called the (transition) isotope shift[1] and can be divided into the finite nuclear-mass shift (MS) and the nuclear volume or field shift (FS)

$$\delta\nu_{\text{IS}}^{AA'} = \delta\nu_{\text{MS}}^{AA'} + \delta\nu_{\text{FS}}^{AA'}. \tag{6.2}$$

---

[1]We can also introduce a level isotope shift, which is the difference in total binding energy of an atomic level in two isotopes. In the following we will use the term isotope shift synonymously with the transition isotope shift.

6.2.1 Mass Shift

The mass shift is connected to the change of the kinetic energy of the nuclear motion in the center-of-mass (CM) frame when additional neutrons are added to the nucleus. For this we can write in a non-relativistic approach

$$E_{\text{kin,nuc}} = \frac{\vec{P}_{\text{nuc}}^2}{2M_{\text{nuc}}}, \tag{6.3}$$

with M_{nuc} the mass of the nucleus. Since the center-of-mass of the atom is per definition at rest in the atoms rest frame, we can replace the nuclear momentum by the negative of the total electron momenta

$$\vec{P}_{\text{nuc}} = -\sum_{i=1}^{N} \vec{p}_i \tag{6.4}$$

resulting in

$$\frac{\vec{P}_{\text{nuc}}^2}{2M_{\text{nuc}}} = \frac{(\sum \vec{p}_i)^2}{2M_{\text{nuc}}} = \frac{1}{2M_{\text{nuc}}} \left(\sum_i \vec{p}_i^2 + \sum_{i \neq j} \vec{p}_i \cdot \vec{p}_j \right). \tag{6.5}$$

This is the total energy of the nuclear motion for a specific electronic state. If one of the electrons is excited into a different state, the nuclear motion must adapt to the new electron momenta and the kinetic energy might change. The corresponding energy must be delivered from the absorbed photon and this gives rise to a small change of the transition energy.[2] The isotope shift is caused by the difference of this energy due to the different masses of the two isotopes. This can be summarized as

$$\delta v_{\text{MS}}^{AA'} = \frac{M_A - M_{A'}}{M_A M_{A'}} (K_{\text{NMS}} + K_{\text{SMS}}). \tag{6.6}$$

The so-called "normal mass shift" (NMS) is the part of the shift that arises from the change of the $\vec{p}^2$-term in (6.5), representing the active electron and can be easily evaluated by replacing the electron mass m_e with the reduced mass of the system. This leads to

$$K_{\text{NMS}} = m_e v. \tag{6.7}$$

The "specific mass shift" (SMS) is caused by the change of the electron correlation terms $\vec{p}_i \cdot \vec{p}_j$. For K_{SMS} there is no analytical solution, it can only be calculated numerically by solving electron-correlation integrals that are notoriously difficult to evaluate. This is already a challenge in a two-electron system and has so far being

---

[2] Momentum conservation requires also that the atom acquires linear momentum in the absorption process. While this can usually be neglected for heavy atoms in allowed dipole transitions, the contribution has to be considered for the lightest elements.

solved accurately and in full detail only for up to three electrons as will be discussed below.

To obtain a size estimate of the mass shift, we consider first the Lyman-α line in hydrogen. Using Eq. (6.7) we obtain the mass shift constant $K_{\text{NMS}} = 1.4 \cdot 10^3$ GHz amu and with the masses of the proton and the deuteron from Eq. (6.6) a mass shift of $\delta \nu_{\text{MS}}^{1,2} = 672.8$ GHz. From the functional dependence on the atomic mass it is clear that this huge contribution decreases roughly as $1/A^2$ with nuclear mass and becomes rather small for heavier elements. In atoms with more than one electron, the specific mass shift has to be considered. Even though there is no analytical solution, a few rules of thumb have been derived [19]:

- For a usual alkaline-like $s \to p$ transition the specific mass shift is expected to be rather small, usually less than the normal mass shift. A notable exception of this rule is the $2s \to 2p$ resonance transition in Be$^+$, where the specific mass shift is larger than the normal mass shift and has the opposite sign.
- Transitions involving d or f states can have specific mass shifts that are several times larger than the normal mass shift and have a positive or a negative sign.

6.2.2 Field Shift

The second part of the isotope shift is related to the finite nuclear size (FNS) and is the interesting part from the nuclear physics point of view. For a point-like nucleus, the electrons experience a true Coulomb potential that approaches $-\infty$ at the center. But for an extended nucleus, the potential deviates from the $1/r$-law within the nuclear volume and acquires a finite value at the nuclear center. Thus, electronic levels that have wave functions with a finite probability density inside the nuclear volume, $|\Psi(0)|^2 \neq 0$, increase in energy since the electron is less strongly bound in this region. This is depicted in the center of Fig. 6.1. The contribution of the finite nuclear size effect to the total binding energy of an atomic level

$$E_{\text{FNS}} = \frac{Ze^2}{6\varepsilon_0} \langle r_c^2 \rangle |\Psi(0)|^2 \qquad (6.8)$$

is proportional to the electron density at the nucleus and the nuclear mean-square charge radius

$$\langle r_c^2 \rangle = \frac{1}{Ze} \int \rho_c(r) r^2 \, dV, \qquad (6.9)$$

where $\rho_c(r)$ is the nuclear charge density normalized to the charge of the nucleus, i.e. $\int \rho_c(r) \, dV = Ze$. If it is possible to determine the total field shift of an electronic state E_{FNS} and to calculate $|\Psi(0)|^2$, the mean-square nuclear charge radius can be determined. Unfortunately, E_{FNS} is experimentally only accessible for hydrogen-like atoms where it contributes particularly to the 1s Lamb shift and it can be extracted from a comparison of the measured binding energy of the electron with corresponding quantum electrodynamical (QED) calculations.

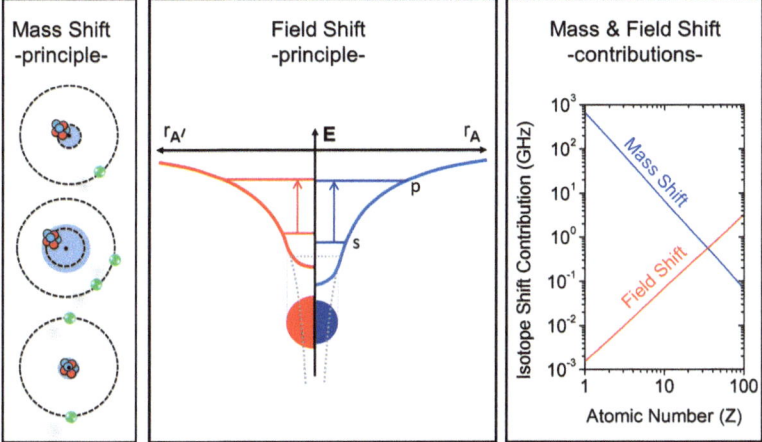

Fig. 6.1 *Left*: Representation of the normal mass shift (NMS, *top*) in a one-electron system and the specific mass shift (SMS) in a two-electron system. If the correlation of the electrons is such that they are preferably close in space, the nuclear center-of-mass motion is large, whereas it is much smaller if the electrons prefer to be far apart and on opposite sides of the nucleus. *Center*: Origin of the field shift. With increasing size of the nucleus, the electrostatic potential deviates already at larger r from the pure Coulomb potential (*dotted line*). Thus, the level energies of the bound electrons change and this change is particularly large for s-electrons having a finite probability for being inside the nucleus. The *dotted horizontal line* represents the level energy of the s-electron for a point-like nucleus, which is lifted in the two isotopes with mass numbers A (*right*) and A' (*left*). The field shift is represented by the different length of the *blue* and the *red arrow*. *Right*: Schematic of the contribution of field shift and mass shift in GHz to the overall isotopic shift, drawn as a function of the atomic number

Laser spectroscopy is able to determine energy differences between two atomic states with very high accuracy. The contribution of the FNS effect to the transition frequency arises from the difference of the electron density at the nucleus $|\Psi(0)|^2$ between the initial (i) and the final state (f) of the transition

$$\delta \nu_{\text{FNS},i \to f} = \frac{Ze^2}{6h\varepsilon_0} \langle r_c^2 \rangle \left(\Delta |\Psi(0)|^2 \right)_{i \to f} \quad (6.10)$$

with

$$\left(\Delta |\Psi(0)|^2 \right)_{i \to f} = |\Psi_f(0)|^2 - |\Psi_i(0)|^2. \quad (6.11)$$

A measurement of the *absolute transition frequency* in an atom would thus allow one the determination of $\langle r_c^2 \rangle$ of its nucleus provided that the transition frequency for a point-like nucleus can be calculated with sufficient accuracy.[3] So far this is only possible for hydrogen-like systems due to difficulties in QED calculations that can currently be solved rigorously only for one-electron systems. An extraction of the

---

[3] This approach is used in the analysis of the spectra of K_α lines of muonic atoms to extract absolute nuclear charge radii.

absolute charge radius based on a laser spectroscopic determination of the transition frequency was so far only possible for the proton based on the $1s$ Lamb shift [20]. A recent result for the proton charge radius obtained from the laser spectroscopic determination of the $2s - 2p$ lamb shift in muonic hydrogen [21] deviates from those of hydrogen and elastic electron scattering by about $5 - 7\sigma$. This constitutes the so-called proton-radius puzzle and is a very interesting topic but not related to radioactive isotopes. For more details we refer to the recent review [22].

If we finally compare the transition frequency of two isotopes, the field shift contribution to the transition isotope shift is

$$\delta \nu_{\text{FS}}^{AA'} = \frac{Ze^2}{6h\varepsilon_0} \Delta |\Psi(0)|^2 \left(\langle r_c^2\rangle^{A'} - \langle r_c^2\rangle^{A}\right)$$

$$= \frac{Ze^2}{6h\varepsilon_0} \Delta |\Psi(0)|^2 \, \delta\langle r^2\rangle^{AA'} = F \, \delta\langle r_c^2\rangle^{AA'}, \quad (6.12)$$

where we have dropped the index of the corresponding atomic transition i → f and introduced the field shift constant F. In principle, F exhibits also a small isotopic dependence, caused by the relativistic correction to the wave function at the origin— i.e. inside the nucleus—and the mass-polarization term $(\vec{p}_i \cdot \vec{p}_j)$. However, its variation along a chain of isotope is usually sufficiently small to be neglected. Only at very high-accuracy it might have to be included as was the case in the determination of the charge radius of lithium isotopes: While the total contribution of relativistic effects to F was calculated to be on the order of 10^{-3}, the variation between the isotopes is only on the 10^{-5} level [23]. Equations (6.2), (6.6) and (6.12) can be written in summary

$$\delta \nu_{\text{IS}}^{AA'} = (K_{\text{NMS}} + K_{\text{SMS}}) \frac{M_A - M_{A'}}{M_A M_{A'}} + F \, \delta\langle r_c^2\rangle^{AA'}. \quad (6.13)$$

In order to get an idea about the size of the field shift in the Ly-α line, we can use the hydrogenic wave function of the $1s$ level and determine $|\Psi(0)|^2 = (\pi a_0^3)^{-1}$ with the Bohr radius a_0. For the field shift constant this results in a value of $F = 1.8$ MHz/fm^2. Using the root-mean-square (RMS) charge radii of hydrogen [24] and deuterium [25] from elastic electron scattering, we obtain

$$\delta \nu_{\text{FS}}^{1,2} = 1.8 \, \frac{\text{MHz}}{\text{fm}^2} \cdot \left(0.895^2 - 2.128^2\right) \text{fm}^2 = -6.7 \, \text{MHz}. \quad (6.14)$$

Thus the field shift is only 10 ppm of the mass shift of 670 GHz in this transition.

For an estimation of the functional dependence of $\delta \nu_{\text{FS}}^{A,A+1}$ we can use the liquid-drop (LD) model for a nucleus of radius $R = r_0 \sqrt[3]{A}$ with $r_0 = 1.2$ fm and a constant charge density of $\varrho_c = Ze/(4\pi R^3/3)$

$$\langle r_c^2\rangle_{\text{LD}} = \frac{4\pi}{Ze} \int_0^R \varrho_c r^4 dr = \frac{3}{5} R^2 = \frac{3}{5} r_0^2 A^{2/3}. \quad (6.15)$$

6 Nuclear Charge Radii of Light Elements and Recent Developments

For small variations of A the charge radius changes according to

$$\delta\langle r_c^2\rangle_{\text{LD}} = \frac{3}{5}r_0^2(A+\delta A)^{2/3} - \frac{3}{5}r_0^2 A^{2/3} \approx \frac{2}{5}r_0^2 A^{-1/3}\delta A, \qquad (6.16)$$

while the electron density at the nucleus increases with Z^2. In total this leads to an increase of the field shift roughly proportional to $Z^2/\sqrt[3]{A}$. The approximate dependency of the mass shift and the field shift on the atomic number is represented on the right in Fig. 6.1. For light elements the mass shift by far dominates the field shift, while they are of approximately the same size around $Z = 38$ and for heavier nuclei the field shift supersedes the mass shift. Please note that due to the opposite signs[4] of the mass shift and the field shift, the total isotope shift vanishes around the crossing point.

It should be noted that the equations given above are a very good approximation for light isotopes, because the probability density can be considered constant with the value $|\Psi(0)|^2$ across the nuclear volume. For heavier isotopes $\delta\langle r_c^2\rangle^{AA'}$ must be replaced by the so-called nuclear parameter $\Lambda^{AA'}$ and higher radial moments have to be considered

$$\Lambda^{AA'} = \delta\langle r_c^2\rangle^{AA'} + \frac{C_2}{C_1}\delta\langle r_c^4\rangle^{AA'} + \frac{C_3}{C_1}\delta\langle r_c^6\rangle^{AA'} + \cdots. \qquad (6.17)$$

The Seltzer coefficients C_2/C_1, etc. are tabulated in [26]. Relativistic corrections and screening effects have to be considered for the electron wavefunctions in heavier atoms, but we do not want to dwell in this further at this point because in the following we will mostly be concerned with charge radii in light and medium-heavy elements.

6.2.3 Evaluation of Mass Shift and Field Shift Constants

Summarizing the isotope shift as

$$\delta\nu_{\text{IS}}^{AA'} = \delta\nu_{\text{MS}}^{AA'} + F\,\delta\langle r_c^2\rangle^{AA'} \qquad (6.18)$$

it becomes apparent that we have to determine $\delta\nu_{\text{MS}}^{AA'}$ and F in order to extract the change in the mean-square charge radius $\delta\langle r_c^2\rangle^{AA'}$ from a measured isotope shift $\delta\nu_{\text{IS}}^{AA'}$. It would be straightforward if *ab-initio* atomic structure calculations could provide these two values. However, already for two-electron systems such calculations are very demanding due to the correlation terms and are only possible in a rigorous way for the lightest systems with up to three electrons as will be discussed

---

[4]The (normal) mass shift leads to an increase of binding energy for the heavier nucleus while the increasing size weakens the binding.

below. With increasing electron number, the number of possible configurations increases and becomes even dramatically large if additional shells are opened as is the case in the transition metals (d-shells) and even worse for the lanthanides and actinides (f-shells and d-shells). Here, many overlapping and nearly degenerated configurations exist. Moreover, QED corrections become more important with increasing Z. Even for a "simple" electronic transition like the $4s\,^2S_{1/2} \to 4p\,^2P_{1/2}$ transition in the alkaline-like Ca$^+$ ion, specific mass shift calculations require large model spaces and several thousands of configuration state functions have to be used in order to obtain a value that has some reliability. Combined with the NMS—being easily and accurately calculable—the total mass shift factor has typically a relative accuracy of about 15–30 %. The field shift constant in this system can be calculated more reliably and relative accuracies on the 10 % level are often estimated [27].

In cases where one has to rely solely on theoretical calculations, the systematic uncertainties of those should be considered. But often it is possible to separate both terms in (6.18) based on nuclear charge radii of stable isotopes determined with other techniques or to use semi-empirical approaches to determine F.

6.2.3.1 Determination of Mass Shift and Field Shift Constants with a King-Plot

A separation of mass and field shift contribution is possible if at least three isotopes have known charge radii, independently determined by other techniques. Therefore (6.13) is multiplied by the inverse mass-scaling factor to obtain

$$\delta v_{\text{IS}}^{AA'} \frac{M_A M_{A'}}{M_A - M_{A'}} = K_{\text{MS}} + F \frac{M_A M_{A'}}{M_A - M_{A'}} \delta \langle r_c^2 \rangle^{AA'} \qquad (6.19\text{a})$$

$$\widetilde{\delta v}_{\text{IS}}^{AA'} = K_{\text{MS}} + F \widetilde{\delta \langle r_c^2 \rangle}^{AA'}, \qquad (6.19\text{b})$$

where $\widetilde{x}$ represents the mass-scaled variable x. Using known isotope masses, one can easily calculate $\widetilde{\delta v}_{\text{IS}}^{AA'}$ from the measured isotopes shifts and $\widetilde{\delta \langle r_c^2 \rangle}^{AA'}$ from the known charge radii. Plotting then $\widetilde{\delta v}_{\text{IS}}^{AA'}$ as a function of $\widetilde{\delta \langle r_c^2 \rangle}^{AA'}$ will result in a straight line with slope F and ordinate crossing at K_{MS}. This kind of presentation is known as "King-Plot". An example is shown in Fig. 6.2 for the Cd isotopes in the $5s\,^2S_{1/2} \to 5p\,^2P_{3/2}$ transition. Here, the charge radii from muonic atoms (see below) tabulated in [28] were used to calculate the modified isotope shifts plotted on the abscissa. The small error bars in y and x direction are based on the statistical uncertainty from laser spectroscopy and the muonic radii, respectively, whereas the larger ones in x direction include the systematic uncertainty of the nuclear charge radii. It is instructive to note that the even isotopes from ^{106}Cd to ^{112}Cd cluster at one point in the King-Plot. This is always the case for isotopes that show a regular behavior of the charge radius with $\delta \langle r^2 \rangle^{AA'} \sim (A - A')$. Therefore, it is of importance for this technique to have information on the charge radii of several isotopes and that these do not show such a regular behavior.

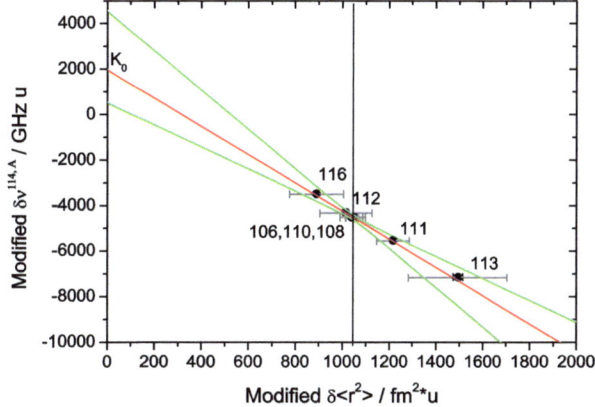

Fig. 6.2 Modified King plot for stable cadmium isotopes in the $5s\,^2S_{1/2} \rightarrow 5p\,^2P_{3/2}$ transition in Cd$^+$. *Small error bars* reflect the statistical uncertainties whereas the *large horizontal error bars* are due to systematic uncertainties. A linear fit allows to determine the mass shift constant from the crossing with the y axis and the slope is determined by the field shift constant F. The 1σ confidence band is also shown

X-ray emission spectroscopy of muonic atoms is a technique that has been applied for a long time and has delivered charge radii for the majority of stable isotopes. Since the muon is 208 times heavier than the electron, it is bound closer to the nucleus and its probability density inside the nucleus is much larger than for an electron with the same principal quantum number. The most reliable information on nuclear charge radii from muonic atom X-ray spectra is obtained from $2p_{1/2,3/2} \rightarrow 1s_{1/2}$ (K$_\alpha$-) transitions. Contrary to the electronic case, the wavefunction cannot be assumed to be constant inside the nucleus due to the much larger probability density. Instead, the Dirac equation of the muon in the nuclear field is numerically solved for the upper and the lower state of the X-ray transition assuming an analytical nuclear charge distribution, usually the two-parameter Fermi distribution

$$\rho_N(r) = \frac{\rho_0}{1 + e^{(r-c)/a}}, \quad (6.20)$$

where ρ_0 is the central nuclear density, c is a size parameter and a is related to the skin thickness t. The latter is usually defined as the radial extension of the region in which the density drops from 90 % to 10 % of the central density, which refers to $t = 4a \ln 3$. The calculated energy difference between the two electronic states corresponds to the X-ray energy and the nuclear charge distribution is now modified in order to obtain agreement with the experimental spectrum. The RMS radius of the nucleus under investigation can be evaluated directly from the parameterized distribution. However, it turns out that the value for $\sqrt{\langle r^2 \rangle}$ is strongly model-dependent since it changes considerably as t is varied or alternative distributions are used. It was shown that the potential difference between the two muonic states connected by the K$_\alpha$ transition can be well approximated by the analytical expression $Br^k e^{-\alpha r}$, where B, k and α are fitting constants and that the corresponding expectation value

$$\langle r^k e^{-\alpha r}\rangle = \frac{4\pi}{Ze} \int \rho_N(r) r^k e^{-\alpha r} r^2 \, dr, \tag{6.21}$$

called the Barrett moment, is largely insensitive to the details of the assumed nuclear density distribution [29], e.g. the skin thickness, and therefore called *model-independent*. Hence, an *equivalent* Barrett radius $R_{k\alpha}$ is introduced, by the implicit relation

$$\langle r^k e^{-\alpha r}\rangle = \frac{3}{R_{k\alpha}^3} \int_0^{R_{k\alpha}} r^k e^{-\alpha r} r^2 \, dr. \tag{6.22}$$

$R_{k\alpha}$ is thus the radius of a sphere of constant charge density that has the same Barrett moment as the nucleus under investigation. Usually the limited knowledge of the nuclear polarization correction is reported as the dominant systematic error of the result. However, it is in most cases not clearly discussed or not even known, how much the polarization correction might change along a chain of stable isotopes. Hence, the term "model-independent" has to be taken with care. This leads to difficulties concerning error propagation in the King plot.

Finally it should be mentioned that there is also another usage of Eqs. (6.19a), (6.19b), namely the extraction of isotope shift parameters in one transition if they are known in another transition of the ion or atom. This is often useful if these can be estimated to good accuracy in a particular transition but this transition does not fulfill the requirements for on-line spectroscopy of rare isotopes, it might lack efficiency for example. In that case both transitions can be measured for stable isotopes—here efficiency is usually not an issue—

$$\widetilde{\delta v}_{IS,1}^{AA'} = K_{MS,1} + F_1 \delta\langle\widetilde{r_c^2}\rangle^{AA'} \tag{6.23}$$

$$\widetilde{\delta v}_{IS,2}^{AA'} = K_{MS,2} + F_2 \delta\langle\widetilde{r_c^2}\rangle^{AA'} \tag{6.24}$$

and the unknown charge radii can be eliminated in one equation, e.g.

$$\frac{1}{F_1}\left(\widetilde{\delta v}_{IS,1}^{AA'} - K_{MS,1}\right) = +\delta\langle\widetilde{r_c^2}\rangle^{AA'} \tag{6.25}$$

$$\widetilde{\delta v}_{IS,2}^{AA'} = \left(K_{MS,2} - \frac{F_2}{F_1} K_{MS,1}\right) + \frac{F_2}{F_1} \widetilde{\delta v}_{IS,1}^{AA'}. \tag{6.26}$$

Thus, a plot of the mass-scaled isotope shift in one transition against that in the other transition should exhibit a straight line with slope F_2/F_1 and ordinate $K_{MS,2} - (F_2/F_1)K_{MS,1}$. Equation (6.26) can also be used to check the internal consistency of isotope shifts in different transitions. Of course, also this technique relies on the existence of at least three stable isotopes or measurements of short-lived isotopes that are produced in larger quantities in more than one transition.

6.2.3.2 Semi-empirical Determination of the Field-Shift Factor

Since the field shift factor is proportional to $\Delta|\Psi(0)|^2$ it can be extracted from other observables that are also sensitive to this probability density. Therefore either the energy of ns states ($n =$ principal quantum number of the state) or the hyperfine splitting for an s state in an isotope with well-known nuclear magnetic moment can be used. Rydberg states in many-electron systems show a deviation from the Rydberg formula that is usually taken care of by introducing the so-called quantum defect ξ. The corresponding formula for the energy of the Rydberg atom is then

$$E_n = E_I - \frac{R_M}{(n-\xi)^2}, \qquad (6.27)$$

where E_I is the ionization energy of the element, R_M the mass-reduced Rydberg constant for the respective isotope with mass M. The quantum defect ξ arises from the fact that the valence electron has some probability being inside the shell of the other electrons and therefore feels more of the unscreened charge of the nucleus. It varies with angular momentum and is largest for s electrons due to the missing centrifugal barrier. For high-lying states the quantum defect is almost constant, but it varies slightly for lower states. Thus, the probability density at the nucleus can be extracted by the Goudsmit-Fermi-Segrè formula

$$|\Psi(0)|^2_{ns} = \frac{1}{\pi a_0^3} \frac{Z_i Z_a^2}{n_a^3} \left(1 - \frac{d\xi}{dn}\right), \qquad (6.28)$$

where a_0 is the Bohr radius, $Z_i = Z$ for s electrons and $Z_a = 1$ for neutral atoms, $Z_a = 2$ for singly charged ions etc. $n_a = n - \xi$ is the effective quantum number after subtraction of the quantum defect ξ. To apply this formula, one must generate a table of level energies starting from a level at low n. From the energies one obtains the effective quantum numbers

$$n_a = \sqrt{\frac{R_\infty}{E'_{ns}}} = \sqrt{\frac{R_\infty}{E_I - E_{ns}}} \qquad (6.29)$$

and thus the variation of the quantum defect with principal quantum number $\xi(n)$. In order to calculate the term in parentheses in Eq. (6.28), one can use the procedure described by Kopfermann [30]

$$\frac{d\xi}{dn} = \frac{\frac{d\xi}{dE'_{ns}}}{\frac{d\xi}{dE'_{ns}} - 2\frac{n_a}{E'_{ns}}}, \qquad (6.30)$$

where the derivative $\frac{d\xi}{dE'_{ns}}$ is determined from a linear or quadratic fit to the data of $\xi(E'_{ns})$ and the derivative is taken at a low value of n.

Another possibility to determine the probability density $|\Psi(0)|^2_{ns}$ is through the hyperfine splitting factor. The a factor of an ns electron for example is given by [30]

$$a_{ns} = \frac{8\pi}{3} hcR_\infty \alpha^2 a_0^3 |\Psi(0)|^2_{ns} F_r(j, Z_i)(1 - \varepsilon_{BR})(1 - \varepsilon_{BW})\frac{\mu_I}{I\mu_B}.$$

R_∞ is the Rydberg constant in cm^{-1}, μ_B the Bohr magneton, μ_I is the nuclear magnetic moment and α is the fine-structure constant. F_r is a radial integral, ε_{BR} and ε_{BW} are the Breit-Rosenthal and Bohr-Weisskopf correction for the finite nuclear charge and nuclear magnetic moment distribution, respectively. Measuring the hyperfine splitting in such a state allows to extract $|\Psi(0)|^2_{ns}$. If possible, both techniques should be applied in order to check the consistency of their results and might then also be compared to *ab-initio* calculations.

6.2.3.3 Ab-initio Calculations of Mass Shift and Field Shift Constants

Mass-shift and field shift constants can in principle be obtained from atomic structure calculations. However, these many-body calculations must reliably treat all electron-electron correlations. As discussed above, this becomes dramatically complex for many-electron systems, but for few-electron systems (not more than three), high-accuracy calculations are feasible nowadays. For two electrons at positions $\vec{r}_1$ and $\vec{r}_2$ bound to a nucleus at position $\vec{R}_0$ the stationary Schrödinger equation including the nuclear motion reads

$$H\Psi(\vec{r}_1, \vec{r}_2) = \left[-\frac{\hbar^2}{2M}\vec{\nabla}_0^2 - \frac{\hbar^2}{2m_e}(\vec{\nabla}_1^2 + \vec{\nabla}_2^2)\right. \tag{6.31}$$

$$\left. + \frac{1}{4\pi\varepsilon_0}\left(-\frac{Ze}{|\vec{r}_1|} - \frac{Ze}{|\vec{r}_2|} + \frac{e^2}{|\vec{r}_1 - \vec{r}_2|}\right)\right]\Psi(\vec{r}_1, \vec{r}_2)$$

$$= E\Psi(\vec{r}_1, \vec{r}_2). \tag{6.32}$$

The last term of the potential energy, representing the electron-electron repulsion term, makes the Hamiltonian nonseparable and, thus, the Schrödinger equation cannot be solved exactly. In order to apply numerical methods, it is usually transformed into the CM frame and relative coordinates, thus removing the explicit appearance of R_0. One obtains

$$H = -\frac{\hbar^2}{2m_e}\left(\vec{\nabla}_1^2 + \vec{\nabla}_2^2 + \frac{\mu}{M}\vec{\nabla}_1 \cdot \vec{\nabla}_2\right) + \frac{1}{4\pi\varepsilon_0}\left(-\frac{Ze^2}{|\vec{r}_1|} - \frac{Ze^2}{|\vec{r}_2|} + \frac{e^2}{|\vec{r}_1 - \vec{r}_2|}\right), \tag{6.33}$$

with the additional mass-polarization term $(\mu/M)\vec{\nabla}_1 \cdot \vec{\nabla}_2$, representing the correlation of electron momenta as discussed above. Here $\mu = m_e M/(m_e + M)$ is the electron reduced mass and M the mass of the nucleus. The Hamiltonian is divided into the dominant nuclear-mass independent term H_0 and the mass polarization term

$$H = H_0 + H' = H_0 + \eta H_{MP} \tag{6.34}$$

with the perturbation parameter $\eta = \mu/M$ and

$$H_{\text{MP}} = -\frac{\hbar^2}{2m_e}\vec{\nabla}_1 \cdot \vec{\nabla}_2. \tag{6.35}$$

The solution to this many-body problem is a kind of textbook example and therefore we will just roughly sketch the way it is handled. For more details see, e.g. Refs. [23, 31, 32].

This problem is treated in perturbation theory using the following approach: First, approximate solutions for H_0 are constructed using a variational approach in an appropriate basis set. Here, a basis already suggested by Hylleraas in 1929 is usually used [33]. It provides a set of functions that are explicitly correlated, since they include, for example in a three-electron system, products of powers of $r_1^{j_1}$, $r_2^{j_2}$, $r_3^{j_3}$, $r_{12}^{j_{12}}$, $r_{13}^{j_{13}}$, and $r_{23}^{j_{23}}$. The constructed solution of H_0 must have been converged to at least 10 digits in order to provide the required accuracy in the next steps. Based on the calculated wave functions, perturbation theory is used to evaluate the contribution of the mass polarization term H', as well as relativistic and QED contributions. Lowest order relativistic corrections are of order α^4 (including the α^2 already existing in the Rydberg constant) while QED corrections start with the order of α^5. This leads to an expansion of the energy of each state in a double power series of η and of α

$$\begin{aligned}E(\alpha, \eta) &= mc^2\alpha^2\left[\mathscr{E}^{(2,0)} + \eta\mathscr{E}^{(2,1)} + \eta^2\mathscr{E}^{(2,2)}\right] \\ &+ mc^2\alpha^4\left[\mathscr{E}^{(4,0)} + \eta\mathscr{E}^{(4,1)}\right] + mc^2\alpha^5\left[\mathscr{E}^{(5,0)} + \eta\mathscr{E}^{(5,1)}\right] \\ &+ mc^2\alpha^6\left[\mathscr{E}^{(6,0)} + \eta\mathscr{E}^{(6,1)}\right] + mc^2\alpha^7\mathscr{E}^{(7,0)} + \cdots \\ &+ \frac{Ze^2}{6\varepsilon_0}\langle r_c^2\rangle\sum_i\langle\delta^3(\mathbf{r}_i)\rangle,\end{aligned} \tag{6.36}$$

where the η-dependent terms are responsible for the mass shift and the last term is the finite nuclear size contribution. Here, $\delta^3(\mathbf{r}_i)$ is the three-dimensional δ-function for the coordinate of the i-th electron and its expectation value $\langle\delta^3(\mathbf{r}_i)\rangle$ is the corresponding electron probability density at the nucleus.

Currently, QED and relativistic contributions are calculated up to the order of α^6 and α^7 terms are approximated using hydrogenic wavefunctions. Some contributions and their dependence on α are listed in Table 6.1. Calculating all terms provides the level energies from which the total transition energies can be determined. In principle, this would allow for an absolute determination of $\langle r_c^2\rangle$ by a comparison with the measured transition frequency. However, the theoretical uncertainty of the transition energy is dominated by mass(η)-independent QED corrections of order α^5 and is on the same order as the finite size effect (≈ 10 MHz). Fortunately, the mass-independent terms cancel in the calculation of the isotope shift and, thus, $\delta\nu_{\text{MS}}$ can be calculated to much higher precision. While two-electron atoms have been treated with this approach since about 1992 [35, 36], calculations in three-electron systems

Table 6.1 Contributions to the mass shift term in isotope shift measurements listed as a function of the fine-structure constant α and the electron reduced-mass to atomic mass ratio $\eta = \mu/M$

Contribution to	Term	Dependence
$\mathscr{E}^{(2,0)}$	Nonrelativistic energy	$Z^2\alpha^2$
$\mathscr{E}^{(2,1)}$	Mass polarization	$Z^2\alpha^2(\mu/M)$
$\mathscr{E}^{(2,2)}$	Second-order mass polarization	$Z^2\alpha^2(\mu/M)^2$
$\mathscr{E}^{(4,0)}$	Relativistic corrections	$Z^4\alpha^4$
$\mathscr{E}^{(4,1)}$	Relativistic recoil	$Z^4\alpha^4(\mu/M)$
$\mathscr{E}^{(5,0)}$	Anomalous magnetic moment	$Z^4\alpha^5$
$\mathscr{E}^{(2,0)}$	Hyperfine structure	$Z^3\alpha^2 g_I \mu_0^2$
$\mathscr{E}^{(3,0)}$	Lamb shift	$Z^4\alpha^5 \ln\alpha + \cdots$
$\mathscr{E}^{(3,1)}$	Radiative recoil	$Z^4\alpha^5 \ln\alpha (\mu/M)$
–	Finite nuclear size	$Z^4\alpha^2 \langle r_c^2 \rangle$

have reached the required accuracy just at the turn of the last century [37]. Since then, the accuracy of the calculations was increased by about two orders of magnitude [23]. As an example, the different contributions from the mass-dependent terms to the isotope shift of ^{11}Li relative to ^{6}Li are listed in Table 6.2 [23]. Not included in the corrections discussed above are nuclear polarization contributions. Nuclei which have a large polarizability—indicated by a large low-lying E1 dipole strength in the nuclear excitation spectrum—can be influenced by the electric field of the atomic shell. This, in turn, causes a contribution to the level energies and therefore to the isotope shift. The corresponding Feynman diagram is shown in Fig. 6.3. This effect cannot be neglected for ^{11}Li [38] and is particularly strong for ^{11}Be [32], the nucleus with the largest dipole strength of all known nuclei.

Using the calculated mass shifts $\delta\nu_{\text{MS}}^{6,A}$, the change in the RMS charge radius between the isotope with mass number A and the reference isotope (in this example: ^{6}Li) can be obtained from the measured isotope shifts $\delta\nu_{\text{IS}}^{6,A}$ using the relation

$$\delta\langle r_c^2\rangle^{6,A} = \frac{\delta\nu_{\text{IS}}^{6,A} - \delta\nu_{\text{MS}}^{6,A}}{F}. \tag{6.37}$$

In order to calculate the RMS charge radius $\langle r_c \rangle$ the charge radius of at least one isotope in the isotopic chain must have been determined by a different technique. In the case of lithium the two stable isotopes, ^{6,7}Li, were investigated by elastic electron scattering in the 1960's and 70's. A recent analysis of the world scattering data showed that the ^{6}Li scattering data is more reliable than that of the more abundant isotope ^{7}Li and a charge radius of $R_c(^6\text{Li}) = 2.59(4)$ fm was extracted [39]. This allows to determine the charge radii of all lithium isotopes according to

$$R_c(^A\text{Li}) = \sqrt{\langle r_c^2\rangle(^6\text{Li}) + \frac{\delta\nu_{\text{IS}}^{6,A} - \delta\nu_{\text{MS}}^{6,A}}{F}}. \tag{6.38}$$

Table 6.2 Contributions to the mass shift $\delta\nu_{\text{MS,Theory}}^{6,11}$ of ^{11}Li relative to ^{6}Li in the $2s\,^2S_{1/2} \rightarrow 3s\,^2S_{1/2}$ transition. The mass-dependent terms are calculated using the mass M listed in the first row for ^{11}Li and for ^{6}Li $M = 7.016\,003\,425\,6(45)$ amu. The unit of the electronic factor F is MHz/fm^2. All other values are in MHz. To demonstrate the degree of agreement between the independent calculations by Yan & Drake and Puchalski & Pachucki, those values which are slightly differing between the groups are listed

Term	^{11}Li
M (amu)	11.04372361(69)[a]
μ/M	36559.1754(27)[b]
$(\mu/M)^2$	−4.7619
$\alpha^2\,\mu/M$	0.0550[c]
	0.0537(4)[d]
$\alpha^3\,\mu/M$	−0.1548(21)
$\alpha^4\,\mu/M$	−0.0215(63)[c]
	−0.0268(90)[d]
ν_{pol}	0.039(4)
Total	36554.323(9)[c]
	36554.325(9)[d]
F	−1.5703(16)

[a][34]
[b]Uncertainties for this line are dominated by the nuclear mass uncertainty
[c]Calculation by Puchalski and Pachucki
[d]Calculation by Yan and Drake

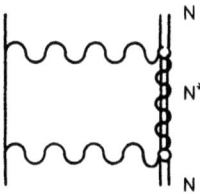

Fig. 6.3 Feynman diagram representation of the nuclear polarizability correction. Photon exchange (*wave*) with the electron (*solid line*) leads to a dipole excitation of the nucleus N into an excited state N*

The field-shift factor F required in this calculation can also be obtained from the constructed wavefunctions using

$$F = \frac{Ze^2}{6\varepsilon_0} \sum_{i=1}^{N} [\langle \delta^3(\mathbf{r}_i)\rangle_{\text{f}} - \langle \delta^3(\mathbf{r}_i)\rangle_{\text{i}}], \qquad (6.39)$$

with the expectation values of the δ-function ($=$ electron probability density) $\langle \cdot \rangle_{i,f}$ in the initial (i) and final (f) state of the transition. Please note that this also includes changes of the core electron density induced by the transition of the valence electron.

6.3 Nuclear Theory: Charge Radii Variations Along Isotopic Chains

In this section, we will shortly discuss possible reasons for changes of the nuclear charge radius along an isotopic chain. These changes are driven, amongst other effects, by the increase in nuclear volume, deformation, or clustering of the nuclei.

6.3.1 Spherical Nuclei

For spherical nuclei that have equally distributed protons and neutrons, the volume occupied by the protons increases proportional to the mass number A. As it has been discussed in Sect. 6.2.2, Eq. (6.16), this leads to a liquid-drop model expectation for the "standard isotope shift" (adding one neutron to a nucleus) of

$$\delta \langle r_c^2 \rangle_{\text{sph}} \approx \frac{\mathrm{d} \langle r_c^2 \rangle_{\text{sph}}}{\mathrm{d}A} \delta A = \frac{2}{5} \frac{\delta A}{\sqrt[3]{A}} r_0^2. \qquad (6.40)$$

Instead of the simple liquid-drop model radius one can use the radius from the more sophisticated finite-range droplet model by Myers et al. [40, 41], which gives a much better approximation of the standard isotope shift [42]. For heavy nuclei, the term $A^{-1/3}$ is almost constant along an isotopic chain and the increase in charge radius is therefore approximately linear. This is illustrated in Fig. 6.4 which shows changes of the mean-square charge radii for elements in the region of the doubly magic nucleus ^{208}Pb, which is expected to be spherical as all doubly magic nuclei. The straight lines that are drawn in Fig. 6.4 are the corresponding expected variations according to the droplet model. All elements in this region have a series of isotopes that indeed follow this trend to a very good approximation. However, there are clear deviations from this linear behavior towards neutron-deficient isotopes and at the $N = 126$ shell closure. The neutron-deficient isotopes show a more or less sudden increase in charge radius. This was first discovered for the mercury isotopes [43], where a large odd-even staggering was observed for isotopes lighter than ^{186}Hg. Most recently the behavior of the neutron-deficient polonium isotopes was studied [44] and a relatively smooth and early deviation from the straight line compared to the other elements in this region was observed. The corresponding increase in charge radius is related to a change from a spherical to a deformed one as will be discussed in the next section. The behavior at the $N = 126$ shell closure is still not fully understood and will be a topic for investigations in the years to come.

6 Nuclear Charge Radii of Light Elements and Recent Developments

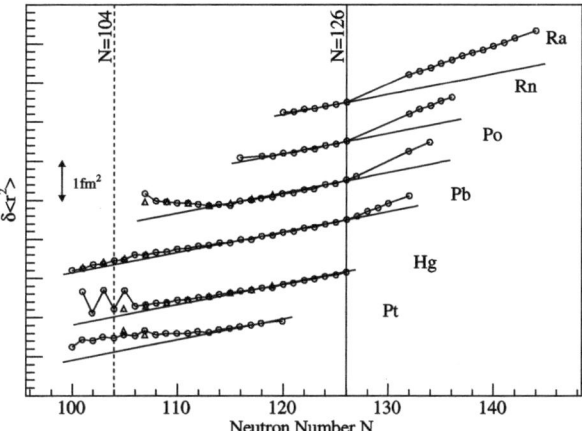

Fig. 6.4 Changes of the mean-square nuclear charge radii in the lead region. A constant offset was used for all isotopes of one element in order to separate the data points of the different elements, which would otherwise lay on top of each other. Thus, the $\delta\langle r_c^2\rangle$ axis is only to be taken relative. The *straight lines* represent the prediction from the spherical finite-range droplet model [42]. The kink at the $N = 126$ shell closure is clearly visible. Neutron-deficient isotopes of most elements exhibit also a deviation from this line due to deformation effects. Figure taken and modified from [44] including data from [45]

6.3.2 Nuclear Deformation

A spherical nucleus of constant density exhibits the smallest $\langle r_c^2\rangle$ of all nuclei with identical density and volume. In order to estimate the change in mean-square charge radius with increasing deformation, a quadrupolar deformed nucleus with sharp edge at radius

$$R_{\text{def}} = R_0\bigl(1 + \beta_2 Y_{20}(\theta)\bigr)/N \tag{6.41}$$

is considered. Here, $Y_{20}(\theta)$ is a spherical harmonic function, β_2 the deformation parameter and N is introduced for volume normalization. The corresponding mean-square charge radius is calculated to be

$$\langle r^2\rangle_{\text{def}} = \langle r^2\rangle_{\text{sph}}\left(1 + \frac{5}{4\pi}\langle\beta_2^2\rangle\right) \tag{6.42}$$

and its change in lowest order can be separated into a volume and a shape term

$$\delta\langle r^2\rangle_{\text{def}} = \delta\langle r^2\rangle_{\text{sph}} + \frac{5}{4\pi}\langle r^2\rangle_{\text{sph}}\delta\langle\beta_2^2\rangle. \tag{6.43}$$

This formula can be generalized for nuclei deformed in a more complicated way by replacing $\beta_2 Y_{20}$ with a sum over all relevant spherical harmonics $\beta_i Y_{i0}$. A typical quadrupole deformation is $\beta_2 = 0.3$. The second term becomes $5\langle r^2\rangle_{\text{sph}} 0.3^2/4\pi \approx 0.04\langle r^2\rangle_{\text{sph}}$. Even though 4 % does not seem to be much, it is a huge effect compared to the relative size of the volume effect for a spherical nucleus generated by a single neutron, obtained from the standard isotope shift in Eq. (6.16) and Eq. (6.15)

$$\frac{\delta \langle r_c^2 \rangle_{\text{sph}}^{A,A-1}}{\langle r_c^2 \rangle_{\text{sph}}} = \frac{1}{3A}, \qquad (6.44)$$

which is already below the 1 % level for masses above 30 amu and for example in the lead region constitutes only a 0.16 % effect. This demonstrates the sensitivity of $\delta \langle r_c^2 \rangle$ and thus the isotope shift to nuclear-shape changes.

A deformed nucleus with spin shows also a change in the electronic structure caused by the electric hyperfine structure. Referring to the lecture note by Neugart and Neyens [16] on nuclear moments, it should be noted that the hyperfine splitting is also sensitive to the deformation parameter $\langle \beta_2 \rangle$. While charge radius changes are only an indication for a variation of $\langle \beta_2^2 \rangle$, its sign can be directly extracted from the hyperfine structure. A comparison between $\langle \beta_2 \rangle$ from the hyperfine structure and $\sqrt{\langle \beta_2^2 \rangle}$ can reveal the nature of the deformation. A nucleus is statically deformed if the minimum of its energy appears at a finite deformation $\langle \beta_2 \rangle$. If the minimum is relatively flat, *i.e.* the slope around the minimum is small, the shape-restoring forces are weak and the nucleus is "soft" against deformation. Contrary, a steep minimum indicates a stiff nucleus. In a soft nucleus collective vibrations can appear that do not contribute to the static deformation since their time average is zero. However, they do cause a rise of the mean-square nuclear charge radius, since the expectation value of $\langle \beta_2^2 \rangle$ is different from zero. This can be used to separate the contribution from the static deformation $\propto \langle \beta_2 \rangle^2$ and the dynamic contribution to the nuclear charge radius, by simply writing

$$\underbrace{\langle \beta_2^2 \rangle}_{\text{charge-radius}} = \underbrace{\langle \beta_2 \rangle^2}_{\text{static}} + \underbrace{\langle \beta_2^2 \rangle - \langle \beta_2 \rangle^2}_{\text{dynamic}}. \qquad (6.45)$$

This identity has to be interpreted as follows: The left side is the information from the charge radius measurement and the first term on the right the expected contribution of the static deformation obtained from the hyperfine structure to the change in mean-square charge radius. If the two do not fit together, the second term must be different from zero and constitutes the contribution of a dynamic deformation. Thus, if a change appears in $\delta \langle r_c^2 \rangle$ that is considerably larger than expected using $\langle \beta_2 \rangle^2$ obtained from the hyperfine structure, one of the isotopes must have a sizable amount of dynamic deformation. This has been observed, *e.g.* in the yttrium isotopes around neutron number $N = 60$ [46], a part of the nuclear chart that is called the "region of sudden onset of deformation". In the isotope ^{88}Y for example, $\delta \langle \beta_2 \rangle^2$ extracted from the B-factor in the hyperfine structure is more than 10 times smaller then the corresponding $\delta \langle \beta_2^2 \rangle$ required to relate the increase in charge radius to the deformation parameter. A similar behavior is observed for yttrium isotopes with $N < 60$, whereas isotopes with $N \geq 60$ fulfill the relation $\delta \langle \beta_2^2 \rangle \approx \delta \langle \beta_2 \rangle^2$. It should be noted that this dynamical deformation can also contribute to nuclei that

6 Nuclear Charge Radii of Light Elements and Recent Developments

are on average spherical. It is for example used in [42] to explain the kink of $\delta\langle r_c^2\rangle$ at the magic numbers $N = 50$ and $N = 82$.

6.3.3 Clustering and Halos in Light Nuclei

Many light nuclei can be described in a way that nucleons merge to smaller entities, called clusters. A more detailed description and background of this topic is given by Martin Freer in this volume under the title "Clustering in light nuclei; from the stable to the exotic" [47].

The most important clusters in these descriptions are the α-cluster (^{4}He^{++}), the deuteron ($d = {}^2$H$^+$), and the triton ($t = {}^3$H$^+$). The nucleus ^{6}Li for example is built from an α-cluster and a deuteron, similarly a triton and an α make up ^{7}Be. An indication that there is some truth in this picture is the fact that one needs less energy to release a deuteron from ^{6}Li than a single proton or neutron. Similarly, the beryllium isotopes with $A \geq 9$ are composed of two α-clusters and additional neutrons. This clustering can be regarded as the reason for the non-existence[5] of ^{8}Be, since there is no "glue" that "connects" the two α's. In ^{9}Be a single neutron provides sufficient binding but the potential minimum is rather flat according to Fermionic Molecular Dynamic Calculations [48]. The "steepness" of the minimum increases for ^{10}Be and, thus, the average distance of the α's is reduced resulting in a smaller charge radius. The determined charge radii along the lithium and beryllium isotopic chains are shown in Fig. 6.5. The strong trend towards smaller charge radii with increasing neutron number is obvious. However, a striking difference is observed for ^{11}Li and ^{11}Be: the charge radius suddenly increases for these isotopes, which are known as "halo" isotopes. A halo nucleus consists of a compact nuclear core with the usual nuclear matter density that is surrounded by a dilute cloud of neutrons [49]. This is illustrated for the case of ^{11}Li in Fig. 6.6(a). The halo is formed by two neutrons and the RMS *matter* radius obtained from the wavefunction of the halo-neutrons is comparable to the matter radius of the stable nucleus ^{208}Pb. The increase in charge radius from ^{9}Li to ^{11}Li can be largely attributed to a pronounced recoil effect of the ^{9}Li core-nucleus in ^{11}Li. The momenta of the two neutrons can be described in the so-called T-system as indicated in Fig. 6.6(b). Assuming a charge radius of the ^{9}Li core-nucleus in ^{11}Li to be identical with that of the free ^{9}Li nucleus, the charge radius of ^{11}Li arising purely from the motion of the ^{9}Li-core in the center-of-mass system can be written as

$$R_c({}^{11}\text{Li}) = \sqrt{R_c^2({}^9\text{Li}) + R_{c\text{-CM}}^2}, \qquad (6.46)$$

with $R_{c\text{-CM}} = \frac{2}{11} R_{c\text{-nn}}$ being the distance between the core and the center of mass. From this formula it is obvious that the change in the mean-square charge radius is

---

[5]To be more specific: ^{8}Be is unbound and does only exist as a resonance in the continuum with a width of about 6 eV corresponding to a half-life of approximately 10^{-16} s.

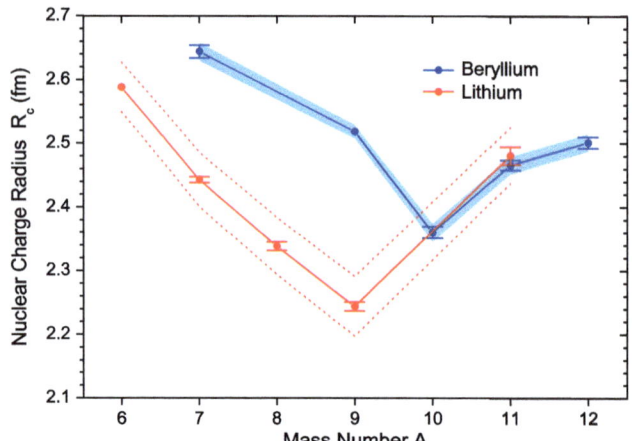

Fig. 6.5 Nuclear charge radii of lithium and beryllium isotopes obtained from isotope shift measurements [50, 51]. *Error bars* are based on the isotope shift uncertainty only. The additional systematic uncertainty caused by the reference charge radius uncertainty is indicated by the *dashed lines* (Li) and the *shaded area* (Be), respectively

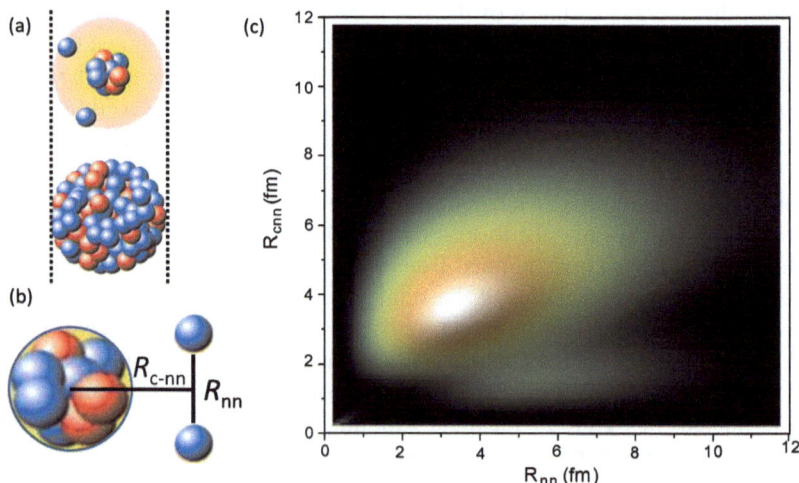

Fig. 6.6 (a) Illustration of the uncommon structure of a halo nucleus for the example [11]Li. The wave function of the two neutrons outside the [9]Li core nucleus expands to large radii and exhibits an RMS *matter* radius comparable to that of the heaviest stable nuclei like [208]Pb illustrated below. (b) The so-called T-system is characterized by the distances between the core and the center-of mass of the two neutrons $R_{c\text{-}nn}$ and the two halo neutrons R_{nn}. In this basis, the nuclear observables accessible by laser spectroscopy—*i.e.* charge radius, electric quadrupole moment and magnetic dipole moment—are calculated easiest. (c) [11]Li correlation density in the T-system. See text for more information. (Figure (c) reprinted from N.B. Shulgina et al.: [11]Li *structure from experimental data*, Nucl. Phys. A **825**, 175–199, http://dx.doi.org/10.1016/j.nuclphysa.2009.04.014.©2009 Elsevier)

6 Nuclear Charge Radii of Light Elements and Recent Developments

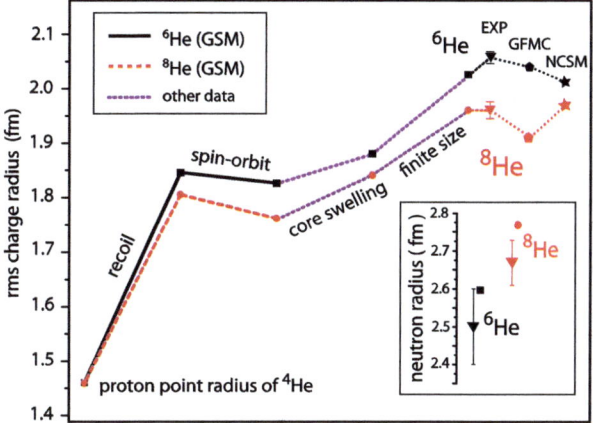

Fig. 6.7 Different contributions to the charge radius of ^{6}He (*solid line, squares*) and ^{8}He (*dashed line, dots*) as calculated in the Gamow Shell-Model (GSM) and comparison with experimental charge radii (EXP) and *ab-initio* calculations using Greens-Function's Monte-Carlo calculations (GFMC) and the No-Core Shell Model (NCSM). The *inset* shows theoretical (GSM) and experimental RMS neutron radii. Figure taken from [54]. (Reprinted with permission from Phys. Rev. C **84** 84051304(R) Copyright 2011 American Physical Society)

just the mean-square distance between the core and the center of mass

$$R_{\text{c-CM}}^2 = R_c^2(^{11}\text{Li}) - R_c^2(^9\text{Li}) = \delta\langle r_c^2\rangle. \quad (6.47)$$

Using the experimental values for ^{9}Li and ^{11}Li as obtained in [39] results in $R_{\text{c-nn}} = 6.1$ fm. The distance between the two neutrons, denoted R_{nn}, can be extracted from two-neutron interferometry data obtained in nuclear breakup reactions and results in $R_{\text{nn}} = 6.6 \pm 1.5$ fm [52]. According to a three-body model of ^{11}Li that has been optimized to agree with all experimental observables of ^{11}Li, the RMS distances between the core and the center of mass of the two neutrons $R_{\text{c-nn}} = 5.55$ fm and between the two halo-neutrons $R_{\text{nn}} = 6.69$ fm are almost equal [53]. Consequently, the ^{11}Li correlation density in the T-system obtained from this model, plotted in Fig. 6.6(c), shows a peak at about 4 fm in both distances. A small contribution of a cigar-like configuration, *i.e.* with the two neutrons on opposite sides of the core nucleus, is observable as a weak cloud at $R_{\text{nn}} \approx 6$ fm and $R_{\text{c-nn}} \approx 2$ fm.

There are several subtle effects also contributing to the size of the charge radius of such a halo nucleus. They are deconvoluted in Fig. 6.7 for the two-neutron and four-neutron halo nuclei ^{6}He and ^{8}He. The radii of these nuclei were calculated in the Gamow Shell-Model [54], based on the point-proton radius of ^{4}He. The point-proton radius is the radius of the proton distribution assuming the protons (and neutrons) being point-like particles. Similar as in ^{11}Li, the additional neutrons in the halo cause a motion of the α-cluster in the center-of-mass system. This leads to an increase of the charge radius, which is slightly smaller for ^{8}He than for ^{6}He due to the more "balanced" configuration of the four neutrons in ^{8}He compared to the two neutrons in ^{6}He. Additionally, the anomalous magnetic moment of the halo neutrons orbiting in the vicinity of the α cluster induces an electric charge density which contributes with a negative sign to the charge density in the nuclear center. This so-called spin-orbit effect is stronger in ^{8}He than in ^{6}He and since its contribution is negative the effect even enhances the difference between ^{6}He and ^{8}He. The size of the core-swelling contribution, caused by a small structural change in the central α-cluster, amounts to roughly 5 % and 7 % increase of $R_c(\alpha)$ in ^{6}He and ^{8}He [54], respectively, according to *ab-initio* Green's-Function Monte-Carlo calculations. Finally, the finite size of the proton and the neutron must be taken into account.

All these contributions can be added to the point-proton radius of a nucleus $\langle r_p^2 \rangle$ according to

$$\langle r_c^2 \rangle = \langle r_p^2 \rangle + \left(R_p^2 + \frac{3\hbar^2}{4M_p^2 c^2} \right) + \frac{N}{Z} R_n^2 + \langle r_{so}^2 \rangle + \langle r_{mec}^2 \rangle \tag{6.48}$$

with $R_p = 0.8775(51)$ fm [55, 56] being the charge radius of a proton, $R_n = -0.1161(22)$ fm [55, 56] the neutron charge radius and M_p the mass of the proton. More subtle effects are also included in Eq. (6.48), namely the Darwin-Foldy term (second term within the parentheses), accounting for the "Zitterbewegung" of the proton due to virtual particle-antiparticle pairs that surround the 'bare' proton. Even a hypothetical point proton, would thus acquire a mean-square charge radius of $3\hbar^2/(2M_p c)^2 = 0.033$ fm^2. The spin-orbit term mentioned above is denoted $\langle r_{so}^2 \rangle$ and additionally one has to take into account meson-exchange currents $\langle r_{mec}^2 \rangle$ between the nucleons, which also contribute to the charge radius.

The most prominent neutron-halo nuclei are ^{6,8}He, ^{11}Li, and ^{11}Be and the laser spectroscopic determination of their charge radii has been achieved within the last decade, based on new experimental techniques presented in the next section and the *ab-initio* atomic structure calculations discussed above. Additionally, charge radii measurements of light neon isotopes are presented since ^{17}Ne is the only proton-halo candidate that has been addressed by laser spectroscopy so far. It should be mentioned that the investigation of halo nuclei is a very active field of research, theoretically as well as experimentally. Here we address only laser spectroscopy, but high-precision Penning-trap mass measurements of these exotic nuclei have contributed considerably to the results discussed here since they were absolutely essential for the mass shift calculations [18]. We refer to some recent review articles on halo nuclei for further studies, *e.g.*, [57, 58].

6.4 Measuring Charge Radii of Halo Isotopes

6.4.1 The Challenge of Halo Nuclei

During the last decade tremendous progress was achieved in the determination of the ground state properties of the lightest elements from hydrogen up to beryllium by laser spectroscopy studies at on-line facilities. This region of the nuclear chart is of great interest since the high-accuracy data from laser spectroscopic investigations provide important benchmarks for *ab-initio* nuclear structure theories. Such calculations, performed, *e.g.*, with the Quantum Monte-Carlo or the No-Core Shell Model approach can only be carried out for the lightest nuclei up to ^{12}C. Moreover, this region is the realm of halo nuclei as discussed above.

In the lightest elements, the mass shift exceeds the field shift contribution to the nuclear charge radius by four to five orders of magnitude. Hence, the isotope shift has to be measured with an accuracy of 1–10 ppm. For helium and lithium the field shift contribution to the isotope shift is on the order of 1 MHz whereas the mass shift between ^{8}He and ^{4}He is about 65 GHz. Hence, the isotope shift has to be measured to an accuracy of about 100 kHz, which corresponds to the Doppler shift of near-UV light at 389 nm experienced by an atom traveling at a velocity of only 4 cm/s. This has to be compared with the recoil velocity of a ^{4}He atom after the absorption or emission of such a single UV photon, which is about 26 cm/s and therefore already 6 times as large. This, in combination with the fact that the halo nuclei are produced only in minute quantities and—once produced—decay after a time considerably less than a second, is the challenge of laser spectroscopy on halo nuclides. So far radioactive isotopes of helium, lithium and beryllium have been studied and for each element a dedicated spectroscopic technique was required. For helium, the atoms were cooled down to low temperatures in a magneto-optical trap (MOT) and kept at this temperature during the spectroscopic investigations [59, 60], whereas for lithium a two-photon excitation was chosen that is in first order Doppler-free and therefore independent of the atomic motion [61]. Contrary, for the beryllium isotopes an approach turned out to be appropriate that actually facilitates the Doppler shift of fast atoms instead of avoiding it [50, 51]. These different techniques will be briefly described in the following sections. Additionally, beryllium ions have been captured in a radio-frequency trap, laser cooled and investigated using an microwave-optical double-resonance technique [62–64]. Previously, radioactive lithium and beryllium isotopes were studied with a combination of laser spectroscopy and nuclear magnetic resonance with β-asymmetry detection in order to extract the nuclear moments [65].

6.4.2 Helium: Spectroscopy on Cold and Trapped Atoms

The isotope ^{6}He was the first halo isotope for which a nuclear charge radius was determined by laser spectroscopy [59]. ^{6}He was produced at the ATLAS accelerator, Argonne, in a transfer reaction of a ^{7}Li beam on carbon: ^{7}Li (^{12}C, ^{13}N) ^{6}He. Later at GANIL, Caen, spallation of a ^{13}C beam impinging on a hot graphite target was used to produce ^{6}He and ^{8}He simultaneously at rates of about 5×10^7 and 1×10^5 atoms/s, respectively [60]. Singly charged ions were produced and delivered into a low-radiation area, where the beam was stopped in a thin hot graphite foil and released as neutralized helium. In both cases, the gaseous reaction products were collected with a turbomolecular pump and the exhausted material mixed with krypton carrier gas and fed into a gas discharge cell. The gas discharge was used for an excitation of the helium atoms into the metastable $1s\,2s\,^3S_1$ state, because laser excitation from the ground state would require a laser with a wavelength of $\lambda < 53$ nm, which is not available. Out of the metastable state, which has a lifetime of several hours, laser excitation can be performed with infrared light at 1083 nm into the $1s\,2p\,^3P_{0,1,2}$ states or into the $1s\,3p\,^3P_{0,1,2}$ states in the near ultraviolet region at 389 nm. Both transitions were employed for the measurement of the charge radius. The infrared transition served for laser cooling, capturing and trapping, and the ultraviolet transition was applied for the isotope shift measurement. The principle of laser cooling and trapping will be briefly described in the following paragraphs.

During the absorption process of a photon by an atom, energy and momentum conservation must be obeyed. The energy of the photon is used to excite an electron in the atom into an energetical higher orbital and the photon momentum $\vec{p} = \hbar\vec{k}$ leads to a change of the velocity of the atom. When we consider first an atom with mass M at rest in the laboratory frame absorbing a photon, the momentum conservation requires that the atom acquires a velocity of $v = \hbar k/M = h/\lambda M = h\nu/Mc$ or $\beta = v/c = E_\gamma/Mc^2$. A ^{6}He atom excited with an ultraviolet photon ($E_\gamma \approx 3$ eV) has therefore a recoil velocity of roughly $v \approx 3/6 \times 10^{-9} c = 5 \times 10^{-10} c \approx 15$ cm/s. Please note that this velocity change of the ^{6}He atom—caused by a single photon—leads to a Doppler shift of approximately $\Delta v_{\text{Doppler}} \approx v_0 \beta \approx 8 \times 10^{14} \cdot 5 \times 10^{-10} = 400$ kHz for the next photon that will be absorbed. Thus, the velocity of the helium atoms must be very well under control to avoid large systematic uncertainties due to Doppler shifts. However, the change of the velocity can also be employed to control the motion of the atom. If a steady stream of photons is directed against an atom in motion and the frequency of the photons is in resonance with an allowed and fast transition of the atom, photons will be repetitively absorbed from one direction and the atom is gradually slowed down. After each absorption, the atom must of course get rid of the excitation energy and therefore another photon is emitted. But since the spontaneous emission is on average isotropic, the net momentum transferred to the atom in a large number of cycles is given by

$$\Delta \vec{p}_{\text{atom}} = \sum_{i=1}^{N} \hbar \vec{k}_{\text{laser}} + \underbrace{\sum_{i=1}^{N} \hbar \vec{k}_{\text{emission}}}_{\approx 0} = N \hbar \vec{k}_{\text{laser}}. \qquad (6.49)$$

Consequently, the laser exerts a force on the atoms in the direction of the laser beam. The condition for this scenario is that the laser frequency stays in resonance during the process. This can be maintained either by quickly tuning the laser frequency or adjusting the atom's resonance frequency using external fields. The first case leads to the deceleration of a bunch of atoms for which the resonance condition is fulfilled whereas other atoms are not cooled. Thus, only a part of the beam is decelerated. If the atomic resonance is tuned along the interaction length, *e.g.*, by applying a varying magnetic field and employing the Zeeman effect in the atomic transition, all

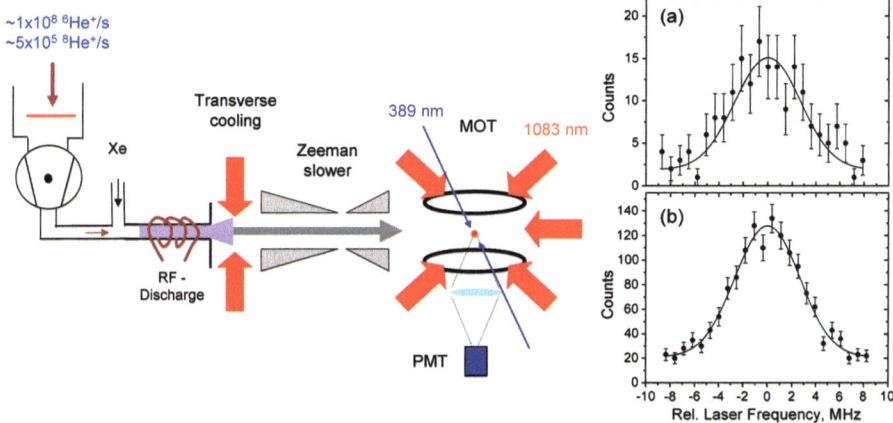

Fig. 6.8 Experimental setup for the isotope shift measurements of He isotopes (*left*) as explained in the text. The *upper plot* (**a**) at the *right* shows the very first spectrum recorded solely with the first ^{8}He atom in the MOT obtained within 0.4 s. The *lower figure* (**b**) shows an integrated spectrum over 30 atoms, resulting in a line center fitting uncertainty of 110 kHz and a $\chi^2 = 0.87$ assuming a simple Gaussian profile. Figure modified from [18], ©The Royal Swedish Academy of Sciences. Reproduced by permission of IOP Publishing. All rights reserved

atoms can be addressed with the laser beam. This is the way how it was used in the helium isotope shift measurement.

The experimental setup is depicted in the left part of Fig. 6.8. It shows the production process on the left and the gas discharge cell, from which the atoms are released at an average beam velocity of about 1000 m/s into the Zeeman slower. At the entrance of the Zeeman slower, laser beams intersect the atomic beam perpendicularly for transversal cooling. The Zeeman slower is a solenoid with a magnetic field along its axis that varies in a way that the atomic resonance condition is maintained while the atoms are being slowed down and therefore experience a strongly decreasing Doppler shift of the laser light. This effect is compensated by a decreasing magnetic field, that leads to a smaller Zeeman splitting. Within a well-designed Zeeman slower, the atoms can be cooled down to very small velocities. Atoms leaving the slower are cold enough that they can be captured in the shallow potential of a magneto-optical trap.

The operating principle of a MOT is described in detail for example in [66, 67]. Briefly summarized it works like this: six red-detuned laser beams are oriented along the axes of a cartesian coordinate system in three pairs of counterpropagating beams. This provides a frictional force on the atoms and slows them down (optical molasse), but does not provide trapping since the force has no spatial dependence. In order to trap the atoms in the center of the trap the laser beams are circularly polarized and a quadrupolar magnetic field is applied that rises linearly from the center of the trap, produced *e.g.* by a pair of anti-Helmholtz coils. The magnetic field shifts the m_F magnetic substate levels in the atom and leads to a position-dependent absorption probability. With the right choice of circularly polarized light and red detuned lasers, the atoms come into resonance with a laser beam only if they drift towards this laser out of the trap center. In this way, a restoring force is realized and the atom can be captured and cooled down to the Doppler limit

$$kT_{\text{Doppler}} = \frac{\hbar \Gamma}{2}, \qquad (6.50)$$

where Γ is the decay rate of the excited state in the two-level system. It corresponds to typical temperatures in the range of a few 100 µK. Trapping times are limited by collisions with residual gas atoms that transfer sufficient momentum to the atom to leave the shallow trap potential. Another loss mechanism that is generally present is the interaction between the trapped atoms. However, in the helium experiments usually only a single atom was kept within the trap and therefore this process can be neglected.

The MOT in the helium experiment was operated in a capture-mode with the cooling lasers turned on until an atom was detected inside the trap by the strong rise of scattered laser light. The fluorescence rate of a single atom reached a signal-to-noise ratio of about 10 after an integration time of about 50 ms. This increase was the trigger for starting the spectroscopy mode until the captured atom was lost from the trap either by radioactive decay or a collisional loss. Trapping rates were on the order of 20,000 ^{6}He atoms/h and 30 ^{8}He atoms/h.

A problem for precision spectroscopy is the presence of the cooling laser during the spectroscopy process. The electronic levels of an atom in a laser beam are slightly shifted due to the influence of the varying electric field. This effect is called AC-Stark shift and increases linearly with intensity. It leads to a shift of the resonance transition and—since the intensity varies over the cross section of the laser beam—is usually also accompanied by a broadening of the observed line shape, called AC-Stark broadening. The relatively intense fields of the cooling laser in a MOT can therefore have a substantial influence on the spectroscopic result if both lasers are applied simultaneously. On the other hand, turning off the cooling laser during spectroscopy, leads to systematic cooling or heating processes correlated with the detuning of the spectroscopy laser. This will also lead to drastic systematic variations of the resonance lineshape. The Argonne group therefore applied a spectroscopy scheme that avoided both processes. The cooling laser was indeed switched off during spectroscopy but only for a very short period of time—typically a few µs—just enough for one excitation of the atom. After this single scattering process, the cooling laser is turned on and cools the atom before the spectroscopy beam is applied again. Additionally, the frequency of the spectroscopy laser is rapidly scanned during this switching, in such a way that spectroscopy is performed alternately on the blue and the red-detuned side of the resonance.

A spectrum obtained from a single captured ^{8}He atom within 0.4 s observation time is shown on the right part of Fig. 6.8. Extensive studies have been carried out in order to investigate and to avoid all systematic shifts. For example the power of the counterpropagating beams of the spectroscopy laser have to be very well balanced. It turned out that the main systematic uncertainty is the exact trapping location of the various isotopes within the trapping volume. If these locations are not exactly at trap center and therefore zero magnetic-field, a small Zeeman shift can arise. This has been conservatively estimated to be less than 30 kHz for the ^{6}He–^{4}He isotope shift and 45 kHz for ^{8}He–^{4}He. An important check for systematic uncertainties was provided by studying all three fine-structure transitions $1s\,2s\,^3S_1 \to 1s\,3p\,^3P_{0,1,2}$ in ^{6}He and two transitions in ^{8}He. After subtracting the calculated mass shifts from the observed isotope shifts, which varied by about 250 kHz for the different transitions, excellent agreement was reached for the field shifts. This also demonstrates the internal consistency of the atomic structure calculations. Finally it should be mentioned that the photon recoil effect has to be taken into account in order to obtain the correct isotope shifts. The absorption process must ensure energy and momentum conservation. An atom at rest will therefore move after the absorption of the photon with a velocity

$$v_{\text{recoil}} = \frac{p_\gamma}{M} = \frac{E_\gamma}{Mc}, \qquad (6.51)$$

which requires the initial photon to carry the required amount of energy

$$E_{\text{kin}} = \frac{p_\gamma^2}{2M} = \frac{h^2 v_0^2}{2Mc^2} \qquad (6.52)$$

additionally to the pure transition energy $h v_0$. Thus, the resonance is shifted by

$$\Delta v_{\text{recoil}} = \frac{h v_0^2}{2Mc^2} \qquad (6.53)$$

and the difference between the corresponding shifts for ^{6}He and ^{8}He has to be taken into account when calculating the isotope shifts. This is about 170 kHz and therefore several times larger than the final uncertainty. More details on the helium isotope shift measurements and its interpretation written for a broader audience can be found in [68].

6.4.3 Lithium: Doppler-Free Two-Photon Spectroscopy on Thermal Atoms

In the case of lithium, a trapping technique cannot be employed even though lithium can be very well trapped in magneto-optical traps. But the short ^{11}Li half-life of only 8.4 ms is prohibitive for this approach. To obtain high efficiency and simultaneously sufficient laser spectroscopic resolution to determine the isotope shift with an accuracy of 100 kHz, a combination of Doppler-free two-photon excitation with efficient resonance ionization and ion detection of the laser excited atoms was applied. Measurements were performed at GSI, Darmstadt, (for ^{8}Li, $T_{1/2} = 838$ ms and ^{9}Li, $T_{1/2} = 178.3$ ms) and at TRIUMF, Vancouver, for ^{8,9}Li and the two-neutron halo isotope ^{11}Li. A simplified scheme of the setup is shown in Fig. 6.9. At both facilities,

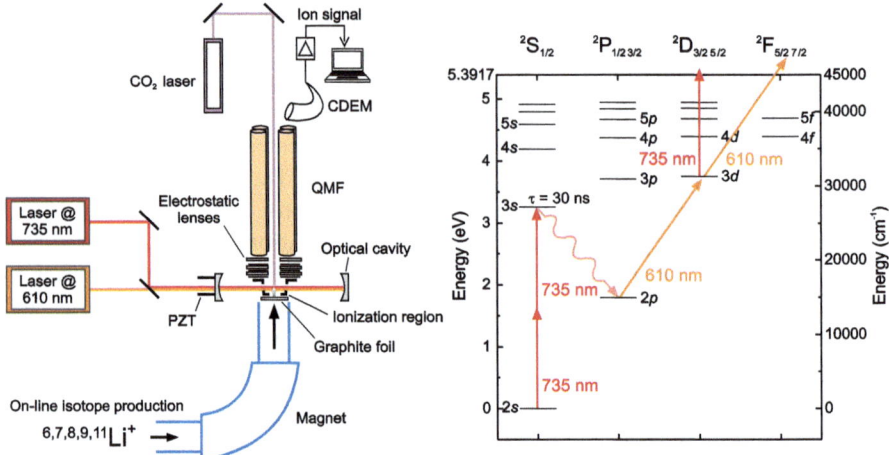

Fig. 6.9 Simplified experimental setup for the lithium spectroscopy (*left*) and level scheme with the two-photon excitation and following resonance ionization ladder for lithium atoms. See text for a detailed description. Figure taken from [23]. (Reprinted with permission from Phys. Rev. A **83** 012516 Copyright 2011 American Physical Society)

the lithium isotopes are produced as singly charged ions, mass separated in a magnetic dipole field with ion beam energies of approximately 40 keV and transported to the experiment, which is sufficiently far away from the hot and highly radioactive source.

To perform high-resolution laser spectroscopy on atomic samples, the ion beam must be stopped, neutralized and transformed into the gaseous state. This was efficiently realized by stopping the ions in a thin graphite foil. To ensure quick release from the foil, it is heated to about 2000 K with 4 W radiation of a CO_2 laser. The laser and the ion beam are focused to a spot size of about 1 mm on the foil. Measurements of the release process showed that the implanted and neutralized ions leave the foil within a few 100 μs, considerably faster than the half-life of ^{11}Li and that the surface ionization probability is sufficiently low in the range of 10^{-4}.

Released from the foil, the thermal atoms do not form an atomic beam but appear as a dilute and hot gas. Thus, laser spectroscopy must be performed very close to the foil surface to achieve a reasonable overlap with the atom cloud and a Doppler-free technique must be employed to reach the required accuracy. Two-photon spectroscopy has been used for the spectroscopy of the $1s \to 2s$ transition in hydrogen over many years with ever increasing resolution and precision (see *e.g.* [69–71] and references therein). In this technique the atom absorbs two photons from counter-propagating laser beams. In the rest frame of the atom, moving with a velocity $\vec{v}$ in the laboratory frame along the laser direction, the frequencies of the two laser beams are shifted according to the Doppler formula

$$\omega'_{1,2} = \frac{\omega_0 \pm \vec{k}\cdot\vec{v}}{\sqrt{1-v^2/c^2}} \approx \omega \pm \vec{k}\cdot\vec{v} + O\left(\frac{v^2}{c^2}\right). \quad (6.54)$$

Hence, if the atom absorbs two photons from the two counterpropagating beams, the resonance condition reads

$$\hbar(\omega'_1 + \omega'_2) = 2\hbar\omega_0 + O\left(\hbar\omega_0 \frac{v^2}{c^2}\right) = (E_f - E_i)/\hbar \qquad (6.55)$$

and the first-order Doppler shift cancels: the combined energy of the two photons becomes independent of the velocity and can be tuned to the resonance energy of the atom between two states i and f. For thermal atoms, the higher order shifts contribute on the level of a few kHz and this can be estimated and corrected with sufficient precision to obtain an accuracy of about 100 kHz as required for the final isotope shift. An important advantage of this technique compared to other Doppler-free methods like saturation spectroscopy is that all atoms of the gaseous ensemble contribute to the resonance signal, whereas in saturation spectroscopy only atoms within a selected velocity class can contribute to the signal. Selection rules for a two-photon transition are those of two combined E1 transitions [72]. Most important is that the two states must be of the same parity, hence $\Delta\ell = 0, \pm 2$, moreover $\Delta J = 0, \pm 1, \pm 2$ and similar for the total angular momentum $\Delta F = 0, \pm 1, \pm 2$. However, in the case of an $s_{1/2} \to s_{1/2}$ transition as it is used in lithium, only $\Delta F = 0$ is allowed due to angular momentum conservation.

For lithium, the two-photon spectroscopy was combined with resonance laser ionization to guarantee both, sufficient accuracy and high detection sensitivity. Resonance ionization spectroscopy [73] was shown to be an extremely sensitive and selective method for ultra-trace analysis. The excitation and ionization scheme is presented on the right in Fig. 6.9. Lithium atoms in the $2s\, ^2S_{1/2}$ ground state are excited by two-photon absorption at 735 nm to the $3s\, ^2S_{1/2}$ excited state which decays with a lifetime of 30 ns to the $2p\, ^2P_{1/2,3/2}$ states. A second laser at 610 nm is used to excite the $2p\, ^2P_{1/2,3/2} \to 3d\, ^2D_{3/2,5/2}$ transition. Finally, the $3d$ states can be laser ionized by either the 735 nm or the 610 nm laser. The laser-generated ions can be collected and detected with an efficiency very close to unity, which makes resonance ionization superior to fluorescence detection. A specialty in this excitation scheme is the $3s \to 2p$ spontaneous decay that decouples the high-precision spectroscopy in the two-photon transition from the resonance ionization in the $2p \to 3d \to \text{Li}^+$ ladder, where maximum efficiency is required to probe the successful $2s \to 3s$ excitation. Performing resonance ionization directly out of the $3s$ state via the $3s \to 5p \to \text{Li}^+$ ladder leads to large AC-Stark broadening in the $2s \to 3s$ transition and renders an accurate determination of the transition frequency impossible. In the $2p \to 3d$ transition a strong AC-Stark shift also occurs, but in this case it is even helpful. As mentioned before, the Doppler-free two-photon excitation is in first order independent of the atoms velocity, whereas the regular E1 (single-photon) transition used for the ionization experiences a Doppler shift and the 610 nm laser would therefore select a single velocity class and diminish the photoionization signal. However, due to the strong laser field, the linewidth of the transition is so much broadened that all atoms are excited along the $2p \to 3d$ transition independent of their velocity. Actually, the broadening obtained with the applied laser power was so strong that all three $2p\, ^2P_{1/2,3/2} \to 3d\, ^2D_{3/2,5/2}$ fine-structure transitions were driven simultaneously.

The excitation ladder shown in Fig. 6.9 requires a sophisticated laser system. First, high intensities are required for the non-linear process of two-photon excitation as well as for an efficient non-resonant laser ionization. This could be easily achieved using pulsed lasers, but these do not provide the required accuracy. Thus, a combination of continuous wave (CW)-lasers with resonant enhancement in a passive optical resonator was applied. Laser light for the two-photon excitation at 735 nm was produced by a titanium:sapphire (Ti:Sa) laser (≈ 1 W) and 610-nm light for the $2p \rightarrow 3d$ transition provided by a dye laser. Both laser beams were simultaneously enhanced in a two-mirror resonator placed in such a way that its focus lies in the interaction region. This provided sufficient intensity for an efficient two-photon excitation and ionization. For precise frequency control, the Ti:Sa laser was stabilized relative to a power-amplified diode laser that was in turn stabilized to a hyperfine component in a transition in molecular iodine $^{127}I_2$. The locking chain includes also the enhancement cavity, which is locked to the Ti:Sa laser frequency in order to maintain the resonance condition and the dye laser is then stabilized to the enhancement resonator in such a way that the resonance with the $2p \rightarrow 3d$ transition of the isotope under investigation is always ensured, while the Ti:Sa laser is scanned across the two-photon resonance.

Ions created by resonance laser-ionization are extracted from the laser interaction region with a negative extractor voltage and focused into a commercial quadrupole mass spectrometer (QMS) with electrostatic lenses. Ions transmitted through the QMS rod system are focused with an exit lens and detected with a continuous dynode electron multiplier (CDEM) detector. Mass suppression between two neighboring isotopes was tested with ions of the stable isotopes ^{6,7}Li produced by surface ionization in the hot carbon catcher and after optimization a suppression factor $>10^8$ was routinely achieved. The dark count rate of the CDEM detector was about 10^{-2}/s.

When using CDEM detectors for detection of radioactive ions, one has to consider that the ions are usually implanted into the surface of the sensitive region. Once these ions decay, the decay products can efficiently trigger another ion event. This leads to artificial detection efficiencies above 100 % but depending on the lifetime of the implanted ions the second signal is delayed relative to the ion implantation. For laser spectroscopy that can cause cross-talk between different channels with different laser frequencies. When the laser frequency is changed on a timescale that is fast compared to the half-life of the isotopes, the decay of those ions implanted during resonance will contribute to later channels and obscure the resonance line shape. To avoid these effects, the laser was slowly scanned across the resonance, collecting data at each frequency for a few seconds and then interrupting the incoming ion beam by using a fast kicker behind the mass separator. Afterwards, the remaining decay signals were recorded into the same frequency channel for about 5 half-lives of the respective isotope before the laser frequency was changed to the next channel and the ion beam turned on again.

Figure 6.10 shows a resonance profile obtained for ^{11}Li. The narrow Doppler-free components are labeled with their respective F quantum numbers, according to the selection rules discussed above (only transitions with $\Delta F = 0$ are allowed). Each

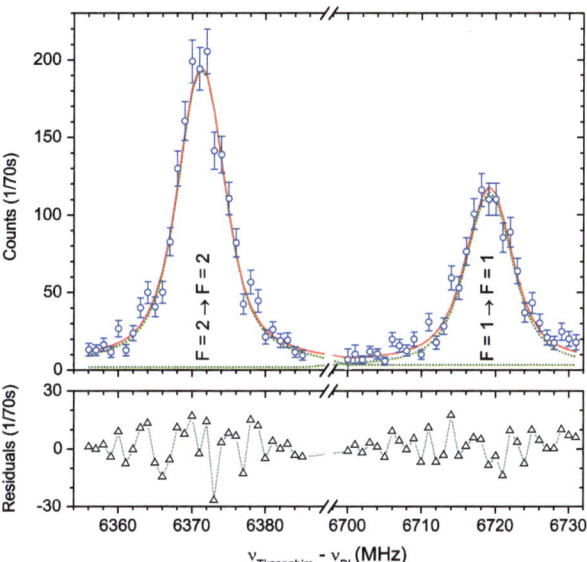

Fig. 6.10 Typical spectrum of ^{11}Li as a function of the two-photon transition frequency. Two hyperfine lines are observed due to the $\Delta F = 0$ selection rule for such a transition. Voigt profiles are fitted to the experimental data points and the residuals of the fit are shown below the spectrum. (Reprinted with permission from Phys. Rev. A **83** 012516 Copyright 2011 American Physical Society)

of the two peaks was fitted with a Voigt profile. Lorentzian and Gaussian linewidths of the Voigt profile were constraint to be equal for both hyperfine structure components.

To calculate the isotope shift, resonance positions of the individual hyperfine components obtained from the fit, must be converted into center-of-gravity (cg) frequencies. The energy shift of the hyperfine state with angular momentum F relative to the J-level energy is given by

$$E_{\text{HFS}} = \frac{A}{2} C_F = \frac{A}{2} \big[F(F+1) - J(J+1) - I(I+1) \big] \quad (6.56)$$

with the Casimir factor C_F and the magnetic dipole hyperfine constant A. It should be noted that in first-order perturbation theory, the hyperfine structure cg coincides with the unperturbed J-level energy and can be calculated from the two hyperfine resonances in the $2s_{1/2} \to 3s_{1/2}$ transition of lithium according to

$$\nu_{\text{cg}} = \frac{C_F \nu_{F'} - C_{F'} \nu_F}{C_F - C_{F'}}, \quad (6.57)$$

where ν_F is the transition frequency of the respective $F \to F$ transition. For ^{7,9,11}Li with nuclear spin $I = 3/2$, this leads to $\nu_{\text{cg}} = \frac{5}{8}\nu_2 + \frac{3}{8}\nu_1$. The obtained frequencies were corrected for AC-Stark shift contributions by measuring the transitions at different laser powers and extrapolating back to zero laser intensity. This could be performed for all isotopes besides ^{11}Li, which had too low statistics for a measurement at low power. Thus, the isotope shift of ^{11}Li was determined relative to a measurement of the reference isotope ^{6}Li with the same power conditions.

All isotope shifts of the complete lithium isotopic chain were measured to an accuracy of 100 kHz or better and the charge radii plotted in Fig. 6.5 [39] and discussed in the theory part were obtained. The isotope shift between the stable isotopes ^{6}Li and ^{7}Li was determined by various groups in a number of transitions in the neutral system as well as in the singly charged ion Li$^+$. With the exception of the $2s\,^2S_{1/2} \rightarrow 2p\,^2P_{1/2,3/2}$ (D1,D2) transitions in neutral lithium a reasonable agreement was obtained for the change of the mean-square charge radius extracted from these measurements and with the result from elastic electron scattering [23]. It should be noted that for this extraction the mass shift calculations for the respective transitions were required. Only the results from the D1 and D2 lines fluctuated strongly and were mutually inconsistent. This was recently resolved and ascribed to quantum interference (cross damping) in the unresolved hyperfine structure of these transitions [74, 75]. The latest results bring the D-line measurements into full agreement with the other transitions and elastic electron scattering. This provided a very important check of the consistency and reliability of the atomic structure mass shift calculations.

Finally it should be noted that the uncertainty of the absolute charge radii of the lithium isotopes is much larger than those for helium. This is due to the relatively large uncertainty of the reference radius of ^{6}Li. For this isotope, new and improved measurements of elastic electron scattering or muonic atoms spectroscopy could considerably reduce this uncertainty.

6.4.4 Beryllium and Neon: High-Accuracy Measurements with Fast Beams of Ions

6.4.4.1 Collinear Fast Beam Spectroscopy

Collinear laser spectroscopy (CLS), also called collinear fast-beam laser spectroscopy (CFBLS) was developed in order to perform high resolution laser spectroscopy on short-lived isotopes to investigate nuclear ground-state properties at ISOL (isotope separation on-line) facilities. It requires ion beams with low emittance and of typically 30–60 keV beam energy. Since such beams are readily available at ISOL facilities, they are often called ISOL-type beams in contrast to the high-energy (at least several MeV/u, up to GeV/u), large emittance beams at in-flight facilities. In CLS, the ion beam is superimposed with a laser beam in collinear (parallel) or anticollinear (antiparallel) geometry. A typical experimental setup at an ISOL or IGISOL facility is depicted in Fig. 6.11. It consists of a target-ion source combination, where short-lived isotopes are first produced and then ionized by various processes. Surface ion sources, plasma ion sources as well as resonance laser ionization are usually applied for the ionization. The ion source is at a high positive potential (typical 30–60 keV), such that the ions extracted are accelerated towards ground potential. The ions are then mass-separated in a magnetic sector field and

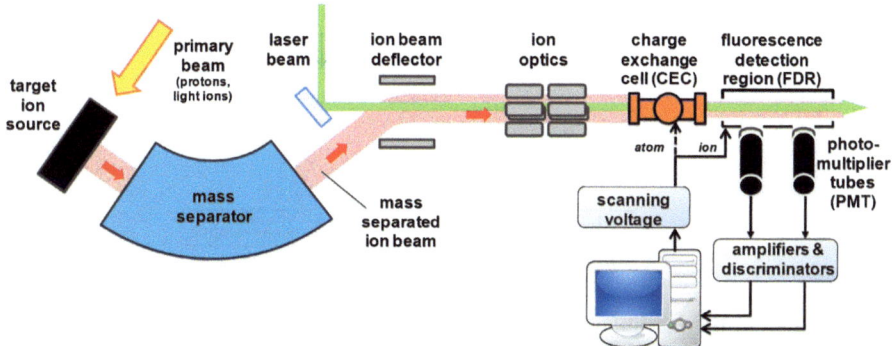

Fig. 6.11 Principle of classical CLS. Radioactive isotopes are produced by bombardment of a target with a high-energy primary beam either in a target container (*e.g.* at CERN-ISOLDE) or within a gas cell (*e.g.* at the JYFL-IGISOL). After extraction and mass separation the ion beam is superimposed with a collimated laser beam using electrostatic deflector plates. Additional ion optical systems match the ion beam profile for maximum overlap with the laser beam. The ions are either neutralized in a charge exchange cell (CEC) or directly studied in the fluorescence detection region (FDR). When the scanning voltage applied to the CEC or the FDR results in a beam velocity that fulfills the resonance condition of an atomic transition with the Doppler-shifted light, the laser induced fluorescence in the FDR is detected by photomultiplier tubes (PMTs) and recorded after signal processing in the data acquisition system

transported to the collinear laser spectroscopy beamline, where an electrostatic deflector is used to superimpose the ion beam with a laser beam either in collinear or in anticollinear geometry.

The electrostatic acceleration has two consequences for laser spectroscopy: First, it leads to a large Doppler shift of the resonance frequency ν_0 of the ion according to

$$\nu = \nu_0 \frac{\sqrt{1-\beta^2}}{1-\beta\cos\theta} \tag{6.58}$$

with the ion velocity in terms of the speed of light $\beta = v/c$ and the angle between the ion beam and the laser beam direction θ. For exact collinear $(+, \theta = 0)$ or anticollinear $(-, \theta = \pi)$ geometry, this simplifies to

$$\nu_\pm = \nu_0 \sqrt{\frac{1\pm\beta}{1\mp\beta}}. \tag{6.59}$$

The laser frequency must therefore be blue-shifted for collinear $\nu_{\text{coll}} = \nu_+$ and red-shifted for anticollinear $\nu_{\text{anticoll}} = \nu_-$ excitation. Please note that different isotopes are accelerated to different velocities (at constant beam energy). These "artificial isotope shifts" must be considered in the analysis. The second effect of the static acceleration is the longitudinal kinematic compression during the acceleration, which leads to a strong reduction of the velocity spread of the ion beam. This was analyzed by Kaufman [8] in a simple approach: the velocity difference $\Delta v = \alpha$ between an ion that starts with a velocity $v_1 = \alpha$ inside the ion source and an ion that is initially at rest is reduced after acceleration by the static potential difference U to

$$\Delta v' = \sqrt{\frac{2eU}{m}}\left(1 - \sqrt{1 + \frac{m\alpha^2}{2eU}}\right) \tag{6.60}$$

$$\approx \sqrt{\frac{2eU}{m}}\frac{m\alpha^2}{4eU} = \alpha\frac{1}{2}\sqrt{\frac{m\alpha^2}{2eU}} = \alpha\frac{1}{2}\sqrt{\frac{\Delta E_{\text{source}}}{eU}}. \tag{6.61}$$

In the last step, the formula was generalized to an energy distribution $\Delta E_{\text{source}} = \frac{m\alpha^2}{2}$ of the ions before acceleration. Kaufmann assumed a thermal ensemble with a Maxwell-Boltzmann distribution with energy width $\Delta E_{\text{source}} = kT$ and obtained the reduction factor

$$R = \frac{\Delta v'}{\Delta v} = \frac{1}{2}\sqrt{\frac{kT}{eU}}. \tag{6.62}$$

However, additional contributions to the initial energy-width in the ion source have to be considered. In a directly heated ion source for example, a voltage drop along the body of the source is unavoidable and leads to an additional energy width since the ionization occurs along a certain depth of the source. This leads to larger remaining Doppler widths. In practice, residual Doppler widths on the order of 50–100 MHz are usually obtained in CLS and the line profile can be described sufficiently well by a Doppler or a Voigt profile.

Once the ion beam and the laser beam are superimposed, additional ion beam optics is usually present in order to shape and steer the ion beam. In the figure, only a quadrupole doublet is shown as a representative of such devices. Afterwards the ion beam enters the so-called charge-exchange cell (CEC) [76]. Some alkaline vapor is generated inside the CEC by heating a small amount of solid alkaline metal. The ion beam passing through the vapor is neutralized through ion-atom charge-exchange reactions. This allows then spectroscopy on fast atoms. Alternatively spectroscopy can be performed directly on the ions, either by removing the CEC from the beamline or not operating it. The CEC is followed by the fluorescence detection region (FDR). The CEC-FDR distance should be as small as reasonably possible to avoid optical pumping into dark states. For the same reason, interaction with the laser light must be avoided along the beam pipe if spectroscopy is performed on the ions. Therefore, an additional potential is applied to the FDR or the CEC, such that the ions are slightly decelerated or accelerated when entering it.

A variable voltage applied to the FDR or the CEC has the additional advantage that ion Doppler tuning can be applied: Varying the potential results in a variation of the ion velocity and therefore of the Doppler-shifted laser frequency that the ion experiences in its rest frame. Thus the laser frequency can be fixed in the laboratory system and does not need to be scanned. Stabilizing a laser at a fixed frequency is much easier than scanning it reproducibly in a reliable manner, whereas the voltage can be scanned easily with high accuracy and reproducibility. A rough estimation of the voltage-to-frequency conversion factor for an ion with mass M is given by the differential Doppler shift

$$\frac{\partial \nu}{\partial U} = \frac{e\nu}{\sqrt{2eU\,Mc^2}} \qquad (6.63)$$

obtained from the non-relativistic Doppler shift.

The principle of collinear laser spectroscopy requires the extraction of the isotope shift from the Doppler shift and therefore the acceleration voltage U must be known to a reasonable accuracy. Usually a 10^{-4} measurement is sufficient, but this depends on the mass region. Essential is here the second derivative of the Doppler-shifted frequency, which can be calculated from (6.63) as

$$\frac{\partial^2 \nu_L}{\partial U\,\partial A} = -\frac{e\nu_L}{\sqrt{2eUM}}\frac{1}{2A}, \qquad (6.64)$$

with A being the mass number of the respective isotope. For the beryllium resonance transition at $\lambda_L = c/\nu_L = 313$ nm, an acceleration voltage of 50 kV and $M = 10$ amu, the differential Doppler shift is about 30 MHz/V and the double differential shift is 1.5 MHz V^{-1} amu^{-1}. Thus, a relative uncertainty of 10^{-4} in the acceleration voltage corresponds to an uncertainty in the artificial isotope shift of ± 7.5 MHz for neighboring isotopes and more than ± 20 MHz for the isotope shift between ^{12}Be and the stable reference isotope ^{9}Be. An accuracy compared to that reached for the lighter elements He and Li is thus not possible with standard collinear laser spectroscopy.

An alternative to CLS are measurements in a Paul trap. These where first discussed as a possible approach for isotope shift measurements on beryllium ions [62, 77]. However, it turned out that the field-shift factor is about an order of magnitude larger for Be$^+$ than for the neutral systems of helium and lithium [78]. This is easily understandable since the single electron in the $2s$ shell experiences a much stronger binding and thus a larger probability density at the nuclear site. A reduction of the required accuracy compared to the previous work was thus found to be tolerable and collinear laser spectroscopy seemed to be possible if—instead of only isotope shifts—the *total transition frequency* is measured with about 10^{-9} accuracy as will be discussed below.

6.4.4.2 Classical Collinear Laser Spectroscopy at Its Accuracy Limits: Neon Isotopes

For a long time, neon isotopes have been the lightest elements for which nuclear charge radii were determined by collinear laser spectroscopy. This was achieved by combining several techniques that allowed for a very sensitive detection of the laser resonance condition as well as for an accurate determination of the isotope shift as required for the light elements. A lecture note on the technique of optical pumping followed by state-selective ionization can be found in [16]. We will shortly summarize it for two reasons: firstly, it is partially similar to the technique applied

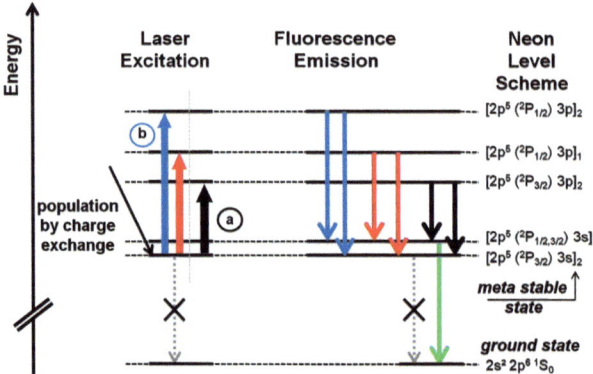

Fig. 6.12 Excitation and fluorescence emission paths within the neon level scheme as applied in CLS. For spectroscopy purposes of radioactive isotopes the excitation path, marked (**a**) has been applied. When Ne spectroscopy was used for voltage calibration of the ISOLDE acceleration potential, the paths marked with (**b**) were simultaneously driven by a single laser being retroreflected at the exit port of the beamline

later to beryllium and, secondly, the isotope ^{17}Ne is so far the only proton-halo candidate that has been investigated by laser spectroscopy.

The neon isotopes produced at ISOLDE were ionized in a plasma source and delivered to the beamline of the collinear laser spectroscopy experiment COLLAPS. Here, charge exchange in a sodium-filled CEC resonantly populated the metastable $[2p^5(^2P^o_{3/2})3s]_2$ level[6] as shown on the left in Fig. 6.12. The total angular momentum of $J = 2$ of this state has the consequence that it is metastable since the decay into the atomic neon $2p^6\,^1S_0$ ground state is forbidden in the electric dipole approximation due to the difference of $2\hbar$ in angular momentum. However, a transfer back to the ground state becomes possible if the atoms are excited by a laser along the $[2p^5(^2P^o_{3/2})3s]_2 \to [2p^5(^2P^o_{3/2})3p]_2$ transition (marked "a" in Fig. 6.12). From the excited state, the atom can return to the ground state by a sequential $[2p^5(^2P^o_{3/2})3p]_2 \to [2p^5(^2P^o_{3/2})3s]_1 \to 2p^6\,^1S_0$ decay. Thus, resonant laser excitation leads to a depopulation of the metastable state. This can be probed extremely sensitive by taking advantage of the different cross sections for collisional ionization: While the metastable state has an ionization energy of only about 5 eV, the ground state is bound much more tightly with an ionization energy of almost 22 eV. The metastable state is therefore easily ionized if the atom collides with another atom in a second gas cell located behind the laser interaction region. By deflecting the ions behind the gas cell they can be separated from the remaining atoms and the ratio of the two beam intensities is a measure for the resonance condition.

The second interesting point is the way the beam energy was determined. The potential distribution inside the plasma ion source does not allow for a very precise

---

[6]This state designation means that the 5 electrons in the $2p$-shell couple to a $^2P_{3/2}$ state. The spin of the valence electron in the $3s$ shell can then couple to a total J of either 1 $(3/2 - 1/2)$ or 2 $(3/2 + 1/2)$. The subscript outside of the square bracket represents the J value.

knowledge of the ion beam energy and introduces a large systematic uncertainty of the isotope shift. To resolve this problem, a beam energy calibration was performed using a peculiarity of the neon excitation scheme: The transition frequencies for the $[2p^5(^2P^o_{3/2})3s]_2 \rightarrow [2p^5(^2P^o_{1/2})3p]_2$ and the $[2p^5(^2P^o_{3/2})3s]_2 \rightarrow [2p^5(^2P^o_{1/2})3p]_1$ transitions (marked "b" in Fig. 6.12) coincide in the laboratory frame if the first transition is excited anticollinearly and the second one collinearly at a beam energy of 61,758.77 eV. Using a single laser beam that is retroreflected at the end of the beamline both transitions are excited at the same time if the condition

$$\nu_L = \sqrt{\nu_1 \nu_2} \tag{6.65}$$

is fulfilled [79]. Since ν_1 and ν_2 were well known from literature, the required laser frequency ν_L could be simply calculated. With this condition the beam energy becomes

$$eU = mc^2 \frac{(\sqrt{\nu_1} - \sqrt{\nu_2})^2}{2\sqrt{\nu_1 \nu_2}}. \tag{6.66}$$

Practically, the condition was not exactly fulfilled and the two resonances appeared separated by a few volt, which could be easily taken into account in the beam energy calibration. With these preparations, the isotope shifts of $^{17-28}$Ne were measured and the charge radii extracted. The results are discussed with respect to a possible two-proton-halo in ^{17}Ne in [80] and in terms of clustering, deformation, shell closures and disappearance of magic numbers in [81]. ^{17}Ne exhibits by far the largest charge radius within the neon chain and this is attributed to a tail in the proton density distribution. A two-proton halo outside a core of ^{15}O can only develop with a significant admixture of s^2-character to the d^2-orbitals.[7] Since the charge radius is very sensitive to these mixing ratios it is an excellent benchmark for theoretical models and according to fermionic molecular dynamics calculations the s^2 contribution is about 40 % [80].

6.4.4.3 Frequency-Comb Based Measurements: Beryllium Isotopes

In order to obtain the charge radius of beryllium isotopes with an accuracy of about 1 % and taking into account the field shift factor of $F = -17.02$ MHz/fm^2 [32, 78] for beryllium, the isotope shift must be determined with an accuracy of about 1 MHz or better.

As described in Sect. 6.4.4.1, beryllium is very sensitive to the exact acceleration voltage because the double-differential isotopic shift according to Eq. (6.64) is 1.5 MHz V^{-1} amu^{-1} at a beam energy of 50 keV and therefore rather large. Assuming a typical voltage uncertainty of $\delta U/U = 10^{-4}$, the voltage-based uncertainty in the isotopic shift between the stable isotope ^{9}Be and ^{12}Be is 22.5 MHz. This exceeds the requested accuracy for beryllium charge radii measurements by more

---

[7] A one-proton halo would require a core of ^{16}F, which is unbound.

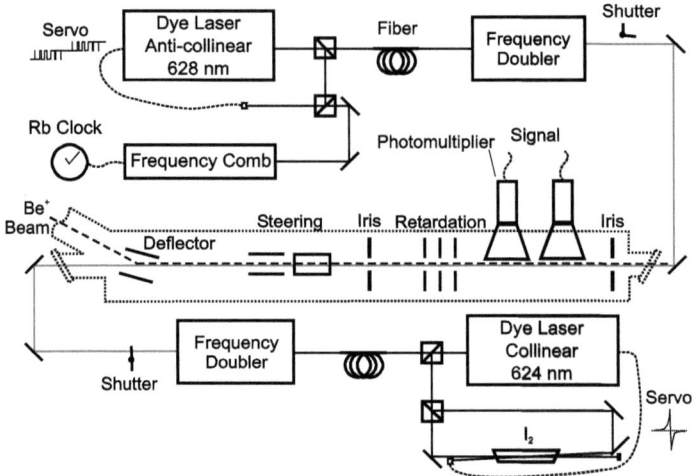

Fig. 6.13 Experimental layout of the frequency-comb based CLS on beryllium isotopes. Two laser systems applied for quasi-synchronous laser excitation on collinear and anticollinear geometry, chopped in fast sequence by mechanical shutters. Figure taken from [50]. (Reprinted with permission from Phys. Rev. Lett. **102**, 062503 Copyright 2009 American Physical Society)

than an order of magnitude, which is why classical CLS could not be applied for this study. However, CLS was used before in order to polarize a beryllium ion beam and to perform β-asymmetry detected nuclear magnetic resonance (β-NMR) measurements [16]. From these, the magnetic dipole and electric quadrupole moments of ^{11}Li as well as the magnetic moment of ^{11}Be were determined for example.

The approach used for isotope shift measurements is partially related to the studies in neon described in the previous section. Again (quasi-)simultaneous excitation in collinear and anticollinear geometry is applied. Deviating from Eq. (6.65), where two different transitions have been probed by one laser beam, here two laser beams with frequencies ν_{coll} and ν_{anticoll} are overlapped with the ion beam in order to excite the same transition in opposite geometries. Both transitions from the ionic ground state into the fine-structure dublett of the excited $2p$ state ($2s\,^2S_{1/2} \to 2p\,^2P_{1/2,3/2}$) at about 313 nm were probed. The *absolute transition frequency* in the ions restframe ν_0 was determined for each isotope with high accuracy using the relation

$$\nu_0 = \sqrt{\nu_{\text{coll}} \cdot \nu_{\text{anticoll}}}. \tag{6.67}$$

Therefore the laser frequency had to be determined with an accuracy of at least $\delta\nu/\nu = 10^{-9}$, which is much higher than in standard CLS experiments.

The setup used to accomplish this task is shown in Fig. 6.13. A continuous-wave ring dye laser at about 628 nm was applied for anticollinear laser spectroscopy and stabilized with a phase-lock to a commercial compact fiber-laser based frequency comb. The laser light was then transported via 20-m long fibers to a second harmonic generation (SHG) device close to the beamline and after frequency doubling to a wavelength of about 314 nm directed along the beamline with a laser intensity of approximately 5 mW. A second ring dye-laser at a wavelength of about 624 nm was stabilized to a molecular transition in iodine. This wavelength stabilization was again repeatedly cross-referenced to the frequency comb, thus transferring the frequency comb's accuracy to the iodine setup. The corresponding laser light was again transported by fibers to a second SHG device close to the vacuum beamline, where light at 312 nm was produced with similar laser power but this time collinearly superimposed with the 50-keV Be$^+$ ion beam from ISOLDE.

In order to clearly separate the collinear spectrum from the anticollinear spectrum, fast shutters were implemented in both laser paths, which in fast sequence blocked alternately one of the laser beams. In a first experiment the isotopic shifts and corresponding charge radii of ^{7,9,10,11}Be were determined with accuracy better than 1.5 MHz which translates to less than 1 % uncertainty in the RMS nuclear charge radius [50]. While the laser wavelength control and laser beam alignment contributed only with about 500 kHz systematic uncertainty to the final uncertainty of the isotope shift, the dominating uncertainty in the total charge radius arises from the experimental uncertainty of the reference charge radius in ^{9}Be.

In this first run, ^{12}Be was not accessible due to the low production yield and the high but customary laser straylight background. At least at the end of the first run the isobaric beam contamination on mass 12 was observed to be very low. This was the motivation for a second beamtime after the implementation of an ion-photon-coincidence technique, as will be described below in Sect. 6.5.2 [51]. Therefore the ion beam has been deflected out of the laser beam axis behind the optical detection region and detected on a secondary electron multiplier. This signal has been fed in a coincidence device together with the delayed signal of two photomultiplier tubes (PMT). This upgrade of classical optical fluorescence CLS with background-suppression and state-of-the-art laser control, thus enabled a precise measurement of the charge radius of ^{12}Be with better than 1 MHz accuracy, which is unsurpassed for such a light element in CLS. This was achieved even with the boundary condition of a weak ion yield of only about 8000 ions per proton pulse, imping statistically every 2–3 seconds onto the ISOLDE target.

The charge radii of the beryllium isotopes are plotted together with those of lithium in Fig. 6.5. The increase from ^{10}Be to ^{11}Be is caused by the single halo neutron in ^{11}Be and can in first-order be ascribed to the core-recoil effect. An average distance of 7.7 fm between the ^{10}Be core and the halo neutron can be extracted from this simple picture [50]. Adding another neutron to ^{11}Be results in the nucleus ^{12}Be, which has a two-neutron separation energy of $S_{2n} \approx 3.7$ MeV and is therefore not expected to be a halo nucleus. This is supported by the experimentally determined matter radius of ^{12}Be of 2.59(6) fm that is considerably smaller than the 2.73(5) fm of ^{11}Be [82]. The charge radius exhibits the opposite trend and increases further. This is ascribed to a promotion of the neutrons expected in the standard shell model to populate the $p_{1/2}$ state into a mixture of sd orbitals. According to fermionic molecular dynamics calculations $(sd)^2$ states contribute to approximately 70 % to the total wavefunction. This indicates clearly the disappearance of the $N = 8$ magic number as discussed in [51].

6.5 Further Developments in Collinear Laser Spectroscopy

6.5.1 Isotope Shift Determinations Using β-Asymmetry Detection

The optical detection of the fluorescence photons from fast ion or atom beams with a scanning voltage applied to the optical detection setup or the CEC, respectively, can be considered the classical approach of collinear laser spectroscopy. However, this approach has a limited sensitivity and requires typically beams of at least 10^4–10^5 particles per second. Sensitivity is of course a critical issue in the detection of exotic short-lived nuclei and the production rates get very small further away from the valley of stability. Thus, increasing sensitivity has been a continuous effort over all the years in CLS and led to the development of many specialized techniques. A very successful direction was the detection of charged particles instead of or in combination with single-photon detection. Examples are resonance ionization combined with CLS [14], state-selective charge-exchange applied for earth-alkaline ions [83] and state-selective ionization for rare-gases [84]. Beta-asymmetry detection has previously been used to determine nuclear moments [16] but not for isotope shift measurements. This was only recently achieved. Another technique that is more generally applicable and has enabled CLS further away from the valley of β-stability is the usage of cooled and bunched ion beams. Both methods will be discussed in the following sections.

Beta-asymmetry detection after optical pumping has been used previously to determine nuclear magnetic moments and nuclear quadrupole moments as described in detail in [16]. Recently, this technique has also been used for the first time to extract nuclear charge radii. The setup used at ISOLDE (CERN) for these measurements in the Mg chain is schematically shown in Fig. 6.14. A polarized ion beam is obtained by applying circularly polarized light ($\sigma^{\pm}$) to the $3s\,^2S_{1/2} \to 3p\,^2P_{1/2}$ transition in Mg$^+$. After several excitation-relaxation processes, the atoms are transferred into the m_F substate with the maximum projection along the laser direction as depicted in the inset of Fig. 6.14. To maintain this polarization, a weak longitudinal magnetic field is applied along the beam pipe. The nuclear polarization achieved by optical pumping is then decoupled from the electron shell in a strong magnetic field (Paschen-Back regime) before the ion is implanted into a MgO crystal. Typical implantation depths at 40–60 keV are of the order of a few 10 nm, deep enough to avoid surface effects in a sufficiently clean crystal. When the laser light is applied at the resonance frequency, the polarization will build up along the beamline and the β-decay of the implanted ions will exhibit an anisotropy in the emission of the positrons/electrons from the β-decay. This asymmetry can be detected using β-telescopes located between the crystal and the magnetic pole-shoes and is used as a probe for the resonant pumping process. The intensity distribution of the emitted electrons or positrons is given by the projection of the β-particle velocity $\vec{v}$ on the spin $\vec{I}$ of the polarized nuclei [85]

$$I(\Theta_{eI}) = 1 + A\frac{\vec{I}}{|\vec{I}|} \cdot \vec{v}/c = 1 + A\frac{v}{c}\cos\Theta_{eI}, \quad (6.68)$$

6 Nuclear Charge Radii of Light Elements and Recent Developments

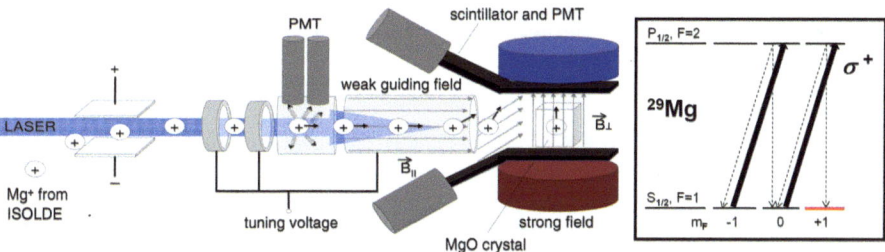

Fig. 6.14 Experimental setup using collinear laser spectroscopy for optical pumping of Mg isotopes with subsequent β-asymmetry detection. Laser induced optical pumping among the m_F-states (*right*) inside a magnetic guiding field leads to a polarized ion beam. After implantation inside a suitable crystal inside a strong magnetic field, the decay asymmetry is observed by two opposite sets of scintillators coupled to photomultiplier tubes (PMT). Applying an RF-field by some external coils (not shown in figure) can then destroy the polarization of the ions inside the crystal and provides a nuclear magnetic resonance signal. *Left* picture taken from [18], ©The Royal Swedish Academy of Sciences. Reproduced by permission of IOP Publishing. All rights reserved

with a parameter $A = a_\beta P_I$ linked to the degree of nuclear polarization P_I and the β-decay asymmetry parameter a_β, depending on the change of the nuclear spin I during the decay. Θ_{eI} is the angle between the electron's direction of flight and the spin axis. The experimental observable is then the β-asymmetry defined as

$$a = \frac{N^\uparrow - N^\downarrow}{N^\uparrow + N^\downarrow}, \qquad (6.69)$$

with the count rates $N^\uparrow$, $N^\downarrow$ measured above and below the crystal, respectively. The observed asymmetry as a function of laser frequency is shown together with standard CLS fluorescence resonance profiles (green) in Fig. 6.15 as dark and grey shaded peaks for ^{21,29,31}Mg.

To determine the g-factor or the electric quadrupole moment of the nucleus, nuclear magnetic resonance can be performed on the polarized nuclei. Therefore, the laser is tuned to a frequency that provides a strong asymmetry signal and a radio-frequency is applied to drive transitions between different m_F substates in order to depolarize the sample. This has previously been applied to study—amongst others—nuclear spins and moments of lithium, beryllium and magnesium isotopes [16]. But here we will discuss the usage of the asymmetry signal itself for an isotope shift measurement. Therefore, a lineshape must be fitted to the asymmetry spectrum that depends on the polarization obtained in the optical pumping process and is also affected by the transition into the strong magnetic field. The lineshape is modelled using a rate equation approximation for the optical pumping process and solving the system of differential equations for the population of the individual m_F states with a Runge-Kutta algorithm. Here the experimental laser intensity and the transition strength between the hyperfine components have to be taken into account. The adiabatic transition into the strong-field region corresponds to a movement along the levels of the Breit-Rabi diagram from the $|F, m_F\rangle$ to the $|m_J, m_I\rangle$ regime. The

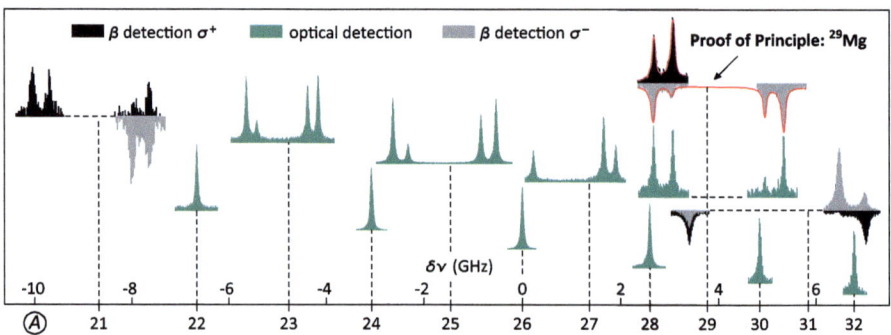

Fig. 6.15 Spectral lines of singly charged Mg isotopes in the D1 transition [86]. The (Doppler–tuning) frequency scale is relative to the reference isotope ^{26}Mg. The *dotted lines* represent the center-of-gravity position for the respective isotope as indicated. Standard fluorescence (*green*) spectroscopy was applied for the magnesium isotopes $^{22-30}$Mg and ^{32}Mg, where the isotopes ^{30,32}Mg with lower yields have been recorded by ion-photon coincidence detection. Only β-asymmetry detection (*grey* and *black*) was possible for the isotopes ^{21,31}Mg. The applicability of this method is demonstrated by a compatibility cross-check with data obtained from fluorescence spectra at isotope ^{29}Mg, for which both spectra—optical and β-asymmetry—are shown. (Reprinted with permission from Phys. Rev. Lett. **108** 042504 Copyright 2012 American Physical Society)

occupation numbers in the $|F, m_F\rangle$ system resulting from the rate equations for optical pumping are transformed into the corresponding $|m_I\rangle$ states and the nuclear spin polarization P_I can be obtained, which determines the amplitude of the β-decay asymmetry observed in the experiment. The experimental spectra are then fitted by varying the hyperfine structure parameters, including the center-of-gravity of the hyperfine structure, which determines the isotopes shift.

Isotope shifts were measured with CLS for all Mg isotopes along the complete sd-shell (^{21}Mg–^{32}Mg) using various detection techniques as presented in Fig. 6.15 taken from [86]. While the isotopes that were produced with sufficient yields were detected in the standard optical way, the isotopes ^{21}Mg and ^{31}Mg had production rates much too low for resonance fluorescence detection. Thus, these isotopes were investigated using the β-asymmetry detection. In order to ensure the compatibility of optical and β-asymmetry detection, the isotope ^{29}Mg was detected with both techniques. This isotope was produced in sufficiently large quantities of about 10^6 ions/s and its half-live of 1.3 s is short compared to the time constants for spin-relaxation processes in the MgO crystal. The isotope shifts obtained with the two techniques agreed within 2σ which could be caused by slightly different experimental conditions or by a statistical fluctuation.

The more exotic even isotopes ^{30,32}Mg were optically detected using an ion-photon coincidence and restricting the observation time to about 3 half-lives of the respective isotope after the proton pulse impinged onto the ISOLDE target.

A schematic view of a collinear laser spectroscopy setup summarizing some of the different detection methods is shown in Fig. 6.16.

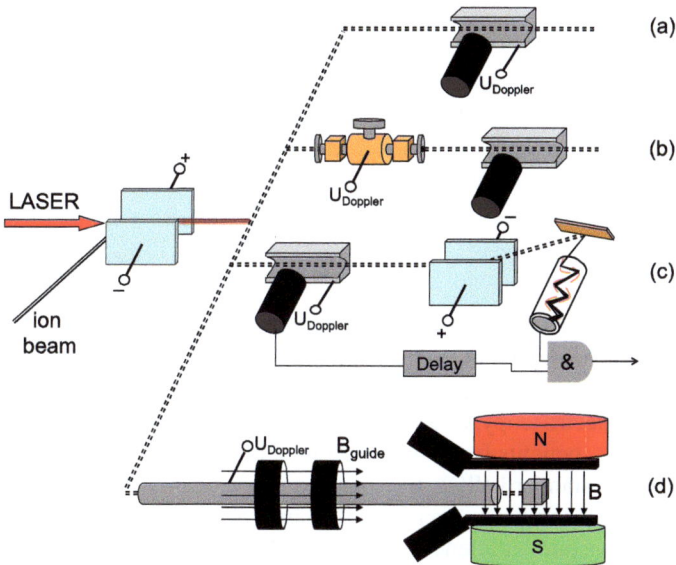

Fig. 6.16 Overview of various detection methods for CLS. In the classical approach (**a**) the scanning voltage ($U_{Doppler}$) is directly applied to the FDR. For spectroscopy of atoms (**b**) the ions are neutralized in an alkaline-vapor loaded charge-exchange cell (CEC) and the atoms velocity is scanned by applying the voltage to the CEC instead of the FDR. If the detection limit in ion spectroscopy has to be increased, the setup in (**a**) can be upgraded to a photon-ion-coincidence setup (**c**) by detecting the ions in a secondary electron multiplier (SEM) and registering only coincidences of delayed PMT and SEM signals. For very short-lived radioactive species the β-asymmetry detection can be applied (**d**). Optical pumping is performed with circularly polarized light in a magnetic guiding field and the spin polarized beam is implanted in a crystal, entangled between two opposite scintillator-PMT arrays in a strong magnetic field

6.5.2 Photon-Ion Coincidence Detection

Laser straylight produces a considerable amount of background on the PMTs in the classical optical detection scheme, limiting the sensitivity at very low yields of radioactive isotopes. This background can be reduced if the photomultiplier signal is gated with a signal of a particle detector downstream the beamline [87], as shown in line (c) of Fig. 6.16. This coincidence technique ensures that only those photons get accepted that appear at a time when an atom or ion was definitely passing through the optical detection region. Therefore, the photon signal is delayed by electronic or digital means for the time required for the particle at the corresponding beam energy to fly from the optical detection region to the particle detector. The flight time can be determined using *e.g.* a time to digital converter or a multi-channel scaler (and an isotope with sufficiently large production rate). An appropriate way to delay the photon signal has to be used to avoid dead-time losses. The width of the gate has to be chosen according to the length of the optical setup. An increase in signal-to-noise of a factor of 1600 has been demonstrated in past experiments [87] and recently it has been applied in the detection of ^{12}Be and ^{30,32}Mg as presented above. This technique is limited by the existence of strong isobaric contaminations, which can hardly be suppressed in many regions of the nuclear chart.

An inverse approach to reduce background from scattered light is to bunch the ions and let them pass the detection region as a short bunch. In this special case all fluorescence events occur during a short period of time corresponding to the length of the ion bunch.

6.6 Towards the Limits: Improving Sensitivity with Cooled and Bunched Ion Beams

Collinear laser spectroscopy, as all other low-energy experiments on short-lived radioactive isotopes, benefits from ion beams with high brilliance. To reduce the emittance beyond the standard ISOL-like beam quality, beam cooling is beneficial for CLS since the laser-ion beam-overlap can be improved, laser straylight reduced and the ion beam transport efficiency increased. This is particularly crucial for weak exotic beams with yields of only a few hundred ions per second. Various cooling mechanisms have been applied for ions in traps and storage rings, such as stochastic, electron, resistive, sympathetic, buffer-gas and laser cooling. Cooling is usually a statistical process that requires multiple interactions of the ion or atom to be cooled with the cooling medium. An ion beam can therefore either be guided multiple times through a cooling device like in a storage ring, where cooling is applied for very short times on each turn, or it can be slowed down and even stopped in a trap in order to obtain sufficient interaction time. In this regard buffer gas cooling has been established as the most universal cooling mechanism for short-lived radioactive isotopes and has been applied at several facilities.

The first buffer-gas cooling in a linear radio-frequency quadrupole (RFQ) was proposed and applied by Douglas and French [88]. Since then this technique has been established as standard technique for ion beam emittance improvement at various radioactive on-line facilities, for example at ISOLDE/CERN [89–91], IGISOL/University of Jyväskylä [92], LEBIT/MSU [93], and ISAC/TRIUMF [94]. Additional to the radio-frequency potential during the buffer gas cooling, the RFQ structure can be segmented in order to apply a DC offset potential. This idea was first proposed at McGill University in 1997 [95], applied for mass measurement by Herfurth and coworkers [89] and later for collinear laser spectroscopy by Nieminen et al. [92, 96]. In CLS the application of the gas-filled segmented RFQ leads to a suppression of background and increase in sensitivity.

6.6.1 Principle of a Radio-Frequency Quadrupole

A linear RFQ is a structure formed by four rods which are oriented as shown in Fig. 6.17. In order to store the ions inside a linear RFQ, the ion beam must be

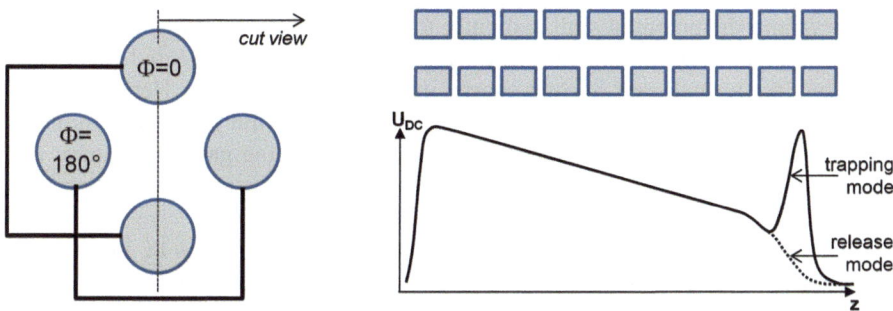

Fig. 6.17 Schematic cut-drawings perpendicular to the symmetry axis (*left*) and side view (*right*) of a segmented RF-quadrupole. The potential shown below displays the DC offset potential in trapping mode (*solid line*) and for the extraction of the cooled ion bunch (*dotted line*)

decelerated first. Thus, the RFQ is installed on a high-voltage platform to which a potential similar to that of the ion source is applied. On the left side of Fig. 6.17, a simplified cut-drawing perpendicular to the symmetry axis is shown and on the right a cut through the central symmetry axis as indicated in the left figure. Opposite rods are connected to the same radio-frequency (RF) potential, while adjacent rods are shifted in phase by 180°. In conventional quadrupole mass filters the rods are not segmented and a single DC-potential is applied at each of the four rods, whereas for RFQ coolers and bunchers each rod is segmented and individual DC potentials can be applied to each segment.

Inside the RFQ the ions are confined by the electric field generated by the voltages $U(t, i)$ applied to the rods,

$$U(t, i) = U_{RF} \cdot \cos(\omega t + \phi) + U_{DC, i}, \tag{6.70}$$

where the first term is the quadrupolar radio-frequency field of a two-dimensional Paul-trap [97] and the second term in Eq. (6.70) describes the DC-offset potential of the individual RFQ-segment i of the RFQ structure. Recently, a first RFQ has been realized and applied for laser spectroscopy, which is driven by a digital square-wave excitation [94] instead of the cos-like function given in Eq. (6.70). A typical DC-potential trend is shown on the right in Fig. 6.17. The solid line represents the applied offset-potentials during the accumulation and cooling time. Additional to the transverse confinement of the ions created by the quadrupole fields of the RFQ, the longitudinal potential well allows for the axial trapping.

For cooling the RFQ structure is filled with a chemically inert buffer gas like helium at typical pressures of the order 10^{-2}–10^{-1} mbar. To minimize charge-exchange processes the buffer gas should have a high ionization potential. Through the interaction of the ions with the buffer gas the ions are slowed down and dissipate energy by collisions with the gas atoms inside the RFQ structure. Thus the ions transversal oscillation amplitude in the quadrupole field is reduced, which inherently leads to reduced mass separation properties of the RFQ. Moreover, the axial kinetic energy decreases during the passage through the structure. The entrance potential of the RFQ as shown in Fig. 6.17 is chosen in such a way that the ions can just overcome the potential wall and enter the RFQ, but are captured between the second potential wall at the exit of the RFQ and the entrance potential as soon as they have lost a small part of their energy in buffer gas collisions. Hence the ions are multiply reflected and successively cooled down into the potential minimum. The final kinetic energy of the ions is limited by the cooling gas temperature. For a detailed description and simulations of the gas-ion interaction for cooling in a segmented RFQ see for example [89, 98] and references therein.

Using fast switches, the potential of the last RFQ segments can be lowered after a certain cooling time t_{cool}, which creates a potential as indicated by the dotted line in the right of Fig. 6.17. The ions can then leave the RFQ as a bunch and are again accelerated to the initial energy reduced by the energy dissipated in the cooling process. Depending on the accumulation and cooling time, the ions form a short bunch with strongly reduced energy spread. In some cases [96] the remaining energy spread of the ion bunch can be reduced down to 10^{-5} of the total acceleration energy, which makes the general velocity compression of CLS, as described in Sect. 6.4.4.1, to some degree redundant. Moreover, if a fast reacceleration of the ions leaving the RFQ becomes dispensable, a slower ion beam might provide a longer laser-ion interaction in the fluorescence detection region. The resulting higher signal intensity/detection efficiency for CLS has not been tested yet and is to be investigated.

6.6.2 Applications of Ion Bunchers in CLS

The main background source of classical CLS with optical fluorescence detection is of continuous nature. It is typical detector noise of the PMTs and the usually dominant laser straylight created inside the vacuum beamline and scattered onto the PMTs. Typical rates are of the order of a few 1000 to 10,000 counts per second, covering the desired fluorescence signal of weak ion beams. By gating the optical detection count rate synchronized with the ion bunch passing the PMT in the detection region, this continuous background is suppressed. The suppression factor S can be simply estimated by the ratio of the sampling and cooling time T and the bunch length t_{gate} as

$$S = \frac{T}{t_{gate}}. \qquad (6.71)$$

The timing scheme and a typical time structure of a cooled bunch is demonstrated in Fig. 6.18. The opening of the RFQ's axial confinement is much shorter than the accumulation time t_{cool} and the cycle is then repeated with a periodicity time T. The right half of Fig. 6.18 displays a typical time-of-flight (TOF) structure of a bunch of ^{101}Cd ions, recorded by fluorescence photons [99] at the COLLAPS setup. The abscissa displays the time relative to the opening of the RFQ and thus the ions time of flight from the RFQ towards the COLLAPS optical detector. A time window t_{gate} of only 2 µs (depicted by the dotted lines in Fig. 6.18) can be used as coincidence gate

6 Nuclear Charge Radii of Light Elements and Recent Developments

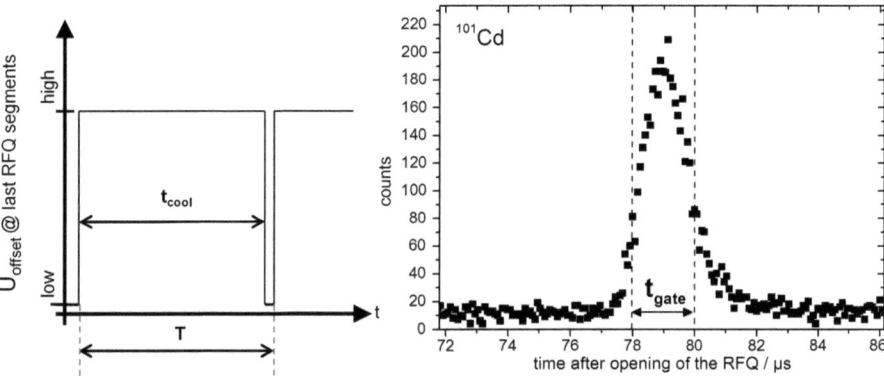

Fig. 6.18 Schematic timing cycle of the DC-potential of the last RFQ electrodes (*left*) and the time structure of a ^{101}Cd ion bunch (*right*) obtained at ISCOOL (ISOLDE). The bunch structure was observed by resonance fluorescence detection at the COLLAPS setup and exhibits a temporal width of about 2 μs after a cooling time of $t_{cool} = 50$ ms [99]

for the laser spectroscopy data recording. For the TOF spectrum shown in the right half of figure Fig. 6.18, which was recorded with a cooling time of $t_{cool} = 50$ ms $\approx T$ in the ISCOOL cooler [90, 91] at ISOLDE, a gating window of $t_{gate} = 2$ μs yields, according to Eq. (6.71), a background suppression factor of $S \approx 25,000$. Practically, a slightly larger time window was used during the measurements in order to be insensitive to small variations of the flight time. With this technique exotic ion beams with very low ion yields came into reach for CLS. Figure 6.19 shows laser spectra of ^{174}Hf from the pioneering work of Nieminen and co-workers [96], which were simultaneously recorded gated and ungated at the IGISOL facility. The cooling time was 500 ms, applying a gate window of $t_{gate} = 28$ μs. Achieving similar background suppression as given in the example shown in Fig. 6.19, Nieminen et al. extrapolated a so far unsurpassed lower detection limit for classical CLS with optical detection of ≈ 50 ions/s.

It should be mentioned that this technique for background suppression is ineffective against ion-beam induced backgrounds, *e.g.* collision-induced residual gas fluorescence. In the case of beams with only a few ions of the desired species, this background is typically dominated by the much more abundant isobaric contamination in the ion beam. However, ion-beam induced background is typically several orders lower than the laser straylight and therefore negligible. It may become relevant if a charge exchange reaction is used on a beam with large isobaric contamination since longer-lived states populated in the CEC can contribute to the background.

At the IGISOL facility a series of refractory elements were studied with bunched ion beams so far: zirconium [100], titanium and hafnium [101], cerium [102], yttrium [46], molybdenum [103], scandium [104] and ytterbium [105]. CLS in combination with the ISCOOL RFQ at ISOLDE was applied and has been completed for gallium [106] and copper [107, 108]. The successful introduction of this technique at the TITAN facility at TRIUMF was recently demonstrated by laser spectroscopy

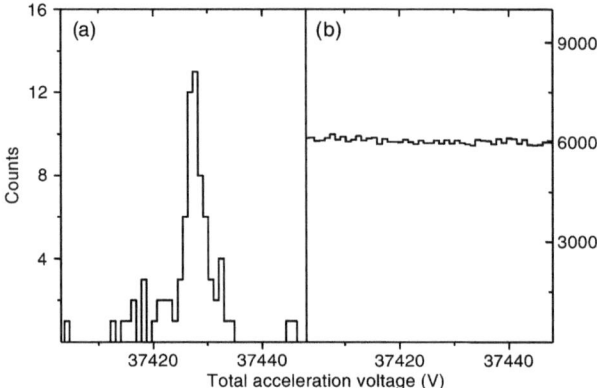

Fig. 6.19 Demonstration of bunched-beam laser spectroscopy. Fluorescence spectrum of ^{174}Hf studied in the $ds^2\,^2D_{3/2} \rightarrow dsp\,^2D_{5/2}$ transition applying a laser wavelength of 301.3 nm. Both spectra were simultaneously recorded and accumulated over 25 minutes at an ion yield of 1300 ions/s. (**a**) gated with $t_{gate} = 28$ μs and $t_{cool} = 500$ ms and (**b**) without gating. (Reprinted with permission from Phys. Rev. Lett. **88** (2002) 094801 Copyright 2002 American Physical Society)

on radioactive rubidium isotopes [109]. A small selection of physics cases will be exemplified here.

In recent years laser spectroscopy with bunched and cooled beams was applied at JYFL to study the sudden onset of deformation in the region around $Z \approx 40$ and $N \approx 60$ and at ISOLDE to investigate the behavior of copper and gallium isotopes above the $N = 28$ shell closure. As an example, studies of neutron-rich gallium isotopes will be briefly discussed, which would have not been possible without the sensitivity increase by the RFQ technique: According to the shell model, the ground-state spin of odd gallium isotopes should be determined by a proton hole in the $\pi p_{3/2}$ state as indicated in Fig. 6.20(a) while the $f_{5/2}$-orbital is not populated. However, experimentally an inversion between these states is found "between" ^{79}Ga and ^{81}Ga. This was determined by fitting the observed hyperfine structures of ^{79,81}Ga depicted in Fig. 6.20(b) for different spin values [110]. The results show a clear signature for a spin $I = 3/2$ for ^{79}Ga, as for almost all other odd isotopes of gallium that were studied (with the exception of ^{75}Ga, which has a collective $I = 1/2^-$ ground state [110]) and $I = 5/2$ for ^{81}Ga. This change can be explained by the tensor force between neutrons and protons induced by pion exchange [111, 112]. The monopole component of this force leads to an attraction of states which have $\vec{l}$ and $\vec{s}$ coupled in the same way, i.e. either to $j_> = l + s$ or to $j_< = l - s$, whereas it is repulsive between proton and neutron states $j_<$ and $j_>$. Since the induced shift of the single-particle energy of a level j caused by the monopole interaction increases linearly with the number of nucleons in the interacting state j', it can induce considerable changes along a chain of isotopes (or isotones). The observed inversion in gallium is caused by the filling of the $\nu g_{9/2}$ subshell starting with ^{71}Ga. The $\nu g_{9/2}$ level is of type $j_<$ thus the increasing number of neutrons reduces the energy of the $\pi f_{5/2}$ and increases the energy of the proton in the $\pi p_{3/2}$ level ($j_>$) until the

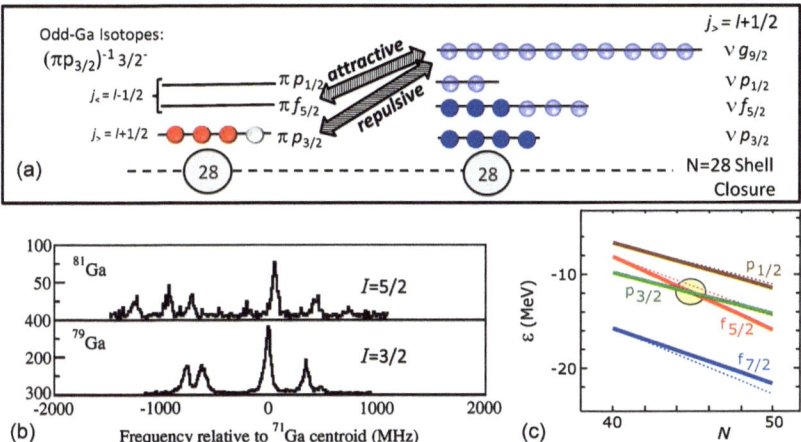

Fig. 6.20 (a) Population of nuclear orbitals of gallium isotopes according to the standard shell model. Indicated is the influence of the monopole moment of the tensor force acting between neutrons in the $\nu\, g_{9/2}$-orbital and protons in the $\pi\, p_{3/2}$, and $\pi\, f_{5/2}$ orbitals. (b) Hyperfine splitting of ^{79}Ga and ^{81}Ga. The spectra must be fitted with different spins ($I = 3/2$ for ^{79}Ga and $I = 5/2$ for ^{81}Ga) in order to get consistent results. (c) Calculation of proton single-particle energies of $^{68-78}$Ni as a function of neutron number [112]

$\pi f_{5/2}$ becomes the ground state. For illustration this is shown in Fig. 6.20(c) where the calculated proton single-particle energies for $^{68-78}_{28}$Ni are plotted as a function of neutron number [112]. The dotted lines represent calculations without the tensor force (central force only) and the solid lines include the tensor force. The strong reduction of the $\pi f_{5/2}$ energy is clearly visible and the crossing with the $\pi p_{3/2}$ orbital appears around $N = 45$.

Besides the increase of sensitivity with bunched ion beams, the accumulation and cooling time in the RFQ provides a possibility for measuring lifetimes of nuclear isomers as an additional spin-off: In a few cases the laser spectroscopy pattern of a well mass-separated ion beam provides more lines than expected from a single isotopic species. This fact can then be attributed to the simultaneous presence of radioactive isotopes in the ion beam not only in the nuclear ground state but also to a certain amount in one or more isomeric states. In order to distinguish among these nuclear states, the cooling and therewith the retention time of the radioactive ions inside the RFQ can be varied and prolonged. Thus the observed transition lines which are caused by the shorter-lived species will become weaker, as they are decaying during the cooling time and will be suppressed in the observed spectrum.

This allows for an identification of the lifetime difference of the components and in conjunction with a known ground-state life-time may be used to measure or at least to estimate a life-time of the isomeric state. However, it should be noted, that an unambiguous identification of the ground state and the isomeric states is not possible solely by laser spectroscopy if no additional information about the ground state is available (*e.g.* spin, lifetime etc.). In that case only mass measurements or radioactive decay spectroscopy can provide the required information. But laser resonance ionization may provide means to know which ions are injected into the trap or delivered to the decay-spectroscopy setup, as demonstrated, e.g., in [113].

In case of rather long-lived states no influence of the constituents can be observed by varying cooling times. Consequently the longest applied cooling time, which did not show any decay losses in the RFQ, can then serve at least as an estimate for the lower limit of the states lifetimes. This was for example conducted for ^{80}Ga and provided a lower lifetime limit for the isomer ^{80,m}Ga of $T_{1/2} \geq 200$ ms [110].

6.6.3 Optical Pumping in the Cooler and Buncher

The atomic state delivered from the ion source—in most cases the ionic ground state—may present a bottleneck which hampers laser spectroscopy for various reasons:

1. In some cases laser excitation from the ground state (GS) of the ion into the first excited state (FES) is technically hard to access with high-resolution CW lasers due to a required wavelength in the deep ultraviolet regime. Wavelengths below 250 nm are hard to realize using higher harmonics generation. Even though CW frequency-quadrupling has been recently applied to generate 215 nm for laser spectroscopy on cadmium [99], its technical complexity makes it inconvenient for routine operation.
2. Some elements suffer from weak oscillator strengths of their ground-state transitions. From this it follows that the transition can only be weakly driven by reasonable laser powers and that the FES has either a weak branching ratio back into the ground state or a relatively long life-time. Both cases strongly reduce the fluorescence signals that can be obtained while the ions are passing through the optical detection region. Charge exchange to the neutral atom might also not be the best option if it is non-resonant and results in a wide distribution of populated states.
3. Not only technical but also physical constraints make certain transitions from the ground state inappropriate for spectroscopic studies of nuclear moments. Especially $J = 0 \rightarrow J = 1$ transitions have the drawback that the corresponding hyperfine structure splitting will provide at maximum 3 lines, from which a simultaneous and independent determination of the magnetic dipole moment μ_I, the spectroscopic electric quadrupole moment Q_0, isotope shift $\delta\nu_{IS}^{AA'}$ and the nuclear spin I is not possible. For this purpose transitions from metastable, states with a higher J value are preferred.

In all of these three cases a laser-based optical manipulation of the ions might offer an appropriate tool to use transitions that are better suited for CLS. Ideally this manipulation takes place during the cooling and accumulation time in a gas-filled RFQ, as described above, since this provides sufficient laser-ion interaction time and a well-matched ion confinement for efficient geometrical overlap of the ions with the laser beam. Centerpiece of this manipulation is the excitation by high-power

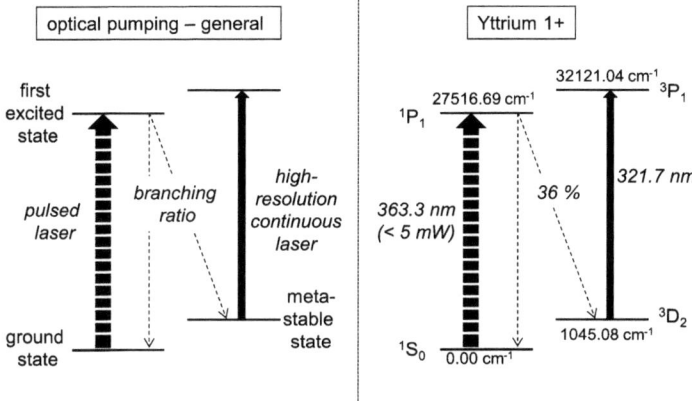

Fig. 6.21 Shown *on the left side* is the general working principle of optical pumping in the RFQ applying pulsed lasers for population of a new meta-stable starting point for high resolution laser spectroscopy. As an example application, the optical pumping and spectroscopy scheme for singly charged yttrium ions [114] is shown *on the right*

(pulsed) lasers[8] from the GS and—with the aid of an supportive branching-ratio—subsequent relaxation into a meta-stable state that serves as a new "ground state" GS* for laser spectroscopy. This process is illustrated in Fig. 6.21 and generally known as one instance of *optical pumping*. From this metastable state spectroscopy can then be performed with appropriate transitions for laser excitation.

The laser system which is ideally suited for optical pumping differs from the spectroscopy laser system in many ways as it has to meet the following technical requirements:

- the laser system should be pulsed, as (opposite to CW lasers) this enables simple and efficient higher harmonic generation of the second, third and fourth harmonic due to the high energy density in the laser pulse
- the repetition rate ν_{rep} should be sufficiently high ($\nu_{\text{rep}} \gg t_{\text{cool}}^{-1}$). Otherwise the ions can not run through several excitation cycles for optical pumping before the bunch is being extracted out of the RFQ.
- for efficient laser excitation the laser linewidth should be matched to the spectroscopic linewidth of the ion ensemble in the RFQ. Although the natural linewidth of these allowed dipole transitions in the ion is usually of the order of a few MHz only, the buffer gas interaction inside the RFQ leads to a collision-broadening and results in a linewidth of typically several GHz.

In pioneering experiments at the IGISOL-facility a Ti:Sa laser system with a repetition rate of $\nu_{\text{rep}} = 10$ kHz [115] was used, which was originally designed for laser-ion-source applications [116]. The efficiency of optical pumping was demonstrated by Cheal and coworkers for example in a test measurement using Y^+ ions

---

[8]High-power lasers in this respect means lasers with higher output power than typically available for CW high-resolution laser spectroscopy.

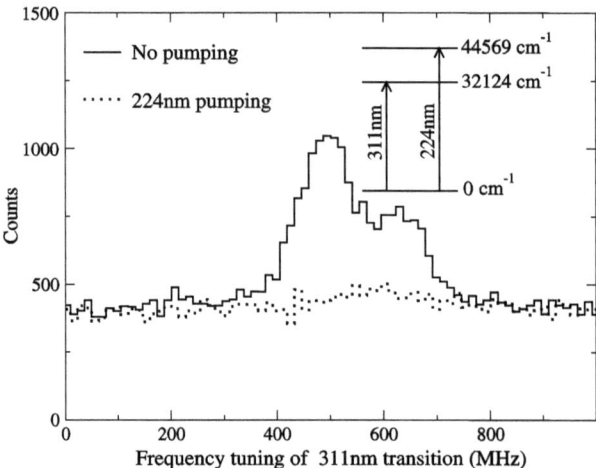

Fig. 6.22 Effect of optical pumping on the ground state population of a cooled yttrium ion bunch at the IGISOL facility [117]. The *solid line* represents a CW laser spectroscopic resonance in the 311-nm transition from the ground state as it is depicted in the *inset*. The *dotted line* represents the response after applying optical pumping along the 224 nm transition. The pumping process apparently transfers the ground state population efficiently into other levels and a fluorescence signal cannot be observed anymore. (Reprinted with permission from Phys. Rev. Lett. **102**, 222501 Copyright 2009 American Physical Society)

[117]. The result is depicted in Fig. 6.22. When exciting the yttrium ions from the ground state with classical collinear laser spectroscopy and using a CW laser at a wavelength of 311 nm, the hyperfine resonance pattern represented by the solid line was observed. The response after optical pumping with the pulsed Ti:Sa laser system at 224 nm is represented by the dotted line. It shows that the optical pumping process with this pulsed laser system is very efficient and apparently the ground state population is almost completely removed.

A number of refractory transition elements has been studied so far utilizing the optical pumping method. In the case of manganese the *challenging wavelengths* for the transitions from the ground state were circumvented by optical pumping with frequency quadrupled pulsed laser light at 231 nm. Subsequent high-resolution laser spectroscopy was performed at 295 nm starting from the $3d^5\,4s$ state at 9473 cm^{-1} above the ions ground state [118]. A very *weak oscillator strength* has been overcome in the case of niobium by optically pumping with laser light at 286 nm again with pulsed lasers and performing nuclear laser probing with CW laser light at 291 nm starting from the $4d^3\,5s$ level at 2357 cm^{-1} [117]. In the case of niobium as well as in the on-line studies of ^{100}Y an unambiguous *spin determination* was hampered by the $J=0 \rightarrow J=1$ transition from the ground state [114]. Applying the optical pumping scheme with the adjacent spectroscopy transition shown in detail on the right side of Fig. 6.21, the nuclear spin of the isomer ^{100,m}Y has been determined and in combination with γ-spectroscopic data the ground state configuration was identified.

The optical pumping method has a high potential and can be extended to other transition metals and probably even beyond. Up to now this technique was only applied at the IGISOL facility, but can be transferred to other RFQ cooler and buncher installations as long as they provide optical access. Studies of manganese isotopes including optical pumping in ISCOOL are currently planned for example at ISOLDE/CERN [119].

6.7 Future Prospects

Laser spectroscopy of radioactive nuclides has now been performed in quite a substantial part of the chart of nuclei [17, 18]. However, the most interesting regions close to the proton and neutron drip lines have been reached in only a few cases. In addition, information is scarce in the region of refractory elements around iron ($Z = 26$) because of the lack of efficient production schemes at ISOL facilities. Similarly, there are gaps in the region of refractory elements around molybdenum ($Z = 42$) and tungsten ($Z = 74$).

A major challenge in the upcoming years will be the coupling of CLS stations to the next generation *in-flight* facilities. In contrast to the ISOL-type ion production mechanism, the in-flight method forms a radioactive ion beam from fragments of a nuclear reaction of relativistic projectiles with atoms in a thin target. The demanding task is the transformation of the resulting relativistic fragment beam with the inherent huge emittance and energy-spread into a narrow collimated ion beam with narrow energy distribution and low emittance, so that it can be efficiently used for CLS.

The transformation of the in-flight fragment ion beam is accomplished in dedicated ion catcher devices, so-called "*gas-cells*". These vacuum chambers are filled with noble gas (typically helium) at pressures of the order of a few tens to several hundreds of mbar and are internally equipped with different versions of drift and guidance electrodes outside a stopping region inside the gas. The ions are entering the gas cell after being slowed down by the passage through previous solid-state energy degraders and create a large number of electron-ion pairs in the gas (in the order of 1 pair per 10–100 eV energy loss of the initial ion in high purity helium gas, depending on experimental conditions). Common to most[9] of the gas cells are (i) the inside lining of the gas cell walls with so-called radio-frequency carpets [120] which keep the heavy ions away from the walls and (ii) the extraction of the buffer gas cooled ions at the transition from the high pressure regime to the vacuum through a thin exit nozzle, in some cases carried out as a radio-frequency orifice.

The individual technical realizations of these gas cells are diverse. Based on the successful operation at the ATLAS facility, a conventional cylindrical gas cell [121] operated at room temperature has been implemented at the CARIBU facility [122]

---

[9]Exception here are small cells as used, *e.g.*, for ion-guide applications at Jyväskylä or Louvain-la-Neuve.

at Argonne National Laboratory. A speciality here is that the radioactive isotopes are not injected but directly created inside the gas cell by spontaneous fission of a macroscopic amount of ^{252}Cf. For the BECOLA laser spectroscopy experiment [123] at the Michigan State University the radioactive ions produced in-flight in the A1900 fragment separator (and in the future from the FRIB accelerator) are moderated in a gas cell as well. Upstream of the BECOLA beamline the ions will be either injected into a conventional gas cell similar to the one at Argonne National Laboratory or into an inverted[10] cyclotron [124]. The latter has the advantage of decoupling the region of where the most parasitic space charges are created from the region where the thermalized ions are finally stopped and extracted. Due to the spiral shape of the ions trajectory in the magnetic field, even long flight paths before stopping are feasible in the cyclotron. This allows for comparably lower buffer gas pressures inside the structure with all its advantages such as shorter transport time of the cooled ions and higher applicable potentials in the gas [124].

The LASPEC collinear laser spectroscopy experiment [125, 126] will be placed at the low-energy branch of FAIR's Super-Fragment-Separator and supplied from a linear gas cell. Deviating from all of the other gas cells mentioned above, this one is operated at a cryogenic temperature of about 60–80 K [127]. This provides higher gas densities (for helium: 250 mbar at 60 K $\approx$ 0.2 mg/cm^3 provides a similar stopping power as 1200 mbar pressure at room temperature [127]) and therefore the cell is more compact than conventional cells at room temperature. The cryogenic operation of the He gas additionally removes the necessity of a gas purification system aiming for a 1-ppb impurity level or even better, because contaminations to the He buffer gas or outgassing from the electrode materials are just frozen out. First tests have recently been successfully conducted at the current fragment separator at GSI [128].

Applying gas cells will extend the range of accessible isotopes for laser spectroscopy further away from the region of β-stability towards the so-called r-process path, as well as into regions that are currently not accessible. The final limitation besides the limited yield from the in-flight production process is the individual extraction time out of the gas cell. Exotic species will only be available if the time span required for extraction and reacceleration is shorter than or at least comparable to the lifetime of the exotic ion.

Moreover, a variety of new CLS experiments have recently started or are under preparation at on-line facilities. In combination with the installation of the TITAN-facility [129] at TRIUMF, the local CLS station—previously mostly applied for the generation of optically polarized beams of ^{8}Li for nuclear magnetic resonance studies—gained new momentum [109]. Within the Spiral-2 upgrade at DESIR/GANIL there will be a designated beamline for a CLS station called LUMIERE [130], which is currently under preparation.

Future developments of CLS in the near future will further focus on increasing sensitivity and—for special purposes—accuracy. The increase in sensitivity will be

---

[10]"Inverted" as it decelerates instead of accelerates the ions, as in typical applications of cyclotrons.

derived from the ongoing evolution of the methodical developments discussed in the previous sections. There are two starting-points for an increase in accuracy:

- *Laser wavelength control and measurement by means of modern quantum optics.* The laser wavelength stabilization can be implemented with a modern, compact fibre-laser based frequency comb, as demonstrated for the CLS measurement along the isotopic chain of beryllium [51] at COLLAPS. With this technique the laser frequency can be controlled even during longer experimental runs or can even be reproduced after a longer experimental break (*e.g.* between two experimental campaigns) with better than 1 MHz accuracy and/or repeatability. The TRIGA-LASER setup [131], which is a prototype of a part of the LASPEC setup, applies this technique routinely. At the BECOLA experiment the implementation of this technique is intended as well.
- *Improvement of the acceleration-potential measurement.* This second aspect for increasing the accuracy of CLS addresses the measurement of the initial acceleration potential of the ion source or RFQ, if applicable, as well as the high voltage of the Doppler-tuning voltage applied either to the FDR or the CEC. With typical equipment at the various on-line facilities, these voltages can be determined with an accuracy and reproducibility of about 100 ppm at best. This represents a limit, especially for light elements as discussed above.

This uncertainty in the acceleration potential had already a verifiable impact in the past: In a sequence of experiments on magnesium ions at ISOLDE, a discrepancy in the isotope shift measurements was observed which could be attributed to the use of different high voltage measurement systems. After a recalibration of both devices [132] with one of the world most precise voltage-dividers with ppm accuracy [133], the results of the individual beamtimes have been brought in agreement with each other [132] as well as (within systematical and statistical uncertainties) with independent high-accuracy ion-trap laser spectroscopy on stable magnesium isotopes as reported in [134].

The utilization of a mobile voltage divider with stabilized environmental conditions, which is aiming for a 10-ppm accuracy and stability and an applicability up to 100 kV is currently under development at the TRIGA-Laser experiment and the Technical University of Darmstadt. In case of a successful completion of this development, a few mobile devices of this type, which are than more sophisticated than any commercially available solution, can be transported to different on-line facilities for high-voltage measurement and monitoring during an on-line beamtime or in order to assure identical conditions among two consecutive experimental campaigns.

6.8 Conclusion

After the proposal of collinear laser spectroscopy in 1976 [8] and its first realization in 1977 [9], a variety of laser spectroscopy stations have been initiated and operated at different places all over the world. Nevertheless, throughout the last three decades only the collinear laser spectroscopy experiment at the IGISOL facility and the COLLAPS experiment at ISOLDE/CERN have been and still are the work horses for laser spectroscopy on a broad spectrum of radioactive isotopes. Affected by the impact and success of these studies and the continuous technical development, partially described in this lecture, the advent of the next generation on-line radioactive ion beam facilities initiated a renaissance of collinear spectroscopy all over the world.

References

1. A. Michelson, Recent advances in spectroscopy, in *Nobel Lecture*, December 12 (1907)
2. H.C. Urey, F.G. Brickwedde, G.M. Murphy, Phys. Rev. **39**, 164 (1932)
3. H. Schüler, T. Schmidt, Z. Phys. A **94**, 457 (1935)
4. H.B.G. Casimir, *On the Interaction Between Atomic Nuclei and Electrons* (Teyler's Tweede Genootschap, Haarlem, 1936)
5. J. Pinard, H. Stroke, Adv. At. Mol. Opt. Phys. **51**, 273 (2005)
6. H.H. Stroke, Hyperfine Interact. **171**, 3 (2006)
7. G. Huber, C. Thibault, R. Klapisch, H.T. Duong, J.L. Vialle, J. Pinard, P. Juncar, P. Jacquinot, Phys. Rev. Lett. **34**, 1209 (1975)
8. S.L. Kaufman, Opt. Commun. **17**, 309 (1976)
9. K.-R. Anton, S.L. Kaufman, W. Klempt, G. Moruzzi, R. Neugart, E.-W. Otten, B. Schinzler, Phys. Rev. Lett. **40**, 642 (1978)
10. B. Schinzler, W. Klempt, S.L. Kaufman, H. Lochmann, G. Moruzzi, R. Neugart, E.-W. Otten, J. Bonn, L. Von Reisky, K.P.C. Spath, J. Steinacher, D. Weskott, Phys. Lett. B **79**, 209 (1978)
11. W. Klempt, J. Bonn, R. Neugart, Phys. Lett. B **82**, 47 (1979)
12. A.C. Mueller, F. Buchinger, W. Klempt et al., Nucl. Phys. A **403**, 234 (1983)
13. V.N. Fedosseev, Y. Kudryavtsev, V.I. Mishin, Phys. Scr. **85**, 058104 (2012)
14. C. Schulz, E. Arnold, W. Borchers, W. Neu, R. Neugart, M. Neuroth, E.W. Otten, M. Scherf, K. Wendt, P. Lievens, Y.A. Kudryavtsev, V.S. Letokhov, V.I. Mishin, V.V. Petrunin, J. Phys. B **24**, 4831 (1991)
15. T.E. Cocolios, H.H. Suradi, J. Billowes, I. Budincevic, R.P. de Groote et al., Nucl. Instrum. Meth. B **317**, 565 (2013)
16. R. Neugart, G. Neyens, Nuclear moments. Lect. Notes Phys. **700**, 135 (2006)
17. B. Cheal, K.T. Flanagan, J. Phys. G **37**, 113101 (2010)
18. K. Blaum, J. Dilling, W. Nörtershäuser, Phys. Scr. T **152**, 014017 (2013)
19. K. Heilig, A. Steudel, *Atomic Data and Nuclear Data Tables* (1974)
20. T. Udem, A. Huber, B. Gross, J. Reichert, M. Prevedelli, M. Weitz, T.W. Hänsch, Phys. Rev. Lett. **79**, 2646 (1997)
21. R. Pohl, A. Antognini, F.o. Nez, F.D. Amaro, F.o. Biraben et al., Nature **466**, 213 (2010)
22. R. Pohl, R. Gilman, G.A. Miller, K. Pachucki, Annu. Rev. Nucl. Part. Sci. **63**, 175 (2013)
23. W. Nörtershäuser, R. Sanchez, G. Ewald, A. Dax, J. Behr et al., Phys. Rev. A **83**, 012516 (2011)
24. I. Sick, Phys. Lett. B **576**, 62 (2003)
25. I. Sick, D. Trautmann, Phys. Lett. B **375**, 16 (1996)
26. E.C. Seltzer, Phys. Rev. **188**, 1916 (1969)
27. B. Cheal, T.E. Cocolios, S. Fritzsche, Phys. Rev. A **86**, 042501 (2012)
28. G. Fricke, K. Heilig, in *Nuclear Charge Radii, Landolt-Börnstein (Numerical Data and Functional Relationships in Science and Technology)*, vol. 20, ed. by H. Schoppe (Springer, Berlin, 1994)
29. R.C. Barret, Phys. Lett. B **33**, 388 (1970)

30. H. Kopfermann, *Nuclear Moments* (Academic Press, New York, 1958)
31. G.W.F. Drake, W. Nörtershäuser, Z.-C. Yan, Can. J. Phys. **83**, 311–325 (2005)
32. M. Puchalski, K. Pachucki, Phys. Rev. A **78**, 052511 (2008)
33. E.A. Hylleraas, Z. Phys. **54**, 347 (1929)
34. M. Smith, M. Brodeur, T. Brunner, S. Ettenauer, A. Lapierre, R. Ringle, L. Ryjkov, F. Ames, P. Bricault, G.W.F. Drake, P. Delheij, D. Lunney, F. Sarazin, J. Dilling, Phys. Rev. Lett. **101**, 200501 (2008)
35. G.W.F. Drake, Z.-C. Yan, Phys. Rev. A **46**, 2378 (1992)
36. G.W.F. Drake, Adv. At. Mol Phys. **31**, 1 (1993)
37. Z.-C. Yan, G.W.F. Drake, Phys. Rev. A **61**, 022504 (2000)
38. M. Puchalski, A.M. Moro, K. Pachucki, Phys. Rev. Lett. **97**, 133001 (2006)
39. W. Nörtershäuser, T. Neff, R. Sanchez, I. Sick, Phys. Rev. C **84**, 024307 (2011)
40. W.D. Myers, W.J. Swiatecki, Ann. Phys. **55**, 395 (1969)
41. W.D. Myers, W.J. Swiatecki, Ann. Phys. **84**, 186 (1974)
42. W.D. Myers, K.-H. Schmidt, Nucl. Phys. A **410**, 61 (1983)
43. J. Bonn, G. Huber, H.-J. Kluge, E.-W. Otten, Phys. Lett. B **38**, 308 (1972)
44. T.E. Cocolios, W. Dexters, M.D. Seliverstov, A.N. Andreyev, S. Antalic et al., Phys. Rev. Lett. **106**, 052503 (2011)
45. M.D. Seliverstov, T.E. Cocolios, W. Dexters, A.N. Andreyev, S. Antalic, A.E. Barzakh, B. Bastin, J. Büscher, I.G. Darby, D.V. Fedorov, V.N. Fedoseyev, K.T. Flanagan, S. Franchoo, S. Fritzsche, G. Huber, M. Huyse, M. Keupers, U. Köster, Yu. Kudryavtsev, B.A. Marsh, P.L. Molkanov, R.D. Page, A.M. Sjødin, I. Stefan, J. Van de Walle, P. Van Duppen, M. Venhart, S.G. Zemlyanoy, Phys. Lett. B **719**, 362 (2013)
46. B. Cheal, M.D. Gardner, M. Avgoulea, J. Billowes, M.L. Bissell, P. Campbell, T. Eronen, K.T. Flanagan, D.H. Forest, J. Huikari, A. Jokinen, B.A. Marsh, I.D. Moore, A. Nieminen, H. Penttilä, S. Rinta-Antila, B. Tordoff, G. Tungate, J. Äystö, Phys. Lett. B **645**, 133 (2007)
47. M. Freer, Clustering in light nuclei; from the stable to the exotic, see Chap. 1 of this volume
48. M. Záková, Z. Andjelkovic, M.L. Bissell, K. Blaum, G.W.F. Drake, C. Geppert, M. Kowalska, J. Krämer, A. Krieger, T. Neff, R. Neugart, M. Lochmann, R. Sánchez, F. Schmidt-Kaler, D. Tiedemann, Z.-C. Yan, D.T. Yordanov, C. Zimmermann, W. Nörtershäuser, J. Phys. G **37**, 055107 (2010)
49. J. Al-Khalili, in *The Euroschool Lectures on Physics with Exotic Beams, Vol. I*, ed. by J. Al-Khalili, E. Roeckl. Lecture Notes in Physics, vol. 651 (Springer, Heidelberg, 2004), p. 77
50. W. Nörtershäuser, D. Tiedemann, M. Zakova, Z. Andjelkovic, K. Blaum et al., Phys. Rev. Lett. **102**, 062503 (2009)
51. A. Krieger, K. Blaum, M.L. Bissell, N. Frömmgen, Ch. Geppert, M. Hammen, K. Kreim, M. Kowalska, J. Krämer, T. Neff, R. Neugart, G. Neyens, W. Nörtershäuser, Ch. Novotny, R. Sanchez, D.T. Yordanov, Phys. Rev. Lett. **108**, 142501 (2012)
52. F.M. Marques, M. Labiche, N.A. Orr, J.C. Angélique, L. Axelsson et al., Phys. Lett. B **476**, 219 (2000)
53. N.B. Shulgina, B. Jonson, M.V. Zhukov, Nucl. Phys. A **825**, 175 (2009)
54. G. Papadimitriou, A.T. Kruppa, N. Michel, W. Nazarewicz, M. Płoszajczak, J. Rotureau, Phys. Rev. C **84**, 051304 (2011)
55. P.J. Mohr, B.N. Taylor, D.B. Newell, Rev. Mod. Phys. **84**, 1527 (2012)
56. J. Beringer, J.-F. Arguin, R.M. Barnett, K. Copic, O. Dahl et al., Phys. Rev. D **86**, 010001 (2012)
57. K. Riisager, Phys. Scr. T **152**, 014001 (2013)
58. I. Tanihata, H. Savajols, R. Kanungo, Prog. Part. Nucl. Phys. **68**, 215 (2013)
59. L.-B. Wang, P. Mueller, K. Bailey, G.W.F. Drake, J.P. Greene, D. Henderson, R.J. Holt, R.V.F. Janssens, C.L. Jiang, Z.-T. Lu, T.P. O'Connor, R.C. Pardo, K.E. Rehm, J.P. Schiffer, X.D. Tang, Phys. Rev. Lett. **93**, 142501 (2004)
60. P. Mueller, I.A. Sulai, A.C.C. Villari, J.A. Alcantara-Nunez, R. Alves-Conde, K. Bailey, G.W.F. Drake, M. Dubois, C. Eleon, G. Gaubert, R.J. Holt, R.V.F. Janssens, N. Lecesne,

Z.T. Lu, T.P. O'Connor, M.G. Saint-Laurent, J.C. Thomas, L.B. Wang, Phys. Rev. Lett. **99**, 252501 (2007)
61. R. Sanchez, W. Nörtershäuser, G. Ewald, D. Albers, J. Behr et al., Phys. Rev. Lett. **96**, 033002 (2006)
62. T. Nakamura, M. Wada, K. Okada, A. Takamine, Y. Ishida, Y. Yamazaki, T. Kambara, Y. Kanai, T.M. Kojima, Y. Nakai, N. Oshima, A. Yoshida, T. Kubo, S. Ohtani, K. Noda, I. Katayama, V. Lioubimov, H. Wollnik, V. Varentsov, H.A. Schuessler, Phys. Rev. A **74**, 052503 (2006)
63. K. Okada, M. Wada, T. Nakamura, A. Takamine, V. Lioubimov et al., Phys. Rev. Lett. **101**, 212502 (2008)
64. A. Takamine, M. Wada, K. Okada, T. Nakamura, P. Schury, T. Sonoda, V. Lioubimov, H. Iimura, Y. Yamazaki, Y. Kanai, T. Kojima, A. Yoshida, T. Kubo, I. Katayama, S. Ohtani, H. Wollnik, H. Schuessler, Eur. Phys. J. A **42**, 369 (2009)
65. W. Geithner, S. Kappertz, M. Keim, P. Lievens, R. Neugart et al., Phys. Rev. Lett. **83**, 3792 (1999)
66. E.L. Raab, M. Prentiss, A. Cable, S. Chu, D.E. Pritchard, Phys. Rev. Lett. **59**, 2631 (1987)
67. S. Chu, *Nobel Lectures, Physics 1996–2000* (World Scientific, Singapore, 2002), p. 122
68. Z.-T. Lu, P. Mueller, G.W.F. Drake, W. Nörtershäuser, S. Pieper, Z.-C. Yan, Rev. Mod. Phys. **85**, 1383 (2013)
69. T.W. Hänsch, S.A. Lee, R. Wallenstein, C. Wieman, Phys. Rev. Lett. **34**, 307 (1975)
70. A. Huber, T. Udem, B. Gross, J. Reichert, M. Kourogi, K. Pachucki, M. Weitz, T.W. Hänsch, Phys. Rev. Lett. **80**, 468 (1998)
71. C.G. Parthey, A. Matveev, J. Alnis, B. Bernhardt, A. Beyer et al., Phys. Rev. Lett. **107**, 203001 (2011)
72. G. Grynberg, B. Cagnac, Rep. Prog. Phys. **40**, 791 (1977)
73. V.S. Letokhov, Comments At. Mol. Phys. **7**, 93 (1977)
74. C.J. Sansonetti, C.E. Simien, J.D. Gillaspy, J.N. Tan, S.M. Brewer, R.C. Brown, S. Wu, J.V. Porto, Phys. Rev. Lett. **107**, 023001 (2011)
75. R.C. Brown, S.J. Wu, J.V. Porto, C.J. Sansonetti, C.E. Simien, S.M. Brewer, J.N. Tan, J.D. Gillaspy, Phys. Rev. A **87**, 032504 (2013)
76. A. Klose, K. Minamisono, Ch. Geppert, N. Frömmgen, M. Hammen, J. Krämer, A. Krieger, C.D.P. Levy, P.F. Mantica, W. Nörtershäuser, S. Vinnikova, Nucl. Instrum. Methods A **678**, 114 (2012)
77. M. Žáková, C. Geppert, A. Herlert, H.-J. Kluge, R. Sánchez, F. Schmidt-Kaler, D. Tiedemann, C. Zimmermann, W. Nörtershäuser, Hyperfine Interact. **171**, 189–195 (2006)
78. Z.C. Yan, W. Nörtershäuser, G.W.F. Drake, Phys. Rev. Lett. **100**, 243002 (2008)
79. W. Geithner, K.M. Hilligsoe, S. Kappertz, G. Katko, M. Keim, S. Kloos, G. Kotrotsios, P. Lievens, K. Marinova, R. Neugart, L. Vermeeren, S. Wilbert, Hyperfine Interact. **127**, 117 (2000)
80. W. Geithner, T. Neff, G. Audi, K. Blaum, P. Delahaye, H. Feldmeier, S. George, C. Guenaut, F. Herfurth, A. Herlert, S. Kappertz, M. Keim, A. Kellerbauer, H.J. Kluge, M. Kowalska, P. Lievens, D. Lunney, K. Marinova, R. Neugart, L. Schweikhard, S. Wilbert, C. Yazidjian, Phys. Rev. Lett. **101**, 252502 (2008)
81. K. Marinova, W. Geithner, M. Kowalska, K. Blaum, S. Kappertz, M. Keim, S. Kloos, G. Kotrotsios, P. Lievens, R. Neugart, H. Simon, S. Wilbert, Phys. Rev. C **84**, 034313 (2011)
82. I. Tanihata, T. Kobayashi, O. Yamakawa, S. Shimoura, K. Ekuni, K. Sugimoto, N. Takahashi, T. Shimoda, H. Sato, Phys. Lett. B **206**, 592 (1988)
83. L. Vermeeren, P. Lievens, A. Buekenhoudt, R.E. Silverans, J. Phys. B **25**, 1009 (1992)
84. W. Borchers, E. Arnold, W. Neu, R. Neugart, K. Wendt, G. Ulm, the ISOLDE Collaboration, Phys. Lett. B **216**, 7 (1989)
85. E.J. Konopinski, Annu. Rev. Sci. **9**, 99 (1959)
86. D.T. Yordanov, M.L. Bissell, K. Blaum, M. De Rydt, Ch. Geppert, M. Kowalska, J. Krämer, K. Kreim, A. Krieger, P. Lievens, T. Neff, R. Neugart, G. Neyens, W. Nörtershäuser, R. Sanchez, P. Vingerhoets, Phys. Rev. Lett. **108**, 042504 (2012)

87. D.A. Eastham, P.M. Walker, J.R.H. Smith, J.A.R. Griffith, D.E. Evans, S.A. Wells, M.J. Fawcett, I.S. Grant, Opt. Commun. **60**, 293 (1986)
88. D.J. Douglas, J.B. French, J. Am. Soc. Mass Spectrom. **181**, 27 (1998)
89. F. Herfurth, J. Dilling, A. Kellerbauer, G. Bollen, S. Henry, H.-J. Kluge, E. Lamour, D. Lunney, R.B. Moore, C. Scheidenberger, S. Schwarz, G. Sikler, J. Szerypo, Nucl. Instrum. Methods A **469**, 254 (2001)
90. A. Jokinen, M. Lindroos, E. Molin, M. Petersson, Nucl. Instrum. Methods B **204**, 86 (2003)
91. H. Franberg, P. Delahaye, J. Billowes, K. Blaum, R. Catherall, F. Duval, O. Gianfrancesco, T. Giles, A. Jokinen, M. Lindroos, D. Lunney, E. Mane, I. Podadera, Nucl. Instrum. Methods B **266**, 4502 (2008)
92. A. Nieminen, J. Huikari, A. Jokinen, J. Äystö, P. Campbell, E.C.A. Cochrane, Nucl. Instrum. Methods A **469**, 244 (2001)
93. S. Schwarz, G. Bollen, D. Lawton, P. Lofy, D.J. Morrissey, J. Ottarson, R. Ringle, P. Schury, T. Sun, V. Varentsov, L. Weissman, Nucl. Instrum. Methods B **204**, 507 (2003)
94. T. Brunner, M.J. Smith, M. Brodeur, S. Ettenauer, A.T. Gallant, V.V. Simon, A. Chaudhuri, A. Lapierre, E. Mane, R. Ringle, M.C. Simon, J.A. Vaz, P. Delheij, M. Good, M.R. Pearson, J. Dilling, Nucl. Instrum. Methods A **676**, 32 (2012)
95. T. Kim, Ph.D. Thesis, McGill University, Montreal (1997, unpublished)
96. A. Nieminen, P. Campbell, J. Billowes, D.H. Forest, J.A.R. Griffith, J. Huikari, A. Jokinen, I.D. Moore, R. Moore, G. Tungate, J. Äystö, Phys. Rev. Lett. **88**, 094801 (2002)
97. W. Paul, H. Steinwedel, Z. Naturforsch. A **8**, 448 (1953)
98. M.D. Lunney, R.B. Moore, Int. J. Mass Spectrom. **190/191**, 153 (1999)
99. M. Hammen, Spins, moment and radii of Cd isotopes, Dissertation, Johannes Gutenberg-Universität Mainz (2013). http://ubm.opus.hbz-nrw.de/volltexte/2013/3590/pdf/doc.pdf
100. P. Campbell, H.L. Thayer, J. Billowes, P. Dendooven, K.T. Flanagan, D.H. Forest, J.A.R. Griffith, J. Huikari, A. Jokinen, R. Moore, A. Nieminen, G. Tungate, S. Zemlyanoi, J. Äystö, Phys. Rev. Lett. **89**, 082501 (2002)
101. P. Campbell, A. Nieminen, J. Billowes, P. Dendooven, K.T. Flanagan, D.H. Forest, Yu.P. Gangrsky, J.A.R. Griffith, J. Huikari, A. Jokinen, I.D. Moore, R. Moore, H.L. Thayer, G. Tungate, S.G. Zemlyanoi, J. Äystö, Eur. J. Phys. A **15**, 45 (2002)
102. B. Cheal, M. Avgoulea, J. Billowes, P. Campbell, K.T. Flanagan, D.H. Forest, M.D. Gardner, J. Huikari, B.A. Marsh, A. Nieminen, H.L. Thayer, G. Tungate, J. Äystö, J. Phys. G **29**, 2479 (2003)
103. F.C. Charlwood, K. Baczynska, J. Billowes, P. Campbell, B. Cheal, T. Eronen, D.H. Forest, A. Jokinen, T. Kessler, I.D. Moore, H. Penttilä, R. Powis, M. Rüffer, A. Saastamoinen, G. Tungate, J. Äystö, Phys. Lett. B **674**, 23 (2009)
104. M. Avgoulea, Yu.P. Gangrsky, K.P. Marinova, S.G. Zemlyanoi, S. Fritzsche et al., J. Phys. G **38**, 025104 (2011)
105. K.T. Flanagan, J. Billowes, P. Campbell, B. Cheal, G.D. Dracoulis, D.H. Forest, M.D. Gardner, J. Huikari, A. Jokinen, B.A. Marsh, R. Moore, A. Nieminen, H. Penttilä, H.L. Thayer, G. Tungate, J. Äystö, J. Phys. G **39**, 125101 (2012)
106. B. Cheal, E. Mane, J. Billowes, M.L. Bissell, K. Blaum et al., Phys. Rev. Lett. **104**, 252502 (2010)
107. K.T. Flanagan, P. Vingerhoets, M. Avgoulea, J. Billowes, M.L. Bissell et al., Phys. Rev. Lett. **103**, 1425012 (2009)
108. P. Vingerhoets, K.T. Flanagan, M. Avgoulea, J. Billowes, M.L. Bissell et al., Phys. Rev. C **82**, 064311 (2010)
109. E. Mane, A. Voss, J.A. Behr, J. Billowes, T. Brunner, F. Buchinger, J.E. Crawford, J. Dilling, S. Ettenauer, C.D.P. Levy, O. Shelbaya, M.R. Pearson, Phys. Rev. Lett. **107**, 212502 (2011)
110. B. Cheal, J. Billowes, M.L. Bissell, K. Blaum, F.C. Charlwood et al., Phys. Rev. C **82**, 051302(R) (2010)
111. T. Otsuka, T. Suzuki, R. Fujimoto, H. Grawe, Y. Akaishi, Phys. Rev. Lett. **95**, 232502 (2005)
112. T. Otsuka, Phys. Scr. T **152**, 014007 (2013)

113. J. Van Roosbroeck, C. Guenaut, G. Audi, D. Beck, K. Blaum et al., Phys. Rev. Lett. **92**, 112501 (2004)
114. K. Baczynska, J. Billowes, P. Campbell, F.C. Charlwood, B. Cheal, T. Eronen, D.H. Forest, A. Jokinen, T. Kessler, I.D. Moore, M. Rüffer, G. Tungate, J. Äystö, J. Phys. G **37**, 10510 (2010)
115. A. Nieminen, I.D. Moore, J. Billowes, P. Campbell, K.T. Flanagan, Ch. Geppert, J. Huikari, A. Jokinen, T. Kessler, B. Marsh, H. Penttilä, S. Rinta-Antila, B. Tordoff, K.D.A. Wendt, J. Äystö, Hyperfine Interact. **162**, 39 (2005)
116. Ch. Geppert, P. Bricault, R. Horn, J. Lassen, Ch. Rauth, K. Wendt, Nucl. Phys. A **746**, 631c (2004)
117. B. Cheal, K. Baczynska, J. Billowes, P. Campbell, F.C. Charlwood, T. Eronen, D.H. Forest, A. Jokinen, T. Kessler, I.D. Moore, M. Reponen, S. Rothe, M. Rüffer, A. Saastamoinen, G. Tungate, J. Äystö, Phys. Rev. Lett. **102**, 222501 (2009)
118. F.C. Charlwood, J. Billowes, P. Campbell, B. Cheal, T. Eronen, D.H. Forest, S. Fritzsche, M. Honma, A. Jokinen, I.D. Moore, H. Penttilä, R. Powis, A. Saastamoinen, G. Tungate, J. Äystö, Phys. Lett. B **690**, 346 (2010)
119. B. Cheal, J. Billowes, M.L. Bissell, P. Campbell, T.E. Cocolios, V.N. Fedoseyev, K.T. Flanagan, D.H. Forest, K.M. Lynch, B. Marsh, I.D. Moore, G. Neyens, T.J. Procter, M.M. Rajabali, M. Reponen, S. Rothe, G. Tungate, accepted proposal to the ISOLDE committee CERN-INTC-2010-073/INTC-P-286; http://cds.cern.ch/record/1298606/files/INTC-P-286.pdf
120. S. Masuda, K. Fujibayashi, K. Ishida, in *Advances in Static Electricity*, Auxilia Brussels, (1970), p. 384
121. G. Savard, J. Clark, C. Boudreau, F. Buchinger, J.E. Crawford et al., Nucl. Instrum. Methods B **204**, 582 (2003)
122. G. Savard, S. Baker, C. Davids, A.F. Levand, E.F. Moore, R.C. Pardo, R. Vondrasek, B.J. Zabransky, G. Zinkann, Nucl. Instrum. Methods B **266**, 4086 (2008)
123. K. Minamisono, G. Bollen, P.F. Mantica, D.J. Morrissey, S. Schwarz, in *Proceedings from the Institute for Nuclear Theory*, vol. 16 (2007), p. 180
124. G. Bollen, D.J. Morissey, S. Schwarz, Nucl. Instrum. Methods A **550**, 27 (2005)
125. D. Rodriguez, K. Blaum, W. Nörtershäuser, M. Ahammed, A. Algora et al., Eur. Phys. J. Spec. Top. **183**, 1 (2010)
126. W. Nörtershäuser, P. Campbell (the LaSpec collaboration), Hyperfine Interact. **171**, 149 (2006)
127. M. Ranjan, S. Purushothaman, T. Dickel, H. Geissel, W.R. Plass, D. Schäfer, C. Scheidenberger, J. Van de Walle, H. Weick, P. Dendooven, Europhys. Lett. **96**, 52001 (2011)
128. W.R. Plaß, T. Dickel, S. Purushothaman, P. Dendooven, H. Geissel et al., Nucl. Instrum. Methods B **317**, 457 (2013)
129. J. Dilling, P. Bricault, M. Smith, H.-J. Kluge, TITAN collaboration, Nucl. Instrum. Methods B **204**, 492 (2003)
130. M. Lewitowicz, J. Phys. **312**, 052014 (2011)
131. J. Ketelaer, J. Krämer, D. Beck, K. Blaum, M. Block, K. Eberhardt, G. Eitel, R. Ferrer, Ch. Geppert, S. George, F. Herfurth, J. Ketter, Sz. Nagy, D. Neidherr, R. Neugart, W. Nörtershäuser, J. Repp, C. Smorra, N. Trautmann, C. Weber, Nucl. Instrum. Methods A **594**, 162 (2008)
132. A. Krieger, Ch. Geppert, R. Catherall, F. Hochschulz, J. Krämer, R. Neugart, S. Rosendahl, J. Schipper, E. Siesling, Ch. Weinheimer, D.T. Yordanov, W. Nörtershäuser, Nucl. Instrum. Methods A **632**, 23 (2011)
133. Th. Thümmler, R. Marx, Ch. Weinheimer, New J. Phys. **11**, 103007 (2009)
134. V. Batteiger, S. Knünz, M. Herrmann, G. Saathoff, H.A. Schüssler, B. Bernhardt, T. Wilken, R. Holzwarth, T.W. Hänsch, Th. Udem, Phys. Rev. A **80**, 022503 (2009)

Chapter 7
The Nuclear Energy Density Functional Formalism

T. Duguet

7.1 Introduction

7.1.1 Generalities

Low-energy nuclear physics aims at addressing several fundamental, yet only partially answered, questions. Among those are (i) *how do neutrons and protons bind inside a nucleus and what are the limits of existence of the latter regarding its mass, neutron-proton imbalance, angular momentum... ?* (ii) *How to explain the complex phenomenology of nuclei starting from elementary two-, three-... A-nucleon (AN) interactions?* (iii) *How do the latter interactions eventually emerge from quantum chromodynamics (QCD)?* Such questions have numerous ramifications and implications such that partial answers to them continuously impact other fields of physics (e.g. astrophysics, tests of the Standard Model). In spite of over eighty years of theoretical and experimental studies, low-energy nuclear physics remains an open and difficult problem. While extensive progress has been made, an accurate and universal description of low-energy nuclear systems from first principles is still beyond reach.

The first difficulty resides in the inter-particle interactions at play. Strong inter-nucleon interactions relevant to describing low-energy phenomena must be modelled within the non-perturbative regime of the gauge theory of interacting quarks and gluons, i.e. QCD. Within such a frame, nucleons are assigned to spin and isospin $SU(2)$ doublets such that they are 4-component fermions interacting in various configurations stemming from invariances of the problem, e.g. they interact through

---

T. Duguet (✉)
CEA-Saclay DSM/Irfu/SPhN, 91191 Gif sur Yvette Cedex, France
e-mail: thomas.duguet@cea.fr

T. Duguet
NSCL and Department of Physics and Astronomy, Michigan State University, East Lansing, MI 48824, USA

central, spin-orbit, tensor, quadratic spin-orbit ... couplings. In addition to its complex operator structure, the $2N$ force produces a weakly-bound neutron-proton state (i.e. the deuteron) in the coupled $^3S_1$–$^3D_1$ partial waves and a virtual di-neutron state in the $^1S_0$ partial wave. Associated large scattering lengths, together with the short-range repulsion between nucleons make the nuclear many-body problem highly non-perturbative. In addition to such difficulties, the treatment of $3N, 4N$ ... interactions in a theory of point-like nucleons is unavoidable. This has become clear over the last fifteen years as one was aiming at a consistent understanding of (i) differential nucleon-deuteron cross-sections [1], (ii) the under-estimation of triton and light-nuclei binding energies [2], (iii) the Tjon line [3], (iv) the violation of the Koltun sum rule [4] and (v) the saturation of symmetric nuclear matter [5, 6] in connection with the Coester line problem [7, 8].

The second difficulty stems from the nature of the system of interest. Most nuclei (i.e. those with masses typically between 10 and 350) are by essence intermediates between few- and many-body systems. As a result (i) most nuclei are beyond theoretical and computational limits of ab-initio techniques that describe the interacting system from basic AN forces, while (ii) finite-size effects play a significant role, which prevents statistical treatments. Furthermore, a unified view of low-energy nuclear physics implies a coherent description of small- and large-amplitude collective motions, as well as of closed and open systems, e.g. of the structure-reaction interface that is mandatory to understand spontaneous and induced fission, fusion, nucleon emission at the drip-line . . . .

The study of the atomic nucleus aims at accessing its ground-state (mass, radius, deformation and multipolar moments . . .) and excited-states (single-particle, vibrational, shape and spin isomers, high-spin and super-deformed rotational bands . . .) properties as well as the various decay modes between them (nuclear, electromagnetic and electroweak), together with reaction properties (elastic and inelastic scattering, transfer and pickup, fusion . . .). This is to be achieved for systems over the nuclear chart, i.e. not only for the nearly 3100 observed nuclei [9] but also for the thousands that are still to be discovered. In that respect, a cross-fertilization between theoretical and experimental studies is topical, with the advent of (i) a new-generation of radioactive-ion-beam (RIB) facilities producing very short-lived systems with larger yields, and (ii) high-sensitivity and high-selectivity detectors allowing measurements with low statistics. Upcoming facilities based on in-flight fragmentation, stopped and reaccelerated beams or a combination of both are going to further explore the nuclear chart towards the limits of stability against nucleon emission, the so-called neutron and proton drip-lines. The study of highly neutron-rich nuclei will help understand the astrophysical nucleosynthesis of about half of the nuclei heavier than iron through the conjectured r-process. The access to nuclei with a large neutron-over-proton ratio has already started to modify certain cornerstones of nuclear structure, e.g. some of the "standard" magic numbers are significantly altered while others (may) appear [10]. When adding even more neutrons, the proximity of the Fermi energy to the particle continuum gives rise to exotic phenomena, such as the formation of light nuclear halos [11, 12] with anomalously large extensions [13, 14] or the existence of di-proton emitters [15, 16]. In addition to reaching out to the most exotic nuclei, experiments closer to the valley of stability still provide critical information. For instance, precise mass measurements using Penning traps [17] or Schottky spectrometry [18] not only refine and extend mass difference formulæ [19] to better understand nuclear structure properties, e.g. pairing correlations, but also contribute to testing the standard model of particle physics, e.g. recent mass measurements have helped refine the validation of the unitarity of the Cabibbo-Kobayashi-Maskawa (CKM) flavour-mixing matrix [20]. Eventually, other limits of existence are of key importance, e.g. the quest for superheavy elements and the conjectured island of stability beyond the $Z = 82$ magic number [21]. In addition to the quoted references, we refer the interested reader to Vols. 1–3 of this series that contain many contributions relevant to the topics alluded to just above.

7.1.2 Nuclear Structure Theory

In such a context, the challenge of contemporary nuclear structure theory is to describe, in a controlled[1] and unified manner, the entire range of nuclei along with the equation of state of extended nuclear matter, from a fraction to few times nuclear saturation density and over a wide range of temperatures. All such properties find an interesting outcome in the physics of neutron stars and supernovae explosions as well as in the nucleosynthesis of heavy elements as already alluded to above.

7.1.2.1 Ab Initio Methods

While bulk properties of nuclei can be roughly explained using macroscopic approaches such as the liquid drop model (LDM) [22, 23], microscopic techniques are the tool of choice for a coherent description of static and dynamical nuclear properties. This leads to defining the class of so-called *ab-initio* methods that consists of solving, as exactly as possible, the nuclear many-body problem expressed in terms of elementary $2N$, $3N$, $4N$... interactions. For three- and four-nucleon systems, essentially exact solutions of the Faddeev or Yakubowski equations can be obtained using realistic vacuum forces [2, 24, 25]. Likewise, Green's function Monte-Carlo (GFMC) calculations [26, 27] provide a numerically exact description of nuclei up to carbon starting from local $2N$ and $3N$ vacuum forces, although such a method already faces huge numerical challenges for ^{12}C. Complementary ab-initio methods allow the treatment of nuclei up to $A \approx 16$, e.g. (i) the no-core shell model (NCSM) [28] that projects the interacting problem on a truncated harmonic oscillator model space or (ii) lattice effective field theory (LEFT) [29] that propagates

---

[1]The notion of "controlled" description refers to the capability of estimating uncertainties of various origins in the theoretical method employed.

nucleons as point-like particles on lattice sites interacting via pion exchanges and multi-nucleon operators.

In the last ten years, a breakthrough has occurred that renders possible the ab-initio calculation of double closed-shell nuclei, along with those in their immediate vicinity, with masses up to $A \approx 60$ on the basis of realistic $2N$ and $3N$ interactions. Three methods have been developed in order to move in this direction. First is Coupled-cluster (CC) theory [30, 31], which constructs the correlated ground-state from a product state using an exponentiated cluster expansion, truncated to B-body operators (typ. $B \sim 2\text{–}3$). Second, self-consistent Green's function (SCGF) theory [32, 33] computes the approximate *dressed* one-body Green's function describing the propagation of a nucleon propagating within the correlated medium. Last but not least, in-medium similarity renormalization group (IMSRG) method [34, 35] proceeds to the decoupling of a finite-density reference state from excitations built on top of it via a sequence of infinitesimal renormalization group transformations. The frontier in the development of such ab-initio many-body methods is not only to push calculations to higher masses but also to extend them to truly open-shell systems. Decisive steps are taken in this direction for SCGF [36, 37], IMSRG [38] and CC [39] theories. This is meant to extend the reach of ab-initio calculations from a few tens to several hundreds of mid-mass nuclei.

7.1.2.2 The Configuration Interaction Method

Accessing even heavier systems requires more drastic approximations to the interacting many-body problem. Part of the physics that cannot be treated explicitly is accounted for through the formulation and use of so-called *in-medium interactions*. The configuration interaction (CI) model [40], i.e. shell model (SM), constructs a model space within which valence nucleons interact through an effective interaction that compensates for high-lying excitations outside that model space as well as for excitations of the core that are not treated explicitly. Even though such an effective interaction can be constructed starting explicitly from elementary interactions [41], certain combinations of two-body matrix elements[2] need to be slightly refitted to experimental data within the chosen model space (sd, pf...) to correct for the so-called monopole part of the interaction. Based on the conjectures that wrong monopoles originate from the omission of the $3N$ force in the starting vacuum Hamiltonian [43], the non-empirical SM based on diagrammatic techniques[3] from renormalized $2N$ and $3N$ interactions is currently being revived [44] and shows promising results [45, 46]. Eventually, spectroscopic properties can be described

---

[2] In the sd shell for example, it is necessary to (slightly) refit about 30 combinations of two-body matrix elements in order to reach about 140 keV root mean square error on nearly 600 pieces of spectroscopic data [42].

[3] The adjective "diagrammatic" refers to many-body methods relying on the use of Feynman or Goldstone diagrams.

with high accuracy using refitted effective interactions [40, 42]. Still, improved accuracy is needed in the SM to use nuclei as laboratories for fundamental symmetries, e.g. to provide the matrix elements needed for the search of neutrinoless double-beta decay [47].

7.1.2.3 The Nuclear Energy Density Functional Method

Last but not least, the theoretical tool of choice for the microscopic and systematic description of medium- and heavy-mass nuclei is the energy density functional (EDF) method [48, 49], often referred to as "self-consistent mean-field and beyond-mean-field methods". Such method has been empirically adapted from well-defined wave-function- and Hamiltonian-based approaches. Based on a relativistic or a non-relativistic framework, the EDF method aims at providing, within one consistent frame, (i) the detailed and complete description of specific nuclei of interest, (ii) systematic trends over a large set of nuclei and (iii) trustful extrapolations in the region of the nuclear chart where experimental data are and will remain unavailable. Thanks to a favourable numerical scaling, the EDF method is indeed amenable to systematic studies of systems with large numbers of nucleons, independent of their expected shell structure. The idealized *infinite nuclear matter* system relevant to the description of compact astrophysical objects such as neutron stars is accessible to EDF calculations as well.

A fundamental aspect of the method is that it relies heavily on the concept of spontaneous breaking and restoration of symmetries. As such, the nuclear EDF method is intrinsically a two-step approach,

1. The first step is constituted by the so-called single-reference EDF (SR-EDF) implementation, originally adapted from the symmetry-unrestricted Hartree Fock Bogoliubov (HFB) method by using a *density-dependent* effective Hamilton "operator" [50]. Later, the approximate energy was formulated directly as a possibly richer functional of one-body density matrices computed from a symmetry-breaking HFB state of reference. The power of the approach relies on its ability to parametrize the bulk of many-body correlations under the form of a functional of one-body density (matrices) while authorizing the latter to break symmetries dictated by the underlying Hamiltonian in order to account for static collective correlations. It is however difficult, if not impossible, to capture in this way the subsequent dynamical correlations associated with good symmetries and quantum collective fluctuations.
2. It is thus the goal of the second step, carried out through the multi-reference (MR) extension of the SR-EDF method, to grasp such long-range correlations. The MR-EDF implementation has been adapted from the generator coordinate method (GCM) performed in terms of symmetry-projected HFB states [51]. Within the EDF context, the MR step necessitates a prescription to extend the SR energy functional[4] associated with a single auxiliary state of reference to the

---

[4] I.e. the density-dependence of the effective Hamilton operator in the traditional formulation.

Fig. 7.1 *Upper panel*: halo parameter δR_{halo} [57] extracted for nearly five hundreds (predicted) spherical nuclei using the SLy4 [58] Skyrme parametrization. *Lower panel*: halo parameter δR_{halo} computed for drip-line chromium isotopes. The halo parameter δR_{halo} quantifies in a model-independent fashion the contribution of the halo structure to the nuclear radius [57]. The *colour scale* refers to a length indicated in Fermi. Large discrepancies in the prediction of the drip-line position and in the extracted halo parameter are obtained from the selection of parametrizations used. Taken from Ref. [59]

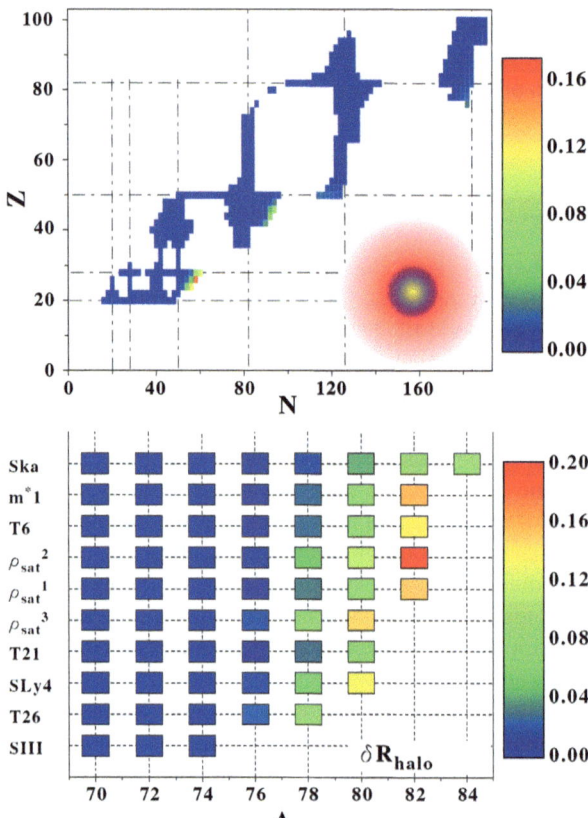

non-diagonal energy kernel associated with a pair of reference states. Although constraints based on physical requirements have been worked out that limit the number of possible prescriptions [52], no first-principle approach to the formulation of such an extension exists today. Although this could have simply remained an academic issue with no measurable consequence, it has been realized recently that the lack of rigorous roots of the EDF method, and in particular of its MR implementation, is responsible for problematic pathologies [53–56].

Modern parametrizations of the nuclear EDF, i.e. Skyrme, Gogny, or relativistic energy functionals, provide a good description of ground-state properties and, to a lesser extent, of spectroscopic features of known nuclei. Still, as of today, EDF parametrizations are phenomenological as they rely on empirically-postulated functional forms whose free coupling constants are adjusted on a selected set of experimental data. This raises questions regarding (i) the connection between currently used EDF parametrizations and elementary AN forces, which is neither explicit nor qualitatively transparent, and regarding (ii) the predictive power of extrapolated EDF results into the experimentally unknown territory. Their lack of microscopic foundation often leads to parametrization-dependent predictions away from known data, i.e. to significant systematic errors, and makes difficult to design systematic improvements. Such a feature is illustrated in Fig. 7.1 for a particular observable of interest related to the prediction of halo structures and the location of the neutron drip-line in medium-mass nuclei [57, 59]. Some systematic limitations of existing EDFs have been empirically identified [60–63] over the last decade that relate to their (too) simple analytical representations and to the biases in their adjustment procedure, as well as to the lack of a solid microscopic foundation. Fuelled by interests in controlled extrapolations of nuclear properties in isospin, density, and temperature, efforts are currently being made to develop energy functionals with substantially reduced errors and improved predictive power. One possible path forward focuses on empirically improving the analytical form and the fitting procedure of existing phenomenological functionals [60, 61, 64–69].

In order to improve on the limitations alluded to above and make EDF calculations truly reliable, several routes must be followed in the future. On the one hand, a better understanding of the foundations of the method and an explicit connection to elementary inter-nucleon interactions must be realized. On the other hand, empirically adjusted parametrizations must rely on advanced fitting and statistical analysis techniques.

7.1.3 Goal of the Present Lecture Notes

The present lecture notes focus on the theoretical foundations of the nuclear energy density functional method. As such, they do not aim at reviewing the status of the field, at covering all possible ramifications of the approach or at presenting recent achievements and applications. For standard reviews that cover the connection to empirical data, we refer the reader to Refs. [48, 49]. In order to achieve our goal within the limits of the present document, the following choices are made in the following

1. the historical perspective is bypassed,
2. the presentation is limited to the non-relativistic framework,
3. time-dependent implementations of the method are not discussed,
4. the Skyrme family of parametrizations is used for illustration,
5. only the *full fledged* multi-reference formalism is discussed,[5]
6. applications are only shown to illustrate points of the formal discussion.

The objective is to provide a *modern* account of the nuclear EDF formalism that is at variance with *traditional* presentations that rely, at one point or another, on a *Hamiltonian-based* picture. The latter is not general enough to encompass what the nuclear EDF method represents as of today. Specifically, the traditional Hamiltonian-based picture does not allow one to grasp the difficulties associated

---

[5] Approximations such as the quasi-particle random phase approximation or the Schroedinger equation based on a collective (e.g. Bohr) Hamiltonian are only mentioned in passing; see Sect. 7.5.7.

with the fact that currently available parametrizations of the energy kernel $E[g', g]$ at play in the method do not derive from a genuine Hamilton operator, would the latter be effective. As such, a key point of the presentation provided below is to demonstrate that the MR-EDF method can indeed be formulated in a *mathematically* meaningful fashion even if $E[g', g]$ does *not* derive from a genuine Hamilton operator. In particular, the restoration of symmetries can be entirely formulated without making *any* reference to a projected state, i.e. within a genuine EDF framework [70]. However, and as will be illustrated below, a mathematically meaningful formulation does not guarantee that the formalism is sound from a *physical* standpoint. We will eventually mention at which price the latter can be ensured as well.

7.2 Prelude

7.2.1 Reference States and Bogoliubov Transformation

The EDF method builds on the *effective* description of a nucleus made of an ensemble of quasi-particles moving independently in their self-created average field(s). As such, the approach relies on the use of product states of Bogoliubov type, which are nothing but a generalization of Slater determinants. To define such many-body states, let us introduce an arbitrary single-particle basis $\{|i\rangle\}$ of the one-body Hilbert space $\mathcal{H}_1$, where $\{i\}$ collects all spatial, spin and isospin quantum numbers necessary to define a given state. Basis states relate to particle creation operators through

$$a_i^\dagger |0\rangle = |i\rangle, \quad (7.1)$$

with $\{a_i, a_j^\dagger\} = \delta_{ij}$. Associated single-particle wave-functions are given by $\psi_i(\vec{r}\sigma\tau) \equiv \langle \vec{r}\sigma\tau | i \rangle$, where σ (τ) denotes the z component of the spin (isospin) $1/2$ nucleon. From there, fully paired Bogoliubov vacua are defined as

$$|\Phi^{(g)}\rangle = \prod_\mu \beta_\mu^{(g)} |0\rangle, \quad (7.2)$$

and carry a collective label g whose definition and meaning will be specified in Sect. 7.2.3.1. Quasi-particle creation and annihilation operators satisfy $\{\beta_\mu^{(g)}, \beta_\nu^{(g)\dagger}\} = \delta_{\mu\nu}$ and relate to particle operators through the so-called Bogoliubov transformation

$$\beta_\mu^{(g)} = \sum_i U_{i\mu}^{(g)*} a_i + V_{i\mu}^{(g)*} a_i^\dagger, \quad (7.3a)$$

$$\beta_\mu^{(g)\dagger} = \sum_i V_{i\mu}^{(g)} a_i + U_{i\mu}^{(g)} a_i^\dagger. \quad (7.3b)$$

7 The Nuclear Energy Density Functional Formalism

Matrices $U^{(g)}$ and $V^{(g)}$, respectively made out of vectors $\mathbf{U}_\mu^{(g)}$ and $\mathbf{V}_\mu^{(g)}$ defined on $\mathscr{H}_1$, combine to make up the matrix representation of the Bogoliubov transformation [51]

$$\mathscr{W}^{(g)} \equiv \begin{pmatrix} U & V^* \\ V & U^* \end{pmatrix}^{(g)} \tag{7.4}$$

whose unitarity provides four identities

$$U^{(g)}U^{(g)\dagger} + V^{(g)*}V^{(g)T} = 1, \tag{7.5a}$$

$$U^{(g)*}V^{(g)T} + V^{(g)}U^{(g)\dagger} = 0, \tag{7.5b}$$

$$U^{(g)\dagger}U^{(g)} + V^{(g)\dagger}V^{(g)} = 1, \tag{7.5c}$$

$$U^{(g)T}V^{(g)} + V^{(g)T}U^{(g)} = 0. \tag{7.5d}$$

Fully paired Bogoliubov states $|\Phi^{(g)}\rangle$ are denoted as "vacua" in the sense that they are annihilated by the set of quasi-particle annihilation operators, i.e.

$$\beta_\mu^{(g)}|\Phi^{(g)}\rangle = 0 \quad \forall \mu. \tag{7.6}$$

Such a notion generalizes the physical vacuum $|0\rangle$, which is annihilated by the set of particle annihilation operators $\{a_i\}$, and Slater determinants that are annihilated by the set of operators $\{a_p, a_h^\dagger\}$, where p (h) denote unoccupied (occupied) single-particle states. Furthermore, Bogoliubov states $|\Phi^{(g)}\rangle$ break particle-number symmetry, i.e. as opposed to Slater determinants they are not eigenstates of the particle (neutron or proton) number operator N. Still, states defined through Eq. (7.2) carry an even number-parity quantum number, i.e. they are linear combinations of eigenstates of N corresponding to even number of particles only. As such, they are appropriate to the description of even-even nuclei. In a more general setting, one may consider Bogoliubov states obtained by performing an even number of quasi-particle excitations on top of a fully paired vacuum or by performing an odd number of such excitations to access odd number-parity states appropriate to the description of odd nuclei [71, 72]. In such a situation, reference states carry an additional label, besides g, to denote the set of quasi-particle excitations that characterizes them.

7.2.2 Elements of Group Theory

The nuclear EDF method relies heavily on breaking and restoring symmetries of the *underlying*, i.e. realistic, nuclear Hamiltonian. As of today, state-of-the-art calculations typically take advantage of breaking translational, rotational and particle-number symmetries, while only restoring the last two. There also exists few calculations treating (solely) the restoration of linear momentum [73]. In order to tackle such a key aspect of the method, let us introduce basic elements of group theory.

We consider the symmetry group $\mathscr{G}$ of the nuclear Hamiltonian H. Because it is the case for the most relevant symmetries, we consider $\mathscr{G}$ to be a continuous, possibly non-abelian, compact Lie group $\mathscr{G} = \{R(\alpha)\}$ parametrized by a set of r real parameters $\alpha \equiv \{\alpha_i \in D_i; i = 1, \ldots, r\}$ defined over a domain of definition $D_{\mathscr{G}} \equiv \{D_i; i = 1, \ldots, r\}$. We thus have $[R(\alpha), H] = 0$ for any $R(\alpha) \in \mathscr{G}$. The invariant measure on $\mathscr{G}$ is defined as $dm(\alpha)$ and its volume is given by

$$v_{\mathscr{G}} \equiv \int_{D_{\mathscr{G}}} dm(\alpha). \tag{7.7}$$

Next, we introduce the set of infinitesimal generators $\vec{C} = \{C_i; i = 1, \ldots, r\}$ that make up the Lie algebra and in terms of which any transformation $R(\alpha)$ of the group can be expressed via an exponential map $R_{\vec{C}}(\alpha)$.

We further consider irreducible representations (Irreps) $S_{ab}^\lambda(\alpha)$ of the group labelled by eigenvalues λ of the Casimir operator Λ. Irreducible representations of dimension d_λ are spanned by states that are also eigenstates of one of the generators, e.g. C_1. Indices "a" and "b" in $S_{ab}^\lambda(\alpha)$ refer to the d_λ corresponding eigenvalues. The unitarity of the Irreps, together with the combination law of two successive transformations, can be read off

$$\sum_c S_{ca}^{\lambda*}(\alpha') S_{cb}^\lambda(\alpha) = \sum_c S_{ac}^\lambda(-\alpha') S_{cb}^\lambda(\alpha) = S_{ab}^\lambda(\alpha - \alpha'), \tag{7.8}$$

where arguments $-\alpha$ and $\alpha - \alpha'$ symbolically denote parameters of transformations $R^{-1}(\alpha)$ and $R^{-1}(\alpha')R(\alpha)$, respectively. Additionally, the orthogonality of the Irreps reads

$$\int_{\mathscr{G}} dm(\alpha) S_{ab}^{\lambda*}(\alpha) S_{a'b'}^{\lambda'}(\alpha) = \frac{v_{\mathscr{G}}}{d_\lambda} \delta_{\lambda\lambda'} \delta_{aa'} \delta_{bb'}. \tag{7.9}$$

Any function $f(\alpha)$ defined on $D_{\mathscr{G}}$ can be decomposed over the Irreps of the group according to

$$f(\alpha) \equiv \sum_{\lambda ab} f_{ab}^\lambda S_{ab}^\lambda(\alpha), \tag{7.10}$$

which defines the set of expansion coefficients $\{f_{ab}^\lambda\}$.

Later on, we wish to apply above considerations to two groups of particular interest, i.e. the abelian group $U(1)$ associated with particle-number symmetry and the non-abelian group $SO(3)$ associated with rotational symmetry. The relevant elements and equations at play for each of these two cases can be deduced from above using correspondence Table 7.1. In the case of $U(1)$, decomposition (7.10) of a function $f(\varphi)$ defined on $D_{U(1)} = [0, 2\pi]$, i.e. its Fourier expansion, reads

$$f(\varphi) \equiv \sum_m f^m e^{im\varphi}. \tag{7.11}$$

7 The Nuclear Energy Density Functional Formalism

Table 7.1 Characteristics of $SO(3)$ and $U(1)$ relevant to the present study. The gauge angle parametrizing $U(1)$ is $\varphi \in [0, 2\pi]$ whereas Euler angles parameterizing $SO(3)$ are $\Omega \equiv (\alpha, \beta, \gamma) \in [0, 4\pi] \times [0, \pi] \times [0, 2\pi]$. One-dimensional Irreps of $U(1)$ are labeled by $m \in \mathbb{Z}$ whereas $(2J+1)$-dimensional Irreps of $SO(3)$ are labeled by $2J \in \mathbb{N}$ and are given by the so-called Wigner functions $\mathscr{D}^J_{MK}(\Omega)$ [74], where $(2M, 2K) \in \mathbb{Z}^2$ with $-2J \leq 2M, 2K \leq +2J$

$\mathscr{G}$	α	$dm(\alpha)$	$v_\mathscr{G}$	$\vec{C}$	Λ	C_1	$R_{\vec{C}}(\alpha)$	$S^\lambda_{ab}(\alpha)$	d_λ
$U(1)$	φ	$d\varphi$	2π	N	N^2	–	$e^{iN\varphi}$	$e^{im\varphi}$	1
$SO(3)$	α, β, γ	$\sin\beta\, d\alpha\, d\beta\, d\gamma$	$16\pi^2$	$\vec{J}$	J^2	J_z	$e^{-i\alpha J_z}e^{-i\beta J_y}e^{-i\gamma J_z}$	$\mathscr{D}^J_{MK}(\Omega)$	$2J+1$

Similarly, the decomposition of a function $f(\Omega)$ defined on $D_{SO(3)} = [0, 4\pi] \times [0, \pi] \times [0, 2\pi]$ over Irreps of $SO(3)$ reads

$$f(\Omega) \equiv \sum_{JMK} f^J_{MK} \mathscr{D}^J_{MK}(\Omega), \tag{7.12}$$

where $\mathscr{D}^J_{MK}(\Omega)$ denotes the so-called Wigner function [74].

7.2.3 Collective Variable and Symmetry Breaking

7.2.3.1 Order Parameters

Whenever $|\Phi^{(g)}\rangle$ breaks a symmetry of the nuclear Hamiltonian, it does not carry the associated symmetry quantum number(s). The three main symmetries considered here lead to loosing good total linear momentum $\vec{P}$, total angular momentum (J^2, J_z) and neutron/proton N/Z quantum numbers. Doing so, $|\Phi^{(g)}\rangle$ acquires non-zero order parameters, i.e. one per broken symmetry, which we group under the generic notation $g \equiv |g|e^{i\alpha} \equiv \langle \Phi^{(g)}|G|\Phi^{(g)}\rangle$, where G is an appropriate operator whose average value in a symmetry conserving state is zero. The norm $|g|$ of the order parameter tracks the extent to which $|\Phi^{(g)}\rangle$ breaks the symmetry, i.e. its "deformation", whereas the phase $\alpha = \text{Arg}(g)$ characterizes the orientation of the deformed body with respect to the chosen reference frame.[6] In the present study, order parameters associated with the breaking of translational, rotational and particle-number symmetries should be specified. As only the latter two are effectively restored in state-of-the-art calculations, Table 7.2 provides the order parameters used to track the breaking of $U(1)$ and $SO(3)$ symmetries. As $|g|$ must be zero/non-zero for good/broken symmetry states, the anomalous density[7] κ^{gg} (see Eqs. (7.19b)–(7.19c)) is a good candidate for $U(1)$. For $SO(3)$, one uses multipole moments $\rho_{\lambda\mu}$ of the matter density distribution $\rho^{gg}_0(\vec{r})$ (see Eqs. (7.26a)–(7.26f))

---

[6] For certain symmetries, e.g. $SO(3)$, the phase α collects in fact several angles. See Table 7.1 for two relevant examples.

[7] Although it can be done rigorously, we do not state explicitly here the definition of the norm of κ.

Table 7.2 Norm and phase of the order parameters associated with broken $U(1)$ and $SO(3)$ symmetries

$\mathcal{G}$	$\|g\|$	$\alpha = \mathrm{Arg}(g)$
$U(1)$	$\|\kappa\|$	φ
$SO(3)$	$\rho_{\lambda\mu}\ (\lambda > 2J)$	α, β, γ

with $\lambda > 2J$ [75]. As for $U(1)$ the phase $\alpha = \mathrm{Arg}(g)$ provides the orientation φ of κ^{gg} in gauge space, while for $SO(3)$ it gives the orientation $\Omega \equiv (\alpha, \beta, \gamma)$ of the deformed density distribution in real space.

7.2.3.2 Symmetry-Breaking Reference State

Eventually, states $|\Phi^{(g)}\rangle$ that are typically dealt with in state-of-the-art calculations can be written in full glory as

$$\left|\Phi^{(\rho_{\lambda\mu}\Omega;\|\kappa_p\|\varphi_p;\|\kappa_n\|\varphi_n)}\right\rangle \equiv R_{\bar{J}}(\Omega) R_N(\varphi_n) R_Z(\varphi_p) \left|\Phi^{(\rho_{\lambda\mu}0;\|\kappa_p\|0;\|\kappa_n\|0)}\right\rangle, \quad (7.13)$$

where the breaking of $U(1)$ appears once for protons (φ_p) and once for neutrons (φ_n). Equation (7.13) indicates that the state corresponding to a finite value of the phase α, i.e. to a given orientation of the "deformed" body, can be obtained from the one at $\alpha = 0$ through the application of the rotation operator

$$\left|\Phi^{(g)}\right\rangle \equiv R(\alpha)\left|\Phi^{(\|g\|0)}\right\rangle. \quad (7.14)$$

7.3 Energy and Norm Kernels

The basic inputs to the nuclear EDF method take the form of the so-called off-diagonal energy and norm kernels

$$E[g', g] \equiv E\left[\langle\Phi^{(g')}|, |\Phi^{(g)}\rangle\right], \quad (7.15a)$$

$$N[g', g] \equiv \langle\Phi^{(g')}|\Phi^{(g)}\rangle, \quad (7.15b)$$

that define quantities associated with two product states $|\Phi^{(g)}\rangle$ and $|\Phi^{(g')}\rangle$ possibly carrying different values of the order parameters.

7.3.1 Norm Kernel

The definition of the norm kernel in Eq. (7.15b) is fully explicit and does not pose any problem. However, the actual computation of both its phase and its norm has posed a great challenge to nuclear theorists over the years. It is only recently that a method to compute $N[g', g]$ unambiguously in terms of Pfaffian was proposed [76]. This constitutes a rather involved technical discussion that goes beyond the scope of the present lecture notes. We refer the interested readers to Refs. [76–80].

7.3.2 Energy Kernel

The energy kernel $E[g', g]$ is postulated under the form of a general, possibly complicated, functional of $|\Phi^{(g')}\rangle$ and $|\Phi^{(g)}\rangle$. Such a feature lies at the heart of the EDF approach as a way to effectively sum up the bulk of many-body correlations. Having no a priori knowledge of the most appropriate functional, one must at least constrain it to fulfil a minimal set [52, 81] of basic properties.

The first requirement states that transforming both $|\Phi^{(g')}\rangle$ and $|\Phi^{(g)}\rangle$ via any element $R(\alpha'') \in \mathscr{G}$ must leave the kernel invariant, i.e.

$$E[\langle\Phi^{(g')}|R^\dagger(\alpha''), R(\alpha'')|\Phi^{(g)}\rangle] = E[\langle\Phi^{(g')}|, |\Phi^{(g)}\rangle], \quad (7.16)$$

which is equivalent to demanding that the kernel only depends on the *difference* of phases of the order parameters labelling the two states, i.e.

$$E[|g'|\alpha', |g|\alpha] = E[|g'|0, |g|\alpha - \alpha']. \quad (7.17)$$

Such a property is necessary and sufficient to ensure later on that the energy is real and independent of the reference frame.

Other requirements relate to the behaviour of the kernel in the limit where $|\Phi^{(g')}\rangle$ and $|\Phi^{(g)}\rangle$ are "close" to each other. In case diagonal and off-diagonal kernels were to be defined through separate means, one must first ensure that they are consistent, i.e. one must ensure that the former is obtained from the latter when taking $|\Phi^{(g')}\rangle = |\Phi^{(g)}\rangle$. Probing the kernel in the vicinity of the diagonal, one further requires that (i) the chemical potentials λ_N and λ_Z obtained through SR calculations are consistent with their extraction from the Kamlah expansion [82] of the particle number restored MR energy and that (ii) the quasi-particle random-phase approximation is recovered from the most general MR scheme whenever $|\Phi^{(g')}\rangle$ and $|\Phi^{(g)}\rangle$ differ harmonically from a common reference state [83, 84]. The latter two requirements are fulfilled [52, 81] if, and only if, $E[\langle\Phi^{(g')}|, |\Phi^{(g)}\rangle]$ does indeed only depend on the bra $\langle\Phi^{(g')}|$ and on the ket $|\Phi^{(g)}\rangle$, as was so far implied by the notation used.

It happens that a sufficient condition for all above properties to be fulfilled is to postulate that the off-diagonal energy kernel is a functional

$$E[g', g] \equiv E[\rho^{g'g}, \kappa^{g'g}, \kappa^{gg'*}], \quad (7.18)$$

in the mathematical sense, of normal and anomalous one-body transition (i.e. off-diagonal) density matrices computed from $\langle\Phi^{(g')}|$ and $|\Phi^{(g)}\rangle$, respectively defined through

$$\rho^{g'g}_{ij} \equiv \frac{\langle\Phi^{(g')}|a^\dagger_j a_i|\Phi^{(g)}\rangle}{\langle\Phi^{(g')}|\Phi^{(g)}\rangle}, \quad (7.19a)$$

$$\kappa^{g'g}_{ij} \equiv \frac{\langle\Phi^{(g')}|a_j a_i|\Phi^{(g)}\rangle}{\langle\Phi^{(g')}|\Phi^{(g)}\rangle}, \quad (7.19b)$$

$$\kappa_{ij}^{gg'*} \equiv \frac{\langle \Phi^{(g')}|a_i^\dagger a_j^\dagger|\Phi^{(g)}\rangle}{\langle \Phi^{(g')}|\Phi^{(g)}\rangle}. \tag{7.19c}$$

One observes that $\rho_{ij}^{g'g*} = \rho_{ji}^{gg'}$, $\kappa_{ij}^{g'g} = -\kappa_{ji}^{g'g}$ and $\kappa_{ij}^{gg'*} = -\kappa_{ji}^{gg'*}$, i.e. the two anomalous densities are antisymmetric whereas the normal density matrix is hermitian whenever $g = g'$.

7.3.3 Pseudo-potential-based Energy Kernel

A particular implementation of the EDF method consists of deriving the EDF kernel from a pseudo Hamiltonian

$$\begin{aligned}
H_{\text{pseudo}} &\equiv \sum_{ij} t_{ij}^{1N\text{pseudo}} a_i^\dagger a_j \\
&+ \left(\frac{1}{2!}\right)^2 \sum_{ijkl} \bar{v}_{ijkl}^{2N\text{pseudo}} a_i^\dagger a_j^\dagger a_l a_k \\
&+ \left(\frac{1}{3!}\right)^2 \sum_{ijklmn} \bar{v}_{ijklmn}^{3N\text{pseudo}} a_i^\dagger a_j^\dagger a_k^\dagger a_n a_m a_l + \cdots, \tag{7.20}
\end{aligned}$$

where $t^{1N\text{pseudo}}$ embodies an effective one-body kinetic energy operator while $\bar{v}_{ijkl}^{AN\text{pseudo}}$ denotes antisymmetrized matrix-elements of a A-body pseudo-potential, i.e. of a A-body effective interaction. The word "pseudo" refers to the fact that operators entering Eq. (7.20) are not the same as the elementary operators entering ab-initio theories; e.g. $v^{AN\text{pseudo}}$ should not be confused with realistic AN interactions. Eventually, H_{pseudo} is only to be seen as a mere intermediary used to generate the fundamental ingredient of the theory, i.e. the off-diagonal energy kernel. In such a context, the latter is computed through

$$\begin{aligned}
E_{\text{pseudo}}[g', g] &\equiv \frac{\langle \Phi^{(g')}|H_{\text{pseudo}}|\Phi^{(g)}\rangle}{\langle \Phi^{(g')}|\Phi^{(g)}\rangle} \tag{7.21a} \\
&= \sum_{ij} t_{ij}^{1N} \rho_{ij}^{g'g} \tag{7.21b} \\
&+ \frac{1}{2} \sum_{ijkl} \bar{v}_{ijkl}^{2N\text{pseudo}} \rho_{ki}^{g'g} \rho_{lj}^{g'g} \\
&+ \frac{1}{6} \sum_{ijklmn} \bar{v}_{ijklmn}^{3N\text{pseudo}} \rho_{li}^{g'g} \rho_{mj}^{g'g} \rho_{nk}^{g'g} + \cdots
\end{aligned}$$

$$+ \frac{1}{4} \sum_{ijkl} \bar{v}_{ijkl}^{2N\text{pseudo}} \kappa_{ij}^{gg'*} \kappa_{kl}^{g'g}$$

$$+ \frac{1}{4} \sum_{ijklmn} \bar{v}_{ijklmn}^{3N\text{pseudo}} \kappa_{ij}^{gg'*} \kappa_{lm}^{g'g} \rho_{nk}^{g'g} + \cdots$$

$$\equiv E_{\text{pseudo}} \left[\rho^{g'g}, \kappa^{g'g}, \kappa^{gg'*} \right], \quad (7.21c)$$

and is indeed a functional of one-body transition density matrices in virtue of the generalized (i.e. off-diagonal) Wick theorem [85]. As long as H_{pseudo} possesses the same symmetries as the underlying nuclear Hamiltonian, Eq. (7.16) is automatically fulfilled for any $R(\alpha'') \in \mathcal{G}$.

7.3.4 Skyrme Parametrization

We now introduce a particular family of EDF parametrizations in view of illustrating some of the points alluded to in the previous section. The Skyrme parametrization[8] is a local energy functional, i.e. it is expressed as a single integral in coordinate space of a local energy density involving a set of local densities derived from the density matrices introduced in Eqs. (7.19a)–(7.19c).

7.3.4.1 Local Densities

Introducing the creation $a^\dagger(\vec{r}\sigma\tau)$ and annihilation $a(\vec{r}\sigma\tau)$ operators in the coordinate representation

$$a(\vec{r}\sigma\tau) \equiv \sum_i \varphi_i(\vec{r}\sigma\tau) a_i, \quad (7.22a)$$

$$a^\dagger(\vec{r}\sigma\tau) \equiv \sum_i \varphi_i^*(\vec{r}\sigma\tau) a_i^\dagger, \quad (7.22b)$$

one obtains the transition density matrices in that representation

$$\rho^{g'g}(\vec{r}\sigma\tau, \vec{r}'\sigma'\tau') \equiv \frac{\langle \Phi^{(g')} | a^\dagger(\vec{r}'\sigma'\tau') a(\vec{r}\sigma\tau) | \Phi^{(g)} \rangle}{\langle \Phi^{(g')} | \Phi^{(g)} \rangle} = \sum_{ij} \varphi_j^*(\vec{r}'\sigma'\tau') \varphi_i(\vec{r}\sigma\tau) \rho_{ij}^{g'g},$$

$$\kappa^{g'g}(\vec{r}\sigma\tau, \vec{r}'\sigma'\tau') \equiv \frac{\langle \Phi^{(g')} | a(\vec{r}'\sigma'\tau') a(\vec{r}\sigma\tau) | \Phi^{(g)} \rangle}{\langle \Phi^{(g')} | \Phi^{(g)} \rangle} = \sum_{ij} \varphi_j(\vec{r}'\sigma'\tau') \varphi_i(\vec{r}\sigma\tau) \kappa_{ij}^{g'g}.$$

---

[8]Coulomb and center-of-mass correction contributions are omitted here for simplicity.

Further considering spin Pauli matrices[9]

$$\sigma_x \equiv \begin{pmatrix} 0 & 1 \\ 1 & 0 \end{pmatrix}, \quad \sigma_y \equiv \begin{pmatrix} 0 & -i \\ i & 0 \end{pmatrix}, \quad \sigma_z \equiv \begin{pmatrix} 1 & 0 \\ 0 & -1 \end{pmatrix}, \quad (7.24)$$

a set of non-local densities containing up to two gradients can be defined

$$\rho_\tau^{g'g}(\vec{r}, \vec{r}') \equiv \sum_\sigma \rho^{g'g}(\vec{r}\sigma\tau, \vec{r}'\sigma\tau), \qquad (7.25a)$$

$$s_{\tau,\nu}^{g'g}(\vec{r}, \vec{r}') \equiv \sum_{\sigma'\sigma} \rho^{g'g}(\vec{r}\sigma\tau, \vec{r}'\sigma'\tau)\langle\sigma'|\sigma_\nu|\sigma\rangle, \qquad (7.25b)$$

$$\tilde{\rho}_\tau^{g'g}(\vec{r}, \vec{r}') \equiv \sum_\sigma 2\bar{\sigma}\kappa^{g'g}(\vec{r}\sigma\tau, \vec{r}'\bar{\sigma}\tau), \qquad (7.25c)$$

$$\tilde{s}_{\tau,\nu}^{g'g}(\vec{r}, \vec{r}') \equiv \sum_{\sigma'\sigma} 2\bar{\sigma}'\kappa^{g'g}(\vec{r}\sigma\tau, \vec{r}'\bar{\sigma}'\tau)\langle\sigma'|\sigma_\nu|\sigma\rangle, \qquad (7.25d)$$

$$\tau_\tau^{g'g}(\vec{r}, \vec{r}') \equiv \sum_\mu \nabla_{\vec{r},\mu} \nabla_{\vec{r}',\mu} \rho_\tau^{g'g}(\vec{r}, \vec{r}'), \qquad (7.25e)$$

$$T_{\tau,\nu}^{g'g}(\vec{r}, \vec{r}') \equiv \sum_\mu \nabla_{\vec{r},\mu} \nabla_{\vec{r}',\mu} s_{\tau,\nu}^{g'g}(\vec{r}, \vec{r}'), \qquad (7.25f)$$

$$\tilde{\tau}_\tau^{g'g}(\vec{r}, \vec{r}') \equiv \sum_\mu \nabla_{\vec{r},\mu} \nabla_{\vec{r}',\mu} \tilde{\rho}_\tau^{g'g}(\vec{r}, \vec{r}'), \qquad (7.25g)$$

$$\tilde{T}_{\tau,\nu}^{g'g}(\vec{r}, \vec{r}') \equiv \sum_\mu \nabla_{\vec{r},\mu} \nabla_{\vec{r}',\mu} \tilde{s}_{\tau,\nu}^{g'g}(\vec{r}, \vec{r}'), \qquad (7.25h)$$

$$j_{\tau,\mu}^{g'g}(\vec{r}, \vec{r}') \equiv -\frac{i}{2}(\nabla_{\vec{r},\mu} - \nabla_{\vec{r}',\mu}) \rho_\tau^{g'g}(\vec{r}, \vec{r}'), \qquad (7.25i)$$

$$J_{\tau,\mu\nu}^{g'g}(\vec{r}, \vec{r}') \equiv -\frac{i}{2}(\nabla_{\vec{r},\mu} - \nabla_{\vec{r}',\mu}) s_{\tau,\nu}^{g'g}(\vec{r}, \vec{r}'), \qquad (7.25j)$$

$$\tilde{j}_{\tau,\mu}^{g'g}(\vec{r}, \vec{r}') \equiv -\frac{i}{2}(\nabla_{\vec{r},\mu} - \nabla_{\vec{r}',\mu}) \tilde{\rho}_\tau^{g'g}(\vec{r}, \vec{r}'), \qquad (7.25k)$$

$$\tilde{J}_{\tau,\mu\nu}^{g'g}(\vec{r}, \vec{r}') \equiv -\frac{i}{2}(\nabla_{\vec{r},\mu} - \nabla_{\vec{r}',\mu}) \tilde{s}_{\tau,\nu}^{g'g}(\vec{r}, \vec{r}'), \qquad (7.25l)$$

where $\vec{\nabla}_{\vec{r}}$ denotes the gradient acting on coordinate $\vec{r}$ while $\bar{\sigma} \equiv -\sigma$. Greek indexes refer to cartesian components of a vector (μ) or a tensor (μ, ν). Densities without Greek index such as $\rho_\tau^{g'g}$, $\tilde{\rho}_\tau^{g'g}$ are scalar densities. Equations (7.25a)–(7.25l) provide non-local matter, spin, pair, pair-spin, kinetic, spin-kinetic, pair-kinetic, pair-

---

[9]Proton/neutron mixing is presently ignored such that $\rho^{g'g}(\vec{r}\sigma\tau, \vec{r}'\sigma'\tau') = \kappa^{g'g}(\vec{r}\sigma\tau, \vec{r}'\sigma'\tau') = 0$ for $\tau \neq \tau'$. This does not correspond to the most general situation [86].

spin-kinetic, current, spin-current, pair-current and pair-spin-current densities for a given isospin projection, respectively.

Eventually, corresponding local densities are trivially obtained through

$$\rho_\tau^{g'g}(\vec{r}) \equiv \rho_\tau^{g'g}(\vec{r},\vec{r}), \qquad s_{\tau,\mu}^{g'g}(\vec{r}) \equiv s_{\tau,\mu}^{g'g}(\vec{r},\vec{r}), \qquad (7.26a)$$

$$\tilde{\rho}_\tau^{g'g}(\vec{r}) \equiv \tilde{\rho}_\tau^{g'g}(\vec{r},\vec{r}), \qquad \tilde{s}_{\tau,\mu}^{g'g}(\vec{r}) \equiv \tilde{s}_{\tau,\mu}^{g'g}(\vec{r},\vec{r}), \qquad (7.26b)$$

$$\tau_\tau^{g'g}(\vec{r}) \equiv \tau_\tau^{g'g}(\vec{r},\vec{r}), \qquad T_{\tau,\mu}^{g'g}(\vec{r}) \equiv T_{\tau,\mu}^{g'g}(\vec{r},\vec{r}), \qquad (7.26c)$$

$$\tilde{\tau}_\tau^{g'g}(\vec{r}) \equiv \tilde{\tau}_\tau^{g'g}(\vec{r},\vec{r}), \qquad \tilde{T}_{\tau,\mu}^{g'g}(\vec{r}) \equiv \tilde{T}_{\tau,\mu}^{g'g}(\vec{r},\vec{r}), \qquad (7.26d)$$

$$j_{\tau,\mu}^{g'g}(\vec{r}) \equiv j_{\tau,\mu}^{g'g}(\vec{r},\vec{r}), \qquad J_{\tau,\mu\nu}^{g'g}(\vec{r}) \equiv J_{\tau,\mu\nu}^{g'g}(\vec{r},\vec{r}), \qquad (7.26e)$$

$$\tilde{j}_{\tau,\mu}^{g'g}(\vec{r}) \equiv \tilde{j}_{\tau,\mu}^{g'g}(\vec{r},\vec{r}), \qquad \tilde{J}_{\tau,\mu\nu}^{g'g}(\vec{r}) \equiv \tilde{J}_{\tau,\mu\nu}^{g'g}(\vec{r},\vec{r}). \qquad (7.26f)$$

Considering neutron-neutron and proton-proton pairing only, densities $\tilde{s}_{\tau,\nu}^{g'g}$, $\tilde{T}_{\tau,\nu}^{g'g}$ and $\tilde{j}_{\tau,\mu}^{g'g}$ are null [86]. We finally introduce the spin-orbit current as the pseudo-vector part of the spin-orbit tensor

$$J_{\tau,\lambda}^{g'g}(\vec{r}) \equiv \sum_{\mu\nu} \varepsilon_{\lambda\mu\nu} J_{\tau,\mu\nu}^{g'g}(\vec{r}). \qquad (7.27)$$

7.3.4.2 Energy Kernel

The basic parametrization of the Skyrme energy kernel is a bilinear local functional built out of the above local densities such that each term may contain up to two gradients and two spin Pauli matrices. It is written as

$$E[\rho^{g'g}, \kappa^{g'g}, \kappa^{gg'*}] \equiv \int d\vec{r}\{\mathcal{E}_\rho^{g'g}(\vec{r}) + \mathcal{E}_{\rho\rho}^{g'g}(\vec{r}) + \mathcal{E}_{\kappa\kappa}^{g'g}(\vec{r})\}, \qquad (7.28)$$

where the term linear in the normal density denotes the effective kinetic energy while the terms bilinear in the normal and anomalous density matrices model the effective nuclear interaction energy. Suppressing the spatial argument $\vec{r}$ for simplicity, the three contributions to the local energy density read

$$\mathcal{E}_\rho^{g'g} = \frac{\hbar^2}{2m} \sum_\tau \tau_\tau^{g'g}, \qquad (7.29a)$$

$$\mathcal{E}_{\rho\rho}^{g'g} = \sum_{\tau\tau'} \Big[C_{\tau\tau'}^{\rho\rho} \rho_\tau^{g'g} \rho_{\tau'}^{g'g} + C_{\tau\tau'}^{\rho\Delta\rho} \rho_\tau^{g'g} \Delta\rho_{\tau'}^{g'g} + C_{\tau\tau'}^{\rho\tau} (\rho_\tau^{g'g} \tau_{\tau'}^{g'g} - \vec{j}_\tau^{g'g} \cdot \vec{j}_{\tau'}^{g'g})$$

$$+ C_{\tau\tau'}^{ss} \vec{s}_\tau^{g'g} \cdot \vec{s}_{\tau'}^{g'g} + C_{\tau\tau'}^{s\Delta s} \vec{s}_\tau^{g'g} \cdot \Delta \vec{s}_{\tau'}^{g'g}$$

$$+ C_{\tau\tau'}^{\rho\nabla J} \big(\rho_\tau^{g'g} \vec{\nabla} \cdot \vec{J}_{\tau'}^{g'g} + \vec{j}_\tau^{g'g} \cdot \vec{\nabla} \times \vec{s}_{\tau'}^{g'g}\big)$$

$$+ C_{\tau\tau'}^{J\bar{J}} \left(\sum_{\mu\nu} J_{\tau,\mu\mu}^{g'g} J_{\tau',\nu\nu}^{g'g} + J_{\tau,\mu\nu}^{g'g} J_{\tau',\nu\mu}^{g'g} - 2\vec{s}_{\tau}^{g'g} \cdot \vec{F}_{\tau'}^{g'g} \right)$$

$$+ C_{\tau\tau'}^{JJ} \left(\sum_{\mu\nu} J_{\tau,\mu\nu}^{g'g} J_{\tau',\mu\nu}^{g'g} - \vec{s}_{\tau}^{g'g} \cdot \vec{T}_{\tau'}^{g'g} \right) + C_{\tau\tau'}^{\nabla s \nabla s} \vec{\nabla} \cdot \vec{s}_{\tau}^{g'g} \vec{\nabla} \cdot \vec{s}_{\tau'}^{g'g} \bigg],$$

(7.29b)

$$\mathcal{E}_{\kappa\kappa}^{g'g} = \sum_{\tau} \bigg\{ C_{\tau\tau}^{\tilde{\rho}\tilde{\rho}} \tilde{\rho}_{\tau}^{gg'*} \tilde{\rho}_{\tau}^{g'g} + C_{\tau\tau}^{\tilde{\tau}\tilde{\rho}} \left(\tilde{\rho}_{\tau}^{gg'*} \tilde{\tau}_{\tau}^{g'g} + \tilde{\tau}_{\tau}^{gg'*} \tilde{\rho}_{\tau}^{g'g} + \frac{1}{2} \vec{\nabla} \tilde{\rho}_{\tau}^{gg'*} \cdot \vec{\nabla} \tilde{\rho}_{\tau}^{g'g} \right)$$

$$+ \sum_{\mu\nu} \left(C_{\tau\tau}^{\tilde{J}\tilde{J}1} \tilde{J}_{\tau,\mu\nu}^{gg'*} \tilde{J}_{\tau,\mu\nu}^{g'g} + C_{\tau\tau}^{\tilde{J}\tilde{J}2} \tilde{J}_{\tau,\nu\nu}^{gg'*} \tilde{J}_{\tau,\mu\mu}^{g'g} + C_{\tau\tau}^{\tilde{J}\tilde{J}3} \tilde{J}_{\tau,\nu\mu}^{gg'*} \tilde{J}_{\tau,\mu\nu}^{g'g} \right) \bigg\}. \quad (7.29c)$$

A key feature of expressions (7.29b) and (7.29c) relates to the fact that local densities are not combined arbitrarily to build the various bilinear terms at play. Given $R(\alpha'') \in \mathcal{G}$, one must characterize the transformation law of each local density induced by the transformation of $\langle \Phi^{(g')} |$ and $| \Phi^{(g)} \rangle$ in order to identify which bilinear combinations can be formed to fulfil Eq. (7.16). Such a procedure must be typically conducted for Galilean transformations, rotations in coordinate, gauge and isospin spaces, as well as for a time-reversal transformation. We refer the reader to Refs. [86, 87] for a detailed discussion regarding the constraints generated by Eq. (7.16) on the diagonal energy kernel $E[\rho^{gg}, \kappa^{gg}, \kappa^{gg*}]$. To give a taste of the constraints at play, let us however exemplify the situation by briefly discussing four transformations of interest.

Fulfilling Eq. (7.16) under Galilean transformations leads to the necessity to *group* several bilinear terms together, i.e. only the sum of terms grouped in between parenthesis in Eqs. (7.29a)–(7.29c) are invariant. This per se reduces the number of free coupling constants entering the EDF kernel. Turning to space rotations, the set of local densities transform according to

$$\rho_{\tau}^{\Omega'-\Omega''\Omega-\Omega''}(\vec{r}) = \rho_{\tau}^{\Omega'\Omega} \left(\mathcal{R}^{-1}(\Omega'')\vec{r} \right) \quad (7.30a)$$

$$\tau_{\tau}^{\Omega'-\Omega''\Omega-\Omega''}(\vec{r}) = \tau_{\tau}^{\Omega'\Omega} \left(\mathcal{R}^{-1}(\Omega'')\vec{r} \right) \quad (7.30b)$$

$$\vec{s}_{\tau}^{\Omega'-\Omega''\Omega-\Omega''}(\vec{r}) = \mathcal{R}^{-1}(\Omega'') \vec{s}_{\tau}^{\Omega'\Omega} \left(\mathcal{R}^{-1}(\Omega'')\vec{r} \right) \quad (7.30c)$$

$$\vdots$$

where $\mathcal{R}(\Omega)$ is the 3-dimensional matrix representation of the rotation, i.e. local densities transform according to their scalar, vector or tensor field character. In order to fulfil Eq. (7.16), densities are combined in Eqs. (7.29a)–(7.29c) such that each bilinear term eventually transforms as a scalar field. As result, integrating over $\vec{r}$ provides a scalar independent of $\mathcal{R}^{-1}(\Omega'')$. Although the realistic nuclear Hamiltonian contains a slight breaking of isospin invariance and of isospin symmetry, only the latter can anyway be characterized in a functional that does not mix protons and neutrons. Enforcing it requires that $C_{nn}^{ff'} = C_{pp}^{ff'}$ and $C_{np}^{ff'} = C_{pn}^{ff'}$. Last but not least,

7 The Nuclear Energy Density Functional Formalism

fulfilling Eq. (7.16) under a rotation in gauge space does not impose any constraint on the part of the EDF kernel that depends on the normal density matrix $\rho^{g'g}$ but imposes that anomalous densities enter under the form of bilinear products of the form $\kappa^{gg'*}\kappa^{g'g}$, which is indeed the case of each term appearing in Eq. (7.29c).

7.3.4.3 Pseudo-potential-based Kernel

Let us now illustrate the pseudo-potential based approach within the Skyrme family of parametrizations. To make the discussion transparent, we simplify it by considering a toy two-body Skyrme pseudo-potential, i.e. the operators considered in Eq. (7.20) are

$$t^{1N\text{pseudo}} \equiv -\frac{\hbar^2}{2m}\delta(\vec{r}_1 - \vec{r}_2)\Delta, \tag{7.31a}$$

$$v^{2N\text{pseudo/toy}} \equiv t_0(1 - P_\sigma)\delta(\vec{r}_1 - \vec{r}_2), \tag{7.31b}$$

where $P_\sigma \equiv (1+\sigma_1\cdot\sigma_2)/2$ is the two-body spin-exchange operator. Further neglecting isospin for simplicity, the EDF kernel computed through Eqs. (7.21a)–(7.21c) can be put under the form

$$E^{\text{toy}}_{\text{pseudo}}\left[\rho^{g'g}, \kappa^{g'g}, \kappa^{gg'*}\right] \equiv \int d\vec{r}\left[\frac{\hbar^2}{2m}\tau^{g'g}(\vec{r}) + A^{\rho\rho}\rho^{g'g}(\vec{r})\rho^{g'g}(\vec{r})\right.$$
$$\left. + A^{ss}\vec{s}^{g'g}(\vec{r})\cdot\vec{s}^{g'g}(\vec{r}) + A^{\tilde{\rho}\tilde{\rho}}\tilde{\rho}^{gg'*}(\vec{r})\tilde{\rho}^{g'g}(\vec{r})\right]. \tag{7.32}$$

In Eq. (7.32), functional coefficients $A^{\rho\rho}$, A^{ss} and $A^{\tilde{\rho}\tilde{\rho}}$ are related to the free parameter t_0 entering the pseudo potential through

$$A^{\rho\rho} = -A^{ss} = \frac{t_0}{2}, \tag{7.33a}$$

$$A^{\rho\rho} = +A^{\tilde{\rho}\tilde{\rho}} = \frac{t_0}{2}, \tag{7.33b}$$

and are thus interrelated.

If we now come back to the generic Skyrme parametrization (7.29a)–(7.29c), it is possible to identify the reduced form that formally matches the above pseudo-potential-based toy functional. It obviously reads

$$E^{\text{toy}}\left[\rho^{g'g}, \kappa^{g'g}, \kappa^{gg'*}\right] \equiv \int d\vec{r}\left[\frac{\hbar^2}{2m}\tau^{g'g}(\vec{r}) + C^{\rho\rho}\rho^{g'g}(\vec{r})\rho^{g'g}(\vec{r})\right.$$
$$\left. + C^{ss}\vec{s}^{g'g}(\vec{r})\cdot\vec{s}^{g'g}(\vec{r}) + C^{\tilde{\rho}\tilde{\rho}}\tilde{\rho}^{gg'*}(\vec{r})\tilde{\rho}^{g'g}(\vec{r})\right], \tag{7.34}$$

and looks indeed formally identical to Eq. (7.32). Still, crucial differences exist between the two. Contrarily to the pseudo-potential-based approach, parameters $C^{\rho\rho}$, C^{ss} and $C^{\tilde{\rho}\tilde{\rho}}$ are not a priori interrelated in the general EDF approach.[10] Such a feature comes from the fact that the functional is postulated rather than computed as the matrix element of an operator. Interrelations between the functional couplings entering a pseudo-potential based EDF kernel are a manifestation of Pauli's principle that is automatically enforced by definition (7.21a). On the contrary, Pauli's principle is violated in the more general approach to the EDF kernel. Let us now try to illustrate such a key point more transparently.

The energy kernel can always be expressed under the generic form (7.21a)–(7.21c), *as long as its dependence on transition densities is polynomial*, which is the case of the above toy functionals. For the local Skyrme parametrization, this is achieved by expanding local densities according to

$$f^{g'g}_{\tau}(\vec{r}) \equiv \sum_{ij} W^{f}_{ji}(\vec{r}\tau) \rho^{g'g}_{ij}, \tag{7.35a}$$

$$\tilde{f}^{g'g}_{\tau}(\vec{r}\tau) \equiv \sum_{ij} W^{\tilde{f}}_{ji}(\vec{r}\tau) \kappa^{g'g}_{ij}, \tag{7.35b}$$

where $W^{f}_{ji}(\vec{r}\tau)$ and $W^{\tilde{f}}_{ji}(\vec{r}\tau)$ can be deduced from the definition of the various local densities at play. In the case of toy bilinear functionals (7.32) and (7.34), one finds

$$W^{\rho}_{ji}(\vec{r}) = \varphi^{\dagger}_{j}(\vec{r}) \varphi_{i}(\vec{r}), \tag{7.36a}$$

$$\vec{W}^{\vec{s}}_{ji}(\vec{r}) = \varphi^{\dagger}_{j}(\vec{r}) \vec{\sigma} \varphi_{i}(\vec{r}), \tag{7.36b}$$

$$W^{\tilde{\rho}}_{ji}(\vec{r}) = \sum_{\sigma} \sigma \varphi_{j}(\vec{r}\sigma) \varphi_{i}(\vec{r}\bar{\sigma}), \tag{7.36c}$$

where $\varphi_i(\vec{r})$ $[\varphi^{\dagger}_i(\vec{r})]$ denotes a spinor with components $\varphi_i(\vec{r}\sigma)$ $[\varphi^*_i(\vec{r}\sigma)]$. With such definitions at hand, the effective two-body matrix elements $\bar{v}^{2N\text{toy}}_{ijkl}$ entering Eq. (7.21b) can be extracted in two different ways, i.e. either focusing on the term proportional to $\rho^{g'g}_{ki} \rho^{g'g}_{lj}$ or focusing on the term proportional to $\kappa^{gg'*}_{ij} \kappa^{g'g}_{kl}$, i.e.

$$\bar{v}^{2N\text{toy}\rho\rho}_{ijkl} \equiv 2 \int d\vec{r} \left[B^{\rho\rho} W^{\rho}_{ik}(\vec{r}) W^{\rho}_{jl}(\vec{r}) + B^{ss} \vec{W}^{\vec{s}}_{ik}(\vec{r}) \cdot \vec{W}^{\vec{s}}_{jl}(\vec{r}) \right]$$

$$= 2 \int d\vec{r} \sum_{\sigma\sigma'} \varphi^*_i(\vec{r}\sigma) \varphi^*_j(\vec{r}\sigma') \left[B^{\rho\rho} \varphi_k(\vec{r}\sigma) \varphi_l(\vec{r}\sigma') \right.$$

$$+ B^{ss} \left(\varphi_k(\vec{r}\bar{\sigma}) \varphi_l(\vec{r}\bar{\sigma}') - \bar{\sigma}\bar{\sigma}' \varphi_k(\vec{r}\bar{\sigma}) \varphi_l(\vec{r}\bar{\sigma}') \right.$$

$$\left. \left. + \sigma\sigma' \varphi_k(\vec{r}\sigma) \varphi_l(\vec{r}\sigma') \right) \right], \tag{7.37a}$$

---

[10] In the case of the present toy functional, the fulfilment of Eq. (7.16) under Galilean transformations does not correlate any of the couplings.

7 The Nuclear Energy Density Functional Formalism

$$\bar{v}_{ijkl}^{2N\text{toy}\kappa\kappa} \equiv 4 \int d\vec{r} B^{\tilde{\rho}\tilde{\rho}} W_{ij}^{\tilde{\rho}*}(\vec{r}) W_{kl}^{\tilde{\rho}}(\vec{r})$$

$$= 4 \int d\vec{r} B^{\tilde{\rho}\tilde{\rho}} \sum_{\sigma\sigma'} \bar{\sigma}\bar{\sigma}' \varphi_i^*(\vec{r}\sigma) \varphi_j^*(\vec{r}\bar{\sigma}) \varphi_k(\vec{r}\sigma') \varphi_l(\vec{r}\bar{\sigma}'), \quad (7.37\text{b})$$

with $B^{ff'} \equiv A^{ff'}$ for Eq. (7.32) and $B^{ff'} \equiv C^{ff'}$ for Eq. (7.34). Such an extraction of effective two-body matrix elements[11] is instrumental to pin down the potential violation of Pauli's principle in the EDF kernel.

7.3.4.4 Spurious Self-interaction and Self-pairing Contributions

In the nuclear EDF framework, Pauli's principle is always satisfied at the level of the individual densities given that one-body density matrices are computed from antisymmetric many-body states (Eqs. (7.19a)–(7.19c)). The violation we now wish to briefly discuss may arise when multiplying several such densities together to build the interaction part of the energy kernel.

The first issue relates to the behaviour of $\bar{v}_{ijkl}^{2N\rho\rho}$ in the particular case where $k = l$ (or $i = j$). Pauli's principle requires such effective matrix elements to be zero given that two nucleons occupy the same single-particle state. It is easy to check that $\bar{v}_{ijkk}^{2\text{toy}\rho\rho} = 0$ in Eq. (7.37a) if, and only if, $B^{\rho\rho} = -B^{ss}$, i.e. if the pseudo-potential-based relationship (7.33a) is satisfied. In the general EDF framework, such interrelations between functional parameters are not enforced and Pauli's principle is violated,[12] e.g. $\bar{v}_{ijkk}^{2N\rho\rho} \neq 0$. Such a violation eventually leads to a contamination of the EDF kernel by spurious self-interaction contributions, i.e. part of the interaction energy originates from individual nucleons interacting with themselves [88, 89]. The self-interaction problem has been extensively studied within DFT for electronic systems and has been shown to contaminate significantly many observables, e.g. ionization energies and, thus, the asymptotic of the electronic density distribution [90].

The self-interaction issue does not concern $\bar{v}_{ijkl}^{2N\kappa\kappa}$. Indeed, such a matrix element is multiplied by $\kappa_{ij}^{gg'*}$ and $\kappa_{kl}^{g'g}$ whose antisymmetry ensures that the corresponding contribution to the energy kernel is anyway zero for $i = j$ and/or $k = l$. However, a second issue relates to the link between $\bar{v}_{ijkl}^{2N\rho\rho}$ and $\bar{v}_{ijkl}^{2N\kappa\kappa}$. Equations (7.21a)–(7.21c) suggests that those two sets of matrix elements should be identical. As a matter of fact, it is straightforward to check that $\bar{v}_{ijkl}^{2N\text{toy}\rho\rho} = \bar{v}_{ijkl}^{2N\text{toy}\kappa\kappa}$ if, and only if, $B^{\rho\rho} = -B^{ss} = B^{\tilde{\rho}\tilde{\rho}}$, i.e. if pseudo-potential-based relationships (7.33a) and (7.33b)

---

[11] The present analysis can be easily extended to trilinear functional terms and effective three-body matrix elements.

[12] This encompasses the intermediate case where the EDF kernel is computed as the matrix elements of a *density-dependent* effective "Hamiltonian". Indeed, in such a case no exchange or pairing term corresponding to the density dependence of the effective vertex appears in the EDF kernel.

are satisfied. In the general EDF framework, such interrelations between functional parameters are not a priori enforced and Pauli's principle is violated, e.g. $\bar{v}^{2N\rho\rho} \neq \bar{v}^{2N\kappa\kappa}$. Such a violation eventually leads to a contamination of the EDF kernel by spurious *self-pairing* contributions. The notion of self-pairing was introduced for the first time in Refs. [54, 55] and generalizes the well-known notion of self-interaction.

Within the nuclear context, the contamination of SR results by self-interaction and self-pairing processes has never been characterized. It thus deserves attention in the future. In Sect. 7.5.8, we will however see that such spurious contributions to the energy kernel have already been understood to be responsible for critical pathologies in MR-EDF calculations.

7.3.4.5 Modern Parametrizations

On the one hand, the bilinear form of the Skyrme parametrization given in Eqs. (7.29a)–(7.29c) constitutes the basis of any modern Skyrme parametrization. On the other hand, none of the modern Skyrme parametrizations strictly corresponds to such a form [48, 54, 55]. The most common departures from it relate to the fact that [48]

1. Couplings $C_{\tau\tau'}^{ff'}$ may further depend on a set of local densities in order to enrich the parametrization and provide more flexibility. Of course, such additional density dependences must not jeopardize Eq. (7.16). Common parametrizations are such that $C_{\tau\tau'}^{\rho\rho}$, $C_{\tau\tau'}^{ss}$ and $C_{\tau\tau}^{\tilde{\rho}\tilde{\rho}}$ depend on the *isoscalar* matter density $\rho_0^{g'g}(\vec{r}) \equiv \rho_n^{g'g}(\vec{r}) + \rho_p^{g'g}(\vec{r})$.
2. Specific couplings might be put to zero for (numerical) convenience, simplicity or because of the difficulty to identify empirical data that can help fix their value unambiguously. Typical examples concern $C_{\tau\tau'}^{JJ}$, $C_{\tau\tau'}^{J\tilde{J}}$, $C_{\tau\tau'}^{\nabla_s \nabla_s}$, $C_{\tau\tau}^{\tilde{t}\tilde{\rho}}$ and $C_{\tau\tau}^{\tilde{J}\tilde{J}1/2/3}$.
3. The form of certain terms might be approximated. This is the case of the so-called exchange term originating from the Coulomb interaction (not shown here) that is usually treated in the Slater approximation.

In the very large majority of cases, such deviations from the strict and complete bilinear form constitute a departure from the pseudo-potential based method, independent of whether or not the bilinear baseline was originally derived from a pseudo potential. Consequently, ad hoc modifications of the EDF parametrizations cause or reinforce a breaking of Pauli's principle and induce pathologies associated with it (see Sects. 7.3.4.4 and 7.5.8). Note that the latter statements apply equally to Gogny or relativistic parametrization of the EDF kernel. Still, most of the enrichments of the analytical form of the Skyrme family of parametrizations have been performed along this line in recent years. With no ambition of being exhaustive, let us mention some of the recent attempts at empirically enriching the Skyrme parametrization in order to improve its global performance and/or overcome a specific limitation. Such developments relate to

1. A dependence of $C^{\tilde\rho\tilde\rho}_{\tau\tau}$ on the scalar-isovector density to better reproduce pairing gaps in neutron-rich nuclei and asymmetric nuclear matter [64, 91–94].
2. A dependence of $C^{\rho\rho}_{\tau\tau'}$ and $C^{ss}_{\tau\tau'}$ on vector-isoscalar and vector-isovector densities to control infinite wavelength spin and isospin instabilities of nuclear matter beyond saturation density [95].
3. An enriched dependence of $C^{\rho\rho}_{\tau\tau'}$ on the scalar-isoscalar density to fully decouple the isoscalar effective mass from the compressibility [96, 97].
4. The pairing part of the EDF derived from a *regularized* zero-range two-body pseudo potential with *separable* Gaussian regulators [91, 98, 99] with the goal to have (i) a way to handle a finite-range pairing vertex that is numerically cost efficient and (ii) the possibility to connect to realistic nuclear forces.
5. A density dependence of $C^{\rho\Delta\rho}_{\tau\tau'}$ to produce a surface-peaked effective mass [100, 101].
6. The use of $C^{\rho\nabla J}_{nn} \neq C^{\rho\nabla J}_{pp}$ to offer more flexibility in the reproduction of spin-orbit splittings [102].

Even more recently, an effort towards the construction of new families of EDF parametrizations that derive strictly from a pseudo potential has emerged. This new trend is motivated by the identification of pathologies in MR-EDF calculations that originate from the breaking of Pauli's principle in any of the existing EDF parametrizations (see Sects. 7.3.4.4 and 7.5.8). Associated on-going developments relate to the construction of

1. A bilinear EDF derived from a zero-range Skyrme-like two-body pseudo potential containing up to six gradient operators [66, 103].
2. The complete bilinear and trilinear EDF derived from zero-range Skyrme-like two- and three-body pseudo potentials containing up to two gradient operators [75, 104].
3. A bilinear EDF derived from a *regularized* zero-range Skyrme-like two-body pseudo potential with up to two gradient operators and Gaussian regulators [105, 106].

7.4 Single-Reference Implementation

The single-reference implementation of the nuclear EDF method exclusively invokes the *diagonal* kernel $E[g, g]$. State $|\Phi^{(g)}\rangle$ is entitled to break as many symmetries of the nuclear Hamiltonian as it finds energetically favourable. That a certain symmetry does break spontaneously usually depends on the number of elementary constituents of the system under consideration (see Sect. 7.4.5). As state $|\Phi^{(g)}\rangle$ acquires a finite order parameter g, the diagonal kernel remains independent of its phase α, as schematically pictured in Fig. 7.2. Such a degeneracy derives trivially from Eq. (7.16). Whenever the system does break the symmetry spontaneously, i.e. whenever the minimal energy is obtained for a non zero value of g, the two-dimensional profile of $E[g, g]$ takes the typical form of a "mexican hat".

Fig. 7.2 Schematic view of the diagonal energy kernel $E[g, g]$ as a function of both the phase and the magnitude of the order parameter associated with a spontaneously broken symmetry

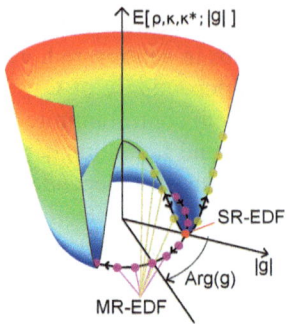

The degeneracy of $E[g, g]$ with respect to α relates to the fact that a spontaneous symmetry breaking at the SR level gives rise to a zero-energy Goldstone mode. One practical consequence is that SR calculations can be performed at any fixed value of α, e.g. $\alpha = 0$.

7.4.1 Equation of Motion

The SR energy is obtained, for a targeted value of $|g|$, through the minimization

$$E^{SR}_{|g|} \equiv \underset{\{|\Phi^{(|g|0)}\rangle\}}{\text{Min}} \{\mathscr{E}_{|g|}\}, \qquad (7.38)$$

within the manifold of (symmetry-breaking) Bogoliubov states. The diagonal energy kernel to be actually minimized reads[13]

$$\mathscr{E}_{|g|} \equiv E[g, g] - \lambda\big[N - \langle\Phi^{(g)}|N|\Phi^{(g)}\rangle\big] - \lambda_{|g|}\big[|g| - |\langle\Phi^{(g)}|G|\Phi^{(g)}\rangle|\big]. \qquad (7.39)$$

The last two terms in Eq. (7.39) introduce Lagrange parameters[14] that are to be adjusted such that the average number of nucleons in $|\Phi^{(|g|0)}\rangle$ is equal to its actual number in the nucleus under study and such that the norm of the order parameter is equal to the desired value $|g|$.

Equations (7.38)–(7.39) lead to solving an equation of motion that takes the form of a constrained Bogoliubov-De Gennes eigenvalue problem[15]

$$\begin{pmatrix} \mathbf{h} - \lambda \mathbf{1} & \mathbf{\Delta} \\ -\mathbf{\Delta}^* & -\mathbf{h}^* + \lambda \mathbf{1} \end{pmatrix}^{(g)} \begin{pmatrix} \mathbf{U} \\ \mathbf{V} \end{pmatrix}^{(g)}_\mu = E^{|g|}_\mu \begin{pmatrix} \mathbf{U} \\ \mathbf{V} \end{pmatrix}^{(g)}_\mu, \qquad (7.40)$$

---

[13] One way to ensure that the minimization is indeed performed within the manyfold of product states consists of adding an additional Lagrange constraint requiring that the *generalized* density matrix [51] $\mathscr{R}$ remains idempotent.

[14] Expressions are given here for linear constraints although practical calculations often rely on quadratic constraints [107].

[15] Depending on the isospin projection τ considered, $\lambda = \lambda_n$ or λ_p.

7 The Nuclear Energy Density Functional Formalism

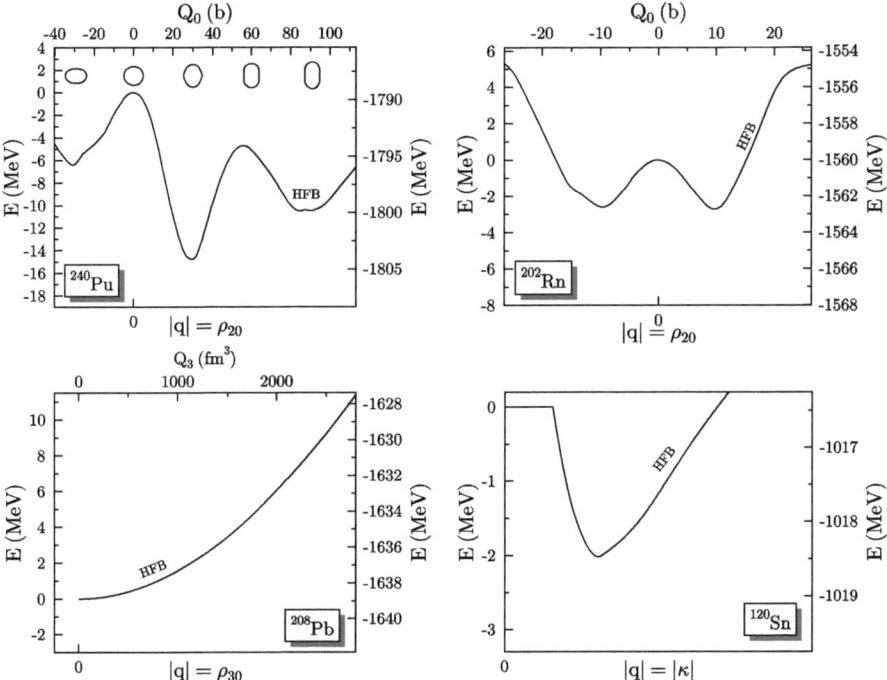

Fig. 7.3 Energy landscapes as a function of the norm of various order parameters [108]. Note that $|q|$ stands for $|g|$ in the figure. *Upper panels*: SR-EDF energy of ^{240}Pu and ^{202}Rn as a function of axial quadrupole deformation ($|g| \equiv \rho_{20}$). *Lower left panel*: SR-EDF energy of ^{208}Pb as a function of axial octupole deformation ($|g| \equiv \rho_{30}$). *Lower right panel*: SR-EDF energy of ^{120}Sn as a function of pairing deformation ($|g| \equiv \|\kappa\|$). Left vertical axes are rescaled with respect to the symmetry conserving, i.e. non-deformed, reference point

which is to be realized iteratively and where the (constrained) one-body fields are defined through functional derivatives of the (modified) diagonal energy kernel

$$\mathbf{h}^{(g)} - \lambda \mathbf{1} \equiv \frac{\delta \mathcal{E}_{|g|}}{\delta \rho^{gg*}}; \quad \mathbf{\Delta}^{(g)} \equiv \frac{\delta \mathcal{E}_{|g|}}{\delta \kappa^{gg*}}. \quad (7.41)$$

The field $\mathbf{h}^{(g)}$ governs the *effective* single-particle motion while the anomalous field $\mathbf{\Delta}^{(g)}$ drives pairing correlations. Explicit expressions of the fields are easily obtained given a specific (e.g. Skyrme) parametrization of the EDF kernel. Equation (7.40) provides the set of quasi-particle energies $E_\mu^{|g|}$ at "deformation" g and the corresponding wave-functions $(\mathbf{U}, \mathbf{V})_\mu^{(g)}$ from which density matrices $\rho^{gg} = \mathbf{V}^{(g)*}\mathbf{V}^{(g)T}$ and $\kappa^{gg} = \mathbf{V}^{(g)*}\mathbf{U}^{(g)T}$, as well as the total energy, can be computed.

The full SR energy landscape, associated with the complete set of reference states $\{|\Phi^{|g|\alpha}\rangle = R(\alpha)|\Phi^{(|g|0)}\rangle; |g| \in [0, +\infty[; \alpha \in D_\mathscr{G}\}$, is generated through repeated calculations performed for various targeted values of $|g|$. The degeneracy of $E[g, g]$ with respect to α makes it unnecessary to solve the equation of motion for $\alpha \neq 0$. As an illustration, Fig. 7.3 displays the energy landscapes associated with various order parameters, i.e. various operators G. First, the energy landscape of ^{240}Pu and ^{202}Rn as a function of axial quadrupole deformation ($|g| \equiv \rho_{20}$) demonstrates that rotational symmetry is spontaneously broken in those nuclei. Second, the energy landscape of ^{208}Pb as a function of axial octupole deformation ($|g| \equiv \rho_{30}$) illustrates that this nucleus is found to remain spherical at the SR-EDF level. Last but not least, the energy landscape of ^{120}Sn as a function of pairing deformation ($|g| \equiv \|\kappa\|$) shows that such a nucleus is superfluid.

The absolute minimum of the SR landscape $E_{GS}^{SR} \equiv \text{Min}_{|g|}\{E_{|g|}^{SR}\}$ provides a first approximation to the ground-state binding energy that incorporates static collective correlations via the breaking of symmetries. Such a solution provides a first approximation to other quantities of interest, e.g. ground-state's charge and matter radii as well as nucleonic density distributions, one-nucleon separation energies and effective single-particle energies (see Sect. 7.4.3), along with individual excitations through an even number of quasi-particle excitations. Using one projection of the angular-momentum vector, e.g. J_x, as the constrain operator gives access to rotational excitations of the nucleus when solving Eq. (7.40) for appropriate values of $\langle \Phi^{(g)} | J_x | \Phi^{(g)} \rangle$. This actually corresponds to using the *velocity* along the phase of the order parameter as a collective degree of freedom.

The full SR landscape provides a richer information. Along the radial direction $|g|$, in particular, the curvature around the minimum characterizes the sensitivity of the system to a change of collective "deformation", whereas the existence of a secondary minimum can be tentatively associated with a shape isomer. Such an analysis is the starting point of the more advanced MR implementation detailed in Sect. 7.5 below.

7.4.2 One-Nucleon Addition and Removal Processes

In the context of SR-EDF calculations, the description of states in the $N \pm 1$ systems rely on Bogoliubov states having the form of one quasi-particle excitations on top of an even number-parity vacuum

$$|\Phi_k^{(g)}\rangle \equiv \beta_k^{(g)\dagger} |\Phi^{(g)}\rangle. \quad (7.42)$$

The even-number parity vacuum being associated with an even-even system, one-nucleon addition and removal energies to final states of the $A \pm 1$ systems are obtained through

$$E_k^{|g|\pm} = \pm\{E[\rho_k^{gg}, \kappa_k^{gg}, \kappa_k^{gg*}] - E[\rho^{gg}, \kappa^{gg}, \kappa^{gg*}]\}$$
$$\mp \lambda\{\langle \Phi_k^{(g)} | N | \Phi_k^{(g)} \rangle - (N \pm 1)\}$$
$$= \lambda \pm E_k^{|g|}, \quad (7.43)$$

where ρ_k^{gg} and κ_k^{gg} denote the density matrices computed from $|\Phi_k^{(g)}\rangle$ [51]. The error associated with the difference between the average number of particles in state

7 The Nuclear Energy Density Functional Formalism

$|\Phi_k^{(g)}\rangle$ and the targeted particle number $N \pm 1$ is compensated for by the last term in the definition of $E_k^{|g|\pm}$. In the perturbative approach (Eq. (7.43)), the chemical potential λ and quasi-particle energies $E_k^{|g|}$ are outputs of Eq. (7.40) solved for the even number-parity vacuum.

Spectroscopic amplitudes associated with the (perturbative) addition and removal of a nucleon are obtained as

$$\langle \Phi_k^{(g)} | a_p^\dagger | \Phi^{(g)} \rangle = U_{pk}^{(g)*}, \tag{7.44a}$$

$$\langle \Phi_k^{(g)} | a_p | \Phi^{(g)} \rangle = V_{pk}^{(g)*}. \tag{7.44b}$$

From these amplitudes, spectroscopic probability matrices are introduced through $\mathbf{S}_k^{(g)+} \equiv \mathbf{U}_k^{(g)} \mathbf{U}_k^{(g)\dagger}$ and $\mathbf{S}_k^{(g)-} \equiv \mathbf{V}_k^{(g)*} \mathbf{V}_k^{(g)T}$ and satisfy, according to Eq. (7.5a), the sum rule

$$\sum_k \mathbf{S}_k^{(g)+} + \sum_k \mathbf{S}_k^{(g)-} = \mathbf{1}. \tag{7.45}$$

Corresponding spectroscopic factors are nothing but the norm of spectroscopic probability matrices and are thus given [57] by

$$SF_k^{(g)\pm} \equiv \text{Tr}_{\mathscr{H}_1}[\mathbf{S}_k^{(g)\pm}]. \tag{7.46}$$

Any inclusion of many-body correlations leads to a fragmentation of the spectroscopic strength associated with one-nucleon addition and removal processes.[16] Within the SR-EDF method, this is the case of static collective correlations that are incorporated via the breaking of symmetries. For example, pairing correlations fragment the strength near the Fermi energy into two peaks belonging, respectively, to addition and removal channels. Similarly, the lifting of the $2j+1$ degeneracy seen at sphericity in the additional/removal spectrum $E_k^{|g|\pm}$ is nothing but the fragmentation of the strength induced by the correlations grasped via the breaking of rotational invariance. Still, this happens at the price of losing good symmetry quantum numbers, which makes difficult to interpret the additional/removal spectrum $E_k^{|g|\pm}$. One must thus await for the MR-EDF description to restore symmetries and achieve a meaningful comparison with experimental data. This will bring further correlations to the description and additional fragmentation of the strength. The latter reveals that separation energies $E_k^{|g|\pm}$ do not target experimental values yet; i.e. absolute values of

---

[16]It is specific to the EDF method to *implicitly* account for correlations via the functional character of $E[g,g]$. As such, one-nucleon separation energies $E_k^{|g|\pm}$ obtained through SR-EDF calculations can be seen as effective centroids of a more fragmented underlying spectrum generated via a theory that explicitly accounts for those correlations.

experimental one-nucleon addition (removal) energies are typically underestimated (overestimated) on purpose by SR-EDF calculations[17] in magic nuclei [48].

7.4.3 Effective Single-Particle Energies

In an ab-initio context, meaningful effective single-particle energies (ESPEs) providing the underlying shell structure relate to the Baranger centroid Hamiltonian. The latter is computed from outputs of the A-body Schroedinger equation through [109, 110]

$$\mathbf{h}^{\text{cent}} \equiv \sum_{\mu \in \mathscr{H}_{A+1}} S_\mu^+ E_\mu^+ + \sum_{\nu \in \mathscr{H}_{A-1}} S_\nu^- E_\nu^-, \tag{7.47}$$

where $\mathscr{H}_{A\pm1}$ denotes the $A \pm 1$ Hilbert space. Specifically, ESPEs are the *eigenvalues* $\{e_p^{\text{cent}}\}$ of the centroid field [109]

$$\mathbf{h}^{\text{cent}} \psi_p^{\text{cent}} = e_p^{\text{cent}} \psi_p^{\text{cent}}, \tag{7.48}$$

and are nothing but barycentre of one-nucleon separation energies weighted by the probability to reach the corresponding $A + 1$ $(A - 1)$ eigenstates through the addition (removal) a nucleon to (from) single-particle state ψ_p^{cent}. As such, they recollect the strength fragmented by many-body correlations.

Let us now transpose the discussion to the context of SR-EDF calculations. Following Baranger, the objective is to build meaningful centroids of the fragmented strength. As discussed above, the only fragmentation of the strength *explicitly* accounted for within the SR-EDF method relates to the breaking of symmetries. Let us illustrate the situation by taking the breaking of particle number and angular momentum as examples. Below, the breaking of the former is explicitly embodied by the Bogoliubov algebra whereas the breaking of the latter is materialized by the labels $|g| \equiv \rho_{\lambda\mu}$ and $\text{Arg}(g) \equiv \Omega$.

As far as gathering the strength fragmented by pairing correlations, one can indeed reach an interesting result [111]. Multiplying the first (second) line of Eq. (7.40) by $\mathbf{U}_k^{(g)\dagger}$ ($\mathbf{V}_k^{(g)\dagger}$) and summing over k, one obtains

$$\sum_k \mathbf{h}^{(g)} \mathbf{U}_k^{(g)} \mathbf{U}_k^{(g)\dagger} + \sum_k \mathbf{\Delta}^{(g)} \mathbf{V}_k^{(g)} \mathbf{U}_k^{(g)\dagger} = \sum_k (\lambda + E_k^{|g|}) \mathbf{U}_k^{(g)} \mathbf{U}_k^{(g)\dagger}, \tag{7.49a}$$

$$\sum_k \mathbf{\Delta}^{(g)} \mathbf{U}_k^{(g)*} \mathbf{V}_k^{(g)T} + \sum_k \mathbf{h}^{(g)} \mathbf{V}_k^{(g)*} \mathbf{V}_k^{(g)T} = \sum_k (\lambda - E_k^{|g|}) \mathbf{V}_k^{(g)*} \mathbf{V}_k^{(g)T}. \tag{7.49b}$$

---

[17]Inaccuracies associated with the quality of empirical EDF parametrizations are responsible for quantitative discrepancies while the present discussion relates to qualitative differences that are built in on purpose.

7 The Nuclear Energy Density Functional Formalism

Adding up both lines, using Eqs. (7.5a) and (7.5b) eventually provides

$$\mathbf{h}^{(g)} = \sum_k \mathbf{S}_k^{(g)+} E_k^{|g|+} + \sum_k \mathbf{S}_k^{(g)-} E_k^{|g|-}, \qquad (7.50)$$

which is analogous to Eq. (7.47) and provides $\mathbf{h}^{(g)}$ with the meaning of a centroid field. The coupling of addition and removal spectroscopic amplitudes via the anomalous field $\mathbf{\Delta}^{(g)}$ in Eq. (7.40) is screened out from the Baranger sum rule. This is an a priori non-trivial result, though straightforward to obtain. Of course, the explicit tackling of pairing correlations does impact the centroid field indirectly via the feedback of such correlations on the normal density matrix and the dependence of $\mathbf{h}^{(g)}$ on the latter. Interestingly, Eq. (7.50) justifies the traditional use by practitioners of the eigenvalues of $\mathbf{h}^{(g)}$ as effective single-particle energies,[18] i.e.

$$\mathbf{h}^{(g)} \psi_p^{(g)} = e_p^{|g|} \psi_p^{(g)}. \qquad (7.51)$$

It is remarkable that Eq. (7.50) could be obtained without making any explicit reference to a Hamilton operator, i.e. within the strict spirit of the EDF method. This is at variance with the standard proof that allows one to connect the centroid field with the static part of the one-nucleon self energy [109, 110].

Sum rule (7.50) only gathers the strength fragmented by correlations associated with the breaking of particle number, not yet the strength fragmented by the breaking of angular momentum. As a matter of fact, $\mathbf{h}^{(g)}$ does break rotational symmetry such that the ESPE spectrum $e_p^{|g|}$ displays the same lifting of the $2j + 1$ degeneracy as $E_k^{|g|\pm}$. Plotted against $|g| = \rho_{20}$, the spectrum $e_p^{|g|}$ takes the form of a so-called Nilsson diagram as is illustrated in Fig. 7.4 for ^{250}Fm. One observes that the minimum of the energy landscape is obtained for a deformation that reflects a compromise between $N = 150$ and $Z \sim 100$ deformed shell gaps in the ESPE spectrum.

One can now go one step further and recollect the strength associated with the breaking of rotational symmetry.[19] To do so, one notices that $\mathbf{h}^{(g)}$, as any one-body operator, transforms under rotation according to[20]

$$\mathbf{h}^{(\rho_{\lambda\mu}\Omega)} = R(\Omega) \mathbf{h}^{(\rho_{\lambda\mu}0)} R^\dagger(\Omega). \qquad (7.52)$$

The fragmented strength is recollected by extracting the monopole, i.e. angular-averaged, part of $\mathbf{h}^{(\rho_{\lambda\mu}\Omega)}$. Expressing Eq. (7.52) in a spherical basis, omitting

---

[18] In view of Eq. (7.50), it thus appears more justified to use eigenvalues of $\mathbf{h}^{(g)}$ as ESPEs rather than its diagonal matrix elements in the basis diagonalizing ρ^{gg}, i.e. the so-called *canonical* basis, as it is often done by practitioners, e.g. see Ref. [57].

[19] Such a procedure can be extended to any subgroup of $\mathscr{G}$.

[20] Equation (7.52) can be recovered by expressing matrices $\mathbf{S}_k^{(g)\pm}$ in a spherical basis $p = n\pi jm\tau$ and by working out how such matrices transform under the rotation of $|\Phi^{(g)}\rangle$ and $|\Phi_k^{(g)}\rangle$.

Fig. 7.4 Energy landscape and effective single-particle energies of ^{250}Fm as a function of axial quadrupole deformation ($|g| = \rho_{20} = \beta_2$) [108]

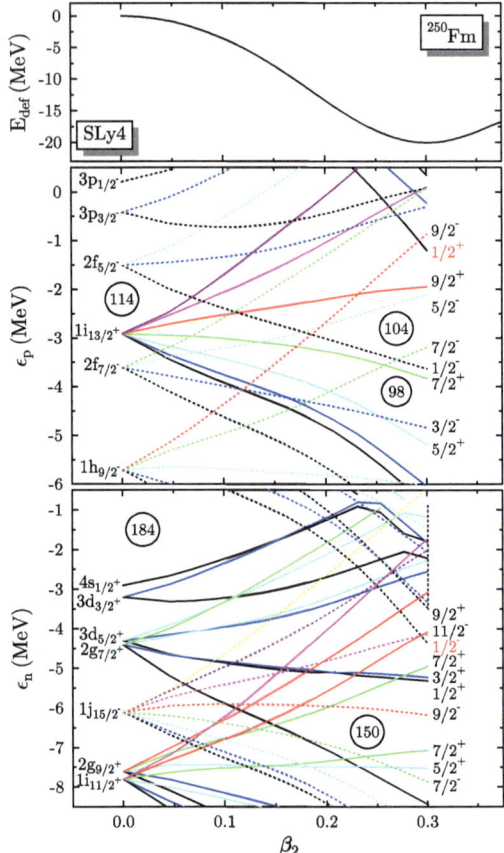

isospin projection and parity[21] quantum numbers for simplicity, as well as using orthogonality relationship (7.9), the monopole operator satisfies [112]

$$h^{\text{mon}[\rho_{\lambda\mu}]}_{njmn'j'm'} \equiv \frac{2J+1}{16\pi^2} \int_{D_\Omega} d\Omega \, \mathscr{D}^{0*}_{00}(\Omega) h^{(\rho_{\lambda\mu}\Omega)}_{njmn'j'm'} \qquad (7.53a)$$

$$= \delta_{jj'}\delta_{mm'} \sum_{m''} h^{(\rho_{\lambda\mu}0)}_{njm''n'jm''}. \qquad (7.53b)$$

Equation (7.53b) demonstrates that $\mathbf{h}^{\text{mon}[\rho_{\lambda\mu}]}$ displays spherical symmetry and is built out of the (j, m) blocks of the deformed operator $\mathbf{h}^{(\rho_{\lambda\mu}0)}$, including an averag-

---

[21] If $\mathbf{h}^{(\rho_{\lambda\mu}0)}$ breaks parity, one further needs to extract the component belonging to the trivial Irreps of C_i, i.e. the inversion center group. Indeed, restoring spherical symmetry does not ensure that parity is a good quantum number, e.g. a $j = 3/2$ single-particle state can be a linear combination of $d_{3/2}$ and $p_{3/2}$ states. Proceeding to such an extraction would deliver a one-body field that is block-diagonal with respect to parity π as well.

ing over the magnetic quantum number m. The monopole field thus extracted carries the deformation label $\rho_{\lambda\mu}$ as a memory of the symmetry breaking field it has been extracted from. Spherical ESPEs gathering the strength of the fragmented spectrum $e_p^{|g|}$ are then obtained through

$$\mathbf{h}^{\text{mon}[\rho_{\lambda\mu}]} \psi_{njm}^{\text{mon}[\rho_{\lambda\mu}]} = e_{njm}^{\text{mon}[\rho_{\lambda\mu}]} \psi_{njm}^{\text{mon}[\rho_{\lambda\mu}]}. \quad (7.54)$$

Equation (7.54) defines the way to extract a spherical effective single-particle energy spectrum out of any SR-EDF calculation. Such a procedure has neither been defined nor used so far.[22] As already mentioned, the above procedure is not limited to $SO(3)$ and can be extended to any broken symmetry.

7.4.4 Equation of State of Infinite Nuclear Matter

Infinite nuclear matter (INM) is an idealized nuclear system that has relevance to the study of several real systems, e.g. the physics of neutron stars or the dynamic of supernovae explosions. The system is made of protons and neutrons and is considered to be homogeneous. The Coulomb interaction between protons is switched off. One is first and foremost interested in computing the equation of state (EOS) of such a system, i.e. its energy per nucleon as a function of its density. This can easily be done at the SR level. Below, we illustrate the procedure at zero temperature on the basis of the strict bilinear Skyrme parametrization introduced in Eqs. (7.29a)–(7.29c). Furthermore, pairing correlations are omitted as they little impact bulk properties such as the EOS. However, one should note that pairing properties, e.g. pairing gaps, of INM are of importance to the physics of neutron stars [113].

7.4.4.1 Definitions

The four basic degrees of freedom characterizing INM are the scalar-isoscalar ρ_0, scalar-isovector ρ_1, vector-isoscalar s_0 and vector-isovector s_1 densities. They can be expressed through neutron and proton as well as spin-up and spin-down densities in the following way

$$\rho_0 = \rho_{n\uparrow} + \rho_{n\downarrow} + \rho_{p\uparrow} + \rho_{p\downarrow}, \quad (7.55a)$$

$$\rho_1 = \rho_{n\uparrow} + \rho_{n\downarrow} - \rho_{p\uparrow} - \rho_{p\downarrow}, \quad (7.55b)$$

$$s_0 = \rho_{n\uparrow} - \rho_{n\downarrow} + \rho_{p\uparrow} - \rho_{p\downarrow}, \quad (7.55c)$$

$$s_1 = \rho_{n\uparrow} - \rho_{n\downarrow} - \rho_{p\uparrow} + \rho_{p\downarrow}, \quad (7.55d)$$

---

[22] Practically speaking, Eqs. (7.53a), (7.53b), (7.54) are particularly trivial to implement in numerical codes that expend deformed solutions out of a spherical, e.g. harmonic oscillator, basis.

such that the inverse relationships read

$$\rho_{n\uparrow} = \frac{1}{4}(1 + I_\tau + I_\sigma + I_{\sigma\tau})\rho_0, \qquad (7.56a)$$

$$\rho_{n\downarrow} = \frac{1}{4}(1 + I_\tau - I_\sigma - I_{\sigma\tau})\rho_0, \qquad (7.56b)$$

$$\rho_{p\uparrow} = \frac{1}{4}(1 - I_\tau + I_\sigma - I_{\sigma\tau})\rho_0, \qquad (7.56c)$$

$$\rho_{p\downarrow} = \frac{1}{4}(1 - I_\tau - I_\sigma + I_{\sigma\tau})\rho_0, \qquad (7.56d)$$

where isospin $I_\tau \equiv \rho_1/\rho_0$, spin $I_\sigma \equiv s_0/\rho_0$ and spin-isospin $I_{\sigma\tau} \equiv s_1/\rho_0$ excesses ($-1 \leq I_i \leq 1$) have been introduced. The typical cases of interest are (i) symmetric nuclear matter ($I_\tau = I_\sigma = I_{\sigma\tau} = 0$), (ii) isospin-asymmetric nuclear matter ($I_\tau \neq 0$), (iii) spin-polarized nuclear matter ($I_\sigma \neq 0$) and (iv) isospin-asymmetric spin-polarized nuclear matter ($I_\tau \neq 0$, $I_\sigma \neq 0$ and $I_{\sigma\tau} \neq 0$).

Infinite nuclear matter being translationally invariant, it is convenient to use a plane wave basis

$$\langle \vec{r}\sigma\tau | \vec{k}\sigma'\tau' \rangle = \varphi_{\vec{k}\sigma'\tau'}(\vec{r}\sigma\tau) = (2\pi)^{-\frac{3}{2}} \exp(i\vec{k} \cdot \vec{r}) \delta_{\sigma\sigma'} \delta_{\tau\tau'}, \qquad (7.57)$$

where $\tau\sigma = \{n\uparrow, n\downarrow, p\uparrow, p\downarrow\}$. Neglecting pairing, the SR state reduces to a Slater determinant obtained by filling individual orbitals $\varphi_{\vec{k}\sigma'\tau'}(\vec{r}\sigma\tau)$ up to the Fermi momentum, i.e. the normal density matrix is diagonal in the plane-wave basis and equal to 1 for states characterized by $|\vec{k}| \leq k_{F,\tau\sigma}$ and 0 otherwise, where $k_{F,\tau\sigma}$ denotes the spin- and isospin-dependent Fermi momentum. The SR state does not carry any non-zero order parameter such that the label g can be dropped in the present section.

Starting from Eq. (7.57), local densities can be computed explicitly. The sum over basis states in Eqs. (7.22a), (7.22b) becomes an integral over the sphere of radius $k_{F,\tau\sigma}$. Eventually, local densities of interest are constant in space and read as

$$\rho_{\tau\sigma} = \int_{|\vec{k}| \leq k_{F,\tau\sigma}} d\vec{k}\, \varphi_{\vec{k}}^*(\vec{r}\sigma\tau) \varphi_{\vec{k}}(\vec{r}\sigma\tau) = \frac{1}{6\pi^2} k_{F,\tau\sigma}^3, \qquad (7.58a)$$

$$\tau_{\tau\sigma} = \int_{|\vec{k}| \leq k_{F,\tau\sigma}} d\vec{k}\, [\vec{\nabla}\varphi_{\vec{k}}^*(\vec{r}\sigma\tau)] \cdot [\vec{\nabla}\varphi_{\vec{k}}(\vec{r}\sigma\tau)] = \frac{3}{20} \frac{2}{3\pi^2} k_{F,\tau\sigma}^5. \qquad (7.58b)$$

With the choice of a Fermi surface centred at $\vec{k} = 0$, current densities vanish $\vec{j}_{q\sigma} = 0$. Also, all gradients of local densities are zero $\nabla_\nu \rho_{q\sigma} = 0$ by construction, as are the pair densities. Using Eqs. (7.56a)–(7.56d), (7.58a) and (7.58b), one relates spin-isospin kinetic densities to spin, isospin and spin-isospin excesses

$$\tau_0 = \tau_{n\uparrow} + \tau_{n\downarrow} + \tau_{p\uparrow} + \tau_{p\downarrow} = \frac{3}{5} c_s \rho_0^{5/3} F_{5/3}^{(0)}(I_\tau, I_\sigma, I_{\sigma\tau}), \qquad (7.59a)$$

$$\tau_1 = \tau_{n\uparrow} + \tau_{n\downarrow} - \tau_{p\uparrow} - \tau_{p\downarrow} = \frac{3}{5} c_s \rho_0^{5/3} F_{5/3}^{(\tau)}(I_\tau, I_\sigma, I_{\sigma\tau}), \qquad (7.59b)$$

7 The Nuclear Energy Density Functional Formalism

$$T_0 = \tau_{n\uparrow} - \tau_{n\downarrow} + \tau_{p\uparrow} - \tau_{p\downarrow} = \frac{3}{5}c_s\rho_0^{5/3}F_{5/3}^{(\sigma)}(I_\tau, I_\sigma, I_{\sigma\tau}), \qquad (7.59c)$$

$$T_1 = \tau_{n\uparrow} - \tau_{n\downarrow} - \tau_{p\uparrow} + \tau_{p\downarrow} = \frac{3}{5}c_s\rho_0^{5/3}F_{5/3}^{(\sigma\tau)}(I_\tau, I_\sigma, I_{\sigma\tau}), \qquad (7.59d)$$

where F-functions [114] are explicated in the Appendix. We further introduce $c_s \equiv (3\pi^2/2)^{2/3}$ and $c_n \equiv (3\pi^2)^{2/3}$.

Last but not least, the results are expressed below in terms of isoscalar $C_0^{ff'}$ and isovector $C_1^{ff'}$ couplings. The latter are related to the couplings in the neutron/proton representation (under the assumption of isospin symmetry) used in Eqs. (7.29a)–(7.29c) through

$$C_0^{ff'} = \frac{1}{2}(C_{\tau\tau}^{ff'} + C_{\tau\bar{\tau}}^{ff'}), \qquad (7.60a)$$

$$C_1^{ff'} = \frac{1}{2}(C_{\tau\tau}^{ff'} - C_{\tau\bar{\tau}}^{ff'}). \qquad (7.60b)$$

The fact that most of the local densities are zero in INM implies that properties will be expressed in terms of a limited number of couplings.

7.4.4.2 Symmetric Nuclear Matter

Symmetric nuclear matter (SNM) is characterized by an equal number of protons and neutrons as well as of spin up and spin down nucleons. Consequently, $\rho_1 = I_\tau = 0$ and $I_\sigma = I_{\sigma\tau} = 0$. Only two local densities ρ_0 and τ_0 subsist, i.e. $\rho_n = \rho_p = \frac{1}{2}\rho_0$ and $\tau_n = \tau_p = \frac{1}{2}\tau_0$, with

$$\rho_0 = \frac{2}{3\pi^2}k_F^3; \qquad \tau_0 = \frac{3}{5}c_s\rho_0^{5/3}. \qquad (7.61)$$

The EOS is obtained from Eqs. (7.29a)–(7.29c) as

$$\frac{E}{A} \equiv \frac{\mathscr{E}_\rho + \mathscr{E}_{\rho\rho}}{\rho_0} = \frac{3}{5}\frac{\hbar^2}{2m}c_s\rho_0^{2/3} + C_0^{\rho\rho}\rho_0^2 + \frac{3}{5}c_sC_0^{\rho\tau}\rho_0^{5/3}. \qquad (7.62)$$

Symmetric nuclear matter presents a stable state such that a minimum energy is obtained for a finite density ρ_{sat}. The pressure of the fluid relates to the first derivative of the EOS with respect to the isoscalar density, which in SNM reads

$$P \equiv \rho_0^2 \left.\frac{\partial E/A}{\partial \rho_0}\right|_A = \frac{2}{5}\frac{\hbar^2}{2m}c_s\rho_0^{5/3} + C_0^{\rho\rho}\rho_0^2 + c_sC_0^{\rho\tau}\rho_0^{8/3}. \qquad (7.63)$$

The equilibrium density ρ_{sat} is obtained as the solution of $P(\rho_{\text{sat}}) = 0$.

The incompressibility of the nuclear fluid relates to the second derivative of the EOS with respect to the isoscalar density and expresses the energy cost to compress the nuclear fluid. It is defined as

$$K \equiv \frac{18P}{\rho_0} + 9\rho_0^2 \frac{\partial^2 E/A}{\partial \rho_0^2}, \quad (7.64)$$

such that at equilibrium

$$K_\infty \equiv 9\rho_0^2 \frac{\partial^2 E/A}{\partial \rho_0^2}\bigg|_{\rho_0=\rho_{\text{sat}}} = -\frac{6}{5}\frac{\hbar^2}{2m} c_s \rho_{\text{sat}}^{2/3} + 6 c_s C_0^{\rho\tau} \rho_{\text{sat}}^{5/3}, \quad (7.65)$$

which needs to be positive for the system to be stable against density fluctuations.

7.4.4.3 Asymmetric Nuclear Matter

In general, INM is characterized by (i) unequal proton and neutron matter densities, i.e. $I_\tau \neq 0$, (ii) a global spin polarization, i.e. $I_\sigma \neq 0$ and (iii) a spin polarization that differs for neutron and proton species, i.e. $I_{\sigma\tau} \neq 0$. The EOS of such a nuclear fluid is given by

$$\frac{E}{A} = \frac{3}{5}\frac{\hbar^2}{2m} c_s F_{5/3}^{(0)}(I_\tau, I_\sigma, I_{\sigma\tau})\rho_0^{2/3} + C_0^{\rho\rho}\rho_0 + C_1^{\rho\rho}\rho_0 I_\tau^2 + C_0^{ss}\rho_0 I_\sigma^2 + C_1^{ss}\rho_0 I_{\sigma\tau}^2$$

$$+ \frac{3}{5}\big[C_0^{\rho\tau} F_{5/3}^{(0)}(I_\tau, I_\sigma, I_{\sigma\tau}) + C_1^{\rho\tau} I_\tau F_{5/3}^{(\tau)}(I_\tau, I_\sigma, I_{\sigma\tau})$$

$$- C_0^{JJ} I_\sigma F_{5/3}^{(\sigma)}(I_\tau, I_\sigma, I_{\sigma\tau}) - C_1^{JJ} I_{\sigma\tau} F_{5/3}^{(\sigma\tau)}(I_\tau, I_\sigma, I_{\sigma\tau})\big] c_s \rho_0^{5/3}.$$

Spin, isospin and spin-isospin symmetry energies are analogues of K_∞ with respect to spin, isospin and spin-isospin excesses, respectively. As such, they characterize the stiffness of the EOS with respect to generating such non-zero excesses. At saturation of SNM, i.e. when $I_\sigma = I_\tau = I_{\sigma\tau} = 0$ and $\rho_0 = \rho_{\text{sat}}$, the three symmetry energies are given by

$$a_\tau \equiv \frac{1}{2} \frac{\partial^2 E_H/A}{\partial I_\tau^2}\bigg|_{I_\sigma=I_\tau=I_{\sigma\tau}=0}$$

$$= \frac{1}{3}\frac{\hbar^2}{2m} c_s \rho_0^{2/3} + C_1^{\rho\rho}\rho_0 + \left[\frac{1}{3}C_0^{\rho\tau} + C_1^{\rho\tau}\right] c_s \rho_0^{5/3}, \quad (7.66a)$$

$$a_\sigma \equiv \frac{1}{2} \frac{\partial^2 E_H/A}{\partial I_\sigma^2}\bigg|_{I_\sigma=I_\tau=I_{\sigma\tau}=0}$$

$$= \frac{1}{3}\frac{\hbar^2}{2m} c_s \rho_0^{2/3} + C_0^{ss}\rho_0 + \left[\frac{1}{3}C_0^{\rho\tau} - C_0^{JJ}\right] c_s \rho_0^{5/3}, \quad (7.66b)$$

7 The Nuclear Energy Density Functional Formalism

$$a_{\sigma\tau} \equiv \frac{1}{2} \frac{\partial^2 E_H/A}{\partial I_{\sigma\tau}^2}\bigg|_{I_\sigma=I_\tau=I_{\sigma\tau}=0}$$

$$= \frac{1}{3}\frac{\hbar^2}{2m}c_s\rho_0^{2/3} + C_1^{ss}\rho_0 + \left[\frac{1}{3}C_0^{\rho\tau} - C_1^{JJ}\right]c_s\rho_0^{5/3}, \qquad (7.66c)$$

and must be positive for the minimum of the EOS to be stable.

Two quantities of interest are intimately connected to the skin thickness of heavy isospin-asymmetric nuclei, i.e. to the difference between their neutron and proton radii. These quantities are the density-symmetry coefficient L

$$L \equiv 3\rho\frac{\partial}{\partial\rho}\left(\frac{1}{2}\frac{\partial^2 E/A}{\partial I_\tau^2}\right)\bigg|_{I_\sigma=I_\tau=I_{\sigma\tau}=0}$$

$$= \frac{2}{3}\frac{\hbar^2}{2m}c_s\rho_0^{2/3} + 3C_1^{\rho\rho}\rho_0 + \left[\frac{5}{3}C_0^{\rho\tau} + 5C_1^{\rho\tau}\right]c_s\rho_0^{5/3}, \qquad (7.67)$$

and the symmetry compressibility

$$K_{sym} \equiv 9\rho^2\frac{\partial^2}{\partial\rho^2}\left(\frac{1}{2}\frac{\partial^2 E/A}{\partial I_\tau^2}\right)\bigg|_{I_\sigma=I_\tau=I_{\sigma\tau}=0}$$

$$= -\frac{2}{3}\frac{\hbar^2}{2m}c_s\rho_0^{2/3} + \frac{10}{3}c_s C_0^{\rho\tau}\rho_0^{5/3} + 10c_s C_1^{\rho\tau}\rho_0^{5/3}. \qquad (7.68)$$

7.4.4.4 Pure Neutron Matter

A particular case of isospin-asymmetric and spin-symmetric nuclear matter is pure neutron matter (PNM) obtained for $I_\tau = 1$ and $I_\sigma = I_{\sigma\tau} = 0$. The EOS of PNM reads

$$\frac{E}{A} = \frac{3}{5}\frac{\hbar^2}{2m}c_n\rho_0^{2/3} + C_0^{\rho\rho}\rho_0 + C_1^{\rho\rho}\rho_0 + \frac{3}{5}c_n C_0^{\rho\tau}\rho_0^{5/3} + \frac{3}{5}c_n C_1^{\rho\tau}\rho_0^{5/3}. \qquad (7.69)$$

7.4.5 Symmetry Breaking and "Deformation"

There are important points to underline regarding the notions of symmetry breaking and "deformation" in finite systems. To do so, let us take rotational symmetry and the deformation of the density distribution as an example. Of course, the discussion conducted below applies to any of the symmetries of interest.

1. The breaking of a symmetry is never quite real in a finite system. Eventually, any quantum state of the nucleus does carry good angular momentum (J, M) such that it is improper to describe it as a wave packet mixing states belonging to different irreducible representations of $SO(3)$, i.e. carrying different values of J.

Only in infinite systems characterized by infinite inertia would the sequence of states belonging to a rotational band be truly degenerate. This makes the symmetry breaking real in infinite systems as it offers the possibility to describe the true ground state as a linear combination of states with different J values. In a finite system, quantum fluctuations associated with finite inertia eventually lift the degeneracy such that good symmetry quantum numbers must eventually be restored.

2. In a finite system, the notion of "deformation" that characterizes the breaking of a symmetry is thus necessarily an artefact associated with an *incomplete* theoretical description. As such, the $J^\pi = 0^+$ ground state of an even-even nucleus is *never* "deformed", given that the density distribution of *any* $J = 0$ quantum state is spherically symmetric. It is only within an incomplete theoretical description such as the SR-EDF method that one may speak improperly of a "deformed" $J^\pi = 0^+$ ground state.[23] Once rotational symmetry is restored, the corresponding density distribution is indeed spherically symmetric.

3. Within, e.g., the SR-EDF method, one notices that the breaking of the rotational symmetry depends on the number of elementary constituents of the even-even nucleus under consideration; i.e. the symmetry does not break in double and single magic nuclei while it breaks in essentially all double open-shell nuclei.[24] This raises an important question. If all $J^\pi = 0^+$ states are eventually equally spherical in front of god, are "spherical" $J^\pi = 0^+$ states more spherical than "deformed" ones!? To rephrase it, one may ask in what way the intermediate artefact of "deformation" tells us anything real about the nucleus under consideration? As a matter of fact, the artefact of ground-state "deformation" does not tell us anything about the ground state but rather about the way the nucleus primarily *excites*. In the case of rotational symmetry, the fact that the ground state comes out to be deformed at the SR-EDF level tells us, at a low theoretical cost, that a rotational band built on top of it should exist. To reverse engineer the statement, any experimental spectrum containing a set of states that can be convincingly ordered as a $J(J+1)$ sequence above the ground state will see the latter being deformed within the (incomplete) SR-EDF description.

---

[23] It is important to underline at this point that the notion of "deformation" differs depending on the angular momentum of the targeted many-body state. This is due to the fact that a symmetry-*conserving* state with angular momentum J does display non-zero multipole moments of the density for $\lambda \leq 2J$ [75]. For example, having a reference state with non-zero quadrupole and hexadecapole moments does *not* characterize a breaking of rotational symmetry if one means to describe a $J = 2$ state. In such a case, one must check multipoles with $\lambda > 4$ (or any odd multipole) to state whether rotational symmetry is broken or not. It happens that product states of the Bogoliubov type usually generate non-zero multipole moments of all (e.g. even) multipolarities as soon as they display a non-zero collective quadrupole moment. As such, they break rotational symmetry independent of the angular momentum of the good-symmetry state one is eventually after.

[24] Of course, the fact that the neutron or proton number is magic is not known a priori but is based on a posteriori observations and experimental facts. In particular, the fact that traditional magic numbers, i.e. $N, Z = 2, 8, 20, 28, 50, 82, 126$, remain as one goes to very isospin-asymmetric nuclei is the subject of intense on-going experimental and theoretical investigations [10].

7 The Nuclear Energy Density Functional Formalism

Table 7.3 Categories of nuclei that tend to break translational, rotational and particle number symmetries as well as associated patterns in their excitation spectrum

	Nuclei	Excitation pattern
Space translation $\vec{a}$	All	Surface vibrations
Gauge rotation φ	All but double magic ones	Energy gap
Space rotation α, β, γ	All but singly-magic ones	Ground-state rotational bands

To conclude, even though the symmetry breaking is fictitious in a finite system it leaves its fingerprint on excitation spectra. Such a connection between the two notions is schematically illustrated in Table 7.3 for the three symmetries of present interest.

7.4.6 Connection to Density Functional Theory?

It has become customary in nuclear physics to assimilate the SR-EDF method, eventually including corrections *a la* Lipkin or Kamlah, with density functional theory (DFT) at play in electronic systems, i.e. to state that the Hohenberg-Kohn (HK) theorem [115] underlays nuclear SR-EDF calculations. This is a misconception as distinct strategies actually support both methods. Whereas the SR-EDF method minimizes the energy with respect to a symmetry-breaking trial density, DFT relies on an energy functional whose minimum must be reached for a local one-body density[25] that possesses *all* symmetries of the actual ground-state density, i.e. that displays fingerprints of the symmetry quantum-numbers carried by the exact ground-state [116]. As a matter of fact, generating a symmetry-breaking solution is known to be problematic in DFT, as it lies outside the frame of the HK theorem, and is usually referred to as the *symmetry dilemma*. To bypass that dilemma and grasp kinematical correlations associated with good symmetries, several reformulations of DFT have been proposed over the years, e.g. see Refs. [117, 118].

Recent efforts within the nuclear community have been devoted to formulating a HK-like theorem in terms of the internal density, i.e. the matter distribution relative to the center of mass of the self-bound system [119, 120]. Together with an appropriate Kohn-Sham scheme [120], it allows one to reinterpret the SR-EDF method as a functional of the internal density rather than as a functional of a laboratory density that breaks translational invariance. This constitutes an interesting route whose ultimate consequence would be to remove entirely the notion of breaking and restoration of symmetries from the EDF approach and make the SR formulation a complete many-body method, at least in principle. To reach such a point though, the work of Refs. [119, 120] must be extended, at least, to rotational and particle-number symmetries, knowing that translational symmetry was somewhat the easy case to deal

---

[25]The scheme can be extended to a set of local densities or to the full density matrix.

with given the explicit decoupling of internal and center of mass motions. Going in such a direction, an interesting formulation was recently proposed that provides the Schroedinger equation based on collective Hamiltonian with a firm ground [121]. This problem deserves significant attention in the future.

7.5 Multi-reference Implementation

In a finite system, quantum fluctuations eventually make the symmetry breaking fictitious such that good symmetries must eventually be restored. From a group theory perspective, the diagonal energy kernel $E[g, g]$ associated with a symmetry breaking state $|\Phi^{(g)}\rangle$ mixes irreducible representations of the symmetry group of interest, and so does E_{GS}^{SR}. The symmetry restoration consists of extracting energies that can be put in one-to-one correspondence with Irreps of the group. In terms of the schematic "mexican-hat" of Fig. 7.2, doing so corresponds to incorporating zero-energy fluctuations along the phase of the order parameter.

Furthermore, fluctuations of $|g|$, i.e. configuration mixing along the radial coordinate of the "mexican-hat", must be considered at the same time. This is well illustrated by Fig. 7.3. On the one hand, the SR energy landscape of ^{240}Pu is stiff in the vicinity of its minimum and well separated from the secondary minimum tentatively associated with a fission isomer. On the other hand, ^{202}Rn is "soft" with respect to axial quadrupole deformation and displays two equally pertinent oblate and prolate minima that are separated by a small barrier of about 2 MeV height. While the SR minimum provides a reasonable picture of what the intrinsic state of ^{240}Pu might be, no single reference state characterized by a fixed value of $|g| = \rho_{20}$ is entitled to do so for ^{202}Rn, i.e. fluctuations in $|g| = \rho_{20}$ are expected to be large a priori.

Within the EDF method, the large amplitude collective motions associated with the fluctuations of both the phase α and the magnitude $|g|$ of the order parameters are accounted for by the multi-reference framework. In doing so, a MR-EDF calculation accesses collective, i.e. "rotational" and "vibrational", excitations while incorporating associated correlations in the ground state. Technically speaking, the MR step invokes the complete set of product states $\{|\Phi^{(|g|\alpha)}\rangle = R(\alpha)|\Phi^{(|g|0)}\rangle; |g| \in [0, +\infty[; \alpha \in D_{\mathcal{G}}\}$ such that the MR energy mixes off-diagonal energy $E[g', g]$ and norm $N[g', g]$ kernels associated with all pairs of states belonging to that set (see below). The restoration of symmetries performed after variation is presently considered, i.e. the states $\{|\Phi^{(|g|0)}\rangle\}$ are determined *prior to* the MR step through repeated SR calculations. A more involved and performing approach consists of determining $|\Phi^{(|g|0)}\rangle$ through the minimization of the symmetry-restored energy $\mathcal{E}_{|g|}^{\lambda}$ defined below, i.e. while including the effect of the fluctuations associated with the restoration of the good symmetry [51].

As mentioned in the introduction, a key aspect of the MR formulation provided below is that it is conducted rigorously from a *mathematical* viewpoint on the basis of a generic EDF kernel $E[g', g]$ that does not necessarily refer to a pseudo Hamilton operator. In particular, the restoration of symmetries is shown to be properly

7 The Nuclear Energy Density Functional Formalism

formulated without making any reference to a projected state [70], which is a necessity in the general EDF context. This however does not guarantee that the MR formalism is sound from a *physical* standpoint as will be illustrated in Sect. 7.5.8.

7.5.1 Symmetry-Restored Kernels

One starts by considering energy and norm kernels as two functions defined over the domain[26] $D_{\mathscr{G}}$ and by decomposing them over the Irreps of $\mathscr{G}$ according to Eq. (7.10), i.e.

$$N\big[|g'|0, |g|\alpha\big] \equiv \sum_{\lambda ab} \mathscr{N}^{\lambda}_{ab}\big[|g'|, |g|\big] S^{\lambda}_{ab}(\alpha), \tag{7.70a}$$

$$E\big[|g'|0, |g|\alpha\big] N\big[|g'|0, |g|\alpha\big] \equiv \sum_{\lambda ab} \mathscr{E}^{\lambda}_{ab}\big[|g'|, |g|\big] \mathscr{N}^{\lambda}_{ab}\big[|g'|, |g|\big] S^{\lambda}_{ab}(\alpha), \tag{7.70b}$$

where the sum runs over all Irreps. Multiplying Eqs. (7.70a), (7.70b) by $S^{\lambda *}_{ab}(\alpha)$, integrating it over the domain of the group and using orthogonality relationship (7.9) allows one to extract the expansion coefficients associated with a specific Irrep, i.e.

$$\mathscr{N}^{\lambda}_{ab}\big[|g'|, |g|\big] = \frac{d_{\lambda}}{v_{\mathscr{G}}} \int_{D_{\mathscr{G}}} dm(\alpha) S^{\lambda *}_{ab}(\alpha) N\big[|g'|0, |g|\alpha\big], \tag{7.71a}$$

$$\mathscr{E}^{\lambda}_{ab}\big[|g'|, |g|\big] \mathscr{N}^{\lambda}_{ab}\big[|g'|, |g|\big]$$
$$= \frac{d_{\lambda}}{v_{\mathscr{G}}} \int_{D_{\mathscr{G}}} dm(\alpha) S^{\lambda *}_{ab}(\alpha) E\big[|g'|0, |g|\alpha\big] N\big[|g'|0, |g|\alpha\big]. \tag{7.71b}$$

The integration over $D_{\mathscr{G}}$ in Eqs. (7.71a), (7.71b) amounts to performing a mixing along the phase of the order parameter in order to lift the degeneracy associated with the fictitious Goldstone mode. As stated earlier, Eqs. (7.70a), (7.70b)–(7.71a), (7.71b) prove that the extraction of the symmetry-restored energy kernel $\mathscr{E}^{\lambda}_{ab}[|g'|, |g|]$ can be rigorously formulated [70] on the basis of a general EDF kernel $E[g', g]$ that satisfies the minimal set of properties introduced in Sect. 7.3, i.e. it is not necessary for such a kernel to derive from a pseudo Hamilton operator (see Sect. 7.5.3 for further discussions). In such a general situation, one cannot and should not invoke a projected state as is (incorrectly) done in standard presentations of the MR-EDF formalism. The above derivation does demonstrate that the projected state can indeed be bypassed without any difficulty.

As $S^{\lambda}_{ab}(0) = \delta_{ab}$ for any λ, setting $\alpha = 0$ into Eqs. (7.70a), (7.70b) provides a sum rule relating symmetry-restored energy and norm kernels to un-rotated symmetry-

---

[26] We take advantage of property (7.16) to fix one of the two phases involved to zero.

breaking kernels, i.e.

$$N[|g'|0,|g|0] = \sum_{\lambda a} \mathcal{N}_{aa}^\lambda[|g'|,|g|], \qquad (7.72a)$$

$$E[|g'|0,|g|0]N[|g'|0,|g|0] = \sum_{\lambda a} \mathcal{E}_{aa}^\lambda[|g'|,|g|]\mathcal{N}_{aa}^\lambda[|g'|,|g|], \qquad (7.72b)$$

where the independence of $\mathcal{E}_{aa}^\lambda[|g'|,|g|]$ and $\mathcal{N}_{aa}^\lambda[|g'|,|g|]$ on a has not been explicitly utilized yet. Exploiting it and particularizing Eqs. (7.72a), (7.72b) to $|g'|=|g|$ provides two sum rules

$$1 = \sum_\lambda d_\lambda \mathcal{N}_{|g|}^\lambda, \qquad (7.73a)$$

$$E_{|g|}^{SR} = \sum_\lambda d_\lambda \mathcal{N}_{|g|}^\lambda \mathcal{E}_{|g|}^\lambda, \qquad (7.73b)$$

the second of which relates, for a given value of $|g|$, the SR energy to the complete set of symmetry-restored energies $\mathcal{E}_{|g|}^\lambda$. In Eqs. (7.73a), (7.73b) simplified notations $\mathcal{E}_{|g|}^\lambda \equiv \mathcal{E}_{aa}^\lambda[|g|,|g|]$ and $\mathcal{N}_{|g|}^\lambda \equiv \mathcal{N}_{aa}^\lambda[|g|,|g|]$ have been used.

First and foremost, sum rule (7.73b) provides a consistency checks in numerical codes used to extract MR energies. However, such a decomposition of the SR energy has shown to be very helpful in pinning down profound issues with the formalism when specifying to $U(1)$ symmetry. Refer to Sect. 7.5.8 for the corresponding discussion.

7.5.1.1 Specification to $U(1)$

Of particular interest is the specification of Eqs. (7.70a), (7.70b)–(7.73a), (7.73b) to the $U(1)$ group, i.e. to particle-number restoration (PNR). Singling out the order parameter $g \equiv \|\kappa\|e^{i\varphi}$ associated with the breaking of nucleon number and omitting the other collective variables at play, one obtains the Fourier decomposition of the kernels

$$N[\|\kappa'\|0,\|\kappa\|\varphi] \equiv \sum_{N\in\mathbb{Z}} \mathcal{N}^N[\|\kappa'\|,\|\kappa\|]e^{iN\varphi}, \qquad (7.74a)$$

$$E[\|\kappa'\|0,\|\kappa\|\varphi]N[\|\kappa'\|0,\|\kappa\|\varphi]$$
$$\equiv \sum_{N\in\mathbb{Z}} \mathcal{E}^N[\|\kappa'\|,\|\kappa\|]\mathcal{N}^N[\|\kappa'\|,\|\kappa\|]e^{iN\varphi}. \qquad (7.74b)$$

From a mathematical viewpoint, the sum in Eqs. (7.74a), (7.74b) runs a priori over all Irreps of $U(1)$, i.e. over both positive *and* negative integers. Following Eqs. (7.71a), (7.71b), one extracts particle-number restored kernels through

$$\mathcal{N}^N[\|\kappa'\|,\|\kappa\|] = \frac{1}{2\pi}\int_0^{2\pi} d\varphi\, e^{-iN\varphi} N[\|\kappa'\|0,\|\kappa\|\varphi], \qquad (7.75a)$$

7 The Nuclear Energy Density Functional Formalism

$$\mathscr{E}^N[\|\kappa'\|,\|\kappa\|]\mathscr{N}^N[\|\kappa'\|,\|\kappa\|]$$
$$= \frac{1}{2\pi}\int_0^{2\pi} d\varphi\, e^{-iN\varphi} E[\|\kappa'\|0,\|\kappa\|\varphi]N[\|\kappa'\|0,\|\kappa\|\varphi]. \quad (7.75b)$$

Setting $\varphi = 0$ into Eqs. (7.74a), (7.74b) provides a sum rule relating particle-number-restored energy and norm kernels to un-rotated particle-number-breaking kernels, i.e.

$$N[\|\kappa'\|0,\|\kappa\|0] \equiv \sum_{N\in\mathbb{Z}} \mathscr{N}^N[\|\kappa'\|,\|\kappa\|], \quad (7.76a)$$

$$E[\|\kappa'\|0,\|\kappa\|0]N[\|\kappa'\|0,\|\kappa\|0] \equiv \sum_{N\in\mathbb{Z}} \mathscr{E}^N[\|\kappa'\|,\|\kappa\|]\mathscr{N}^N[\|\kappa'\|,\|\kappa\|]. \quad (7.76b)$$

Further setting $\|\kappa'\| = \|\kappa\|$ provides two sum rules

$$1 = \sum_{N\in\mathbb{Z}} \mathscr{N}^N_{\|\kappa\|}, \quad (7.77a)$$

$$E^{SR}_{\|\kappa\|} = \sum_{N\in\mathbb{Z}} \mathscr{N}^N_{\|\kappa\|}\mathscr{E}^N_{\|\kappa\|}, \quad (7.77b)$$

the second of which relates, for a given value of $\|\kappa\|$, the SR energy to the whole set of particle-number restored energies $\mathscr{E}^N_{\|\kappa\|}$.

7.5.1.2 Specification to $SO(3)$

Of particular interest is the specification of Eqs. (7.70a), (7.70b)–(7.73a), (7.73b) to the $SO(3)$ group, i.e. to angular-momentum restoration (AMR). Singling out the order parameter associated with the breaking of angular momentum and omitting the other collective variables at play, one obtains the expansion of the kernels

$$N[\rho'_{\lambda\mu}0,\rho_{\lambda\mu}\Omega] \equiv \sum_{JMK} \mathscr{N}^J_{MK}[\rho'_{\lambda\mu},\rho_{\lambda\mu}]\mathscr{D}^J_{MK}(\Omega), \quad (7.78a)$$

$$E[\rho'_\lambda 0,\rho_{\lambda\mu}\Omega]N[\rho'_{\lambda\mu}0,\rho_{\lambda\mu}\Omega]$$
$$\equiv \sum_{JMK} \mathscr{E}^J_{MK}[\rho'_{\lambda\mu},\rho_{\lambda\mu}]\mathscr{N}^J_{MK}[\rho'_{\lambda\mu},\rho_{\lambda\mu}]\mathscr{D}^J_{MK}(\Omega). \quad (7.78b)$$

Following Sect. 7.5.1, one extracts angular-momentum restored kernels through

$$\mathscr{N}^J_{MK}[\rho'_{\lambda\mu},\rho_{\lambda\mu}] = \frac{2J+1}{16\pi^2}\int_{D_{SO(3)}} d\Omega\, \mathscr{D}^{J*}_{MK}(\Omega)N[\rho'_{\lambda\mu}0,\rho_{\lambda\mu}\Omega], \quad (7.79a)$$

$$\mathcal{E}^J_{MK}[\rho'_{\lambda\mu},\rho_{\lambda\mu}]\mathcal{N}^J_{MK}[\rho'_{\lambda\mu},\rho_{\lambda\mu}]$$
$$=\frac{2J+1}{16\pi^2}\int_{D_{SO(3)}}d\Omega\,\mathscr{D}^{J*}_{MK}(\Omega)E[\rho'_{\lambda\mu}0,\rho_{\lambda\mu}\Omega]N[\rho'_{\lambda\mu}0,\rho_{\lambda\mu}\Omega]. \quad (7.79b)$$

Setting $\Omega = 0$ into Eqs. (7.78a), (7.78b) provides a sum rule relating angular-momentum restored energy and norm kernels to un-rotated angular-momentum breaking kernels, i.e.

$$N[\rho'_{\lambda\mu}0,\rho_{\lambda\mu}0] \equiv \sum_{JM}\mathcal{N}^J_{MM}[\rho'_{\lambda\mu},\rho_{\lambda\mu}], \quad (7.80a)$$

$$E[\rho'_{\lambda\mu}0,\rho_{\lambda\mu}0]N[\rho'_{\lambda\mu}0,\rho_{\lambda\mu}0] \equiv \sum_{JM}\mathcal{E}^J_{MM}[\rho'_{\lambda\mu},\rho_{\lambda\mu}]\mathcal{N}^J_{MM}[\rho'_{\lambda\mu},\rho_{\lambda\mu}]. \quad (7.80b)$$

Further setting $\rho'_{\lambda\mu} = \rho_{\lambda\mu}$ provides two sum rules

$$1 = \sum_J (2J+1)\mathcal{N}^J_{\rho_{\lambda\mu}}, \quad (7.81a)$$

$$E^{SR}_{\rho_{\lambda\mu}} = \sum_J (2J+1)\mathcal{N}^J_{\rho_{\lambda\mu}}\mathcal{E}^J_{\rho_{\lambda\mu}}, \quad (7.81b)$$

the second of which relates, for a given value of $\rho_{\lambda\mu}$, the SR energy to the whole set of angular-momentum restored energies $\mathcal{E}^J_{\rho_{\lambda\mu}}$.

7.5.2 Full Fledged MR Mixing

In practice, PNR and AMR are often combined. To make formula bearable, we come back to a generic symmetry group. Starting from the symmetry-restored kernels extracted through Eqs. (7.71a), (7.71b), one mixes the components[27] of the targeted Irrep and further performs the mixing over the norm of the order parameter to define the MR energy through

$$E^{MR}_{\lambda k} \equiv \underset{f^{\lambda k*}_{|g'|a}}{\text{Min}}\left\{\frac{\sum_{|g|,|g'|}\sum_{a,b}f^{\lambda k*}_{|g'|a}f^{\lambda k}_{|g|b}\mathcal{E}^\lambda_{ab}[|g'|,|g|]\mathcal{N}^\lambda_{ab}[|g'|,|g|]}{\sum_{|g|,|g'|}\sum_{a,b}f^{\lambda k*}_{|g'|a}f^{\lambda k}_{|g|b}\mathcal{N}^\lambda_{ab}[|g'|,|g|]}\right\}. \quad (7.82)$$

Mixing coefficients $f^{\lambda k}_{|g|b}$ are determined by solving the Hill-Wheeler equation of motion [122] obtained as a result of minimization (7.82)

$$\sum_{|g|b}\mathcal{E}^\lambda_{ab}[|g'|,|g|]\mathcal{N}^\lambda_{ab}[|g'|,|g|]f^{\lambda k}_{|g|b} = E^{MR}_{\lambda k}\sum_{|g|b}\mathcal{N}^\lambda_{ab}[|g'|,|g|]f^{\lambda k}_{|g|b}. \quad (7.83)$$

---

[27] Such a mixing does not appear in the case of the $U(1)$ group given that its Irreps are of dimension 1.

Equation (7.83) denotes an eigenvalue problem, expressed in a non-orthogonal basis, whose eigen-solution is nothing but the MR energy $E_{\lambda k}^{MR}$. As a matter of fact, Eq. (7.83) provides a complete set of excitation energies $\{E_{\lambda k}^{MR}; k = 0, 1, 2, \ldots\}$ for each value of the symmetry quantum number λ. As such, one accesses the low-lying collective spectroscopy along with associated correlations in the ground state.

7.5.3 Pseudo-potential-based Energy Kernel

In the particular case of a pseudo-potential-based EDF kernel, the MR energy (Eq. (7.82)) can be factorized into a more conventional form invoking a MR *wave function*. The derivation provided below does *not* hold when employing an EDF kernel that does not strictly derive from a pseudo Hamiltonian, e.g. for any of the modern Skyrme, Gogny and relativistic parametrizations. As such, the MR energy $E_{\lambda k}^{MR}$ cannot be expressed in terms of a MR wave-function in the most general EDF context, e.g. when using a *density-dependent* "Hamiltonian". Such a fact is systematically overlooked in standard presentations of the EDF theory, which constitutes a problem given the intimate connection between such a feature and the pathologies alluded to in Sect. 7.5.8.

In virtue of Eq. (7.21a), one can first re-express the symmetry-restored energy and norm kernels (Eqs. (7.71a), (7.71b)) according to

$$\mathcal{N}_{ab}^{\lambda}[|g'|,|g|] = \langle \Phi^{(|g'|0)}|P_{ab}^{\lambda}|\Phi^{(|g|0)}\rangle, \tag{7.84a}$$

$$\mathcal{E}_{ab}^{\lambda}[|g'|,|g|]\mathcal{N}_{ab}^{\lambda}[|g'|,|g|] = \langle \Phi^{(|g'|0)}|H_{\text{pseudo}}P_{ab}^{\lambda}|\Phi^{(|g|0)}\rangle, \tag{7.84b}$$

where the transfer operator is introduced as

$$P_{ab}^{\lambda} \equiv \frac{d_\lambda}{v_\mathscr{G}} \int_{D_\mathscr{G}} dm(\alpha) S_{ab}^{\lambda*}(\alpha) R(\alpha). \tag{7.85}$$

Further considering that $P_{ac}^{\lambda} P_{db}^{\zeta} = \delta_{\lambda\zeta}\delta_{cd}P_{ab}^{\lambda}$ and that $[H_{\text{pseudo}}, P_{ac}^{\lambda}] = 0$, as well as that $P_{ac}^{\lambda} = (P_{ca}^{\lambda})^{\dagger}$, one can finally factorize the full fledged MR energy according to

$$E_{\lambda k}^{MR} \equiv \underset{|\Psi_k^{\lambda c}\rangle}{\text{Min}} \left\{ \frac{\langle \Psi_k^{\lambda c}|H_{\text{pseudo}}|\Psi_k^{\lambda c}\rangle}{\langle \Psi_k^{\lambda c}|\Psi_k^{\lambda c}\rangle} \right\}, \tag{7.86}$$

where the MR *wave-function* is defined by

$$|\Psi_k^{\lambda c}\rangle \equiv \sum_{|g|}\sum_{b} f_{|g|b}^{\lambda k} P_{cb}^{\lambda}|\Phi^{(|g|0)}\rangle, \tag{7.87}$$

and where the mixing coefficients are obtained through Eq. (7.83). In such a context, one recovers the textbook Hamiltonian-based GCM [51] performed along the variable $|g|$ on the basis of symmetry-projected HFB wave-functions.

7.5.4 Other Observables

Other observables besides binding energies and low-lying excitation spectra can be extracted from MR-EDF calculations, once Eq. (7.83) has been solved. Typical quantities of interest are expectation values and transition matrix elements of electromagnetic and electroweak operators. Recently, ground-state density distributions have also been extracted [123, 124] whereas transition densities or pair transfer form factors could be calculated in the future.

The archetypal quantity one wishes to compute is the $B(E2)$ [125]

$$B(E2; J'_{k'} \to J_k) = \frac{1}{2J'+1} \sum_{M=-J}^{+J} \sum_{M'=-J'}^{+J'} \sum_{\mu=-2}^{+2} |\langle \Psi_k^{JM} | Q_{2\mu} | \Psi_{k'}^{J'M'} \rangle|^2, \quad (7.88)$$

where the electric quadrupole moment operator $Q_{2\mu} = e \sum_p r_p^2 Y_{2\mu}(\Omega_p)$ is written for point protons with their bare electric charge e. Independent of whether one uses a pseudo-potential EDF kernel or not, auxiliary observables are computed as matrix elements of bare operators in between MR wave-functions. The latter can always been built according to Eq. (7.87) as soon as Eq. (7.83) is solved to extract $f_{|g|b}^{\lambda k}$. In view of the overall accuracy of the method, the current agreement of computed, e.g., $B(E2)$ or $B(E3)$ values with experimental data is considered to be reasonably good and justifies this common practice. Would the accuracy of the method improve significantly, one could consider going beyond such a paradigm by, e.g., designing density functional kernels for auxiliary observables as well.

In the present context, computing Eq. (7.88) eventually boils down to evaluating the matrix element of a tensor operator, e.g. $Q_{2\mu}$, in between two reference states on which different transition operators are applied. Coming back to our general notations, this corresponds to computing

$$\langle \Phi^{(|g'|0)} | P_{a'c'}^{\lambda'} T_\mu^{\lambda''} P_{ca}^{\lambda} | \Phi^{(|g|0)} \rangle = \frac{2\lambda+1}{2\lambda'+1} (\lambda \lambda'' \lambda' | c \mu c') \sum_{\nu=-\lambda}^{+\lambda} (\lambda \lambda'' \lambda' | a, a-\nu, \nu)$$

$$\times \langle \Phi^{(|g'|0)} | P_{a'\nu}^{\lambda'} T_{a-\nu}^{\lambda''} | \Phi^{(|g|0)} \rangle, \quad (7.89)$$

where the matrix element appearing on the right-hand side can eventually be evaluated, after expanding $P_{a'\nu}^{\lambda'}$ according to Eq. (7.85), on the basis of the generalized Wick theorem [85].

7.5.5 Dynamical Correlations

Let us now summarize the way correlations are incorporated in the nuclear EDF approach. The power of the method relies on (i) the parametrization of the "bulk" of correlations, i.e. the part of the binding energy that varies smoothly with neutron and/or proton numbers, under the form of a functional of the one-body density ma-

7 The Nuclear Energy Density Functional Formalism

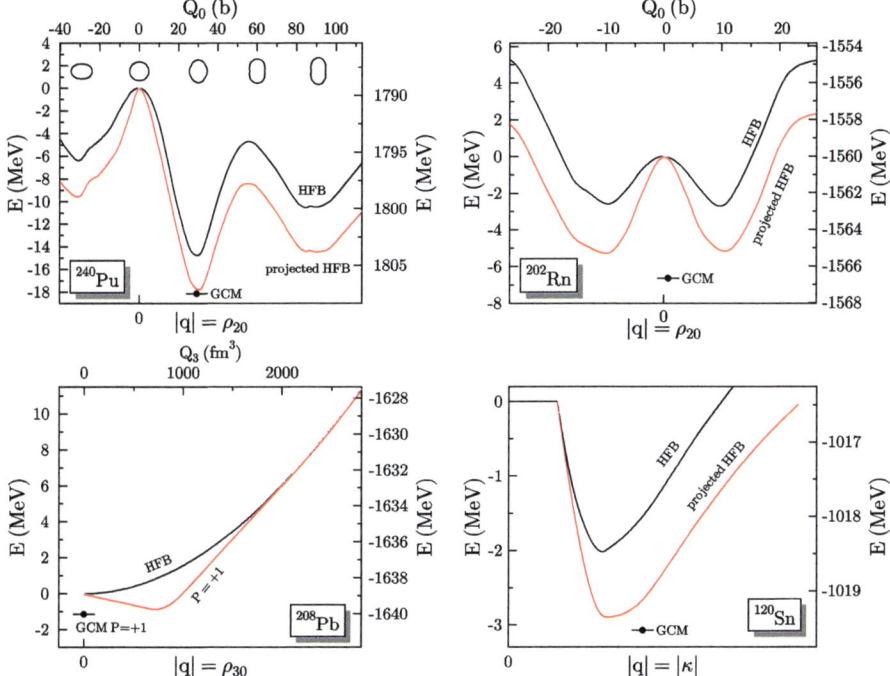

Fig. 7.5 *Upper panels*: energy of ^{240}Pu and ^{202}Rn as a function of the axial quadrupole degree of freedom ($|g| \equiv \rho_{20}$): single-reference calculation (*full black line*), with the added effect of particle number and ($J = 0$) angular momentum restorations (*full red line*) as well as of the shape mixing along $|g| \equiv \rho_{20}$ (*black circle* labelled as "GCM"). *Lower left panel*: energy of ^{208}Pb as a function of the axial octupole degree of freedom ($|g| \equiv \rho_{30}$): single-reference calculation (*full black line*), with the added effect of (positive) parity restoration (*full red line*) and mixing of shapes along $|g| \equiv \rho_{30}$ (*black circle* labelled as "GCM"). *Lower right panel*: energy of ^{120}Sn as a function of the pairing degree of freedom ($|g| \equiv \|\kappa\|$): single-reference calculation (*full black line*), with the added effect of neutron number restoration (*full red line*) and mixing along $|g| \equiv \|\kappa\|$ (*black circle* labelled as "GCM"). *Left vertical axes* are rescaled with respect to the symmetry conserving, i.e. non-deformed, reference point. Please note that $|q|$ stands for $|g|$ in the present figure. Taken from Ref. [108]

trices and on (ii) the grasping of correlations that vary quickly with the filling of nuclear shells through the breaking of symmetries along with the subsequent treatment of the fluctuations of the associated order parameters. Incorporating the second type of correlations within symmetry-conserving approaches, e.g. the CI method, would necessitate tremendous computational efforts in heavy open-shell nuclei.

Of course, the success of the approach eventually relies on the validity of the empirical decoupling between the bulk of correlations and those that are more explicitly accounted for. To some extent, the different scales that characterize these two categories of correlations play in favour of such an empirical decoupling. Let us come back to the four nuclei considered in Fig. 7.3 to illustrate this point. Figure 7.5 separates the binding energy of ^{240}Pu, ^{202}Rn, ^{208}Pb and ^{120}Sn into

Table 7.4 Schematic classification of correlation energies as they naturally appear in the nuclear EDF method. The quantity A_{val} denotes the number of valence nucleons while G_{deg} characterizes the degeneracy of the valence major shell

Correlation energy	Treatment	Scales as	Varies with
Bulk	Summed into EDF kernel	~ 8 A MeV	A
Static collective	Finite order parameter $|g|$	$\lesssim 25$ MeV	$A_{\text{val}}, G_{\text{deg}}$
Dynamical collective	Fluctuations of g	$\lesssim 5$ MeV	$A_{\text{val}}, G_{\text{deg}}$

1. the symmetry conserving SR energy,
2. the symmetry-unrestricted SR energy,
3. the symmetry-restored MR energy,
4. the full fledged MR-EDF energy.

The symmetry conserving SR-EDF result (full black line at $|g| = 0$) provides the "bulk" part of the energy and accounts for, at least, 98 % of the binding energy. Authorizing the breaking of symmetries (absolute minimum of the full black line) does not bring anything to stable double closed-shell nuclei such as ^{208}Pb. However, the spontaneous breaking of rotational symmetry brings up to 20 MeV correlation energy in heavy double open-shell nuclei such as ^{240}Pu, which accounts for about 2 % of the binding. In a transitional nucleus such as ^{202}Rn, the symmetry breaking only accounts for 2 MeV but it signals that such a nucleus should not even be considered at the SR level because of the anticipated large amplitude fluctuations. Superfluidity associated with the breaking of neutron and/or proton numbers typically accounts for 2 MeV in singly-open shell nuclei such as ^{120}Sn. Most important, including pairing is mandatory to describe other observables, e.g. the odd-even mass staggering, individual excitations of even-even nuclei or the moment of inertia of rotating systems. Restoring symmetries (absolute minimum of the full red line) brings in additional correlations, even in nuclei whose SR minimum is symmetry conserving. Typically, restoring angular momentum (^{240}Pu and ^{202}Rn), parity (^{208}Pb) or neutron number (^{120}Sn) add between 1 MeV and 3 MeV correlation energy, depending on how much the symmetry is broken in the first place. Last but not least, the fluctuations of $|g|$ ("GCM" circle) differentiate nuclei that are stiff (i.e. ^{240}Pu, ^{208}Pb, ^{120}Sn) from those that are soft (e.g. ^{202}Rn) with respect to the collective degree of freedom under study. While the correlation energy is of the order of one or two hundreds keV in the former, it can be as large as 1 MeV in the latter. Although the examples discussed here are only illustrative, they are quite representative of the various behaviours one may encounter. Eventually, Table 7.4 recall the various categories of correlations at play and summarizes schematically the scale and the scaling that characterize them. For systematic studies on how correlations impact binding energies and other observables in the context of MR-EDF calculations, see Refs. [126–128].

7.5.6 State-of-the Art Calculations

As of today, full fledged MR-EDF calculations are limited to even-even nuclei. In their most advanced form, they simultaneously restore neutron number, proton number and angular momentum from triaxially deformed Bogoliubov states and further perform the mixing of quadrupole shapes ($|g| = \rho_{2\mu}$ with $\mu = -2, 0, 2$). Such calculations are available for non-relativistic Skyrme [125] and Gogny [129] functionals as well as for relativistic Lagrangians [130]. Still, those cutting-edge calculations are currently limited to light nuclei such that approximations are needed (e.g. limiting oneself to axially deformed shapes) to tackle heavy nuclei. An important effort is also being pursued to restore both good angular momentum and isospin from triaxially deformed Slater determinants [131]. This is relevant to the evaluation of isospin mixing and isospin-breaking corrections to super-allowed β-decay in view of testing the unitarity of the CKM matrix [132]. The versatility of the method also permits to address delicate questions such as the quest of neutrino-less double β-decay to pin down the Dirac or Majorana character of neutrinos [133].

The current forefront corresponds to extending MR-EDF schemes in several (complementary) directions. First and foremost, it is crucial to have the ability to perform MR-EDF calculations of odd-even and odd-odd nuclei. This poses a great technical challenge [72] but will extend the reach of the method tremendously and greatly enhance the synergy with upcoming experimental studies. Along the same line, MR-EDF schemes must be extended such as to include *diabatic* effects [134], i.e. configurations generated through an even number of quasi-particle excitations. This is expected to improve significantly the description of, e.g., the first 2^+ excited state in near-spherical nuclei and to allow a clean description of K isomers. Also of importance is the implementation of the MR method on the basis of references states generated through *cranked* SR calculations, i.e. calculations employing a constraints on $\langle \Phi^{(g)} | J_{x,y,z} | \Phi^{(g)} \rangle \neq 0$ [135–137]. By accounting for Coriolis effects, this is expected to improve moments of inertia that are systematically too low in MR calculations based on uncranked states. Eventually, state-of-the-art calculations should combine quadrupole and octupole degrees of freedom [138] as well as the mixing over $\|\kappa\|$ [139, 140]. The latter also impacts moment of inertia significantly and authorizes the description of pairing fluctuations and pairing vibrations near closed shell, as well as the computation of pair transfer overlap functions.

All such extensions are particularly timely given that upcoming RIB facilities are accessing an increasingly larger number of short-lived atomic nuclei. Among the latter, exotic nuclei with a large neutron excess are likely to require more systematically the inclusion of MR correlations from the outset, i.e. to be less-good "mean-field" nuclei than those located near the valley of β stability.

7.5.7 Approximations to Full Fledged MR-EDF

Several approximations to or variants of the full fledged MR-EDF approach are being pursued with great success. It is beyond the scope of the present lecture notes to review them. Let us however mention the most important ones and refer the reader to recent associated works.

The quasi-particle random phase approximation (QRPA) can be motivated in many different ways, one of which is the approximation of the MR kernels in the limit where $|\Phi^{(g')}\rangle$ and $|\Phi^{(g)}\rangle$ differ harmonically from a common reference state [83, 84]. Quasi-particle random phase approximation, along with its extensions, provides vibrational excitations of various multipolarities and associated ground-state correlations. This includes low-lying states as well as giant resonances. A limitation of such an approximation is its inability to describe violently anharmonic systems undergoing large amplitude motion. There is a significant on-going effort to develop the method in deformed nuclei [141–145] on the basis of complete EDF parametrizations and efficient algorithms [146–148]. This will permit to address many upcoming challenges including the quest of potentially new exotic vibrational modes [149].

Second is the collective (e.g. Bohr) Hamiltonian that can be motivated in two different ways, one of which is the (topological) Gaussian overlap approximation [150–152] of the transition EDF kernels. In practice, however, inertia parameters are not computed from available full fledged MR-EDF calculations. Indeed, the latter are not complete enough at this point in time to compute inertia parameters reliably. Five-dimensional collective Hamiltonians built from non-relativistic Skyrme [153, 154] and Gogny [155, 156] functionals as well as from relativistic Lagrangians [157] are available. Work is currently being pursued to improve on the Inglis-Belyaev moments of inertia and cranking mass parameters by means of Thouless Valentin [158, 159]. Within such a scheme, low-lying collective spectra of heavy even-even nuclei can be computed while including the full quadrupole dynamics.

Last but not least, it is worth mentioning the recent revival of the interacting boson model (IBM) within a microscopic setting, i.e. based on the mapping of triaxial HFB energy landscapes generated from a Gogny functional [160] or a relativistic Lagrangian [161]. Such a method allows the efficient description of low-lying collective spectra of complex heavy nuclei.

As for full fledged MR-EDF calculations, modern accounts of the three above methods are only available for even-even nuclei. Extensions to odd-even and odd-odd nuclei must be envisioned in the future.

7.5.8 Pathologies of MR-EDF Calculations

In spite of the mathematically sound formulation of the MR-EDF method provided above, pathologies were identified under the form of spurious divergences [162, 163] and steps [53] in potential energy curves obtained from PNR calculations. Examples are given in Fig. 7.6 for two different Skyrme parametrizations of the EDF kernel. The occurrence of such anomalies were analysed in details in Refs. [53, 55, 56] and put in connection with non-analyticities of the energy kernel over the complex plane, after performing the continuation $z = e^{i\varphi}$, where φ

7 The Nuclear Energy Density Functional Formalism   341

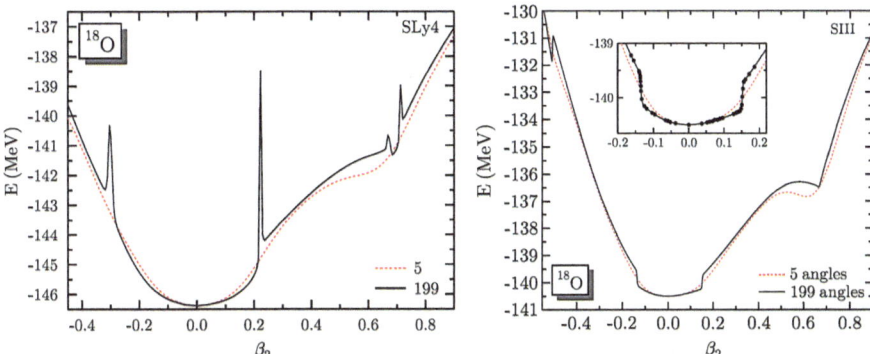

Fig. 7.6 Proton-number restored energy $\mathscr{E}^Z_{\rho_{20}}$ of ^{18}O as a function of the axial quadrupole deformation (β_2 is a dimensionless measure of ρ_{20}) using 5 and 199 discretization points in the integral over the gauge angle (Eqs. (7.75a), (7.75b)). *Left panel*: calculations performed with the SLy4 Skyrme parametrization and a density-independent pairing interaction. *Right panel*: calculations performed with the SIII Skyrme parametrization and a density-independent pairing interaction. Taken from Ref. [55]

denotes the gauge angle characterizing the off-diagonal energy kernel at play (see Sect. 7.3.2). In particular, the problem manifests differently depending on the analytical structure of the EDF kernel [55]. The left panel of Fig. 7.6 is characteristic of the general case where divergences occur whenever a proton and/or neutron single-particle level crosses the Fermi energy [53]. Additionally, the potential energy surface displays finite steps across any such divergence. The right panel of Fig. 7.6 illustrates the particular case of a functional that is strictly bilinear in the density matrices of a given isospin species. In such a situation, no divergence occurs and one is only left with finite discontinuities.

A step towards the formulation of a remedy to the problem was made in Refs. [54–56]. Firstly, the problem was shown to relate to the breaking of Pauli's principle discussed in Sect. 7.3.3. Specifically, spurious contributions associated with self-interaction and self-pairing processes are multiplied with dangerous weights in the *off-diagonal* energy kernel $E[g', g]$, which results in the anomalies illustrated in Fig. 7.6. Secondly, divergences and steps were shown to constitute the visible part of the problem only, i.e. PNR energies are not only contaminated where divergences and steps occur but also *away* from them.

Another striking manifestation of spurious self-interaction and self-pairing processes in PNR calculations was identified in Ref. [55]. Whereas contributions to sum rule (7.77b) corresponding to $N \leq 0$ are zero in the absence of self-interaction and self-pairing, i.e. when working within the pseudo-potential-based approach, non-analyticities of the energy kernel over the complex plane translate into[28] having $\mathscr{N}^N \mathscr{E}^N \neq 0$ for $N \leq 0$. Such a feature is illustrated in Fig. 7.7 for the interaction

---

[28]The overlap kernel being analytical over the complex plane, it is straightforward to prove that $\mathscr{N}^N = 0$ for $N \leq 0$.

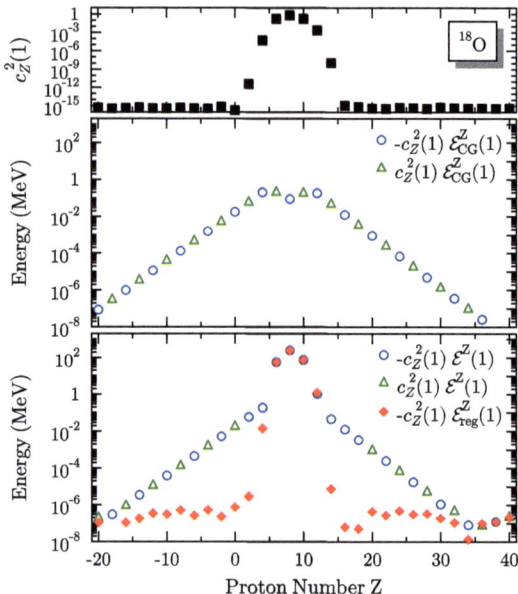

Fig. 7.7 Proton-number-restored kernels as a function of the Z one restores. *Upper panel*: norm kernel $\mathcal{N}^Z$. *Middle panel*: spurious contribution to the (weighted) energy kernels. *Lower panel*: uncorrected $\mathcal{N}^Z \mathcal{E}^Z$ and corrected $\mathcal{N}^Z \mathcal{E}^Z_{\text{REG}}$ proton-number-restored energy kernels. All results are obtained using the same SR state calculated for ^{18}O at a deformation of $\beta_2 = 0.371$. The neutron number is not restored. Taken from Ref. [55]

energy part (i.e. the kinetic energy contribution is omitted) obtained from PNR calculation of ^{18}O. The distribution of absolute values of $\mathcal{N}^Z \mathcal{E}^Z$ as a function of Z does not follow the distribution of the weights $\mathcal{N}^Z$ displayed in the upper panel. Instead, it has a long tail that spreads visibly to $Z = -20$ and $Z = 34$, before it cannot be distinguished from numerical noise anymore. In these tails, $\mathcal{N}^Z \mathcal{E}^Z$ displays alternating signs, which is clearly unphysical.

The fact that PNR calculations do provide non-zero (weighted) energies for negative or null particle numbers is certainly the most illuminating proof that having a mathematically well-founded formalism is necessary but not sufficient to make it physically meaningful, i.e. while mathematics makes sum rule (7.77b) run over all Irreps a priori, physics requires that the expansion coefficients associated with negative integers are zero, which is not guaranteed in general and is not the case for *any* existing modern parametrization of the EDF kernel.

Although most clearly highlighted through PNR calculations, i.e. in calculations realizing the mixing over the gauge angle, pathologies due to the violation of Pauli's principle contaminate *any* type of MR mixing. Figure 7.8 displays the result of a MR-EDF calculation of ^{18}O including both PNR and AMR, and compares it to the result obtained via PNR only.

It is interesting to note at this point that certain approximations to full fledged MR-EDF calculations [48], i.e. calculations based on a collective Hamiltonian or on QRPA, avoid the dramatic pathologies discussed above by bypassing the problem from the outset, i.e. thanks to the approximation to the off-diagonal kernels that define them. However, such methods are not free from less dramatic, i.e. smooth and finite, contaminations associated with the presence of spurious self-interaction and self-pairing in the energy kernel. This question deserves attention in the future.

7 The Nuclear Energy Density Functional Formalism

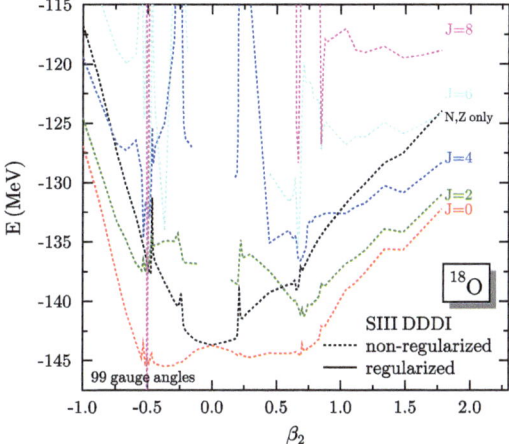

Fig. 7.8 Proton-number- and angular-momentum-restored energies of ^{18}O for various values of J as a function of the axial quadrupole deformation. The integral over the gauge angle (Eqs. (7.75a), (7.75b)) uses 99 discretization points. Calculations are performed with the SIII Skyrme parametrization and a density-dependent pairing interaction. *Solid lines* defined in the legend are not shown in the present figure but are (will be) visible in the original reference [164]. The *curve* labelled with "N, Z only" only performs the restoration of particle number

7.5.9 Towards Pseudo-potential-based Energy Kernels

In order to resolve the difficulties illustrated above, a regularization of the off-diagonal energy kernel was designed for parametrizations that are strictly polynomial in the density matrices [54]. The method was meant to eliminate a posteriori the pathologies contaminating MR-EDF calculations without fully enforcing the Pauli principle from the outset. Exposing the regularization method is beyond the scope of the present document and we refer the reader to Ref. [54] for details. As of today, the regularization method has been implemented not only in pure PNR calculations [55] but also for the most general MR-EDF calculations available [164]. This includes the most advanced ones aiming at the description of odd nuclei [72]. In spite of solving the problem for pure PNR calculations, the regularization method leaves implementations that go beyond it, e.g. calculations mixing PNR and AMR, with unwanted pathologies [164].

As of today, the only viable route to a sound MR-EDF formalism relies on energy kernels that strictly derive from a pseudo-potential [75], i.e. kernels that enforce the Pauli principle from the outset to bypass spurious self-interaction and self-pairing processes. Several efforts [75, 104–106] in this direction are currently being pursued as alluded to in Sect. 7.3.4.5. This constitutes a turning point in the construction of nuclear EDF parametrizations. It is beyond the scope of the present lecture notes to expose such developments. Let us however briefly explain why such a route is not straightforward to follow. As a matter of fact, *none* of the modern, i.e. Skyrme, Gogny or relativistic, parametrizations belong to the category of *strict* pseudo-potential-based EDF kernels. The reason for such a situation is precisely that practitioners have moved away from the strict pseudo-potential-based philosophy throughout the last four decades because of its apparent lack of flexibility and its inability to produce high-quality EDF parametrizations. The challenge is thus to develop pseudo-potentials that are more general than those considered in the past such that they can provide a high-quality phenomenology. The pseudo potentials must however be simple enough for the fit of its free parameters to be meaningfully handled. Several new families of EDF parametrizations strictly deriving from pseudo potentials and allowing for safe MR-EDF calculations can be expected to be published in the coming years.

7.5.10 Towards Non-empirical Energy Kernels

On the longer term, it is mandatory to go beyond the empirical formulation of the nuclear EDF method in order to augment its predictive power. This requires the design of ab-initio many-body methods from which both SR- and MR implementations of the EDF method, i.e. both diagonal and off-diagonal energy functional kernels, can be derived through a set of controlled approximations. This is meant to lead to so-called *non-empirical* energy functionals possessing a link to the underlying nuclear Hamiltonian describing few-body scattering and bound-state observables. The objective is not to replace but rather complement the development of empirical EDFs based on trial and error by combining the predictive character of an ab-initio method with the gentle numerical scaling of EDF calculations. Indeed, while empirical EDFs already achieve an accuracy for known observable that will be difficult, if not impossible, to reach with purely non-empirical functionals, they lack predictive power away from the experimentally known region of the nuclear chart.

The first way to improve on such limitations consists of using "pseudo-data" generated from ab-initio calculations for nuclei located in the experimentally unknown region (i) for the fitting procedure of EDF parametrizations and (ii) to benchmark extrapolations from such EDF parametrizations. In this way, unknown couplings of the empirical EDF parametrization can be "microscopically" constrained. Eventually, the goal is to discriminate between different functional forms. The benefit of such an indirect approach is that any ab-initio method that can provide precise enough benchmarks for the systems and observables of interest can be employed. However, no direct/explicit connection with vacuum interactions is realized such that no specific insight about the *form* of new functional terms that could capture the missing physics is easily gained in this way, i.e. the predictive power of EDF calculations away from the benchmarks remains bound to the quality of the postulated functional form such that improvements still rely on trial and error.

A greater challenge is to connect explicitly the *form* of the energy functional kernel, in addition to the *value* of its couplings, to vacuum nuclear interactions. One is essentially looking for *microscopically-educated guesses*. Ground-breaking, though very incomplete, works in this direction have been undertaken recently [165–170]. Eventually, a fine-tuning of the couplings, within the intrinsic error bars with which they will have been produced, can be envisioned [171]. In this context, microscopically-educated functionals are to be derived through analytical approximations of the ground-state energy computed via a given ab-initio method of reference (preferably the same as the one providing benchmarks for observable quantities). It is a challenging task whose complexity depends on the nuclear Hamiltonian and the many-body method one starts from. In particular, ab-initio methods that are amenable to such a mapping must share certain key features of the nuclear EDF method, the most important of which being the notion of spontaneous symmetry breaking. Let us take the part of the EDF that drives superfluidity as an example, i.e. the part that depends on the anomalous pairing tensor $\kappa_{ij}^{g'g}$ (see Sect. 7.2.3). Such a functional dependence of the EDF kernel exists only because pairing correlations are grasped through the breaking of good particle-number associated with $U(1)$ gauge symmetry. Deriving microscopically-educated EDF kernels can thus only be achieved starting from an ab-initio method that also incorporates pairing correlations through the breaking of $U(1)$ gauge symmetry.

7.6 Conclusions

Very significant advances have been made in the last 15 years within the frame of the nuclear energy density functional method. In doing so, the focus of the field has shifted in several respects, with the consequences that

1. routine applications have moved from SR to MR calculations,
2. one can address, e.g. neutron-rich, nuclei that do not fit the mean-field paradigm,
3. applications are now equally dedicated to ground and excited states,
4. one can provide both

    (a) the detailed quantitative picture of a given system of interest,
    (b) study trends through large-scale MR calculations,

5. advances in the field are bound to making consistent progress regarding

    (a) the foundations of the approach and its formal consistency,
    (b) the rooting of EDFs into basic many-body methods and interactions,
    (c) the building of EDFs from improved fitting protocols,
    (d) the building of EDF parametrizations from enlarged data sets,
    (e) the further development of powerful numerical tools,

    while points (a), (b) and (c) were essentially discarded 15 years ago,
6. applications more strongly impact astrophysics and particle physics.

The field is expected to move forward in these directions in the next 10 years. Most probably, this will be the era of the strong overlapping with emerging ab-initio methods for mid-mass nuclei and of the materialization of powerful numerical tools dedicated to the description of odd-even and odd-odd nuclei. In addition to these already on-going trends, one can expect surprises to emerge that will guide the development of the EDF methods in new directions.

Acknowledgements It is a great pleasure to thank deeply all those I have had the chance to collaborate with on topics related to the matter of the present lecture notes, i.e. B. Avez, M. Bender, K. Bennaceur, P. Bonche, B. A. Brown, P.-H. Heenen, D. Lacroix, T. Lesinski, J. Meyer, V. Rotival, J. Sadoudi, N. Schunck and C. Simenel. I also wish to thank M. Bender for providing me with several of the figures that are used in the present lecture notes.

Appendix: F-Functions

Kinetic densities are expressed in INM in terms of functions $F_m^{(0)}(I_\tau, I_\sigma, I_{\sigma\tau})$, $F_m^{(\tau)}(I_\tau, I_\sigma, I_{\sigma\tau})$, $F_m^{(\sigma)}(I_\tau, I_\sigma, I_{\sigma\tau})$ and $F_m^{(\sigma\tau)}(I_\tau, I_\sigma, I_{\sigma\tau})$ defined through [114]

$$F_m^{(0)} \equiv \frac{1}{4}\big[(1 + I_\tau + I_\sigma + I_{\sigma\tau})^m + (1 + I_\tau - I_\sigma - I_{\sigma\tau})^m$$
$$+ (1 - I_\tau + I_\sigma - I_{\sigma\tau})^m + (1 - I_\tau - I_\sigma + I_{\sigma\tau})^m\big], \quad (7.90a)$$

$$F_m^{(\tau)} \equiv \frac{1}{4}\big[(1 + I_\tau + I_\sigma + I_{\sigma\tau})^m + (1 + I_\tau - I_\sigma - I_{\sigma\tau})^m$$
$$- (1 - I_\tau + I_\sigma - I_{\sigma\tau})^m - (1 - I_\tau - I_\sigma + I_{\sigma\tau})^m\big], \quad (7.90b)$$

$$F_m^{(\sigma)} \equiv \frac{1}{4}\big[(1 + I_\tau + I_\sigma + I_{\sigma\tau})^m - (1 + I_\tau - I_\sigma - I_{\sigma\tau})^m$$
$$+ (1 - I_\tau + I_\sigma - I_{\sigma\tau})^m - (1 - I_\tau - I_\sigma + I_{\sigma\tau})^m\big], \quad (7.90c)$$

$$F_m^{(\sigma\tau)} \equiv \frac{1}{4}\big[(1 + I_\tau + I_\sigma + I_{\sigma\tau})^m - (1 + I_\tau - I_\sigma - I_{\sigma\tau})^m$$
$$- (1 - I_\tau + I_\sigma - I_{\sigma\tau})^m + (1 - I_\tau - I_\sigma + I_{\sigma\tau})^m\big]. \quad (7.90d)$$

Their first derivatives with respect to spin, isospin and spin-isospin excesses are

$$\frac{\partial F_m^{(\tau)}}{\partial I_\tau} = \frac{\partial F_m^{(\sigma)}}{\partial I_\sigma} = \frac{\partial F_m^{(\sigma\tau)}}{\partial I_{\sigma\tau}} = m F_{m-1}^{(0)}, \quad (7.91a)$$

$$\frac{\partial F_m^{(0)}}{\partial I_\tau} = \frac{\partial F_m^{(\sigma)}}{\partial I_{\sigma\tau}} = \frac{\partial F_m^{(\sigma\tau)}}{\partial I_\sigma} = m F_{m-1}^{(\tau)}, \quad (7.91b)$$

$$\frac{\partial F_m^{(0)}}{\partial I_\sigma} = \frac{\partial F_m^{(\tau)}}{\partial I_{\sigma\tau}} = \frac{\partial F_m^{(\sigma\tau)}}{\partial I_\tau} = m F_{m-1}^{(\sigma)}, \quad (7.91c)$$

$$\frac{\partial F_m^{(0)}}{\partial I_{\sigma\tau}} = \frac{\partial F_m^{(\tau)}}{\partial I_\sigma} = \frac{\partial F_m^{(\sigma)}}{\partial I_\tau} = m F_{m-1}^{(\sigma\tau)}, \quad (7.91d)$$

while their second derivatives are

$$\frac{\partial^2 F_m^{(j)}}{\partial I_i^2} = m(m-1) F_{m-2}^{(j)}, \quad (7.92)$$

for any $i, j \in \{0, \tau, \sigma, \sigma\tau\}$. Remarkable values are

$$F_0^{(0)}(I_\tau, I_\sigma, I_{\sigma\tau}) = 1, \qquad F_0^{(i)}(I_\tau, I_\sigma, I_{\sigma\tau}) = 0, \qquad (7.93\text{a})$$

$$F_1^{(0)}(I_\tau, I_\sigma, I_{\sigma\tau}) = 1, \qquad F_1^{(i)}(I_\tau, I_\sigma, I_{\sigma\tau}) = I_i, \qquad (7.93\text{b})$$

and

$$F_m^{(0)}(0, 0, 0) = 1, \qquad (7.94\text{a})$$

$$F_m^{(i)}(0, 0, 0) = 0, \qquad (7.94\text{b})$$

$$F_m^{(\tau)}(0, 1, 0) = F_m^{(\tau)}(0, 0, 1) = 0, \qquad (7.94\text{c})$$

$$F_m^{(\sigma)}(1, 0, 0) = F_m^{(\sigma)}(0, 0, 1) = 0, \qquad (7.94\text{d})$$

$$F_m^{(\sigma\tau)}(1, 0, 0) = F_m^{(\sigma\tau)}(0, 1, 0) = 0, \qquad (7.94\text{e})$$

$$F_m^{(0)}(1, 0, 0) = F_m^{(0)}(0, 1, 0) = F_m^{(0)}(0, 0, 1) = 2^{m-1}, \qquad (7.94\text{f})$$

$$F_m^{(\tau)}(1, 0, 0) = F_m^{(\sigma)}(0, 1, 0) = F_m^{(\sigma\tau)}(0, 0, 1) = 2^{m-1}, \qquad (7.94\text{g})$$

$$F_m^{(0)}(1, 1, 1) = F_m^{(i)}(1, 1, 1) = 4^{m-1}, \qquad (7.94\text{h})$$

where $i \in \{\tau, \sigma, \sigma\tau\}$.

References

1. N. Kalantar-Nayestanaki et al., Rep. Prog. Phys. **75**, 016301 (2012)
2. A. Nogga, H. Kamada, W. Glöckle, Phys. Rev. Lett. **85**, 944 (2000)
3. A. Nogga, S.K. Bogner, A. Schwenk, Phys. Rev. C **70**, 061002 (2004)
4. A. Faessler, S. Krewald, G.J. Wagner, Phys. Rev. C **11**, 2069 (1975)
5. J. Fujita, H. Miyazawa, Prog. Theor. Phys. **17**, 360 (1957)
6. W. Zuo et al., Nucl. Phys. A **706**, 418 (2002)
7. F. Coester et al., Phys. Rev. C **1**, 769 (1970)
8. R. Brockmann, R. Machleidt, Phys. Rev. C **42**, 1965 (1990)
9. A. Sonzogni, NNDC Chart of Nuclides (2007). http://www.nndc.bnl.gov/chart/
10. O. Sorlin, M.-G. Porquet, Prog. Part. Nucl. Phys. **61**, 602 (2008)
11. I. Tanihata et al., Phys. Rev. Lett. **55**, 2676 (1985)
12. M. Fukuda et al., Phys. Lett. B **268**, 339 (1991)
13. P.G. Hansen, B. Jonson, Europhys. Lett. **4**, 409 (1987)
14. A.S. Jensen et al., Rev. Mod. Phys. **76**, 215 (2004)
15. B. Blank, M. Ploszajczak, Rep. Prog. Phys. **71**, 046301 (2008)
16. M. Pfützner et al., Rev. Mod. Phys. **84**, 567 (2012)
17. K. Blaum, Phys. Rep. **425**, 1 (2006)
18. B. Schlitt et al., Hyperfine Interact. **99**, 117 (1996)
19. M. Wang et al., J. Phys. Conf. Ser. **312**, 092064 (2011)
20. I.S. Towner, J.C. Hardy, Rep. Prog. Phys. **73**, 046301 (2010)
21. V. Zagrebaev, A. Karpov, W. Greiner, J. Phys. Conf. Ser. **420**, 012001 (2013)
22. P. Möller et al., At. Data Nucl. Data Tables **59**, 185 (2002)

23. G. Royer, C. Gautier, Phys. Rev. C **73**, 067302 (2006)
24. J.L. Friar et al., Phys. Lett. B **311**, 4 (1988)
25. A. Nogga et al., Phys. Lett. B **409**, 19 (1997)
26. S.C. Pieper, R.B. Wiringa, J. Carlson, Phys. Rev. C **70**, 054325 (2004)
27. S. Pastore et al., arXiv:1302.5091 (2013)
28. P. Navratil et al., J. Phys. G **36**, 083101 (2009)
29. E. Epelbaum et al., Phys. Rev. Lett. **109**, 252501 (2012)
30. G. Hagen et al., Phys. Rev. C **82**, 034330 (2010)
31. S. Binder et al., arXiv:1211.4748 (2012)
32. W.H. Dickhoff, C. Barbieri, Prog. Part. Nucl. Phys. **52**, 377 (2004)
33. A. Cipollone, C. Barbieri, P. Navratil, arXiv:1303.4900 (2013)
34. K. Tsukiyama, S.K. Bogner, A. Schwenk, Phys. Rev. Lett. **106**, 222502 (2011)
35. H. Hergert et al., Phys. Rev. C **87**, 034307 (2013)
36. V. Somà, T. Duguet, C. Barbieri, Phys. Rev. C **84**, 064317 (2011)
37. V. Somà, C. Barbieri, T. Duguet, Phys. Rev. C **87**, 011303 (2013)
38. H. Hergert et al., Phys. Rev. Lett. **110**, 242501 (2013)
39. A. Signoracci, T. Duguet, G. Hagen, unpublished (2013)
40. E. Caurier et al., Rev. Mod. Phys. **77**, 427 (2005)
41. D.J. Dean et al., Prog. Part. Nucl. Phys. **53**, 419 (2004)
42. B.A. Brown, W.A. Richter, Phys. Rev. C **74**, 034315 (2006)
43. A.P. Zuker, Phys. Rev. Lett. **90**, 042502 (2003)
44. T. Otsuka et al., Phys. Rev. Lett. **105**, 032501 (2010)
45. J.D. Holt et al., J. Phys. G **39**, 085111 (2012)
46. J.D. Holt, A. Schwenk, Eur. Phys. J. A **49**, 39 (2013)
47. J.D. Holt, J. Engel, Phys. Rev. C **87**, 064315 (2013)
48. M. Bender, P.-H. Heenen, P.-G. Reinhard, Rev. Mod. Phys. **75**, 121 (2003)
49. T. Niksic, D. Vretenar, P. Ring, Prog. Part. Nucl. Phys. **66**, 519 (2011)
50. J.W. Negele, D. Vautherin, Phys. Rev. C **5**, 1472 (1972)
51. P. Ring, P. Schuck, *The Nuclear Many-Body Problem* (Springer, New York, 1980)
52. L.M. Robledo, Int. J. Mod. Phys. E **16**, 337 (2007)
53. J. Dobaczewski et al., Phys. Rev. C **76**, 054315 (2007)
54. D. Lacroix, T. Duguet, M. Bender, Phys. Rev. C **79**, 044318 (2009)
55. M. Bender, T. Duguet, D. Lacroix, Phys. Rev. C **79**, 044319 (2009)
56. T. Duguet et al., Phys. Rev. C **79**, 044320 (2009)
57. V. Rotival, T. Duguet, Phys. Rev. C **79**, 054308 (2009)
58. E. Chabanat et al., Nucl. Phys. A **635**, 231 (1998)
59. V. Rotival, K. Bennaceur, T. Duguet, Phys. Rev. C **79**, 054309 (2009)
60. T. Lesinski et al., Phys. Rev. C **74**, 044315 (2006)
61. T. Lesinski et al., Phys. Rev. C **76**, 014312 (2007)
62. M. Kortelainen et al., Phys. Rev. C **77**, 064307 (2008)
63. M. Bender et al., Phys. Rev. C **80**, 064302 (2009)
64. J. Margueron, H. Sagawa, K. Hagino, Phys. Rev. C **77**, 054309 (2008)
65. T. Niksic, D. Vretenar, P. Ring, Phys. Rev. C **78**, 034318 (2008)
66. B.G. Carlsson, J. Dobaczewski, M. Kortelainen, Phys. Rev. C **78**, 044326 (2008)
67. S. Goriely et al., Phys. Rev. Lett. **102**, 242501 (2009)
68. M. Kortelainen et al., Phys. Rev. C **82**, 024313 (2010)
69. M. Kortelainen et al., Phys. Rev. C **85**, 024304 (2012)
70. T. Duguet, J. Sadoudi, J. Phys. G, Nucl. Part. Phys. **37**, 064009 (2010)
71. T. Duguet et al., Phys. Rev. C **65**, 014310 (2002)
72. B. Bally et al., Int. J. Mod. Phys. E **21**, 1250026 (2012)
73. R.R. Rodriguez-Guzman, K.W. Schmid, Eur. Phys. J. A **19**, 45 (2004)
74. D.A. Varshalovich, A.N. Moskalev, V.K. Khersonskii, *Quantum Theory of Angular Momentum* (World Scientific, Singapore, 1988)

75. J. Sadoudi, Constraints on the nuclear energy density functional and new possible analytical forms (2011). Université Paris XI, France, http://tel.archives-ouvertes.fr/docs/00/04/49/86/PDF/tel-00001784.pdf
76. L.M. Robledo, Phys. Rev. C **79**, 021302 (2009)
77. L.M. Robledo, Phys. Rev. C **84**, 014307 (2011)
78. B. Avez, M. Bender, Phys. Rev. C **85**, 034325 (2012)
79. M. Oi, M. Takahiro, Phys. Lett. B **707**, 305 (2012)
80. Z.-C. Gao, Q.-L. Hu, Y.S. Chen, arXiv:1306.3051 (2013)
81. L.M. Robledo, J. Phys. G **37**, 064020 (2010)
82. A. Kamlah, Z. Phys. **216**, 52 (1968)
83. B. Jancovici, D.H. Schiff, Nucl. Phys. **58**, 678 (1964)
84. D.M. Brink, A. Weiguny, Nucl. Phys. A **120**, 59 (1968)
85. R. Balian, E. Brézin, Nuovo Cimento **64**, 37 (1969)
86. E. Perlinska et al., Phys. Rev. C **69**, 014316 (2004)
87. J. Dobaczewski, J. Dudek, Phys. Rev. C **52**, 1827 (1995)
88. J.P. Perdew, A. Zunger, Phys. Rev. B **23**, 5048 (1981)
89. N. Chamel, Phys. Rev. C **82**, 061307 (2010)
90. A. Ruzsinsky et al., J. Phys. Chem. **126**, 104102 (2007)
91. T. Duguet, Phys. Rev. C **69**, 054317 (2004)
92. M. Yamagami, Y.R. Shimizu, T. Nakatsukasa, Phys. Rev. C **80**, 064301 (2009)
93. N. Chamel, Phys. Rev. C **82**, 014313 (2010)
94. M. Yamagami et al., Phys. Rev. C **86**, 034333 (2012)
95. J. Margueron, H. Sagawa, J. Phys. G **36**, 125102 (2009)
96. B. Cochet et al., Nucl. Phys. A **731**, 34 (2004)
97. B. Cochet et al., Int. J. Mod. Phys. E **13**, 187 (2004)
98. T. Lesinski et al., Eur. Phys. J. A **40**, 121 (2009)
99. T. Lesinski et al., J. Phys. G **39**, 015108 (2012)
100. M. Zalewski, P. Olbratowski, W. Satula, Phys. Rev. C **81**, 044314 (2010)
101. A.F. Fantina et al., J. Phys. G **38**, 025101 (2011)
102. E. Moya de Guerra, O. Moreno, P. Sarriguren, J. Phys. Conf. Ser. **312**, 092045 (2011)
103. D. Davesne, A. Pastore, J. Navarro, J. Phys. G **40**, 095104 (2013)
104. J. Sadoudi et al., Phys. Scr. T **154**, 014013 (2013)
105. J. Dobaczewski, K. Bennaceur, F. Raimondi, J. Phys. G **39**, 125103 (2012)
106. K. Bennaceur, J. Dobaczewski, F. Raimondi, arXiv:1305.7210 (2013)
107. A. Staszczak et al., Eur. Phys. J. A **46**, 85 (2010)
108. M. Bender, private communication (2013)
109. M. Baranger, Nucl. Phys. A **149**, 225 (1970)
110. T. Duguet, G. Hagen, Phys. Rev. C **85**, 034330 (2012)
111. J. Sadoudi, T. Duguet, unpublished (2013)
112. T. Duguet, unpublished (2013)
113. N. Chamel, S. Goriely, J.M. Pearson, Pairing: from atomic nuclei to neutron-star crusts, in *Fifty Years of Nuclear BCS: Pairing in Finite Systems*, ed. by R. Broglia, W. Zelevinsky (World Scientific, Singapore, 2013), p. 284
114. M. Bender, J. Dobaczewski, J. Engel, W. Nazarewicz, Phys. Rev. C **65**, 054322 (2002)
115. P. Hohenberg, W. Kohn, Phys. Rev. **136**, B864 (1964)
116. H.A. Fertig, W. Kohn, Phys. Rev. A **62**, 052511 (2000)
117. E.K.U. Gross, L.N. Oliveira, W. Kohn, Phys. Rev. A **37**, 2809 (1988)
118. A. Gorling, Phys. Rev. A **47**, 2783 (1993)
119. J. Engel, Phys. Rev. C **75**, 014306 (2007)
120. J. Messud, M. Bender, E. Suraud, Phys. Rev. C **80**, 054314 (2009)
121. T. Lesinski, arXiv:1301.0807 (2013)
122. D.L. Hill, J.A. Wheeler, Phys. Rev. **89**, 1106 (1953)
123. J.-M. Yao et al., Phys. Rev. C **86**, 014310 (2012)
124. J.-M. Yao, H. Mei, Z.P. Li, Phys. Lett. B **723**, 459 (2013)

125. M. Bender, P.-H. Heenen, Phys. Rev. C **78**, 024309 (2008)
126. M. Bender, G.F. Bertsch, P.-H. Heenen, Phys. Rev. C **73**, 034322 (2006)
127. M. Bender, G.F. Bertsch, P.-H. Heenen, Phys. Rev. C **78**, 054312 (2008)
128. L.M. Robledo, G.F. Bertsch, Phys. Rev. C **84**, 054302 (2011)
129. T.R. Rodriguez, J.L. Egido, Phys. Rev. C **81**, 064323 (2010)
130. J.-M. Yao et al., Phys. Rev. C **81**, 044311 (2010)
131. W. Satula et al., Phys. Rev. C **81**, 054310 (2010)
132. W. Satula et al., Phys. Rev. C **86**, 054316 (2012)
133. T.R. Rodriguez, G. Martinez-Pinedo, Phys. Rev. C **85**, 044310 (2012)
134. T. Duguet, Problème à N corps nucléaire et force effective dans les méthodes de champ moyen auto-cohérent (2002). http://tel.archives-ouvertes.fr/docs/00/04/49/86/PDF/tel-00001784.pdf
135. D. Baye, P.-H. Heenen, Phys. Rev. C **29**, 1056 (1984)
136. H. Zdunczuk, J. Dobaczewski, W. Satula, Int. J. Mod. Phys. E **16**, 377 (2007)
137. B. Avez et al., unpublished (2013)
138. J. Meyer et al., Nucl. Phys. A **588**, 597 (1995)
139. M. Bender, T. Duguet, Int. J. Mod. Phys. E **16**, 222 (2007)
140. N.L. Vaquero, T.R. Rodriguez, J.L. Egido, Phys. Lett. B **704**, 520 (2011)
141. S. Peru, H. Goutte, Phys. Rev. C **77**, 044313 (2008)
142. K. Yoshida, Eur. Phys. J. A **42**, 583 (2009)
143. D. Pena Arteaga, P. Ring, arXiv:0912.0908 (2009)
144. C. Losa et al., Phys. Rev. C **81**, 064307 (2010)
145. J. Terasaki, J. Engel, Phys. Rev. C **82**, 034326 (2010)
146. T. Nakatsukasa, T. Inakura, K. Yabana, Phys. Rev. C **76**, 024318 (2007)
147. J. Toivanen et al., Phys. Rev. C **81**, 034312 (2010)
148. P. Avogadro, T. Nakatsukasa, Phys. Rev. C **84**, 014314 (2011)
149. N. Paar, J. Phys. G **37**, 064014 (2010)
150. P.-G. Reinhard, K. Goeke, Rep. Prog. Phys. **50**, 1 (1987)
151. K. Hagino, P.-G. Reinhard, G.F. Bertsch, Phys. Rev. C **65**, 064320 (2002)
152. S.G. Rohozinski, J. Phys. G **39**, 095104 (2012)
153. L. Prochniak et al., Nucl. Phys. A **730**, 59 (2004)
154. L. Prochniak, S.G. Rohozinski, J. Phys. G **36**, 123101 (2009)
155. J. Libert, M. Girod, J.-P. Delaroche, Phys. Rev. C **60**, 054301 (1999)
156. J.-P. Delaroche et al., Phys. Rev. C **81**, 014303 (2010)
157. T. Niksic et al., Phys. Rev. C **79**, 034303 (2009)
158. Z.P. Li et al., Phys. Rev. C **86**, 034334 (2012)
159. N. Hinohara et al., Phys. Rev. C **85**, 024323 (2012)
160. K. Nomura et al., Phys. Rev. C **83**, 014309 (2011)
161. K. Nomura et al., Phys. Rev. C **84**, 014302 (2011)
162. D. Almehed, S. Frauendorf, F. Dönau, Phys. Rev. C **63**, 044311 (2001)
163. M. Anguiano, J.L. Egido, L.M. Robledo, Nucl. Phys. A **683**, 227 (2001)
164. M. Bender et al., unpublished (2013)
165. B. Gebremariam, T. Duguet, S.K. Bogner, Phys. Rev. C **82**, 014305 (2010)
166. B. Gebremariam, S.K. Bogner, T. Duguet, Nucl. Phys. A **851**, 17 (2011)
167. N. Kaiser, Eur. Phys. J. A **45**, 61 (2010)
168. J.W. Holt, N. Kaiser, W. Weise, Eur. Phys. J. A **47**, 128 (2011)
169. J.W. Holt, N. Kaiser, W. Weise, Prog. Part. Nucl. Phys. **67**, 353 (2012)
170. N. Kaiser, Eur. Phys. J. A **48**, 36 (2012)
171. M. Stoitsov et al., Phys. Rev. C **82**, 054307 (2010)

MIX
Papier aus verantwortungsvollen Quellen
Paper from responsible sources
FSC® C105338

If you have any concerns about our products,
you can contact us on
ProductSafety@springernature.com

In case Publisher is established outside the EU,
the EU authorized representative is:
**Springer Nature Customer Service Center GmbH
Europaplatz 3, 69115 Heidelberg, Germany**

Printed by Libri Plureos GmbH
in Hamburg, Germany